Human Thermal Environments

Human Thermal Environments

The effects of hot, moderate, and cold environments on human health, comfort and performance

Second edition

K. C. Parsons

CRC Press
Taylor & Francis Group
Boca Raton   London   New York

CRC Press is an imprint of the
Taylor & Francis Group, an **informa** business

First edition published 1993
Second edition published 2003
by Taylor & Francis
11 New Fetter Lane, London EC4P 4EE

Simultaneously published in the USA and Canada
by Taylor & Francis Inc,
29 West 35th Street, New York, NY 10001

Taylor & Francis is an imprint of the Taylor & Francis Group

British Library Cataloguing in Publication Data
A catalogue record for this book is available from the British Library

Library of Congress Cataloging in Publication Data
Parsons, K. C. (Kenneth C.)
   Human thermal environments: the effects of hot, moderate,
   and cold environments on human health, comfort and
   performance / K. C. Parsons—2nd ed.
      p.   cm.
   1. Temperature—Physiological effect.   I. Title.

QP82.2.T4 P38 2002
612′.01446—dc21   2002072017

ISBN 0–415–23793–9 (pbk)

ISBN 0–415–23792–0 (hbk)

Contents

Preface to the second edition

Since the first edition of *Human Thermal Environments* was published in 1993, the subject has moved into exciting times. While fundamental principles are accepted, knowledge has increased, methods of application are used and some are controversial. From a base in military research and studies of fit, young people, there is now a recognition of the importance of the subject in all aspects of life worldwide. Many of the ideas expressed in the first edition have come to fruition such as the recognition of the six basic parameters (variables) as an essential starting point and the use of computers. The first edition recorded work up until the 1990s and may have stimulated research and application after that. It was particularly important in describing work in the 1980s when the foundations for new approaches were established. Developments that were at an early stage are now accepted. This second edition reports activity and progress into the twenty-first century while maintaining the description of the fundamentals of the subject and its development.

Programmes of European research have been a feature of the 1990s and beyond as the European Community has integrated activities over its member countries, stimulated co-operative research across countries and laboratories, industry, the military, consumers and other interested parties. Globalisation including worldwide electronic communication has had significant effects. New approaches to health and safety requirements have stimulated research and 'useable methods' and risk assessment strategies have become an issue. In heat stress, the predicted heat strain assessment method has been developed, in cold stress working practices in industrial work are being established and the WCI reviewed, and in thermal comfort the *PMV/PPD* index hangs on, as the agenda has moved to behavioural and adaptive approaches. Global databases of the results of collections of laboratory and field studies, made available on the worldwide web, can now allow quality control and research across the world to perform analysis on raw data. There has been an interest in special groups (e.g. people with disabilities) and special environments (e.g. vehicles, chilled ceiling and displacement ventilation). Knowledge of human skin contact with hot, moderate and cold surfaces has been developed, integrated and reported as part of the whole series of International Standards that were first described at an early stage in the first edition and are now fully updated in this second edition. They provide an influential role in determining the direction of the subject and in stimulating research activity. New developments have provided new opportunities but we should not forget that the integrated subject of human thermal environments is based upon responses of people and the same fundamental principles apply. The second edition has introduced new developments since the previous edition while still maintaining a comprehensive description of the essentials of the subject.

All chapters have been revised, most with significant additions and there are new chapters on dehydration and water requirements and on thermal comfort for special populations, special environments and adaptive modelling. Chapters 1–7 present fundamental principles. Chapter 1 describes the six basic variables that make up human thermal environments and now includes the effects of solar radiation. Improvements to the heat balance equation include the effects of clothing ventilation and practical application is enhanced with a description of the thermal audit which will become a fundamental starting point in all human thermal environments assessments. Chapters 2 and 3 describe physiological and psychological responses. The controversial selective brain cooling is discussed as well as developments in behavioural thermoregulation. Chapter 4 is a new chapter which marks the increasingly recognised importance of dehydration and water requirements in responses to human thermal environments. Control of body water, indicators of dehydration and practical recommendations for drinking are provided. Chapter 5 is concerned with measuring methods and now includes personal monitoring systems and specification of physiological instruments. Chapters 6 and 7 are concerned with metabolic heat and clothing. New techniques, including the doubly labelled water method and CO_2 build-up in rooms, are presented as well as developments in clothing ventilation and active clothing. Chapters 8, 9, 10 and 11 cover thermal comfort, heat stress and cold stress, all of which have had significant developments. The adaptive approaches to thermal comfort are described as well as response to specific populations and environments. New developments in heat stress assessment are presented, in particular the new Predicted Heat Strain method that is being proposed as an International Standard. The original research for the Wind Chill Index is now described in detail as it comes under scrutiny and new proposals are developed. Chapter 12 is enhanced with the description of recent studies and approaches, in particular, the importance of the role of distraction in reducing productivity. Chapter 13 is significantly enhanced beyond consideration of hot surfaces with the entirely new sections on contact with moderate and cold surfaces, reflecting much improved knowledge and data in those areas. Chapter 14 reflects a major development in the application of the subject with the publication, revision and development of International Standards, now essential reading for the practitioner. Chapter 15 has also been updated with significant developments. As predicted in the first edition, thermal models have taken their place as an integrated part of the subject's activity and their role in computer aided environmental design is becoming established.

K. C. Parsons
Loughborough, 2002

Acknowledgements to the
second edition

In addition to those acknowledged in the first edition, I would like to thank all those who have supported me with the second edition. The death of Pharo Gagge of the J. B. Pierce Foundation, Yale, USA was a great loss to the subject from which it has yet to recover and his work makes a significant contribution to this book. Ralph Goldman, now retired, has continued to inspire, as have Ole Fanger and Bjarne Olesen from Denmark and Professor Bernhardt Metz from France; Rainer Goldsmith, now retired from the Department of Human Sciences at Loughborough University, influenced us all with his extensive knowledge and experience, as did the late Ernest Hamley who kept me in touch with the history of the subject and always had a story to tell. I am particularly appreciative of the support of Ingvar Holmér from Sweden who has consistently pursued an international agenda for the subject and has stimulated many symposia, research projects and conferences worldwide. I am also grateful to Victor Candas for his perceptive advice and experience concerning human thermal physiology and thermoregulation and to Byron Jones and Elizabeth McCullough from KSU, USA for the high quality of their work and the unstinting way in which they disseminate it to the wider scientific community.

I acknowledge the support of Cathy Bassey, John Fisher, David Fishman and Sina Talal of the BSI and Murielle Gauvin from AFNOR for their work with ISO/TC 159 SC5 and CEN/TC 122 WG11. I also acknowledge Geoff Crockford, Margaret Hanson, Alan Snary, W. R. Withey and Tony Youle from BSI committees. ISO/TC 159 SC5 WG1 has continued to provide a forum for international debate. We have grown old together and I acknowledge the contributions of G. Alfano, V. Candas, G. Crockford, D. Gabay, H.-J. Gebhardt, B. Griefahn, J. Hassi, E. Hartog, G. Havenith, I. Holmér, R. Ilmarinen, B. Kampmann, G. Langkilde, J. Malchaire, B. Olesen, H. Rintamäki, S. Sawada, Y. Tochihara and K. Tsuzuki.

In recent years, I have been grateful for the discussions of developments in thermal comfort with Mike Humphreys and Fergus Nicol of Oxford Brookes University and Dennis Loveday, Ahmed Taki and Genhong Zhou from Loughborough.

The Department of Human Sciences at Loughborough has provided a unique blend of Ergonomics, Psychology and Human Biology, creating a stimulating environment from which to view different perspectives of the subject. The pioneering work of Professor Will Floyd, Brian Shackel and Stuart Kirk as well as that of Peter Jones, Peter Stone, Mike Cooke and others has grown into a significant and productive academic body and I am proud to have carried the banner forward as Head of the Department since 1996. I acknowledge the following academic colleagues: Dr T. Baguley, Dr K. Brooke-Wavell, Dr S. Brown, Dr J. Brunstrom, Professor N. Cameron, Dr J. Cromby, Dr C. Crook, Professor K. Eason, Dr H. Gross, Mrs S. Harker, Dr R. Haslam, Dr G. Havenith, Dr R. Hooper, Professor J. Horne, Dr P. Howarth, Professor M. Lansdale, Dr N. Mansfield, Dr S. Mastana, Dr D. Middleton, Professor K. Morgan, Dr N. Norgan, Dr L. Reyner, Dr E. Rousham and Mr M. Sinclair.

I would also acknowledge the support of members of the Human Thermal Environments Laboratory including George Havenith, Ollie Jay, Shuna Powell, Karen Bedwell, Trevor Cole, Lynda Webb and my PhD students: Julia Scriven, Roger Haslam, Paul Wadsworth, Wen Qi Shen, Mike Neale, Marc McNeil, Lisa Bouskill, Mandy Stirling, Damian Bethea and Simon Hodder.

Hilarie Kemp helped me recover the previous manuscript and kept me on track with the new one. I am grateful to her for extensive support with book production, the enthusiasm and dedication showed, for her careful attention to detail and for finding amusement in parts of the text (some of which was intended). Thankyou also to Pam Taylor for help in the final stages of book preparation.

Finally I would acknowledge the support of my family. Benjamin has graduated in English and is training to be a journalist. Hannah is reading Italian and Samuel is embarking on his secondary school career, with a passion for rugby. Jane, as ever, is there for all of us.

K. C. Parsons
Loughborough
March 2002

Preface to the first edition

Responses to thermal environments play a major role in human existence. All humans have exhibited physiological and behavioural responses to heat and cold, from cave men, through ancient civilizations to the present day. Over the last few centuries, such responses have undergone systematic scientific investigation. In parallel with major scientific discoveries, principles have been established and during the twentieth century these have been formalized into methods used in environmental design and evaluation. Technology has played a supporting role and in recent years the so-called information technology revolution has provided new opportunities for the development and application of well established principles and understanding. For example, in areas of environmental measurement, thermal models and in computer aided environmental design. Much is known and the foundations have been laid; however there is still much to discover in this subject. What is certainly true is that there is greater than ever interest and activity in the area.

The book presents knowledge concerning human responses to hot, moderate, and cold thermal environments for people exposed to air at normal atmospheric pressure. A comprehensive and integrated approach to the subject is taken, defining human thermal environments in terms of six basic parameters: air temperature; radiant temperature; humidity and air velocity of the environment; the clothing worn and the activity of the person. Much of the book is concerned with how these parameters interact to produce physiological and psychological responses and how they, in turn, will affect human health, comfort, and performance.

The underlying principles, derived from the component disciplines of physics, physiology, and psychology are described, as are methods used in the practical assessment of thermal environments. A historical perspective is often taken to enhance description and understanding of current methods and techniques.

In Chapters 1–6, fundamental principles are presented. Chapter 1 defines the six basic parameters and demonstrates how they interact to influence heat exchange between the body and the environment. The analysis of heat transfer leads to the body heat balance equation, a fundamental concept in the rational method for assessing human thermal environments. Chapters 2 and 3 cover physiological and psychological responses. Human thermoregulation and body temperature are described, as are psychological and psycho-physical responses and perspectives. Chapter 4 is concerned with measurement of the basic environmental parameters, physiological and psychological responses. The thermal index is introduced as a fundamental assessment technique. Chapters 5 and 6 are concerned with the principles and estimation of the basic parameters, metabolic heat production, and clothing insulation.

Chapters 7, 8 and 9 consider thermal comfort, heat stress, and cold stress. There has been much activity in this area and recent developments, particularly new rational thermal indices, are described. Interference with activity, performance, and productivity is considered in Chapter 10. A systematic approach is taken and fundamental research reviewed. A performance model is discussed, as well as ways forward, in this area of great economic concern.

Chapters 11, 12 and 13 present topics not previously presented in detail that will be of great interest and influence in the future. Skin reaction on contact with solid surfaces is difficult to study where burns occur, for ethical reasons. Knowledge of temperatures that produce burns is needed however to aid in the design of products. Scientific findings and practical suggestions are presented as a contribution to this highly controversial area. The development of International Standards has been a feature of the last ten years and, with segmented world markets, will continue to be of interest. Chapter 12 describes current developments.

Digital computers and developments in knowledge and techniques have led to the availability of thermal models. Fundamental principles and important and influential models are described. The possibility of computer aided design in this area is demonstrated.

There is no doubt that the practice of this subject is best performed using computers. For example, in recent years my students of climatic ergonomics, physiology, and psychology routinely use software in all aspects of their project and survey work. Sufficient detail is provided in the book to allow the production of software. Source listings of some important computer programs are provided in the appendices. In the first plan of the book a suite of programs was to be included. However, after consideration, I decided to refer to texts where software could be obtained in disk form. This will allow the reader access to easy updates and save errors and laborious typing of source code. All of the equations and models referred to in this book are available on disk for microcomputers from references provided in the text.

K. C. Parsons
Loughborough, 1992

Acknowledgements to the first edition

I am responsible for any errors and some of the inspiration and knowledge presented in this book. I have benefited greatly from the advice, friendship, and openness of many experts throughout the world. These particularly include colleagues from the Department of Human Sciences at Loughborough, those on British, European, and International Standards committees, and fellow members of the Ergonomics Society and of the American Society of Heating, Refrigerating and Air Conditioning Engineers (ASHRAE).

I first met Fred Rohles on an aeroplane between Copenhagen and London. He invited me to give a presentation at an ASHRAE symposium he was organizing in New York. After the symposium, I visited Pharo Gagge at the J. B. Pierce Laboratory, Yale and stayed with Pharo and his wife Edwina in their home. It was a most hospitable, inspirational, and informative visit for which I will always be grateful. I would like to thank Pharo Gagge of the J. B. Pierce Laboratory and also Fred Rohles, Elizabeth McCullough and Byron Jones of the Institute for Environmental Research, Kansas, for their help, advice, and support.

I first attended the ISO Working Group concerned with the Ergonomics of the thermal environment in 1983. The group had been meeting since mid 1970s and the United Kingdom had joined late. Despite some scepticism, but skilfully managed by the Chairman, Gerard Aubertin, standards were, and still are being, produced. The international experts in the group debated many aspects of the subject and, over the years, undoubtedly made a major contribution particularly in stimulating interest and research. I was privileged to be part of this and acknowledge the contribution of members who included Gerard Aubertin, Gaetano Alfano, Ole Fanger, François Grivel, George Havenith, Thomas Hettinger, Ingvar Holmér, Jacques Malchaire, Bjarne Olesen, Jean-Jacques Vogt, and Joseph Yoshida. I would also acknowledge Bernard Metz and Harald Siekmann, colleagues from European Standards committees, and Peter Crabbe, Geoff Crockford, David Fishman, Rod Graves and Ron McCaig from the BSI committee concerned with human response to the thermal environment.

There have been a number of books on aspects of human response to thermal environments and I have in particular often referred to two over the last ten years. These are *Indoor Climate* by D. A. McIntyre and *The stress of hot environments* by D. McK Kerslake. I would acknowledge the contribution of the authors. I would also acknowledge the advice of Ralph Goldman previously from USARIEM in the USA and of Mike Haisman when previously at the APRE in the UK. Mike Haisman provided continued enthusiasm, support, and advice throughout the thermal modelling work carried out with Roger Haslam and the work on the MAPS system with Paul Wadsworth, which integrated thermal models into a computer enquiry handling system.

Colleagues within the Department of Human Sciences have provided support that I acknowledge. These include Rainer Goldsmith, Ernest Hamley, Jim Horne, Stuart Kirk, Brian Shackel, and Peter Stone. I would also like to thank Mary Hewitt for helping type the manuscript, the many students who have produced projects and theses on all aspects of the subject and maintained my interest and the research workers, past and present, in the Human Modelling Group at Loughborough.

The greatest acknowledgement is to my family for their support: Jane, Benjamin, Hannah and Samuel.

Notation

The choice of notation is based on that used by the American Society of Heating, Refrigerating and Air Conditioning Engineers in Chapter 8 of the Handbook Fundamentals (1989). Standard definitions of terms, symbols and units can also be found in ISO 13731 (2002).

a, b	constants used in the calculation of air velocity from kata cooling time	ND
A	activity factor for calculation of pumping effects on clothing insulation	ND
A_{cl}	surface area of clothed body	m^2
A_{cov}	percentage of body surface area covered by a garment	%
A_D	DuBois body surface area	m^2
A_p	projected area of the body	m^2
A_r	effective radiating area of the body	m^2
b_i	thermal penetration coefficient for material i	$J\,s^{-1/2}\,m^{-2}\,K^{-1}$
BM	basal metabolic rate	$W\,m^{-2}$
B	behaviour	ND
B	exponent used in psychophysical power law	ND
c	specific heat capacity	$kJ\,kg^{-1}\,°C^{-1}$
C	convective heat loss per unit area	$W\,m^{-2}$
C_B	weight of subject clothed before exposure	kg
C_A	weight of subject clothed after exposure	kg
$C_{p,bl}$	specific heat capacity of blood	$kJ\,kg^{-1}\,°C^{-1}$
C_{res}	dry respiration heat loss per unit body surface area	$W\,m^{-2}$
$C_{SIG_{sk}}$	effector controlling signal for vasoconstriction	ND
d	diameter of spherical sensor	m
dt_{sk}	clothing insulation index derived from mean skin temperature of subjects when clothed and unclothed	$°C$

DRY	dry heat loss from the skin	$W\,m^{-2}$
D	distance of load carriage	m
D_c	magnitude estimate of cold discomfort	ND
D_w	magnitude estimate of warm discomfort	ND
E	environment	ND
E	evaporative loss per unit body surface area	$W\,m^{-2}$
EE	energy equivalent of burning food in oxygen	$W\,h\,1\,O_2^{-1}$
EST	Burton and Edholm Equivalent Shade Temperature	$°C$
$ESAT$	Burton and Edholm Equivalent Still Air Temperature	$°C$
E_p	predicted evaporation rate	$W\,m^{-2}$
E_{dif}	evaporative loss by diffusion through skin per unit area	$W\,m^{-2}$
E_{max}	maximum evaporative potential per unit area	$W\,m^{-2}$
E_{req}	required evaporative loss per unit area for heat balance	$W\,m^{-2}$
E_{res}	evaporative loss from respiration	$W\,m^{-2}$
E_{rsw}	evaporation of sweat secreted due to thermoregulatory control	$W\,m^{-2}$
E_{sw}	total evaporation rate due to sweating	$W\,m^{-2}$
E_{sk}	total evaporative heat loss from the skin	$W\,m^{-2}$
ET	effective temperature	$°C$
f	STPD reduction factor	ND
f	partial pressure of water vapour in air	mm Hg
f_{cl}	clothing area factor	ND
f_p	projected area factor	ND
F_{pcl}	permeation efficiency factor	ND
F	kata factor	$m\,cal\,cm^{-2}$
FO_2	fraction of oxygen in expired air	ND
FCO_2	fraction of carbon dioxide in expired air	ND
F_{eff}	effective radiation area factor	ND
F_{p-n}	angle factor between a person, p and a surface, n	ND
g	radiant response ratio	ND
G	mass velocity	$kg\,m^{-2}\,s^{-1}$
h	combined heat transfer coefficient	$W\,m^{-2}\,K^{-1}$
h_c	convective heat transfer coefficient	$W\,m^{-2}\,K^{-1}$
h_{cl}	thermal conductance of a clothing ensemble	$W\,m^{-2}\,K^{-1}$
h_e	evaporative heat transfer coefficient	$W\,m^{-2}\,kPa^{-1}$
h_r	radiative heat transfer coefficient	$W\,m^{-2}\,K^{-1}$

h_{fg}	heat of vaporization of water	$kJ\,kg^{-1}$
H	height	m
H	metabolic heat production per unit body surface area	$W\,m^{-2}$
H_{res}	total respiratory heat loss	$W\,m^{-2}$
H_{sk}	total heat loss at the skin	$W\,m^{-2}$
HR	heart rate	beats per min
HR_0	heart rate while at rest	beats per min
HR_m	component of heart rate due to work	beats per min
HR_s	component of heart rate due to static exertion	beats per min
HR_T	component of heart rate due to thermal strain	beats per min
HR_N	component of heart rate due to emotional response	beats per min
HR_e	residual component of heart rate	beats per min
HR_r	heart rate after recovery	beats per min
i_m	moisture permeability index (clothing)	ND
i_{cl}	clothing vapour permeation efficiency	ND
I_{cl}	intrinsic clothing insulation	Clo, $m^2\,{}^\circ C\,W^{-1}$
I_{cle}	effective clothing insulation	Clo, $m^2\,{}^\circ C\,W^{-1}$
I_{EQUIV}	Equivalent clothing insulation	Clo, $m^2\,{}^\circ C\,W^{-1}$
I_{res}	resultant clothing insulation	Clo, $m^2\,{}^\circ C\,W^{-1}$
I_t	total clothing insulation	Clo, $m^2\,{}^\circ C\,W^{-1}$
I_a	thermal insulation of the boundary layer on a nude person	Clo, $m^2\,{}^\circ C\,W^{-1}$
$IREQ$	clothing insulation required	Clo, $m^2\,{}^\circ C\,W^{-1}$
$IREQ_{min}$	clothing insulation required for thermal equilibrium	Clo, $m^2\,{}^\circ C\,W^{-1}$
$IREQ_{Neutral}$	clothing insulation required for thermal comfort	Clo, $m^2\,{}^\circ C\,W^{-1}$
I_{clu}	effective clothing insulation of a garment	Clo, $m^2\,{}^\circ C\,W^{-1}$
$I_{clu,i}$	effective clothing insulation of garment i	Clo, $m^2\,{}^\circ C\,W^{-1}$
I_{clo}	effective clothing insulation in Clo units	Clo
I_{ecl}	resistance of clothing to the transfer of water vapour	$m^2\,kPa\,W^{-1}$
I_{ea}	resistance of air layer to the transfer of water vapour	$m^2\,kPa\,W^{-1}$
k	constant of psychophysical power law	ND
k_s	thermal conductivity for sensible heat	$W\,m^{-2}\,{}^\circ C^{-1}$
K	heat transfer by conduction	$W\,m^{-2}$
K_e	thermal conductivity of water vapour	$W\,m^{-2}\,kPa^{-1}$
L	thermal load per unit body area	$W\,m^{-2}$
LR	Lewis relationship	$K\,kPa^{-1}$

M	metabolic free energy production per unit body area	W m^{-2}
M_p	component of metabolic rate due to posture	W m^{-2}
M_w	component of metabolic rate due to type of work	W m^{-2}
M_m	component of metabolic rate due to whole body movement	W m^{-2}
M_net	metabolic heat load in Givoni/Goldman model	W
M_act	component metabolic energy due to subject's activity	W m^{-2}
M_shiv	component of metabolic heat due to shivering	W m^{-2}
M_rsw	rate at which sweat is secreted	$\text{kg s}^{-1}\,\text{m}^{-2}$
M_bl	blood flow	$\text{kg s}^{-1}\,\text{m}^{-2}$
M_i	metabolic rate representative of activity i	W m^{-2}
Met	metabolic rate symbol used by Nishi and Gagge (1977)	W m^{-2}
NA	weight of nude subject after exposure	kg
NB	weight of nude subject before exposure	kg
O_i	fractional oxygen content of inspired air	ND
O_e	fractional oxygen content of expired air	ND
P_1	pulse rate from 30 seconds to 1 minute during recovery	beats per min
P_2	pulse rate from 1.5 to 2 minutes during recovery	beats per min
P_3	pulse rate from 2.5 to 3 minutes during recovery	beats per min
P_a	partial pressure of water vapour in air	kPa
$P_\text{sk,s}$	saturated water vapour pressure at skin temperature	kPa
P_swb	saturated water vapour pressure at aspirated wet bulb temperature	kPa
P	person	ND
P	measured atmospheric pressure	kPa
Q_lim	heat storage limit	W h m^{-2}
Q_crsk	combined thermal exchange between body core and skin	W m^{-2}
Q_res	total rate of heat loss through respiration	W m^{-2}
Q_sk	total rate of heat loss from the skin	W m^{-2}
r	efficiency of sweating	ND
R	radiative heat loss per unit area	W m^{-2}
R_cl	intrinsic thermal insulation of clothing	$\text{m}^2\,\text{K W}^{-1}$

$R_{e,cl}$	intrinsic evaporative resistance of clothing	$m^2\,kPa\,W^{-1}$
$R_{e,t}$	total evaporative resistance	$m^2\,kPa\,W^{-1}$
R_s	solar load in index of thermal stress	$W\,m^{-2}$
RQ	respiratory quotient	ND
R_{ct}	thermal resistance of clothing, symbol used by Umbach (1988)	$m^2\,°C\,W^{-1}$
R_{et}	water vapour resistance of clothing, used by Umbach (1988)	$m^2\,kPa\,W^{-1}$
RT	recovery time	min
RM	increase in heart rate per unit of metabolic rate	beats per min $m^2\,W^{-1}$
R_{ct}	total thermal resistance of clothing, used by Umbach (1988)	$m^2\,°C\,W^{-1}$
R_{et}	total water vapour resistance of clothing, used by Umbach (1988)	$m^2\,kPa\,W^{-1}$
S	rate of heat storage per unit area	$W\,m^{-2}$
SW_{req}	sweating required. Required sweat rate index	$W\,m^{-2}$
SW	sweating required, used by Givoni (1963)	$W\,m^{-2}$
SW_p	predicted sweat rate	$W\,m^{-2}$
SW_{max}	maximum sweat rate that can be achieved by persons	$W\,m^{-2}$
S	warmth sensation rating	ND
t	recorded temperature of expired air	°C
t_a	air temperature	°C
t_b	mean body temperature	°C
t_{cl}	surface temperature of clothed body	°C, K
t_{con}	contact temperature	°C
t_{cr}	core temperature	°C
t_{dp}	dew point	°C
t_g	globe temperature	°C, K
t_{nwb}	natural wet bulb temperature	°C
t_o	operative temperature	°C
t_{pr}	plane radiant temperature	°C
t_r	mean radiant temperature	°C
t_s	surface temperature	°C, K
t_{ty}	tympanic membrane temperature	°C
t_{db}	dry bulb temperature	°C
t_{wb}	aspirated wet bulb temperature	°C
$t_{b,n}$	mean body temperature for a person in thermal neutrality	°C
t_i	temperature of surface i	°C
t_{ch}	chilling temperature index	°C

t_c	contact temperature	°C
t_{cc}	contact temperature if the solid surface were made entirely of the coating on the skin and corrected for type of contact	°C
t_{res}	Dry Resultant Temperature index	°C
t_h	temperature of hot body	°C
t_i	time on activity i	min
t_{sk}	mean skin temperature	°C
$t_{sk,n}$	mean skin temperature for a person in thermal neutrality	°C
t_w	mean temperature of surrounding walls	°C
t_{sh}	body shell temperature	°C
t_{re}	rectal temperature	°C
t_{ref}	predicted equilibrium rectal temperature in Givoni/Goldman model	°C
t_k	mean temperature of kata thermometer range	°C
T	total time over all activities	min
T_{eq}	equivalent temperature	°C
T_{core}	mean core temperature of thermal model	°C
T_{skin}	mean skin temperature of thermal model	°C
T_{muscle}	mean muscle temperature of thermal model	°C
T_{ceq}	Equivalent contact temperature index	°C
T_{eq}	Equivalent temperature index	°C
T_{pri}	plane radiant temperature in direction i	°C
T_u	local air turbulence	ND
v	air velocity (speed)	$m\,s^{-1}$
V	ventilation rate	$1\,s^{-1}$
VO_2	oxygen uptake	$ml\,kg\,min^{-1}$
V_{ex}	volume of expired air	1
w	skin wettedness	ND
w_{max}	maximum skin wettedness that can be achieved by persons	ND
w_p	predicted skin wettedness	ND
w_{req}	required skin wettedness	ND
W	external mechanical work per unit area	$W\,m^{-2}$
$W_{SIG_{cr}}$	effector controlling signal for vasodilation	ND
W_{SIG_b}	mean body temperature component of sweating controller signal	ND
W	weight of load carried	kg
Wk	symbol for mechanical work, used by Nishi and Gagge (1977)	$W\,m^{-2}$
W_{ct}	comfort rating, used by Umbach (1988)	ND
W_c	comfort rating, used by Umbach (1988)	ND

W_{ex}	mechanical work symbol from Givoni and Goldman (1973)	W
x	average cloudiness in tenths	ND
Y	mechanical efficiency of the body doing work	ND
Y	subjective warmth vote	ND

Greek

α	absorptivity	ND
α	azimuth angle	degree
α	relative mass of skin to core	ND
β	elevation	degree
ε	emissivity	ND
η	mechanical efficiency of work	ND
λ	latent heat of evaporation	$J\,kg^{-1}$
λ	wavelength	μm
μ	viscosity of the air	$kg\,m\,s^{-1}$
Φ	magnitude of stimulus	ND
Φ_0	magnitude of stimulus at absolute perception threshold	ND
σ	Stefan–Boltzmann constant $= 5.67 \times 10^{-8}$	$W\,m^{-2}\,K^{-4}$
ϕ	relative humidity	ND
Ψ	magnitude of sensation	ND

1 Human thermal environments

INTRODUCTION

The sun, at a temperature of around 5500 °C, provides heat to the earth at the top of the atmosphere at a rate of around 1370 W m^{-2} (the Solar Constant – the exact value depending upon the orbit of the earth around the sun). Human existence, along with that of other organisms, depends upon that energy. On the earth, the energy is transferred from place to place and from one form to another hence creating a wide range of environments. The challenge for every person is to successfully interact with his or her local environment. The human body responds to environmental variables in a dynamic interaction that can lead to death if the response is inappropriate, or if energy levels are beyond survivable limits, and it determines the strain on the body as it uses its resources to maintain an optimum state. In the case of the thermal environment this will determine whether a person is too hot, too cold or in thermal comfort.

Air temperature, radiant temperature, humidity and air movement are the four basic environmental variables that affect human response to thermal environments. Combined with the metabolic heat generated by human activity and clothing worn by a person, they provide the six fundamental factors (sometimes called the six basic parameters as they vary in space and time but fixed representative values are often used in analysis) that define human thermal environments. The general, but fundamental point, is that it is the interaction of the six factors to which humans respond. This was shown by Fanger (1970) in his classic book *Thermal Comfort*. It also applies for humans in hot or cold environments. Collins (1983) considers a skier with a child on his back. At the bottom of the slope the adult was hot and sweating. The child suffered hypothermia. Metabolic heat production in the skiing adult had compensated for heat loss driven by high relative air velocity across the body and low air temperatures. The child had been inactive. The adult and the child had experienced very different human thermal environments and hence had very different responses.

The ability to lose heat by the evaporation of sweat is crucial to a person under heat stress. Environmental humidity is therefore of great importance as is the nature of protective clothing. Death and heat illness in military personnel have been attributed to high metabolic heat production, radiant and air temperatures, and relatively impermeable protective clothing. Often however, it is because there is a lack of understanding of the effects of the interaction of the six basic variables.

Often environments are assessed or environmental limits are defined only in terms of air temperature. This is insufficient in many situations, as all other five factors are relevant. For example, thermal comfort in offices or vehicles may be greatly affected by solar radiation and specifying comfort limits in terms of air temperature alone will be inadequate. That is not to say that reasonable assumptions cannot be made. In terms of thermal comfort, it is sometimes reasonable to specify comfort limits in terms of air temperature on the assumption that radiant temperature equals air temperature; there is still air at 50 per cent relative humidity; and that persons in the environment wear 'normal' clothing, and conduct light activity. However, it is better that assumptions are explicitly stated rather than imply that air temperature alone determines thermal comfort.

In specific situations other factors will be influential; for example, a person's posture will greatly affect heat exchange between the body and the environment. Behavioural factors can be important. Motivation, and the degree of acclimatization to heat, can play a major role in determining heat stress casualties in military exercises. The six fundamental factors therefore, provide a minimum requirement for a useful conceptual basis upon which a consideration of human thermal environments can be based.

BASIC PARAMETERS

Temperature

At a molecular level, temperature can be considered as the average kinetic energy (heat) in a body. If heat energy is lost from a body, its temperature will fall and if it flows into a body its temperature will rise. It is a law of thermodynamics that there is a net energy flow from bodies at higher temperatures to bodies at lower temperatures. The temperature of the human body is an important indicator of its condition (comfort, heat or cold stress, performance). Humans are homeotherms and 'attempt' to maintain internal body temperature near to about 37 °C. A deviation of more than a few degrees from this value can have serious consequences. The temperature of the human body will therefore be greatly affected by the temperature of fluids or solids surrounding it as these will influence the heat transfer to and from the body. The human body is commonly surrounded by clothing and then almost entirely by air. Contact may also be with solid surfaces, water (total or partial immersion), other fluids, or even space. The thermodynamic principles of heat transfer will apply in all of these cases. Throughout this book the 'normal' case of a person in air is assumed unless otherwise stated. A 'driving force' for heat transfer between the human body and the surrounding air will therefore be determined by air temperature.

Air temperature (t_a)

For practical convenience air temperature can be defined as *the temperature of the air surrounding the human body which is representative of that aspect of the surroundings which determines heat flow between the human body and the air.*

Naturally, the temperature of the air will vary; heat exchange between bodies is a continuous process. The temperature of the air at a great distance from the human body of interest will not necessarily be representative of that which determines heat flow. The temperature of the air very close to the (clothed) body will also not be representative as this will be influenced by so called 'boundary conditions'; for example, in a 'cold' environment there will be a layer of 'warmer' air surrounding the body.

Radiant temperature

In addition to the influence of air temperature on the temperature of the human body there is also an influence of radiant temperature. Heat is exchanged by radiation between all bodies, and there is a net heat flow from a hot to a cold body by an amount related to the difference between the fourth powers of the absolute temperatures of the two bodies. There is a net flow of radiant energy between the sun and the earth (i.e. no intervening medium is required – it will flow through a vacuum). Thermal radiation is a part of the electromagnetic spectrum that includes X-rays (short wave length), light, and radio waves (long wavelength). A useful way (although the analogy should not be taken too far) of conceptualizing thermal radiation is therefore to relate it to light. In any environment there will be continuous energy exchanges, reflections, absorptions, and so on. As a person moves around a room, the lighting environment (and thermal radiation) may change. At any point in space therefore, there is a unique radiation environment.

McIntyre (1980) describes the concept of the radiation field. At any point in a radiation field there will be a dynamic exchange (e.g. in time and direction) of energy (heat) by radiation. The overall radiation field, in a room for example, can be defined in terms of heat transfer or, more conveniently, in terms of radiant temperatures. In the assessment of humans in thermal environments it is the radiation exchange, at the position of the person in the radiation field, that is important. This can be analysed to increasing levels of complexity. For convenience, radiant temperatures are used, which are defined by McIntyre (1980) as 'the temperature of a black-body source that would give the same value of some measured quantity of the radiation field as exists in reality.' Two radiant temperatures are commonly used to summarize the radiant heat exchange between the human body and the environment. The *mean radiant temperature*, which provides an overall average value and the *plane radiant temperature*, which provides information concerning direction of radiant exchange (and if measured in different directions can give variations in direction about the mean radiant temperature). Plane radiant temperature will be important, for example, where relatively large radiant 'sources' are present in an environment at a specific orientation to the body (e.g. electric fires, steel furnace, heated ceiling, the sun, or a very cold wall).

Mean radiant temperature (t_r)

The *mean radiant temperature* is defined as *the temperature of a uniform enclosure with which a small black sphere at the test point would have the same radiation exchange as it does with the real environment*. The use of the sphere in the definition shows the average in three dimensions. The reference to an equivalent effect in the defined standard environment is a widely used technique. The definition above is generally used and is consistent with the concept of the radiation field. For a sphere, the mean radiant temperature will not depend upon its orientation in the surroundings. For a non-spheroidal shape such as the human body, the concept of *effective mean radiant temperature* is used: 'the temperature of a uniform enclosure with which the test surface would have the same radiation exchange as it does with the real environment.' This will depend upon the orientation of the object in the surroundings.

There has been some inconsistency in defining mean radiant temperature. ISO 7726 (1998) for example, gives the definition: 'The mean radiant temperature is the uniform temperature of an imaginary enclosure in which radiant heat transfer from the human body is equal to the radiant heat transfer in the actual non-uniform enclosure.' This standard and other references (e.g. Fanger, 1970) define mean radiant temperature directly with respect to the human body. The use of a sphere is therefore regarded as an approximation to the mean radiant temperature (still given the symbol t_r). The use of an ellipsoid or cylindrical sensor is then said to be superior as it represents the shape of a standing person more closely than a sphere. The mean radiant temperature will thus depend upon the orientation of the human body in the radiation environment. The mean radiant temperature for the human body can be calculated from the temperature of surrounding surfaces and their orientation with respect to man. Methods of measurement, analysis, and calculation are described in Chapter 5.

Plane radiant temperature (t_{pr})

ISO 7726 (1998) defines *plane radiant temperature* as *the uniform temperature of an enclosure where the radiance on one side of a small plane element is the same as in the non-uniform actual environment.* McIntyre (1980) defines the plane radiant temperature at a point as 'the temperature of a uniform black hemisphere, centred on a plane element with its basal plane in the plane of the element which would give the same irradiance on the element.' (He defines irradiance as the flux per unit area falling on a surface from all directions). The point is that plane radiant temperature is related to radiant exchange in one direction. Both definitions refer to a small plane element at a point. The confusion between definitions relating to either a point or to the human body does not seem to apply. However, the effects of directional radiation will clearly be influenced by the orientation of the body.

The plane radiant temperature can be measured in many orientations: for example, up/down, left/right and front/back. It can also be measured in the direction of a radiation source such as the sun. The projected area factor is estimated as A_p/A_r where A_p is the surface area projected in one direction and A_r is the total radiant surface area. For example, for a standing man the up/down projected area factor is low (0.08). It is higher for left/right orientation (e.g. 0.23) and higher again for front/back (e.g. 0.35). If one could imagine the plane radiant temperature measured on the six faces of a cube then ISO 7726 (1998) gives the following equation as an estimate of mean radiant temperature for a standing man.

$$t_r = \frac{0.08\left(t_{pr1} + t_{pr2}\right) + 0.23\left(t_{pr3} + t_{pr4}\right) + 0.35\left(t_{pr5} + t_{pr6}\right)}{2(0.08 + 0.23 + 0.35)}, \qquad (1.1)$$

where t_{pr1} to t_{pr6} are the plane radiant temperatures in directions: (1) up; (2) down; (3) right; (4) left; (5) front; and (6) back.

This point can also be shown (and values estimated) using photographic techniques (e.g. Fanger, 1970; Underwood and Ward, 1966 – see Figure 1.1). To avoid possible confusion the reader should distinguish between A_p/A_r, the projected area factor in one direction and A_r/A_D, the effective radiation area factor, since the latter is the ratio of the total area of the body 'available' for radiant exchange divided by the total surface area of the body (A_D – see later in this chapter). It will depend upon posture for example, ISO 7933 (1989) gives a value for A_r/A_D, of 0.77 for a standing person, 0.72 sitting and 0.67 crouched. Clearly, the integration of projected area factors over all directions gives A_r. Radiant temperatures are further discussed later in this chapter in relation to radiant heat transfer.

Solar radiation

The principles in the use of radiant temperatures apply equally for indoor and outdoor environments. However, the particular nature of solar radiation in

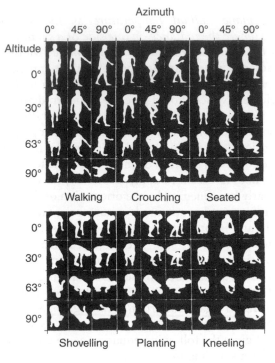

Figure 1.1 A selection of silhouettes of a subject in various postures, corresponding to the areas illuminated by the sun's rays at the angles of altitude and azimuth shown.

Source: Underwood and Ward (1966).

terms of its quality (spectral content) and directional properties requires it to be given special consideration. Radiant heat is part of the electromagnetic spectrum (see Figure 1.2).

Any effects of radiation intensity may be due to both the level of radiation and its wavelength or spectral content. It will also depend upon the direction of the radiation received by the body and the body posture and orientation.

Quantity: solar radiation level

At the mean distance of the earth to the sun; i.e. 1.5×10^{11} m, the irradiance of a surface normal to the solar beam is known as the *solar constant*. Satellite measurements provide a value of 1373 W m^{-2} for the solar constant. Direct solar radiation at the ground however has a maximum value of about 1000 W m^{-2} as some heat is scattered and absorbed by the atmosphere. This is further absorbed by cloud cover, aerosols, and pollution, etc. (see Figure 1.3).

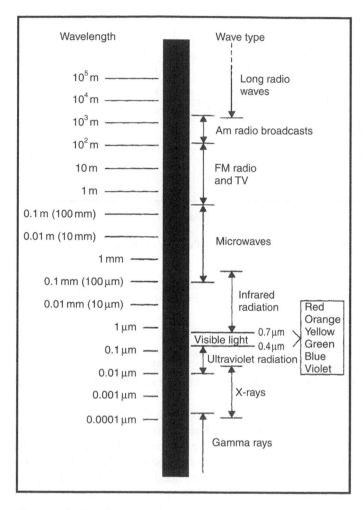

Figure 1.2 The electromagnetic spectrum.

Quality: spectral content of solar radiation

The total energy emitted by the sun (E) is obtained by multiplying the solar constant by the area of a sphere at the earth's mean distance d:

$$E = 4\pi d^2 \times 1373 = 3.88 \times 10^{26}\,\text{W}. \tag{1.2}$$

For black-body radiation the energy emitted obeys Stefan's Law ($E = \sigma T^4$) so:

$$\sigma T^4 = 3.88 \times \frac{10^{26}}{4\pi r^2}. \tag{1.3}$$

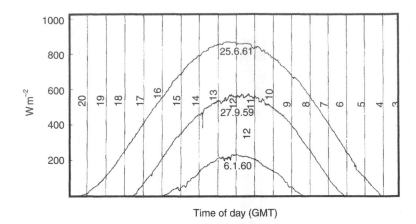

Figure 1.3 Solar radiation on three cloudless days at Rothamsted (52 °N, 0 °W). During the middle of the day, the record tends to fluctuate more than in the morning and evening, suggesting a diurnal change in the amount of dust in the lower atmosphere, at least in summer and autumn. Three recorder charts were superimposed to facilitate this comparison.

Source: Monteith and Unsworth (1990).

The radius of the sun $r = 6.69 \times 10^8$ m.

Stefan–Boltzmann constant $\sigma = 5.67 \times 10^{-8} (\text{W m}^{-2}\,\text{K}^{-4})$.

So from equation (1.3) the temperature of the sun $T = 5770$ K (Monteith and Unsworth, 1990).

The higher the surface temperature of an object the shorter the wavelength of radiation emitted (Wien's law). The solar spectrum therefore has radiation at much shorter wavelengths than terrestrial emission where temperatures are much lower than those of the sun (see Table 1.1 and Figure 1.4). *Note*: The ultra violet spectrum can be subdivided into UVA (400 to 320 nm) which produces tanning in the sun, UVB (320 to 290 nm) which is responsible for skin cancer and vitamin D synthesis, and UVC (290 to 200 nm) which is potentially harmful but is absorbed by the ozone in the stratosphere.

Table 1.1 Distribution of energy in the spectrum of radiation emitted by the sun

Waveband (nm)	Energy %
0–300	1.2
300–400 (ultra violet)	7.8
400–700 (visible/PAR)	39.8
700–1500 (near infrared)	38.8
1500–∞	12.4
	100.0

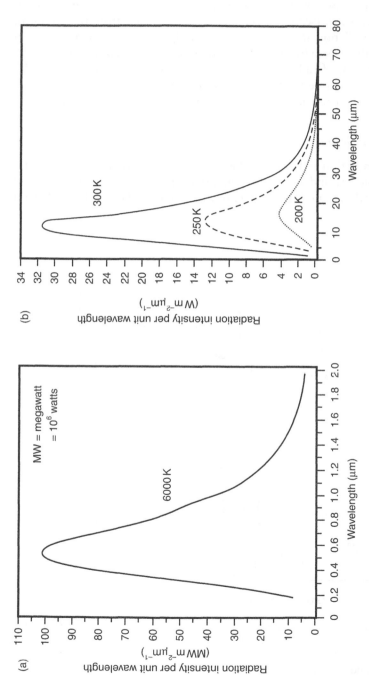

Figure 1.4 Variation in the intensity of black-body radiation with wavelength: (a) $T = 6000\,\mathrm{K}$ (approximately the emission temperature of the sun). (b) $T = 200\,\mathrm{K}$, $250\,\mathrm{K}$ and $300\,\mathrm{K}$ (range of earth emission temperatures). Note the differences in scale.

Source: Neiburger *et al.* (1982).

Solar radiation and the human body

Santee and Gonzalez (1988) identify six radiation terms that are relevant to the thermal responses of the human body. There are three solar radiation terms (direct, diffuse, (scattered sky) and reflected (solar from the ground)), two thermal radiation terms (sky and ground) and the thermal radiation emitted from the person (see Figure 1.5). A commonly held term is the *albedo* that is the amount of radiation reflected from a surface. It is calculated by dividing the amount reflected by the total amount arriving at a surface. Absorption of solar radiation by man is considered by Blazejezyk (2000).

Direct solar radiation

Direct solar radiation arrives from the direction of the solar disc. This varies throughout the day and year due to the earth's rotation on its own axis and its orbit around the sun. With respect to the human body it will also depend upon the orientation and posture of a person. Although direct solar radiation falls across the entire body surface area exposed to direct sunlight, the

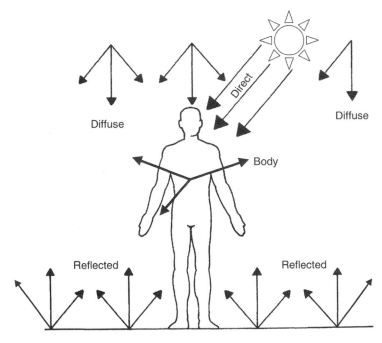

Figure 1.5 The impact of direct, diffuse and reflected solar radiation is dependent on the orientation of the individual relative to the radiation source.

Source: Pandolf *et al.* (1988).

amount of direct solar radiation received is equal to the full radiative intensity normal to the solar beam multiplied by the cross-sectional area of an object normal to the solar rays (A_p). This relationship is known as Lambert's cosine law. Note that A_p values are normal to the direction of the sun. These can be seen in Figure 1.1. Underwood and Ward (1966) give the following equation for A_p for an average man as:

$$A_p = 0.043 \sin \theta_s + 2.997 \cos \theta_s \sqrt{(0.02133 \cos^2 \phi + 0.0091 \sin^2 \phi)} \, \text{m}^2,$$

(1.4)

where
    θ_s = solar elevation angle = angle between sun and horizon
    (Note that the complementary angle to θ_s is θ_z = zenith angle = deviation
    of the sun from vertical)
    ϕ = azimuth angle = orientation of the body relative to the sun
        $\phi = 0$ = facing directly towards or away from the sun
        $\phi = 90$ = sideways (shoulders towards the sun).
(See Santee and Gonzalez, 1988).

Diffuse solar radiation

Diffuse solar radiation describes all other (than direct) scattered radiation received from the blue sky (including the very bright aureole surrounding the sun) and from clouds either by reflection or transmission (Monteith and Unsworth, 1990).

    Diffuse solar, reflected solar, ground thermal and sky thermal radiation are considered as isotrophic, that is the radiation is emitted in all directions from an area source. For a fuller description of solar radiation the reader is referred to Monteith and Unsworth (1990). A description of how to measure solar radiation is provided in Chapter 5.

Temperature charts and scales

Temperature refers to how hot or cold an object or fluid is. It is related to heat content (e.g. temperature as average kinetic energy); however, temperature will also depend upon properties such as specific heat capacity and mass. The thermal sensations of hot or cold are psychological phenomena and, although there are physiological mechanisms in the body which respond to temperature, thermal sensation depends upon such things as previous experience, individual differences, and rates of change of temperature. Humans are therefore not good temperature measuring instruments and cannot provide reliable scales of temperature.

    If there is no net heat flow between two objects in contact, then they are at the same temperature. The steady state level of mercury in a glass thermometer placed in melting ice is the state of the mercury at the same temperature as melting ice. The state of the mercury in boiling water is its state at the temperature of boiling water. This provides the basis of thermometry and temperature measurement scales.

The origins of scales for quantifying temperature and temperature changes derive from human applications. Sanctorius in the early seventeenth century (see Cetas, 1985) developed an air thermometer and used it to measure oral and hand skin temperatures. There are now three temperature scales in common use; the practical working scales of Celsius (sometimes called centigrade) and Fahrenheit, and the theoretical absolute temperature (Kelvin) scale.

Practical working scales (°C, °F)

Working scales of temperature are defined by assigning the temperature of a reproducible known event (e.g. freezing or boiling) arbitrary values (e.g. 0, 100 etc.). The Celsius (°C) and Fahrenheit (°F) scales use two points. The temperature of the freezing point of water is given the value 0 °C and 32 °F. The temperature of the boiling point of water is given the value 100 °C and 212 °F (at 1 atm pressure). The Celsius scale is divided into 100 equal divisions or degrees (1 °C difference) and the Fahrenheit scale into 180 equal divisions (1 °F difference). A simple method of converting from one to the other is by equating 180 °F to 100 °C temperature difference. Therefore, for temperature difference 180 °F = 100 °C so,

$$F = \frac{9}{5}C \quad \text{or} \quad C = \frac{5}{9}F$$

temperature difference, and as 0 °C = 32 °F the scale values convert using

$$F = \frac{9}{5}C + 32 \quad \text{or} \quad C = \frac{5}{9}(F - 32).$$

For example a body temperature of 98.4 °F gives

$$C = \frac{5}{9}(98.4 - 32)$$
$$= 36.9\,°C.$$

A body temperature of 37 °C gives

$$F = \frac{9}{5} \times 37 + 32$$
$$= 98.6\,°F.$$

The Celsius scale is the most widely used throughout the world and is an SI unit; however, the Fahrenheit scale is still used, for example in the USA.

Absolute temperature scale (K)

When pressure is kept constant the volume of a gas varies in proportion to its temperature. For any gas, as temperature is reduced, the volume decreases linearly until it becomes liquid. If the linear relationship is extrapolated to a theoretical zero volume, it is found that the temperature at this point is always −273.15 °C. This is considered to be the lowest temperature possible and is said to be absolute zero or zero degrees Kelvin (0 K). The intervals on the Kelvin scale are equivalent to those on the Celsius scale; so a temperature difference of 1 °C is equivalent to a temperature difference of 1 K. The freezing point of water (0 °C) is 273.15 K and boiling point 373.15 K (i.e. $K = C + 273.15$).

The Kelvin scale is extensively used in scientific work and is correctly used in SI units. The important point in terms of scale properties is that it has an absolute zero, whereas Celsius and Fahrenheit scales have arbitrary zeros. In common practice it is not usual to quote room air temperature or body temperature, for example, as 293.15 K or 310.15 K respectively, 20 °C and 37 °C are more conveniently used.

The International Practical Temperature Scale of 1968 (IPTS-68) is a state of the art approximation to the Kelvin scale. The procedure of the IPTS-68 is to define a number of temperatures at natural fixed points (see Table 1.2) and provide a method of interpolation. A precisely constructed

Table 1.2 Fixed points of the IPTS-68a*

Equilibrium state	Assigned value of IPTS	
	T_{68}(K)	t_{68}(°C)
Triple point of equilibrium hydrogen	13.81	−259.34
Liquid and vapour phase equilibrium of hydrogen at a pressure of 33 330.6 N m^{-2} (25/76 standard atmosphere)	17.042	−256.108
Boiling point of equilibrium hydrogen	20.28	−252.87
Boiling point of neon	27.102	−246.048
Triple point of oxygen	54.361	−218.789
Boiling point of oxygen	90.188	−182.962
Triple point of water[†]	273.16	0.01
Boiling point of water[††]	373.15	100
Freezing point of zinc	692.73	419.58
Freezing point of silver	1235.08	961.93
Freezing point of gold	1337.58	1064.43

Sources: Cetas (1985). From Preston-Thomas (1976).

Notes

* Except for the triple points and one equilibrium hydrogen point (17.042 K), the assigned values of temperature are for equilibrium states at a pressure $\rho_0 = 1$ standard atmosphere (101 325 N m^{-2}). In the realization of the fixed points, small departures from the assigned temperatures will occur as a result of differing immersion depths of thermometers or the failure to realize the required pressure exactly. If allowance is made for these small temperature differences, they will not affect the accuracy of realization of the scale.

† The water used should have the isotopic composition of ocean water.

†† The equilibrium state between the solid and liquid phases of tin (freezing point of tin) has the assigned value of $t_{68} = 231.9681$ °C and may be used as an alternative to the boiling point of water.

platinum resistance thermometer is used for interpolation and any given platinum resistance thermometer is calibrated at the fixed points to determine the corrections required to reproduce the scale of the ideal thermometer. In practice platinum resistance thermometers are calibrated against the fixed points and standards laboratories use them to calibrate other thermometers (Cetas, 1985).

Air velocity (*v*)

Air movement across the body can influence heat flow to and from the body and hence body temperature. The movement of air will vary in time, space and direction. A description of air velocity at a point could therefore be in terms of a time variation in intensity in three orthogonal axes. For convenience the air velocity (or more correctly the scalar quantity, air speed) can be considered to be the 'mean' air velocity intensity over an exposure time of interest and integrated (e.g. square root of the sum of squares of air velocity in each direction) over all directions. Studies of human response, e.g. discomfort due to draught, have shown that variation in air velocity is important. ISO 7730 (1994) suggests that both mean air velocity and the standard deviation of the value should be taken. This also provides so-called *turbulence intensity* that is the ratio of the standard deviation of the air velocity to the mean air velocity.

The air movement (in combination with air temperature) will affect the rate at which warm air or vapour is 'taken' away from the body, thus affecting body temperature. Mean air velocity, as described above, provides a commonly used overall value for representing this effect on the body. More detailed analysis will increase complexity but may have practical value in some situations.

Humidity

If liquid, e.g. water or sweat, is heated by the human body, evaporates to a vapour and is lost to the surrounding environment, then heat has been transferred from the body to the environment and the body is cooled. The 'driving force' for this vapour (or mass) transfer is the difference in mass per unit volume of moist air: that is the difference in absolute humidity (mass concentration or density of water vapour) between that at the skin surface and that in the environment. For convenience, the 'driving force' for heat loss is considered to be the difference in partial vapour pressures between that at the skin and that in the environment. Kerslake (1972) connects the absolute humidity (in kg m^{-3}) and partial vapour pressure (P_a in kPa) by the gas laws and temperature T (in K):

$$\text{absolute humidity} = 2.17 \frac{P_a}{T}.$$

The humidity of the environment is therefore a basic parameter. It can be expressed in a number of forms; however, two are commonly used: relative humidity and partial vapour pressure.

Relative humidity ϕ (often given as a percentage) is the ratio of the prevailing partial pressure of water vapour to the saturated water vapour pressure,

$$\phi = \frac{P_a}{P_{sa}},$$

where the *partial vapour pressure*, $P_a(\mathrm{N\,m^{-2}})$ is the prevailing partial pressure of water vapour in the air.

The basic parameter is the humidity in the environment around the body which is representative of the driving force for heat loss by vapour or mass transfer. The concept of a *saturated vapour pressure* is important. If a kettle of water is boiled, or water is left on an open surface, vapour will transfer from the kettle (or surface) into the air. The water vapour in the air will exert a partial vapour pressure (P_a). If water vapour is continually transferred to the air, eventually a maximum amount will be reached where the air will hold no more. The vapour pressure at this point is termed saturated vapour pressure. The relative humidity at this point will be 100 per cent. The partial vapour pressure in the saturated air will depend upon the temperature of the air: the higher the air temperature, the more vapour the air can hold and hence the greater the partial vapour pressure. The saturated vapour pressure, P_{sa} (mb) at a temperature t (°C) is given by Antoine's equation:

$$P_{sa} = \exp\left(18.956 - \frac{4030.18}{t + 235}\right). \tag{1.5}$$

The various units of pressure used are related as follows:

$$1000\,\mathrm{N\,m^{-2}} = 1\,\mathrm{kPa} = 10\,\mathrm{mb} = 7.52\,\mathrm{mm\,Hg} = 7.52\,\mathrm{Torr}.$$

Another term sometimes used is the dew point: the temperature at which dew would begin to form if the air was slowly cooled (Kerslake, 1972). That is, if water vapour is in air, then if the air is cooled there will be a temperature at which the air cannot 'hold' the water vapour as it has become saturated at that temperature. Dew point temperature (t_{dp}) can be obtained from

$$t_{dp} = \frac{4030.18}{18.956 - 1nP_a} - 235.$$

It is interesting that heat can still be lost from the body, by vapour transfer, to air even if air temperature is below 0 °C. More detailed analysis is however required.

Air temperature, radiant temperature, air velocity, and humidity therefore are the four basic variables that should be quantified (measured/estimated) if one is considering human thermal environments. A derivation of this point, and the role of metabolic heat production and clothing, can be made using an analysis of heat transfer between the body and the environment and consideration of the body heat balance equation.

THE HEAT BALANCE EQUATION FOR THE HUMAN BODY

That the internal temperature should be maintained at around 37 °C dictates that there is a heat balance between the body and its environment. That is, on average, heat transfer into the body and heat generation within the body must be balanced by heat outputs from the body. That is not to say that a steady state occurs, since a steady state involves unchanging temperatures and temperatures within the body and avenues of heat exchange will vary; the point is that for a constant temperature there will be a dynamic balance.

If heat generation and inputs were greater than heat outputs, the body temperature would rise and if heat outputs were greater the body temperature would fall. The heat balance equation for the human body can be represented in many forms. However, all equations have the same underlying concept and involve three types of terms; those for heat generation in the body, heat transfer, and heat storage. The metabolic rate of the body (M) provides energy to enable the body to do mechanical work (W) and the remainder is released as heat (i.e. $M - W$). Heat transfer can be by conduction (K), convection (C), radiation (R), and evaporation (E). When combined together all of the rates of heat production and loss provide a rate of heat storage (S). For the body to be in heat balance (i.e. constant temperature), the rate of heat storage is zero ($S = 0$). If there is a net heat gain, storage will be positive and body temperature will rise. If there is a net heat loss, storage will be negative and body temperature will fall.

The conceptual equation

The conceptual heat balance equation is

$$M - W = E + R + C + K + S$$

i.e. for heat balance ($S = 0$)

$$M - W - E - R - C - K = 0,$$

where $M - W$ is always positive E, R, C and K are rates of heat loss from the body (i.e. positive value is heat loss, negative value is heat gain).

Units

It is important that all of the above terms can be expressed as rates of heat production or loss; this allows a simple addition of gains and losses. The units of the rate of energy gain or loss are energy per second, that is, joules per second ($J\,s^{-1}$) or watts (W). It is traditional (and useful) to 'standardize' over persons of different sizes by using units of watts per square metre of the total body surface area. The units are then $W\,m^{-2}$.

Body surface area

Total body surface area is traditionally estimated from the simplified equation of Dubois and Dubois (1916).
That is

$$A_D = 0.202 \times W^{0.425} \times H^{0.725},$$

where

A_D = Dubois surface area (m^2)
W = Weight of body (kg)
H = Height of body (m).

A standard value of 1.8 m^2 is sometimes used for a 70 kg man of height 1.73 m. It is recognized that A_D provides only an estimate of body surface area. There are more accurate equations, however, A_D is traditionally used and the nature of the heat balance equation means that any error is systematic and unimportant. In addition, objects of the same shape but different size have different heat transfer coefficients. This can be important but is not usually considered.

The practical heat balance equation

The principles and concept of the heat balance equation are described above. For an analysis of heat exchange between the body and the environment, and hence to quantify components of the equation to allow overall calculations (e.g. to make it useful in practice), there are two objectives:

1   The specific avenues of heat production and exchange for the human body must be identified.
2   Equations must be determined for the calculation (or estimate) of heat production and exchange. It is important for practical applications that terms in the equations are those which can be measured (or estimated) e.g. the basic parameters.

Fanger (1970) provides a classic analysis of this type to satisfy one of his necessary (but not sufficient) conditions for thermal comfort. He uses the following heat balance equation:

$$H - E_{dif} - E_{sw} - E_{res} - C_{res} = R + C,$$

where

H = metabolic heat production
E_{dif} = heat loss by vapour diffusion through skin
E_{sw} = heat loss by evaporation of sweat
E_{res} = latent respiration heat loss
C_{res} = dry respiration heat loss.

Note that for 'normal' conditions heat exchange by conduction is often assumed to be negligible. ASHRAE (1989a) gives the following equation of heat balance:

$$M - W = Q_{sk} + Q_{res} = (C + R + E_{sk}) + (C_{res} + E_{res}),$$

where all terms have units of $W\,m^{-2}$ and

M = rate of metabolic energy production
W = rate of mechanical work
Q_{sk} = total rate of heat loss from the skin
Q_{res} = total rate of heat loss through respiration
C = rate of convective heat loss from the skin
R = rate of radiative heat loss from the skin
E_{sk} = rate of total evaporative heat loss from the skin
C_{res} = rate of convective heat loss from respiration
E_{res} = rate of evaporative heat loss from respiration.

Note that

$$E_{sk} = E_{rsw} + E_{dif},$$

where

E_{rsw} = rate of evaporative heat loss from the skin through sweating
E_{dif} = rate of evaporative heat loss from the skin through moisture diffusion.

A practical approach is therefore to consider heat production within the body $(M - W)$, heat loss at the skin $(C + R + E_{sk})$ and heat loss due to respiration $(C_{res} + E_{res})$. The next objective is to quantify components of the heat balance equation in terms of parameters that can be determined (measured or estimated).

Heat production within the body $(M - W)$

Heat production within the body is related to the activity of the person. In general, oxygen is taken into the body (i.e. breathing air) and is transported by the blood to the cells of the body where it is used to burn food. Most of the energy released is in terms of heat. Depending upon the activity, some external work will be performed. Energy for mechanical work will vary from about zero (for many activities) to no more than 25 per cent of total metabolic rate. Methods for the determination and estimation of metabolic rate, which is one of the six basic factors for defining human thermal environments, are described in Chapter 6.

Heat loss at the skin $(C + R + E_{sk})$

Sensible heat loss $(R + C)$

ASHRAE (1997) gives the following derivation:

$$C = f_{cl}h_c(t_{cl} - t_a)$$
$$R = f_{cl}h_r(t_{cl} - t_r)$$
$$C + R = f_{cl}h(t_{cl} - t_o), \tag{1.6}$$

where

$$t_o = \frac{(h_r t_r + h_c t_a)}{h_r + h_c} \quad \text{and} \quad h = h_r + h_c.$$

The actual transfer of heat through clothing (conduction, convection, and radiation) is combined into a single thermal resistance value R_{cl}. So,

$$C + R = \frac{(t_{sk} - t_{cl})}{R_{cl}}. \tag{1.7}$$

Combining equations (1.6) and (1.7) to remove t_{cl} the final form is:

$$C + R = \frac{(t_{sk} - t_o)}{\left(R_{cl} + \frac{1}{f_{cl}h}\right)}, \tag{1.8}$$

where C and R are as defined above and

f_{cl} = clothing area factor. The surface area of the clothed body A_{cl}, divided by the surface area of the nude body A_D (ND)
h_c = convective heat transfer coefficient $(\text{W}\,\text{m}^{-2}\,\text{K}^{-1})$
h_r = linear radiative heat transfer coefficient $(\text{W}\,\text{m}^{-2}\,\text{K}^{-1})$
h = combined heat transfer coefficient $(\text{W}\,\text{m}^{-2}\,\text{K}^{-1})$
t_o = operative temperature (°C)
t_r = mean radiant temperature (°C)
t_a = air temperature (°C)
t_{cl} = mean temperature over the clothed body (°C)
t_{sk} = mean skin temperature (°C)
R_{cl} = thermal resistance of clothing $(\text{m}^2\,\text{K}\,\text{W}^{-1})$.

Air temperature (t_a), mean radiant temperature (t_r) and the thermal resistance of clothing (R_{cl}), are all basic variables or parameters which must be measured (or estimated) to define the environment (see Chapters 5 and 7). Mean skin temperature can be estimated as a constant value (e.g. around

33 °C for comfort and 36 °C under heat stress), or predicted from a dynamic model of human thermoregulation; see Chapter 15. All other values above can be calculated from values of basic parameters – this is discussed later in the section on heat transfer. However, for a seated person Mitchell (1974) gives

$$h_c = 8.3v^{0.6} \quad \text{for } 0.2 < v < 4.0$$
$$h_c = 3.1 \quad\quad \text{for } 0 < v < 0.2,$$

where v is the air velocity in m s^{-1} (i.e. a basic parameter).

The radiative heat transfer coefficient h_r can be given by

$$h_r = 4\varepsilon\sigma \frac{A_r}{A_D} \left[273.2 + \frac{t_{cl} + t_r}{2} \right]^3, \tag{1.9}$$

where

ε = the area weighted emissivity of the clothing body surface (ND)

σ = Stefan–Boltzmann constant, $5.67 \times 10^{-8} \left(\text{W m}^{-2}\,\text{K}^{-4} \right)$

A_r = effective radiative area of the body (m^2).

ε is often assumed to be between 0.95 to 1.0; A_r/A_D can be estimated as 0.70 for a sitting person and 0.73 for a standing person (Fanger, 1967); t_r is a basic parameter; and t_{cl} must be calculated using iteration techniques. ASHRAE (1997) suggests that a value of $h_r = 4.7$ W m^{-2} K^{-1} is a reasonable approximate for 'typical indoor conditions'.

Evaporative heat loss from the skin (E_{sk})

ASHRAE (1997) gives the following equation:

$$E_{sk} = \frac{w \left(P_{sk,s} - P_a \right)}{\left[R_{e,cl} + \frac{1}{f_{cl}h_e} \right]}, \tag{1.10}$$

where

P_a = water vapour pressure in the ambient air (kPa)

$P_{sk,s}$ = water vapour pressure at the skin, normally assumed to be that of saturated water vapour at skin temperature, t_{sk} (kPa)

$R_{e,cl}$ = evaporative heat transfer resistance of the clothing layer (m^2 kPa W^{-1})

h_e = evaporative heat transfer coefficient (W m^{-2} kPa^{-1})

w = skin wettedness (Gagge, 1937). The fraction of wet skin (ND);

h_e is calculated using the Lewis Relation $h_e = LR\,h_c$. This is an important development in the establishment of the body heat balance equation allowing comparison and combination of dry and evaporative heat transfer.

The values P_a and $R_{e,cl}$ are basic parameters. $P_{sk,s}$ is calculated from Antoine's equation (equation 1.5) using the value for mean skin temperature t_{sk} (assumed constant value or from a dynamic model of human thermoregulation (see Chapter 15)). Skin wettedness varies from a value of 0.06 when only natural diffusion of water through the skin occurs (E_{dif}), to 1.0 when skin is completely wet and maximum evaporation occurs (E_{max}).

E_{max} can be calculated by letting $w = 1.0$ in the above equation (1.10). Skin wettedness can be calculated from:

$$w = 0.06 + 0.94\,\frac{E_{rsw}}{E_{max}}, \qquad (1.11)$$

where

$$E_{rsw} = M_{rsw} \times h_{fg}, \qquad (1.12)$$

and

$$h_{fg} = \text{heat of vaporization of water} = 2430\,\text{kJ}\,\text{kg}^{-1} \text{ at } 30\,^\circ\text{C}$$
$$M_{rsw} = \text{rate at which sweat is secreted } (\text{kg}\,\text{s}^{-1}\,\text{m}^{-2}).$$

E_{max} can be determined from the basic parameters and equation (1.10). Skin evaporation and skin wettedness are often calculated from the heat balance equation in terms of wettedness required (w_{req}) or evaporation required (E_{req}), to maintain the body in heat balance. They can be calculated by determining E_{rsw} using the equation (1.12) and

$$M_{rsw} = 4.7 \times 10^{-5}\,W_{SIG_b}\,\exp\left(\frac{W_{SIG_{sk}}}{10.7}\right),$$

where n signifies neutral, and

$$W_{SIG_b} = t_b - t_{b,n} \quad \text{for} \quad t_b > t_{b,n}$$
$$W_{SIG_{sk}} = t_{sk} - t_{sk,n} \quad \text{for} \quad t_{sk} > t_{sk,n},$$

and mean body temperature $t_b = \alpha t_{sk} + (1 - \alpha)t_{cr}$, i.e. a weighted average of skin and core temperature, with weighting α depending upon degree of vasodilation.

$$t_{sk,n} = 33.7\,^\circ\text{C}$$

i.e. mean skin temperature for a person in thermal neutrality.

$$t_{cr,n} = 36.8\,°C$$

i.e. core temperature in thermal neutrality.

Typical values for α are 0.2 for thermal equilibrium while sedentary, 0.1 in vasodilation and 0.33 for vasoconstriction.

The mean body temperature in neutrality (neutral bulk temperature) $t_{b,n}$ is therefore

$$t_{b,n} = (0.2 \times 33.7) + (0.8 \times 36.8)$$
$$= 36.18\,°C.$$

Heat loss from respiration $(C_{res} + E_{res})$

Heat loss from respiration is by 'dry' convective heat transfer due to cool air being inhaled, heated to core temperature in the lungs, and heat transferred in exhaled air to the environment (C_{res}). In addition, inhaled air is moistened (to saturation) by the lungs. When exhaled, therefore, there is a mass (heat) transfer from the body core to the outside environment (E_{res}). ASHRAE (1997) gives the following equation for total respiratory heat loss:

$$C_{res} + E_{res} = [0.0014\,M(34 - t_a) + 0.0173\,M(5.87 - P_a)].$$

A_D is calculated as described above. All other values are basic parameters, with P_a in kPa, M in $W\,m^{-2}$ and t_a in °C.

EXAMPLE CALCULATION

The presentation of the human heat balance equation above is in summary form yet still rather cumbersome. The usefulness of the equation in practice is the ability to identify and calculate values showing the relative contribution of each of the components. A hypothetical example is given below. For example, consider a man standing in a hot environment conducting light work in light clothing.

Basic parameters

For the example the following measurements and estimates have been made to define the human thermal environment:

Air temperature: $t_a = 30\,°C$
Mean radiant temperature: $t_r = 40\,°C$

Relative humidity: $\phi = 60\%$

Air velocity: $v = 0.25\,\mathrm{m\,s^{-1}}$

Metabolic rate (light work): $M = 100\,\mathrm{W\,m^{-2}}$

External work: $W = 0\,\mathrm{W\,m^{-2}}$

Clothing: $R_{cl} = 0.093\,\mathrm{m^2\,°C\,W^{-1}}$

$R_{e,cl} = 0.015\,\mathrm{m^2\,kPa\,W^{-1}}$

It is assumed here that R_{cl} and $R_{e,cl}$ have been measured, or are available, from tables. It may be more convenient to express the insulation value as a basic parameter in Clo. R_{cl} can then be calculated (in $\mathrm{m^2\,°C\,W^{-1}}$) using $1\,\mathrm{Clo} = 0.155\,\mathrm{m^2\,°C\,W^{-1}}$ (i.e. in the example, clothing insulation is $0.6\,\mathrm{Clo} = 0.093\,\mathrm{m^2\,°C\,W^{-1}}$). $R_{e,cl}$ values are not widely available and the example assumes that we know this value. A value of $0.015\,\mathrm{m^2\,kPa\,W^{-1}}$ is however, typical and could be assumed in some cases. Evaporative resistance may also be described by a variety of clothing indices which can be used directly in the heat balance equation or to calculate $R_{e,cl}$, for example i_m or F_{pcl} – see Chapter 7.

Simple calculations

Metabolic heat production $(\mathrm{W\,m^{-2}}) = M - W = 100 - 0 = 100$

$$f_{cl} = 1 + 0.31\,\mathrm{Clo} = 1 + \frac{0.31\,R_{cl}}{0.155} = 1.186.$$

For $t_{sk} = 35\,°\mathrm{C}$ saturated vapour pressure at skin temperature is given by

$$P_{sk,s} = \exp\left(18.956 - \frac{4030.18}{35 + 235}\right)\,\mathrm{mb} = 56.23\,\mathrm{mb} = 5.623\,\mathrm{kPa}.$$

For $t_a = 30\,°\mathrm{C}$ saturated vapour pressure at air temperature is given by

$$P_{sa} = \exp\left(18.956 - \frac{4030.18}{30 + 235}\right)\,\mathrm{mb} = 4.243\,\mathrm{kPa}$$

$$P_a = \phi P_{sa} = 0.6 \times 4.243 = 2.55\,\mathrm{kPa}$$

Assumptions: $t_{sk} = 35\,°\mathrm{C}$; $\varepsilon = 0.95$; $A_r/A_D = 0.77$; $\mathrm{LR} = 16.5\,\mathrm{K\,kPa^{-1}}$; $A_D = 1.8\,\mathrm{m^2}$.

Heat transfer coefficients

$v > 0.2\,\mathrm{m\,s^{-1}}$ therefore,

$$h_c = 8.3 v^{0.6} = 3.61\,\mathrm{W\,m^{-2}\,K^{-1}}$$

$$h_e = 16.5\,h_c = 59.61\,\mathrm{W\,m^{-2}\,kPa^{-1}}$$

Calculating h_r and t_{cl} using iteration (from equation (1.9)):

$$h_r = 4 \times 0.95 \times 0.77 \times 5.67 \times 10^{-8} \times \left(\frac{t_{cl} + 40}{2} + 273.2\right)^3$$

$$= 16.59 \times 10^{-8} \left(\frac{t_{cl} + 40}{2} + 273.2\right)^3$$

$$t_{cl} = \frac{\frac{1}{R_{cl}} t_{sk} + f_{cl}(h_c t_a + h_r t_r)}{\frac{1}{R_{cl}} + f_{cl}(h_c + h_r)}$$

$$= \frac{10.75 \times 35 + 1.186(108.30 + 40h_r)}{10.75 + 1.186(3.61 + h_r)}.$$

Starting with $t_{cl} = 0.0$, and repeatedly evaluating new values for $h_r, t_{cl}, h_r, t_{cl}, \ldots$ until the difference between two consecutive values of $t_{cl} \le 0.01$, for this example provides:

$$h_r = 4.99 \, \text{W m}^{-2} \, \text{K}^{-1}$$
$$t_{cl} = 35.4\,^{\circ}\text{C}.$$

Operative temperature is then given by

$$t_o = \frac{(h_r t_r + h_c t_a)}{(h_r + h_c)}$$

$$= \frac{(4.99 \times 40) + (3.61 \times 30)}{4.99 + 3.61} = 35.8\,^{\circ}\text{C}.$$

Combined heat transfer coefficient

$$h = h_c + h_r$$
$$= 3.61 + 4.99$$
$$= 8.60 \, \text{W m}^{-2} \, \text{K}^{-1}.$$

Calculation of the components of the heat balance equation

$$C + R = \frac{(t_{sk} - t_o)}{\left(R_{cl} + \frac{1}{f_{cl} h}\right)}$$

$$= \frac{35 - 35.8}{\left(0.093 + \frac{1}{(1.186 \times 8.6)}\right)}$$

$$= -4.18 \, \text{W m}^{-2}$$

(i.e. a heat gain).

$$E_{sk} = \frac{w(P_{sk,s} - P_a)}{\left(R_{e,cl} + \frac{1}{f_{cl}h_e}\right)}$$

$$= \frac{w(5.623 - 2.55)}{\left(0.015 + \frac{1}{(1.186 \times 59.61)}\right)}$$

$$= w \times 105.58 \, \mathrm{W\,m^{-2}}.$$

For $w = 1$, $E_{sk} = E_{max}$, so $E_{max} = 105.58 \, \mathrm{W\,m^{-2}}$.

$$\begin{aligned}
C_{res} + E_{res} &= 0.0014 \, M(34 - t_a) + 0.0173 \, M(5.87 - P_a) \\
&= 0.0014 \times 100(34 - 30) + 0.0173 \times 100(5.87 - 2.55) \\
&= 6.3 \, \mathrm{W\,m^{-2}}.
\end{aligned}$$

The heat balance equation then becomes:

$$\begin{aligned}
M - W &= C + R + E_{sk} + C_{res} + E_{res} \\
100 - 0 &= -4.18 + (w \times 105.58) + 6.3.
\end{aligned}$$

Thus a skin wettedness of 0.93 will provide sufficient heat loss at the skin through evaporation. That is, the body will sweat to thermo-regulate and achieve heat balance. For maximum evaporation, E_{max} wettedness is 1. For the above example this gives a value of $E_{max} = 105.58 \, \mathrm{W\,m^{-2}}$. The required wettedness for balance is given by

$$w_{req} = \frac{E_{req}}{E_{max}}.$$

Therefore,

$$E_{req} = w_{req}E_{max} = 0.93 \times 105.58 = 97.88 \, \mathrm{W\,m^{-2}}$$

(i.e. E_{sk} for heat balance).

Using the latent heat of vaporization of water ($22.5 \times 10^{-5} \, \mathrm{J\,kg^{-1}}$) and taking account of the efficiency of sweating, r (e.g. some sweat drips and latent heat is not lost) ISO 7933 uses

$$r = 1 - \frac{w^2}{2}.$$

The sweating required (to provide the evaporation required) can then be calculated as

$$S_{req} = \frac{E_{req}}{r} \, \mathrm{W\,m^{-2}}$$

where an evaporated sweat rate of 0.26 litres (kg) per hour corresponds to a heat loss of $100\,\mathrm{W\,m^{-2}}$. For the above example, to maintain heat balance the body would therefore have to produce 0.454 litres of sweat per hour. That is,

$$r = 1 - \frac{0.93^2}{2}$$
$$\cong 0.56$$
$$S_{\mathrm{req}} = \frac{97.88}{0.56}$$
$$\cong 174.79\,\mathrm{W\,m^{-2}}$$
$$\cong 0.454 \text{ litres per hour.}$$

The above example demonstrates a calculation of the heat balance equation and how it can be used to identify the relative importance of each of the components. Although sweating required in a hot environment is used in the example, requirements for other components can also be calculated: for example, in terms of metabolic rate, air velocity, or the insulation of clothing.

The conceptual basis for the body heat balance equation is well established. However, equations for components of the equation are continually updated from the results of research. Of particular importance has been the theory of heat transfer. Much of the development of the heat balance equation for the human body has been adapted from heat (and mass) transfer theory. An understanding of this theory is useful for a thorough understanding of the body heat balance equation and a brief summary is provided in Appendix.

Heat balance and clothing ventilation

An important aspect of work that has not been considered in the above example is the effects of human movement on the thermal properties of clothing. In hot environments saturated air at skin temperature can be transferred directly from the skin to the air through vents and openings in clothing. In the example provided, one litre of saturated air at skin temperature $(t_a = 35\,^\circ\mathrm{C}, \phi = 100$ per cent) will contain approximately 140 joules of heat. In the environment one litre of air $(t_a = 30\,^\circ\mathrm{C},$ $\phi = 60$ per cent) contains approximately 80 joules of heat. For every litre of air lost directly from the skin – through openings in clothing and replaced by environmental air, $140 - 80 = 60\,\mathrm{J}$ of heat is lost to the environment. (Further details of this calculation are provided in Chapters 5 and 7.)

The amount of heat transferred depends upon a number of factors including the environmental conditions (in very hot environments heat could be transferred to the body), skin condition, the ventilation and dynamic properties of clothing, and the activity of the person. Ventilation of clothing varies from zero (completely sealed) through 1 litre min^{-1} for very low exchange conditions to 60 litres min^{-1} for medium conditions and up to 300 litres min^{-1} for relatively high levels of exchange. Therefore heat loss with air leaving the clothing environment is equal to the heat content of air leaving the ensemble less the heat content when entering $\cong 140$ J/litre $- 80$ J/litre $= 60$ J/litre.

In the above example, if we assume medium levels of exchange (60 litres min$^{-1} = 1$ litre s^{-1}) then 60 W of heat is lost. ($\cong 33.3$ W m^{-2} assuming 1.8 m^2 for body surface area.) If we ignore any interaction with the intrinsic heat transfer properties of clothing (and other heat transfer mechanisms) we can include this as a heat loss in the heat balance equation.

The heat balance equation – including the effects of clothing ventilation – will now be

$$M - W = C + R + E_{sk} + C_{res} + E_{res} + \text{Heat loss due to clothing ventilation.}$$

So,

$$100 - 0 = -4.18 + w \times 105.58 + 6.3 + 33.3,$$

which gives, for heat balance, $w = w_{req} = 0.61$. Required evaporation is now

$$0.61 \times 105.58 = 64.58 \text{ W m}^{-2}.$$

Taking account of sweating efficiency as above, this now corresponds to 0.21 litres of sweat required per hour – a significant reduction from 0.454 litres per hour in the previous analysis.

Heat balance and solar load

For work influenced by solar radiation there will be an additional heat input to the person. On a clear day with full sun the direct, diffuse and reflected radiation can provide an input of about 800 W m^{-2} to the person. On a cloudy day this may be reduced to about 200 W m^{-2}. The amount of energy received by the person by direct solar radiation will depend upon the projected angle (A_p).

Burton and Edholm (1955) suggest that, for a clear sun, the mean radiation incident in a man is equivalent to about 4.6 Mets (where 1 Met $= 58$ W m^{-2}). They provide the following equation for estimating radiation (R)

$$R = 4.6(1 - 0.9x) \times a \text{ Mets,}$$

where R is the solar radiation output; x is the amount of cloud ($10/10 = 1 =$ overcast; $0/10 = 0 =$ clear sky); a is the absorbing power of clothes where $a = 0.88$ for black clothing, 0.57 for khaki and 0.20 for white clothing.

In the above example, we could assume a thin cloud cover of 5/10 and black clothing $a = 0.88$. So,

$$R = 4.6 \times 58(1 - [0.9 \times 0.5]) \times 0.88$$
$$= 129 \, \text{W m}^{-2}.$$

The heat balance equation including the effects of clothing ventilation and solar load will now be

$$M - W = C + R + E_{sk} + C_{res} + E_{res} + \text{Heat loss to clothing ventilation}$$
$$+ \text{Heat 'loss' due to solar load}$$
$$100 - 0 = -4.18 + w\,105.58 + 6.3 + 33.3 - 129,$$

which gives for heat balance $w = w_{req} = 1.83$. If we assume that the maximum wettedness is $w = 1$, then heat balance cannot be achieved. If we put $w = 1$ into the above equation we see that there is an imbalance and that $90 \, \text{W m}^{-2}$ is stored in the body. This would correspond to a rise in body temperature of

$$\frac{\Delta T}{dt} = \frac{90 \times 1.8}{70 \times 3.49 \times 10^3} \, ^\circ\text{C s}^{-1}$$
$$= 6.63 \times 10^{-4} \, ^\circ\text{C s}^{-1}.$$
$$= 0.04 \, ^\circ\text{C min}^{-1}.$$

For a 70 kg man with a surface area of $1.8 \, \text{m}^2$ and specific heat requirements of the body of 3.49 kJ/(kgK). So for a limit of 1 °C rise in body temperature the man could work in the sun for around 25 min. The exposure time may be increased because of the influence of the thermal inertia of the body and clothing; however, it can be seen that the solar load clearly increases the heat stress on a person and emphasizes the importance of shade, weather conditions and colour of clothing.

FROM THERMAL AUDIT TO EDUCATED GUESS: A PRACTICAL APPROACH

The human heat balance equation has been developed from a knowledge of human physiology and heat transfer theory. Equations have been developed that can easily be calculated on a computer. The nature of human behaviour however is such that, however complex the equations become, they will never provide a perfect representation of heat transfer. In practical terms, it is the discipline of systematic analysis of avenues of heat production and transfer that provides a powerful assessment. Parsons (1992a) terms this the 'thermal audit', which he says 'allows the ergonomist to analyse any work situation and "account for" avenues of heat transfer between the human body and its environment.' The implication is that the thermal audit is an essential part of a practical environmental assessment or design; that is, it should always be carried out. Withey and Parsons (1994) suggest that the thermal audit based upon equations is a 'valuable tool when reliable data are available, when the necessary hardware and software tools are at hand, and when the risk assessor has an adequate understanding of the assumptions and calculations'. A qualitative approach is the proposal that provides a systematic basis for thermal risk assessment and risk management. It can be argued that if the assessor has an adequate understanding of the assumptions and principles behind the calculation, a thermal audit can be conducted without resort to complex calculations. This is demonstrated with the example given earlier in the chapter.

Practical problem and the thermal audit

Consider a man standing in a hot environment conducting light work in light clothing. Air temperature is 30 °C and mean radiant temperature is 40 °C. Relative humidity is 60 per cent and there is perceptible light air movement. The clothing is of a light boiler suit type, which is air permeable.

Analysis

Metabolic heat production – light work is around $100 \, \text{W} \, \text{m}^{-2}$.

Dry heat transfer – is driven by the difference between skin temperature and environmental temperature. As air temperature is 30 °C and mean radiant temperature is 40 °C, operative temperature will be about 35 °C which will be very close to skin temperature in a warm to hot environment – so there will only be a small dry heat transfer between the skin and the environment.

Heat loss by evaporation at the skin – as the dry heat transfer will be small, metabolic heat will have to be balanced by evaporation of sweat. The skin will therefore be wet and at a partial vapour pressure close to saturated vapour pressure at skin temperature (35 °C). The driving force for evaporation will be the difference in partial vapour pressures between skin and air. As relative humidity is 60 per cent at an air temperature of 30 °C, then there should be a significant vapour pressure gradient between skin and air. As clothing is air permeable, significant evaporation should be possible even in light air movement. Saturated vapour pressure at skin temperature is around 5.5 kPa and at 30 °C, around 3.5 kPa. So vapour pressure in the air at 60 per cent relative humidity is around 2.5 kPa – the difference providing a driving force of $5.5 - 2.5 = 3$. The evaporative heat transfer coefficient is 16.5 times the convective heat transfer coefficient, so can be estimated from $16.5 \times 8.3 \times \sqrt{v}$ which is around $135\sqrt{v} (\cong 70)$ so total heat transfer will be $70 \times 3 = 210\,\mathrm{W\,m^{-2}}$. Clothing will, however, reduce this value by around 0.5 so maximum heat transfer by evaporation will be around $105\,\mathrm{W\,m^{-2}}$.

Heat transfer by respiration – as the air temperature is $30\,°\mathrm{C}$ and 60 per cent humidity, some heat will be lost by respiration. This is generally small and close to a maximum of $20\,\mathrm{W\,m^{-2}}$. As the air is warm and humid an estimate of respiratory heat loss would be around $10\,\mathrm{W\,m^{-2}}$.

Based upon a knowledge of the principles of the heat balance equation and a rationale of educated guesses, this thermal audit provides the following heat balance equation:

$$\text{Metabolic heat} = \text{dry heat loss} + \text{evaporation heat loss}$$
$$(100) \qquad\qquad (0) \quad + \quad (w\ 105)$$
$$+ \text{respiratory heat loss}$$
$$(10).$$

So, if the body is wet with sweat, a balance can be achieved. This will be the condition of the workers who may be uncomfortable. Dehydration will be an issue. If clothing allows pumping, then evaporative potential will increase, as well as some dry heat loss. This would reduce the required sweating. If solar radiation is present it will depend upon the weather conditions. In full sun and extra load of around $200\,\mathrm{W\,m^{-2}}$ would be added, which would make heat balance impossible and thus allowable exposure times would be limited.

Comparison of the analysis with the method using the equations of heat transfer shows very similar results. The systematic approach and training in the principles of the heat balance equation can provide a realistic thermal audit. This point is taken further in Chapter 10 when the issues are presented on how to represent principles of assessment in a useable form.

2 Human thermal physiology and thermoregulation

INTRODUCTION

The human heat balance equation describes how the body (homeotherm) can maintain an internal body temperature near to 37 °C in terms of heat generation and heat exchange with the environment. In practice, what is achieved is not a steady state (constant temperatures) but a dynamic equilibrium: as external conditions continually change, so the body responds to 'regulate' internal body temperature.

Metabolic heat is produced in the cells of the body, each of which must also maintain homeothermy. This heat is transferred from the cell to its surroundings mainly by conduction, due to thermal gradients between the cell and its surroundings, and by convection, due to movement of extracellular fluids, e.g. blood. There is therefore a dynamic and complex heat transfer between the cells of the human body that will depend upon the thermophysical and physiological properties of the cells, e.g. the thermal conductivity, density and specific heat of the cells and blood perfusion rates. If the body did not lose heat to the environment (i.e. if it were completely 'lagged') then, although there may be heat exchanges within the body, there would be no effective temperature gradient between the body and the environment. Heat would then be stored and body temperature would rise at about 1 °C per hour for a resting person. For most cases, however, there is an effective temperature gradient between the internal body and the human skin. There is a net heat transfer from the cells of the body to the surface of the body where it can be lost to the environment by conduction, convection, radiation and evaporation at the skin surface and the lungs.

The thermal properties of blood, muscle, fat, bone, etc. will therefore be important for internal heat transfer and hence body heat exchange. However, to regulate temperature in a changing environment, this 'passive' system must be controlled by a dynamic system of thermoregulation. Both of these systems are discussed separately below, however, it is important to remember that the body functions as a 'whole' and not as separate component parts.

THE THERMOPHYSICAL PROPERTIES
OF THE HUMAN BODY

Heat is generated within the cells of the body and thermophysical properties of the body will determine its rate of transfer to the surface. The measurement of the thermal properties of biological materials is discussed by Chato (1985). It should be remembered that living materials are continually changing, that blood perfusion will greatly influence these properties and that rate of blood flow will greatly influence heat transfer. For the development of thermal models (see Chapter 15) these properties are required for the passive system of the model. Values used by Stolwijk and Hardy (1977) are shown in Table 2.1.

The average insulation of tissue, over the temperature gradient from the 'core' to the skin of the body was measured as 0.15 Clo units for vasodilated skin, to 0.9 Clo units for vasoconstricted skin using human subjects immersed in water baths (Burton and Bazett, 1936) where 1 Clo $= 0.155\,\mathrm{m^2\,°C\,W^{-1}}$. This can be compared with 0.16 Clo and 0.64 Clo for vasodilated and vasoconstricted skin (Monteith and Unsworth, 1990). Burton and Edholm (1955) also suggest changes of between three and four times in tissue insulation values between vasodilated and vasoconstricted tissues (Hardy and Dubois, 1938; Winslow *et al.*, 1936).

It is important to recognize that the above figures are average values; the reality is more complex. Burton and Edholm (1955) give the example of heat reaching the hand. This will depend upon insulation to flow of heat down the length of the arm, the insulation from the arteries to the distributing subcutaneous vascular bed and the resistance through the skin to the surface.

The effects of blood flow

ASHRAE (1993) describes a single model of blood flow and heat transfer based on that used by Gagge *et al.* (1971). This two-node model (see Chapter 15) has a 'shell' (skin) and a 'core'. Normal blood flow for sedentary activity at thermally neutral body conditions is $1.75\,\mathrm{g\,s^{-1}\,m^{-2}}$. For each 1 °C rise above thermo-neutral body core temperature (36.8 °C) blood flow increases by $56\,\mathrm{g\,s^{-1}m^{-2}}$. For each 1 °C decrease below thermo-neutral body skin temperature (33.7 °C) there is a proportional resistance to blood flow on average. (This resistance depends upon the area of the body, i.e. greater in the hands and feet than in the trunk where it is negligible for example.) For the two-node model the effects of core (t_{cr}) and skin (t_{sk}) temperatures (°C) on blood flow (m_{bl}) are given by

$$m_{bl} = \frac{1}{3600}\left[6.3 + \frac{200\,\mathrm{W_{SIG_{cr}}}}{1 + 0.5\,\mathrm{C_{SIG_{sk}}}}\right], \tag{2.1}$$

Table 2.1 Estimated basal heat production, blood flow for each compartment and thermal conductance between compartments

Segment	Compartment (N)	Volume, cm³	Length, cm	Radius, cm	Average cond. W/(cm °C) ×10⁴	TC (N) W/°C	Basal heat production QB (N) W	Basal blood flow BFB (N) 1 h⁻¹
1 Head	1 Core	3010		8.98	41.8	1.61	14.95	45.00
	2 Muscle	3380		9.32	41.8	13.25	0.12	0.12
	3 Fat	3750		9.65	33.4	16.10	0.13	0.13
	4 Skin	4020		9.88	33.4		0.10	1.44
2 Trunk	5 Core	14 680	60	8.75	41.8	1.59	52.63	210.00
	6 Muscle	32 580	60	13.15	41.8	5.53	5.81	6.00
	7 Fat	39 650	60	14.40	33.4	23.08	2.49	2.56
	8 Skin	41 000	60	14.70	33.4		0.47	2.10
3 Arms	9 Core	2240	112	2.83	41.8	1.40	0.82	0.84
	10 Muscle	5610	112	4.48	41.8	10.30	1.11	1.14
	11 Fat	6580	112	4.85	33.4	30.50	0.21	0.20
	12 Skin	7060	112	5.02	33.4		0.15	0.50
4 Hands	13 Core	260	96	0.93	41.8	6.40	0.09	0.10
	14 Muscle	330	96	1.04	41.8	11.20	0.23	0.24
	15 Fat	480	96	1.27	33.4	11.50	0.04	0.04
	16 Skin	670	96	1.49	33.4		0.06	2.00
5 Legs	17 Core	6910	160	3.71	41.8	10.50	2.59	2.69
	18 Muscle	17 100	160	5.85	41.8	14.40	3.32	3.43
	19 Fat	19 480	160	6.23	33.4	74.50	0.50	0.52
	20 Skin	20 680	160	6.42	33.4		0.37	2.85
6 Feet	21 Core	430	125	1.06	41.8	16.30	0.15	0.16
	22 Muscle	510	125	1.14	41.8	20.60	0.02	0.02
	23 Fat	730	125	1.36	33.4	16.40	0.05	0.05
	24 Skin	970	125	1.57	33.4		0.08	3.00
	25 Central blood	2500						3.00
Total							86.44	285.13

Source: Stolwijk and Hardy (1977).

for $1.4 \times 10^{-4} < m_{bl} < 2.5 \times 10^{-2}\,\text{kg s}^{-1}\,\text{m}^{-2}$,

$$\text{and } W_{SIG_{cr}} = 0 \qquad \text{for} \quad t_{cr} \leq 36.8$$
$$= t_{cr} - 36.8 \quad \text{for} \quad t_{cr} > 36.8,$$

$$\text{and } C_{SIG_{sk}} = 33.7 - t_{sk} \quad \text{for} \quad t_{sk} < 33.7$$
$$= 0 \qquad\qquad \text{for} \quad t_{sk} \geq 33.7,$$

where $1\,\text{kg} = 1$ litre of blood.

The combined thermal exchange between the core and skin (Q_{crsk}) for this two-node model can then be written

$$Q_{crsk} = (K + C_{p,bl} m_{bl})(t_{cr} - t_{sk}), \tag{2.2}$$

where

K = effective conductance between core and skin, $5.28\,\text{W m}^{-2}\,\text{K}^{-1}$

$C_{p,bl}$ = specific heat capacity of blood, $4.187\,\text{kJ kg}^{-1}\,\text{K}^{-1}$.

Also for the two-node model the effect of blood flow on the relative masses of skin and core components can be written as

$$\alpha = 0.0418 + \frac{0.745}{(m_{bl} - 0.585)}, \tag{2.3}$$

where α is the relative mass of skin to core. Mean body temperature t_b is then given by

$$t_b = \alpha t_{sk} + (1 - \alpha)t_{cr}. \tag{2.4}$$

In ASHRAE (1997) the coefficient in equation (2.1) is changed from 200 to 175 and limiting temperatures for $W_{SIG_{cr}}$ and $C_{SIG_{sk}}$ are 37 °C and 34 °C respectively. In equation 2.3 m_{bl} is in $\text{Lh}^{-1}\text{m}^{-2}$ limited from 90 $\text{Lh}^{-1}\text{m}^{-2}$ to 3 $\text{Lh}^{-1}\text{m}^{-2}$. This simple two-node model demonstrates the 'average' change in the thermal properties of tissues with blood flow. It is important to recognize that this is a simplification. In the human body there is much variation across the body, and there are confounding factors. For example, in exercise or shivering, blood flow is increased even when 'attempting' to preserve heat. Burton and Bazett (1936) for example show that greatest tissue resistance varies under cool conditions: for cold conditions shivering causes blood flow, which decreases resistance.

A detailed description of the thermophysical properties of the human body is not available. However what is not in dispute is that the properties are important in defining the so-called passive or controlled part of the human thermal system. To maintain the body within acceptable temperatures a 'controlling' system can be considered to act on the 'controlled' system. This is the system of human thermoregulation.

Human thermoregulation

A most powerful form of human thermoregulation is behavioural; put on or take off clothes, change posture, move, take shelter, etc. The human body also has a physiological system of thermoregulation and it is this that is described below. Both systems continually interact and respond to changing environments in an attempt to ensure human survival and comfort.

Not only does the body maintain internal temperature at around 37 °C; it also maintains cell temperatures all over the body at levels which avoid damage. There are regulating processes which operate within the cells and it is the way in which the whole body regulates all cell temperatures that make up the system of human thermoregulation. Bligh (1985) suggests 'the constancy of every living cell relative to that of its immediate, and also more remote, environments, requires a mass of regulator processes...' and that 'a difficulty in any study of biological homeostasis or regulation is that in even the simplest unit of life, the cell, the stabilizing processes are so numerous and inter-related that it can be difficult to determine what principle underlies the stabilizing function.'

Although the sum of the parts make up the whole (or greater), at the level of the 'whole-body' it can be argued that there is a central regulating system. There have been numerous studies of this system and a detailed discussion is given by Bligh and Moore (1972), Hensel (1981), Bligh (1985) and Gregor and Windhorst (1996). For most practical purposes it is useful to consider a simple engineering analogy. Although, as Hensel (1981) points out, 'analogies from technology may be misleading since the principles actually used by control engineers are only part of the theoretical possibilities of control systems.' The development of the digital computer has increased the practical possibilities; however, the point is still valid. There has been some dispute about what is the regulated variable in thermoregulation. Bligh (1978) considers this and concludes that evidence suggests that temperature or a function of various temperatures is the controlled variable, as opposed to such ideas that body heat content is the controlled variable (Houdas *et al.*, 1972; Houdas and Guicu, 1975) or that rate of heat outflow is adjusted to balance metabolic heat production (Webb *et al.*, 1978). There are numerous proposed system models of human thermoregulation. Four simple models are discussed below. Although they are different in composition, for most practical purposes they are almost identical and can explain human thermoregulatory response. All models recognize that when the body becomes hot it

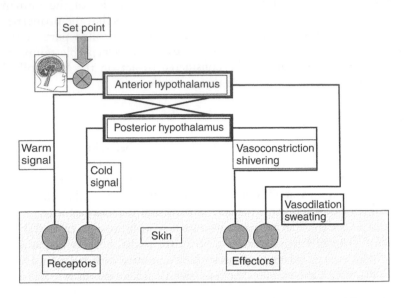

Figure 2.1 Simplified diagram of the thermoregulatory system. The main routes
are shown in heavy lines. Displacement of the brain temperature
above the set point results in vasodilation and sweating: reduction in
skin temperature produces vasoconstriction and shivering. There are
crossed inhibitory connections between the warm and cold systems.

Source: Modified from McIntyre (1980).

loses heat by vasodilation and, if required, sweating. If the body becomes cold
then heat is preserved by vasoconstriction and, if necessary, generated by
shivering. Another fundamental point of agreement is that the primary
control centre for thermoregulation is in the hypothalamus.

The details of how information is sensed by the body and transferred to
the controller, integrated and processed by the controller, and transformed
into effector signals that stimulate effective output, are not known. In
particular the establishment of the so-called 'set point' and how it changes
with factors such as time or exercise is not fully understood.

It is generally accepted that temperature sensors inside the body are
situated in the hypothalamus as well as in the medulla, spinal cord and other
sites. There also appears to be some local control for example by the spinal
cord (Bligh, 1985). There are two types of thermal sensor distributed across
the skin of the body, so-called warm and cold receptors. Signals from these
sensors, as well as from 'core' sensors, are integrated at the hypothalamus.

McIntyre (1980) provides the simplified system shown in Figure 2.1,
which he suggests is sufficient for most practical applications. In the model,
if the brain temperature rises above the set point the anterior hypothalamus
causes vasodilation and sweating. Reduction in skin temperature produces

vasoconstriction and also contributes to shivering. Inhibitory pathways between the anterior and posterior hypothalamus provide 'damping' and prevent the system working against itself and 'unnecessary oscillation'.

Stolwijk and Hardy (1966) provide a closed loop, negative feedback system representation of human thermoregulation (Figure 2.2). In this system, the thermal response is proportional to the difference between the value of the controlled variable (e.g. hypothalamus temperature) and its 'required value' (set point). The constant of proportionality differs for each response and is positive or negative; see, for example, equation (2.1) for blood flow given above and proposed by ASHRAE (1993). A similar system of thermoregulation is used in the thermal models of Stolwijk and Hardy (1977) and Gagge *et al.* (1971).

The strength of the signal (warm or cold) is proportional to the difference between actual core temperature and 36.8 °C, actual skin temperature and 33.7 °C, and actual mean body temperature – from equation (2.4) – and neutral mean body temperature (calculated for $t_{cr} = 36.8$, $t_{sk} = 33.7$ in equation (2.4)). Provision is made for the effects of rate of change of sensor temperature but insufficient empirical data are available to provide a mechanism for predicting the effects. Zhu (2001) has however provided

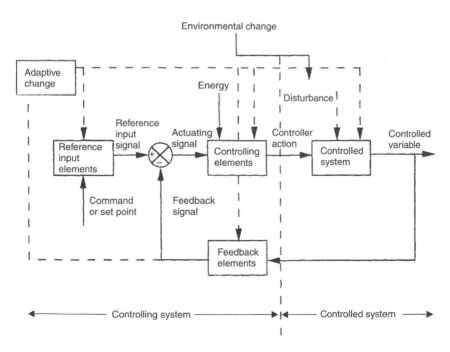

Figure 2.2 Block diagram of closed-loop negative-feedback control system with the addition of adaptive change.
Source: Stolwijk and Hardy (1966).

a mechanism for including rate of change of skin temperature in predicting thermoregulatory responses to transients.

Neuronal models

There are a number of differing views about the structure of the human system of thermoregulation. Bligh (1985) suggests two basic principles: '... the comparison of a signal from a sensor of the regulated variable with a stable reference signal, or the comparison of the signals from two sensors of the regulated variable, having different response characteristics.' Both invoke the concept of a 'set point' and an action related to deviation from the 'null' position. Evidence is insufficient to decide between these two (or other) alternatives, however neuronal models have been hypothesized. Bligh (1979) suggests the model shown in Figure 2.3.

In the model, the input to the hypothalamus (central nervous system (CNS) control) is from skin and core thermosensors and the output is in terms of heat loss effectors (vasocontrol and sweating) and heat production effectors (metabolic heat). The right heart and lungs are regarded as a thermal mixer, although actively controlled heat exchange could be implemented at this site. The blood leaving the left heart is assumed to be at the mean temperature of the circulatory blood. As blood circulates across all core thermosensors it provides feedback from thermoregulatory effectors into core thermosensors. The vasomotor temperature controller is distributed between skin sensors, CNS control, efferent nerves, and sensors.

Within the hypothalamus (CNS control), Bligh (1979) provides a neuronal representation which was derived from the effects on thermoregulation

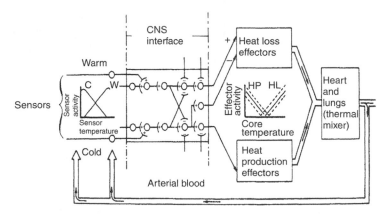

Figure 2.3 A diagrammatic representation of a thermoregulatory system adapted to give expression to the concepts of how the afferent pathways from the thermosensors might relate centrally to the efferent effectors.

Source: Bligh (1985).

of several synoptically active substances when introduced into a lateral cerebral ventriole of a sheep. The open circles represent areas of sensor to effector pathways where the + sign represents a stimulating influence and the − sign an inhibitory influence. It can be seen that the controlling system involves both integration of a number of sensor signals from individual sources and 'cross over' inhibitory connections.

Details of how thermosensors relate to effectors and the inclusion of multiple sensors of temperature distribution governed by passive and controlled heat exchange, storage and generation mechanisms, and including counter-current exchange would lead to a numerical model which could be represented in dynamic form on a digital computer. Although, there has been a great deal of research into heat transfer and the passive system, there seems to have been relatively little on the controlling system, including individual differences and the effects of genetic makeup. There would also appear to be potential for useful research using modern technology and developments in software and control theory.

Hierarchy of systems

If a number of body systems experience strain and resources are limited such that all desirable responses cannot be effected, then any response of the body must be based on a system of 'compromise'. For example, if food is in short supply should it be utilized in shivering to maintain temperature or in hot conditions should water be 'lost' as sweat if the body is already dehydrated? In exercise there is a competition for blood between skin and working muscles and blood distribution must be controlled.

Hensel (1981) notes that in competition between systems, thermoregulation is often relatively powerful. Homeotherms, in the cold, maintain body temperature even during starvation and sweating is often maintained at high rates, even when the body is dehydrated. Blood pressure is also sacrificed to vasodilation (leading to heat syncope). At extremes, therefore, thermoregulation appears to have priority. However, at less extremes small shifts in body temperature are allowed, to maintain other systems. Where resources are required for other vital functions in the body then additional strain will be placed on the system of thermoregulation.

PHYSIOLOGICAL RESPONSES

Thermosensors

Temperature sensitive receptors have been identified in both the skin and hypothalamus. There is also evidence for thermoregulation in the midbrain, medulla oblongata and spinal cord as well as in blood vessels, the abdominal cavity and a number of other sites (Hensel, 1981).

In the skin thermoregulators are free nerve endings widely distributed over and within the epidermis (Weddel and Miller, 1962): 'Their signals are carried by either non-myelinated C fibres or small myelinated A fibres and the main afferent spinal pathway is the lateral spinothalmic tract' (Edholm and Weiner, 1981). Thermoreceptors (skin and central) are either 'warm' or 'cold' types, according to the response to stimuli. Their morphology is not fully understood; however, they respond to both static and dynamic stimuli by a change in the rate of firing of the associated nerve.

The distribution of thermoreceptors in the body, static and dynamic responses, and methods of integration to provide a stimulus for controller action (as well as how they relate to thermal sensation) are discussed in detail by Hensel (1981), McIntyre (1980) and Kenshalo (1970).

Central components

Thermosensors are connected to the hypothalamus by nervous pathways. The existence of the two linked hypothalmic centres was established by placing neurological tension at different levels of the brain and by warming, cooling and electrically stimulating different areas and observing effects (Edholm and Weiner, 1981). The anterior hypothalamus and pre-optic region control heat loss and the posterior hypothalamus is involved with vasoconstriction and shivering. This hypothalamus controller includes responses using a number of efferent nervous pathways. Of particular importance is the sympathetic adrenomedullary system (SAM) which '... provides a widely distributed network of noradrenergic terminals activating

Table 2.2 Six thermoregulatory effector pathways

Pathway	Effect
1 Adrenergic non-medullated nerve fibres of the sympathetic system	Vasoconstriction of skin blood vessels, possible inhibition of vasodilation nerves and pilomotor reaction
2 Sympathetic nerves and artine vasodilator nerves	Cutaneous vasodilation and inhibition of vasoconstrictor tone
3 Ordinary skeletal supply	Shivering
4 Sympathetic supply to the adrenal medulla releases adrenaline and catecholamines	Increases heat production, cardiac output and enhances skin vasoconstriction while dilating muscle blood vessels. Glucose and free fatty acid utilization shivering thermogenesis
5 Sympathetic adreno-medullatory system	Non-shivering thermogenesis
6 Non-medullated (cholinergic) sympathetic nerves	Sweating

Source: Edholm and Weiner (1981).

metabolic and vascular responses and a motor supply to the adrenal medulla for the release of adrenaline into the circulation' (Edholm and Weiner, 1981). Six of the pathways and the effects are described in Table 2.2.

Changes in skin blood flow are also caused by local heating and cooling. With prolonged exposure to heat or cold endocrine systems can cause changes, for example, in response to seasonal variations. Bligh (1985) provides a discussion.

Vasodilation, vasoconstriction and direction of blood flow

The body causes skin vasodilation to increase heat loss and vasoconstriction to reduce heat loss. Cold vasoconstriction still allows some blood flow for the required small amount of oxygen to reach the cells. In the limbs, a 'countercurrent heat exchange' occurs due to constriction of superficial veins so that cool blood from the skin returns along the venae comitans close to the artery, hence gaining heat and returning to the body core. During vasodilation venous blood returns near to the skin hence increasing the availability of heat loss from the skin to the environment.

Arterio-venous anastomoses deep to the skin capillaries can open and reduce the fall in temperature along the length of the artery, hence increasing arterial temperature, raising skin temperature, and increasing heat loss. Conversely, the anastomoses can be closed, hence reducing heat loss. Kerslake (1972) demonstrates the effects of countercurrent heat exchange, arterio-venous anastomoses and return of blood through superficial veins (Figure 2.4).

Cabanac (1995b, 2000) has suggested that there is a mechanism of selective brain cooling (SBC) (and warming) in humans whereby when the brain is hot, blood from the mouth and nasal areas is directed to cool the brain, and when the brain is cool it is directed to preserve the metabolic heat of the brain. Clearly breathing and panting will cool the blood in the head but its type and degree of control and effectiveness is for debate. For example, Jessen and Kuhnen (2000) consider SBC to occur with significance only with animals with a rete mirabile (network of fine arterioles) which includes cats and dogs but not humans.

Piloerection

Piloerection or 'hairs standing on end' occurs when the skin becomes cold and is an attempt to reduce heat loss by maintaining a layer of still air between the body and the environment. As humans have relatively little hair and are often clothed this reaction is usually regarded as making an insignificant contribution to human thermoregulation. It may be however, that in some circumstances this has significance, for example during shivering, in still air environments and as an interactive parameter in determining the thermal insulation of clothing.

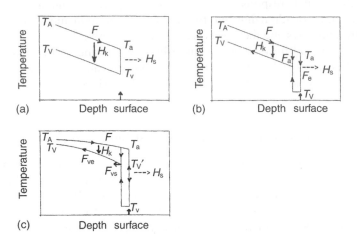

Figure 2.4 Heat loss at the skin surface (H_s) due to blood flow variations.
(a) Simple counter current heat exchange system. Heat regained
between artery and vein (H_k) depends upon the arterial (T_A) and venous
(T_V) blood temperatures and conductivity of tissues between them.
(b) Opening of arterio-venous anastomoses increases blood flow and
reduces temperature difference between artery and vein. This increases
skin heat loss and reduces heat regain from artery to vein. (c) Effect of
returning some blood through a superficial vein (F_{vs}) instead of through
the venus comitans (F_{ve}). Heat loss at the skin is further increased.

Source: Kerslake (1972).

Heat production by shivering

Both skin temperature and core temperature affect the onset of shivering
which can be both voluntary and involuntary. Bligh (1985) describes
shivering as the '... simultaneous asynchronous contraction of the muscle
fibres in both the flexor and exterior muscles'; i.e. activity producing heat
with no net external muscular work. If the body temperature falls then
metabolic rate begins to increase, first due to an increase in muscle tone
(causing stiffness) and then due to shivering. Shivering can vary in intensity
from 'mild' to 'violent' and can greatly increase metabolic heat production
by up to around five times the non-shivering level for short periods. For
a person standing at rest for example, shivering can increase metabolic heat
production from around $70\,\mathrm{W\,m^{-2}}$ to around $200\,\mathrm{W\,m^{-2}}$ or more. High
levels of shivering will tend to be continuous, whereas mild shivering will
'come and go' throughout an exposure. Although shivering is effective in
increasing metabolic heat production and hence reducing a drop in 'core'
temperature, for example, a large amount of the increased heat produced
can be lost to the environment. The role of piloerection may be important

by preserving heat generated and increasing the effectiveness of shivering. In very cold environments or during cold water immersion however, shivering can arrest the fall in body core temperature but it can also increase heat loss to the environment. ASHRAE (1993) gives the following equation (used in the two-node model (see Gagge *et al.*, 1971 – Chapter 15) for calculating additional metabolic heat energy due to shivering (M_{shiv}):

$$M_{shiv} = 19.4 C_{SIG_{sk}} C_{SIG_{cr}}, \quad Wm^{-2} \tag{2.5}$$

where $C_{SIG_{sk}} = (33.7 - t_{sk})$ for skin temperature less than 33.7 °C (otherwise $C_{SIG_{sk}} = 0$) and $C_{SIG_{cr}} = (36.8 - t_{cr})$ for core temperatures less than 36.8 °C (otherwise $C_{SIG_{cr}} = 0$).

The latent metabolic energy M is thus

$$M = M_{act} + M_{shiv},$$

where M_{act} is the metabolic energy due to the subjects' activity. Equation (2.5) assumes that both skin and core temperature must be reduced to stimulate shivering. There is still some debate about this.

Sweating

When the body temperature rises, sweat is secreted over the body to allow cooling by evaporation. There are two types of sweat gland: apocrine glands are found in the armpits and pubic regions, are generally vestigial, and are responsible for the distinctive odour in these regions; eccrine glands are distributed about the body (many on the forehead, neck, trunk, back of forearm and hand, and fewer on thighs, soles and palms). It is the eccrine glands that perform the thermoregulatory function. The control and 'behaviour' of sweat glands has been a subject of many studies. For a review the reader is referred to Kerslake (1972) and Edholm and Weiner (1981). In a hot environment the evaporation of sweat is the dominant method for maintaining a stabilized core temperature (for a given level of metabolic rate – Neilsen (1938), Lind (1963) – see Figure 2.5). This is within the 'prescriptive zone' (WHO, 1969).

Beyond the prescriptive zone, core temperature and skin temperature rise. Vasodilation is further stimulated by sweating and provides the blood supply that carries fluid to the sweat glands. Increase in local skin temperature can increase sweat gland production and stimulate glands that were inactive (Kerslake, 1972).

During acclimatization (e.g. due to repeated exposures to heat stress over a number of days) maximum sweat production is greatly increased and a given submaximal rate is achieved at lower skin or core temperature. In addition, the distribution of sweat production may change. The process of acclimatization can in simple terms be regarded as the 'training' of sweat glands.

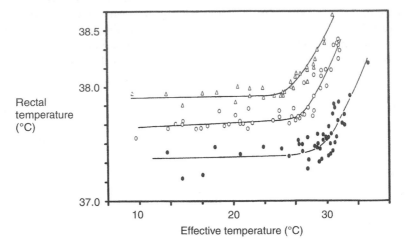

Figure 2.5 Equilibrium rectal temperatures of one subject working at energy expenditures of 180 (●), 300 (○) and 420 (△) kcal h^{-1} in a range of climatic conditions (ET).

Source: Lind (1963).

Hidromeiosis is a reduction in sweat associated with wetting the skin (Kerslake, 1972). This occurs when the skin is in hot, humid conditions or when it is immersed in warm water. Kerslake (1972) provides a comprehensive discussion of hidromeiosis and concludes that it is a local phenomenon, involving sweating and a hydrated and swollen epidermis. The decline in sweating is exponential and tends towards zero; however, recovery is rapid. Edholm and Weiner (1981) suggest that continually wetted skin reabsorbs water, leading to epidermal swelling and poral closure.

Another often quoted reason for decline in sweating (even when body temperature is rising) is due to sweat gland fatigue. It is debatable however, whether this occurs, and Kerslake (1972) concludes that any fatigue is probably '...a small factor in comparison with the effects of local skin temperature and hidromeiosis'.

BEHAVIOURAL THERMOREGULATION

In addition to the active physiological system of thermoregulation, there is an important behavioural response. This is also related to autonomic response where rapid changes in the environment can cause rapid physiological responses (e.g. if part of the skin is rapidly cooled or heated an appropriate autonomic response will be made). The perception of this change in the environment may also produce a behavioural response (e.g. to move away from a cold draught). Behavioural responses greatly affect the human thermal environment. A simple change in posture, orientation towards a heat source, putting on of clothes or movement within the environment can all have significant effects. Over long-term exposures optimum behaviour can be learned (behavioural acclimatization). Working practices for hot or cold environments can provide powerful tools to avoid unacceptable heat stress or cold stress (NIOSH, 1986). The thermal stimuli that lead to some behavioural responses often lead to such intense sensation that the response is difficult to resist voluntarily. For example, vasoconstriction in the cold can lead to such discomfort that behavioural responses are almost automatic. This can be clearly seen where minimally clothed subjects are exposed in a cold chamber. Subjects have to be continually reminded to maintain standard posture otherwise they reduce surface area by clenching fists and bringing arms close to the body.

In addition to relatively transient behavioural responses to thermal conditions there are more long-term responses. Longer term exposure to thermal stress may lead either directly (due to being hot or cold) or indirectly (due to practices involved in avoiding heat or cold stress) to behavioural and other psychological changes (e.g. mood). Rivolier *et al.* (1988), for example, describe experiences in the Antarctic. Clearly donning and doffing clothing to perform even simple body functions can be tedious. These responses can lead in a complex way to many other factors involving organizational dynamics and effectiveness. Other factors will also be important (e.g. wind noise, general stress, isolation). The interaction of the effects, is probably very important but not understood.

Further behavioural responses could also be termed 'technical regulation' (Hensel, 1981). This includes building of shelter and involves designing the environment for human occupancy. Buildings and climatic architecture can be considered as creating micro-climates which, together with clothing, have allowed humans to inhabit the whole of the surface of the earth. Hensel (1981) provides the more 'holistic' system of human temperature regulation shown in Figure 2.6.

An interesting area for consideration is the extent to which humans learn to respond to thermal stimuli, in both a physiological way (e.g. acclimatization) and behavioural way. It is known that humans plan, model the world, and test scenarios; they use memory to utilize previous experience. The role of such a model of a human in holistic (physiological, autonomic and behavioural) thermoregulation has yet to be comprehensively explored.

Understanding of their thermal position, and a perception of what they 'ought' to do based upon it, will influence people's behavioural thermoregulation. A survival expert from Lapland once asked a group of people what they would do if caught out in the snow. Dig a snow hole was a common answer. No! Look for a house, the number of people found dead in snow holes...

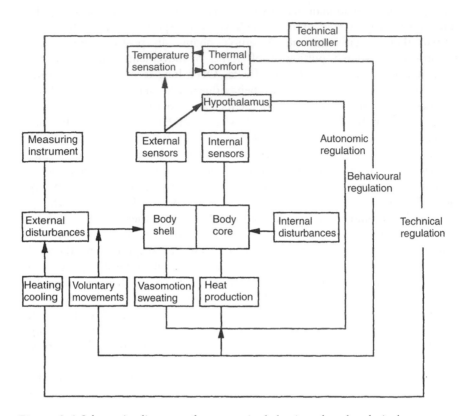

Figure 2.6 Schematic diagram of autonomic, behavioural and technical temperature regulation in man.

Source: Hensel (1981).

BODY TEMPERATURE

The body can be considered conceptually in two parts, a 'core' and a 'shell' (inner and outer). It is core temperature (t_{cr}) which the thermoregulatory system attempts to maintain (around 37 °C) and as part of thermoregulation the shell temperature (t_{sh}) varies. Mean body temperature (t_b) is the average temperature over the whole body (shell and core) and is often taken as a weighted average of shell and core temperature. In practice, there is no such thing as core or shell temperature. They provide only a conceptual convenience. It has been suggested that they be defined as sites on the body. For example, core temperature could be defined as hypothalamus temperature or rectal temperature. Shell temperature could be defined as the average temperature over the skin (mean skin temperature, t_{sk}). This however misses the point. 'Core' and 'shell' temperatures provide convenient concepts. In practice, temperatures vary over the whole of the body. If one is considering a particular area of the body then one should refer to that (e.g. rectal temperature). Experience has shown however, how particular temperatures (rectal, mean skin, oral, aural, etc.) relate to human responses (heat stress, cold stress, comfort), and how one can view the temperature in the conceptual framework of body core and shell.

Core temperature

Core temperature has no definition, as core tissues are not defined. However it is generally considered as inner body temperature or the temperature of the vital organs including the brain. These core tissues are maintained within a narrow range of temperatures by thermoregulation. If core temperature rises or falls, then there are practical consequences for the body in terms of health, comfort and performance. An 'estimate' of 'core' temperature is therefore useful and this is obtained by a number of methods. Each method has its own characteristics and method of useful interpretation associated with it. For example, aural (external auditory meatus) temperature can give information about brain temperature and responds rapidly to changes. It may therefore be useful for investigating thermoregulation and transient conditions; its' sensors, however, can be affected by the external environment. Rectal temperature provides a method little influenced by external environment, it provides a good average value of internal body temperature; however, it is slow to respond to changes and is affected by leg muscle temperature. A fuller discussion of methods of measurement is provided in Chapter 5. A list of common sites are mouth, ear canal, tympanum, oesophagus, abdomen, rectum, and urine (see Chapter 5 for details).

Shell temperature

The temperature of the shell (outer tissues of the body) varies with external environmental conditions and the thermoregulatory state of the body (vasodilated, sweating, etc.). Shell temperature is often taken as the mean skin temperature (t_{sk}) over the body. Local skin temperature can, however, vary from the mean. In the cold for example, feet and hands can have much lower skin temperatures than the trunk or forehead. Local skin temperatures can be important (to health or performance, for example), and should be measured individually (e.g. on hands and feet). Mean skin temperature is commonly calculated by taking a weighted average (according to body mass/area) of temperature taken from a number of body sites. The number of sites required to provide a representative average increases as the body becomes cold and skin temperatures vary across the body. Methods of estimating mean skin temperature are provided in Chapter 5.

Mean body temperature

Mean body temperature (t_b) is the average temperature over the body. It is usually estimated from a weighted average of shell and core temperatures. ASHRAE (1993) gives the following equation:

$$t_b = \alpha t_{sk} + (1 - \alpha)t_{cr},$$

where α is a weighting factor that depends upon skin blood flow (see equations (2.1) and (2.3)). Estimates of α vary from 0.1 to 0.3 for vasodilated and vasoconstricted skin respectively.

3 Psychological responses

INTRODUCTION

Thermal environments greatly influence thermal sensation and human behaviour which are psychological responses. The potentially dramatic consequences of humans being unable to maintain acceptable body temperatures, and the unpleasant thermal sensations associated with cold for example, have ensured that humans have responded to (or anticipated) climatic change as a high priority. This is particularly true where people exist in climates for which they cannot survive without behavioural measures (buildings, clothing, etc.). It is reflected in the language of many cultures, as seen in the rich semantic structure and prevalence of conversation about the weather. Despite this prime importance, psychological research into the effects of thermal environments is in its infancy. There has been a great deal of research into establishing the relationship between the six basic parameters that make up human thermal environments and thermal sensation (warm, cold, etc.), but there is relatively little known about why individuals differ in their responses. Little is known about behavioural responses, what constitutes thermal pleasure, provides a fresh or inspirational thermal environment, or contributes to a convivial social environment, except that thermal environments affect these psychological responses. Rivolier *et al.* (1988) report on a large biomedical study conducted over six months in Antarctica. Behavioural responses (clothing, working practices, shelter) ensured survival of the body in acceptable microclimates. However, other psychological factors became very important. 'The results show the importance of monitoring the whole range of human responses, in particular the early warning signs of psychological stress... if disregarded... would cause severe problems.' Hensel (1981) suggests that thermoreception leading to qualities such as 'warm' or 'cold' is qualitative, derived from sensory experience, and hence cannot be based on physics or physiology. This relationship is shown in Figure 3.1.

The phenomenal manifold contains immediate sensory, affective, and volitive experiences and is concerned with thermal sensation. The physical and physiological manifolds are concerned with an objective world where

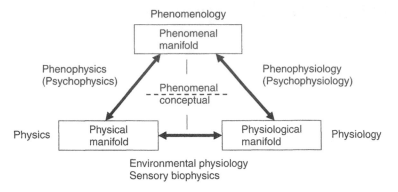

Figure 3.1 Approaches to thermoreception.
Source: Hensel (1981).

sensory qualities are considered as properties of objective things. The correlation between physical or physiological factors and sensory phenomena is the domain of psychophysics and psychophysiology respectively, e.g. the relationship between temperature and warmth, or the relationship between sweating and feelings of stickiness. A further avenue of research is into the correlation between environmental conditions, physiological responses and hence psychological phenomena (behaviour, sensation).

McIntyre (1980) describes the three approaches that have generally been taken to investigate human response to thermal environments and these have also been described in previous chapters. The physical or engineering approach has been to represent man as a 'heat engine' and considers his interaction with the environment as heat exchange mechanisms that are represented in terms of the six basic parameters (air temperature, etc.). Thermal physiology is the study of the mechanics by which the body responds to the thermal environment, e.g. the thermoregulatory responses. Neither physical nor the physiological method can directly predict psychological responses. Psychological studies can investigate psychological phenomena directly and empirical (correlation) studies can be used to relate physical and physiological (e.g. mean skin temperature, sweat rate) factors to psychological responses. An important point about human response, however, is that the three approaches are for convenience of study. Humans simply respond to environments as humans; i.e. involving an integration and interaction of the mechanisms described.

The above discussion demonstrates the role of psychology in response to thermal environments and notes that knowledge of psychological responses is limited. Further developments in this area are likely to be influenced by developments in environmental psychology, e.g. the 'effects of continuous physical settings on various aspects of behaviour' (Ittleson *et al.*, 1974). A classical study was conducted at the Hawthorne laboratories (1924).

The aim was to improve worker productivity by 'improving' physical working conditions. It was found however that, although 'improved' working conditions (heating, lighting, etc.) could improve productivity, there was no simple relationship and social factors dominated. A simple psychological model of a human who predictably responds to environmental stimuli was therefore naïve and limited in application. A similar conclusion could be made about some studies of what are termed 'sick buildings'. Single or combinations of physical environmental factors are not always (often) the cause of dissatisfaction with buildings (Robertson *et al.*, 1985). So what can psychology offer to the assessment of human thermal environments? It is too easy to make the simple statement that 'humans are more complex than that!' and this has led to a lack of interaction between workers in different disciplines. A famous biophysicist who worked in the Second World War once reflected that if there was a problem (e.g. working at high altitudes in aeroplanes) and the physiologist could provide a solution, then the engineers would almost always be able to implement it. If you wanted to complicate the issue however, then you should include a psychologist in the team. The point is that simple models of humans can prove very useful. For example, in the design of a building it is useful to know what environmental conditions, on average, produce thermal comfort. If a percentage of dissatisfaction with the conditions can also be predicted, then this is a bonus. It is of course important to note that the model of the human used has limited application and that an understanding of thermal comfort is not known. However, although more sophisticated knowledge and models are desirable, the knowledge gained from simple models of man is non-trivial in usefulness for application. Canter (1985) suggests that the role of psychology is to provide ways of thinking about problems, and models of what is happening. Much has been achieved in terms of thermal environmental conditions and predicting average thermal sensation. Much can still be gained from use of the models of environmental psychology. A summary of psychological models is provided below and a description of the findings of psychological research in this area is reviewed. Methods of 'measuring' psychological responses depend upon the model used and are described in Chapter 5.

PSYCHOLOGICAL MODELS

A simple stimulus/response model of humans, where for a given physical environment, psychological response (e.g. sensation) can be predicted, is useful for some limited practical applications but clearly naïve. It has however provided the basis for much research into thermal comfort. Canter (1975, 1985) describes three types of psychological model: determinist, interactionist, and transactionist.

The strong *determinist model* is where people are seen as being so constituted that unitary, relatively simple aspects of our surroundings (e.g. physical thermal environment) have specific consequences for what we think, feel and do. This leads to simplistic questions by designers, for example, what temperature will provide comfort? A weaker form of determinism is a model where there are associations, ideas or meanings associated by people with environmental stimuli. This then leads to behaviour. This is termed *behaviourism* and is a major school of psychology. The object is then to identify the 'meanings' which the environmental stimuli will elicit which will be specific to any individual and will change with time.

Canter's *interactionist model* is based on the formula:

$$B = F(P, E),$$

(from Gestalt psychology and Lewin's (1936) field theory), that is, behaviour is a function of the person (P) and the environment (E). This emphasizes that the characteristics the individual person brings to the situation interact with the environment to produce behaviour. In terms of Lewin (1936), behaviour is a function of the total field that existed at the time the behaviour took place. This is elaborated by Canter to what is called weak transactionism and finally strong transactionism (see Figure 3.2).

In strong transactionism 'we should look to the objectives or goals of the individual, and the way in which they are structured and organized by the social process of which that person is a part. These processes in turn give rise to a set of place specific actions and experiences.' Canter (1985) concludes 'the understanding of the meanings that the inner life of people have for these transactions with the outer world is the key to applying psychology.' The reference to Canter (1985) is used to summarize models in environmental psychology. There have been many texts on this subject and for further information the reader is referred to Krasner (1980), Canter (1977), Moos (1976), Lee (1976) and Bonnes and Secchiatoli (1995). The simple point is that most psychological investigations involving the effects of thermal environments have used the 'strong deterministic model'.

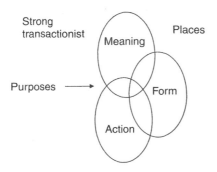

Figure 3.2 Strong transactionist model.
Source: Canter (1985).

Developments from this model would emphasize total, interactive environments, the identification of the involvement of a more sophisticated model of individual human characteristics and that it is the interaction of these in any context that will influence human behaviour.

Which method is used for psychological investigation will depend upon the model adopted. It is often argued by psychologists that the simple deterministic model is inappropriate. However, this depends upon the objective and it should be remembered that all models are approximations. Confusion can be caused by a misunderstanding of objectives between investigators. Information based on a strong deterministic model (e.g. 20 °C air temperature will provide thermal comfort on average), may be sufficient to a designer in the first stages of design. This will not, however, be acceptable to a psychologist attempting to understand the nature of thermal comfort. In addition, the importance of individual characteristics will depend upon the response of interest. Elevated core temperature will affect health in all individuals and skin contact with metal at 100 °C will generally cause pain and burns irrespective of psychological context. The appropriateness of any psychological model, therefore, will depend upon the application to which it is being put. There are numerous other psychological models not presented above; however, two issues in particular are worthy of mention. As the subject considers requirements for thermal comfort, acceptability, satisfaction and so on in a worldwide context the cultural factors become important. The role of climate in an individual's psychological construction is of interest and qualitative techniques involving discourse analysis may be useful in promoting understanding.

THERMAL SENSATION

Thermal sensation is related to how people 'feel' and is therefore a sensory experience and a psychological phenomenon. It is not possible to define sensation in physical or physiological terms. There have, however, been many studies that have correlated physical conditions and physiological response with thermal sensation and it is from these studies that models for predicting the thermal sensation of groups of individuals have been based. These models are widely used by designers and engineers in an attempt to provide comfortable environments (e.g. Fanger, 1970).

It is important to distinguish thermal sensation (how a person feels – warm, neutral, cold, etc.) from affective or value judgements that are related to how a person would like to feel, for example, comparative judgements such as warmer or cooler, or other value judgements such as comfortable or acceptable. A further distinction should be made because of possible semantic confusion. When a person says it is a cold environment, for example, this could mean either that the person feels cold, or that, because of his or her experience, it could be considered or described as cold. In an environment at an air temperature of 5 °C for example, heavily clothed, exercising persons may feel 'hot', but the environment may be described as 'cold'. It is important to recognize that thermal sensation is how the person feels, not how the environment may be described. (This is important when making subject measurements – see Chapter 5.) Although human thermal environments are defined by six basic parameters, it is the integration of the parameters that determines the thermal state of the body (i.e. a person senses the state of his nerve endings, situated about 200 µm below the skin surface), and thermal sensation is determined by the 'thermal state' of the body and not an environmental component (Fanger, 1970).

The study of thermal sensation can be divided into approaches historically determined by the purpose of the study. Physiologists have studied thermoreceptor behaviour and attempted to relate this to thermal sensation. Studies in classical psychophysics have investigated the relationship between physical intensity and psychological sensation, attempting to determine psychophysical laws. Designers and engineers consider whole-body thermal sensation as related to the thermal state of the body or part of it (local sensation). Although the mechanisms are not fully understood, it is generally agreed that the area, position and duration of stimulus on the body, pre-existing thermal state of the body, temperature intensity and rate of change of temperature all affect thermal sensation.

THERMORECEPTION

Nerve endings

After much research, it is now generally agreed that there are no specialized temperature sensors in the skin. There are, however, nerve endings, some of which respond to warm stimuli and some to cold. The nerve endings are distributed differently across the skin. It is thought that there are generally more cold receptors than warm; however, Hensel (1981) notes that studies in this area have used stimuli of very small areas and some stimuli may have been subthreshold to a single or a few fibres, but suprathreshold if a larger area stimulus had been used, because of area summation. Figure 3.3 shows the rate of firing of cold and warm receptors with temperature from which it can be seen that there is no simple monotonic relationship between temperature and rate of firing.

The rate of firing of the receptors depends not only on 'static' temperature but also on rate of change of temperature. In addition, over a range of temperatures from about 29 °C to 37 °C, there will be adaptation (see Figure 3.4). When adapted, there is a difference in threshold temperature for a change in sensation that depends upon adapting temperature and rate of change of temperature (see Figures 3.5, 3.6 and 3.7).

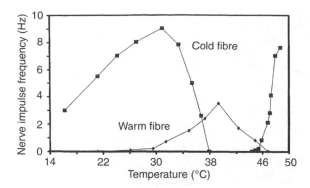

Figure 3.3 The steady-state discharge of a simple cold fibre and a warm fibre in the tongue of a cat. The cold fibre exhibits a paradoxical discharge above 45 °C.

Source: Dodt and Zotterman (1952).

There are, therefore apparent anomalies whereby the skin can feel 'cool' at a temperature greater than that where the skin feels 'warm', because of previous adapting temperature and rate of change of temperature. These phenomena relate to small areas of the skin and also to whole-body sensations. The human body would therefore be unreliable as a temperature-measuring instrument.

For a particular temperature a more intense sensation will be felt, the greater the area is stimulated (see Figure 3.8). This area effect reduces in magnitude as the intensity of the stimulus increases i.e. pain threshold is independent of size of areas stimulated.

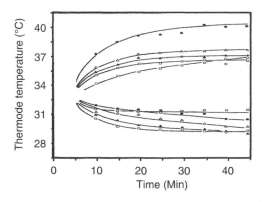

Figure 3.4 The course of adaptation to the temperature of a thermode applied to the forearm. Each of the five subjects was asked to keep the stimulus temperature just perceptibly warm or cool.

Source: Kenshalo (1970).

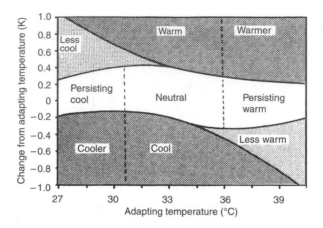

Figure 3.5 Just noticeable changes in forearm skin temperature as a function of
the adapting temperature. Warm and cool thresholds are shown by the
solid curve; the dotted curve is the threshold for a change in sensation.
Source: Kenshalo (1970).

Spatial summation occurs not only at a single site but also over sites
symmetrically placed on the body when simultaneously stimulated. This
does not occur if the sites are not symmetrical and may be explained by
integration of stimuli at the level of the spinal cord. Area sensation may
translate to the whole-body, which is very sensitive to temperature changes;
however, there is some debate and there may be a link to area size. Summation will also depend upon strength of stimulus, and there is evidence of an
interaction between warmth and tactile sensations.

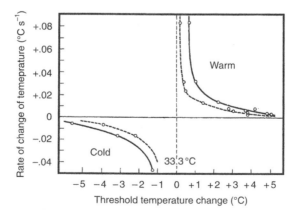

Figure 3.6 Average thresholds (ΔT) of warm and cold sensations on the forearm ($20\,\mathrm{cm}^{-2}$). Dashed lines thresholds; solid lines distinct sensations.
Source: Hensel (1981).

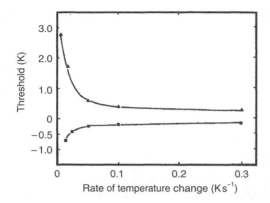

Figure 3.7 The effect of the rate of change of stimulus temperature on warm and cool threshold. The adapting temperature was 32.5 °C on the forearm. The skin is relatively insensitive to the slow rates of temperature change.

Source: Kenshalo (1970).

There is a variation on regional sensitivity for warm and cold stimuli. This is also shown by Nadel *et al.* (1973) who observed increase in thigh sweating, by applying heat to different parts of the body, and obtained a simple index by dividing this value by the area irradiated and the increase in skin temperature. Crawshaw *et al.* (1975) performed a similar experiment but observed decrease in thigh sweating by applying cold stimuli over different areas of the body in an ambient temperature of 37 °C (see Figure 3.9).

The results can be used to provide 'weightings' for formulae for calculating mean skin temperature (t_{sk}) based on body surface area or, possibly

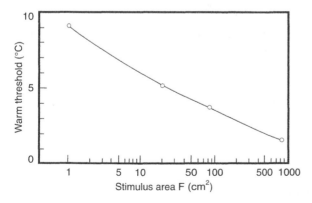

Figure 3.8 Average thresholds (ΔT) of warm sensations on the forearm for linear temperature rises of 0.017 °C s^{-1} as a function of stimulus area (F). Initial temperature 30 °C.

Source: Hensel (1981).

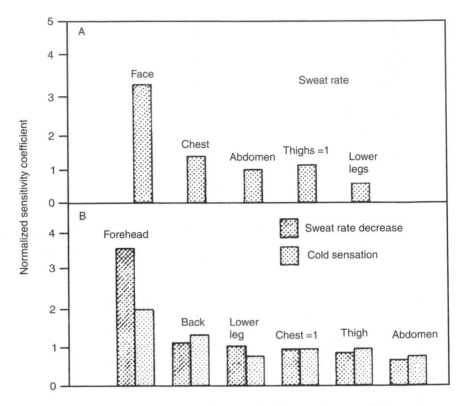

Figure 3.9 Sensitivity coefficients of various body areas: A – sensitivity to warming assessed by reflex sweat rate and B – sensitivity to cooling assessed by decrease in reflex sweat rate and cold sensation.

Source: A – Nadel *et al*. (1973) and B – Crawshaw *et al*. (1975).

more useful for analysing thermal response, thermal drive due to sensitivity to warm or cold stimuli (Table 3.1).

The sensation 'felt' by a person is bipolar in nature (e.g. hot, neutral, cold) and at extremes possibly represents a different quality. Hensel (1981) suggests for example the following sensation conditions:

> painful (cold)
> icy
> cold
> cool
> neutral or indifferent
> lukewarm
> warm
> hot
> painful (hot).

Table 3.1 Number of cold and warm spots per square centimetre in human skin

	Cold spots*	Warm spots†
Forehead	5.5–8.0	
Nose	8.0	1.0
Lips	16.0–19.0	
Other parts of face	8.5–9.0	1.7
Chest	9.0–10.2	0.3
Abdomen	8.0–12.5	
Back	7.8	
Upper arm	5.0–6.5	
Forearm	6.0–7.5	0.3–0.4
Back of hand	7.4	0.5
Palm of hand	1.0–5.0	
Finger dorsal	7.0–9.0	1.7
Finger volar	2.0–4.0	1.6
Thigh	4.5–5.2	
Calf	4.3–5.7	
Back of foot	5.6	
Sole of foot	3.4	

Notes
* Strughold and Porz (1931).
† Rein (1935).

Whether 'hot' is a more intense form of warmth or icy and painful are on the same continuum is debatable. Pain is a very complex sensation. In terms of heat, pain occurs at around 45 °C skin temperature (close to burn threshold), with some adaptation up to 46 °C (Hardy *et al.*, 1952). Pain in the cold can be intense but physical correlations are less clear. It is related to vasoconstriction of blood vessels. At low temperatures however, cold-induced vasodilation increases blood flow with an associated pleasant sensation. The sensation and reporting of pain depend upon a number of individual factors, and these are discussed further in Chapter 13. For a more detailed summary and discussion of thermal sensation, the reader is referred to Kenshalo (1968, 1970, 1979), McIntyre (1980) and Hensel (1981).

Psychophysics

Since the nineteenth century there have been many attempts to establish a psychophysical law which relates physical stimulus and psychological sensation. Human thermal environments are however made up of the interaction of six basic parameters and even if one considers a simple thermoreceptor, or a small area of skin, there is no simple relationship between temperature, rate of firing and possible sensation.

Stevens (1960) proposed that the magnitude of sensation ψ grows as a power function of stimulus magnitude ϕ where

$$\psi = k\phi^B,$$

k is a constant depending upon units and B is the exponent that depends upon sensory mode (visual, noise, thermal, etc.). This is modified to:

$$\psi = k(\phi - \phi_0)^B,$$

for stimuli close to the absolute threshold (ϕ_0).

Stevens *et al.* (1969) considered discomfort caused to a subject lying supine on a bed under a radiant heat source. Equations for the magnitude estimate of discomfort were:

$$D_c = 0.33(22 - t_0)^{1.66},$$

and

$$D_w = 4.1(t_0 - 22)^{0.77},$$

where D_c and D_w are magnitude estimates of discomfort for cold and warmth respectively and t_0 is operative temperature.

Although the curves 'fit' the empirical data, it does not necessarily imply that a power law applies. Hensel (1981) gives three fundamental reasons for why a power law should not, or has not been shown to, apply.

1   It has not been shown that the magnitude estimate is a ratio scale, which would be required for the power law.
2   The stimuli used are arbitrary (temperature, operative temperature, etc.).
3   Stimulus discharge from thermoreceptors is bell shaped.

It is clear that response to 'static' temperature is not a power function; however, response to rate of change of temperature may be. It could be argued that the constant k and exponent B in the law provide a wide degree of freedom for curve fitting especially if the measure of the physical stimulus is arbitrary. How this would be used in practice to predict whole-body sensation is also not clear. A psychophysical law may be useful for predicting the response of local heating or cooling of the skin. A more rational approach, however, would be to consider the six basic parameters which make up human thermal environments, predict the response of the body in these environments and relate the response to thermal sensation.

Whole-body responses

The thermoreceptors provide information regarding temperatures over the body and hence greatly influence thermal sensation. The body 'feels' the state of the thermoreceptors and not the physical environment. Despite this, there has been some success at relating ratings of thermal sensation with physical parameters such as temperature. After much research over many years it is now generally considered that '. . . cold sensation is determined by mean skin temperature. Warm sensation depends initially on skin temperature, then on deep body temperature; warmth discomfort is dependent on skin wettedness' (McIntyre, 1980). It should be emphasized that these are generalizations for 'typical' conditions. There are exceptions. For a clothed person working in a freezer room, for example, local hand or foot temperature may correlate with whole-body cold sensation rather better than mean skin temperature. Chatonnet and Cabanac (1965) demonstrated that even if the skin temperature is high (in a warm bath), a fall in deep body temperature can cause cold sensations.

In transient conditions thermal comfort may be predicted more accurately from air temperature (physical conditions) than from knowledge of skin and body temperatures. Gagge *et al.* (1967) found that thermal sensation changed, immediately air temperature changed, even though there was a delay before skin and body temperature changed. During exercise some sweating is required for comfort and skin temperature would be relatively low. The body is less sensitive to changes in thermal stimuli, being less able to attend due to physical activity and any change in thermal state can be easily accounted for by a small change in sweat rate.

Thermal pleasure is related to a relief from thermal discomfort. It can be considered as a drive for behavioural thermoregulation and is a transient phenomenon. Generally, if a person is too cold a stimulus that causes loss of heat produces 'displeasure' and preservation or gain of heat 'pleasure'. Conversely, if a person is 'too hot', heat gain is perceived as 'unpleasant' and heat loss 'pleasant'. Thermal pleasure does not occur at thermal neutrality. Cabanac (1981) suggests that pleasure is the motivation for a behaviour which tends to restore neutrality. He observes that as human thermoregulation is dynamic, skin temperature is not constant and we are always in a state of warm or cold defence with no interval in between. A state without pleasure is therefore rare.

SEMANTICS, PSYCHOLOGICAL MODELS AND MULTIDIMENSIONAL SCALING

We have seen that it is important to distinguish between more objective ratings such as sensation and affective or evaluative ratings such as comfort and pleasure. In everyday language, however, these dimensions are often confounded, and distinctions are not made. In addition, the richness of semantics for describing thermal environments and responses to them will depend upon individuals, their experiences, their language and culture. When investigating thermal environments therefore, it is sometimes useful to first investigate the psychological dimensions (or constructs), which individuals and groups use to describe their world.

Two commonly used approaches are *semantic differential techniques* (Osgood *et al.*, 1957) and *personal construct theory* methods (repertory grid – Kelly, 1955). The methods which are discussed further in Chapter 5, invoke factor analysis or multi-dimensional scaling techniques to build a psychological model of the way in which the thermal environments are perceived and 'modelled' by the person. Similar stimuli on a particular dimension (or all combinations of dimensions) will be placed close together in the multi-dimensional model of psychological space. This type of method has been used in investigations of other environmental stimuli, e.g. visual environment (Boyce, 1981), but has not been widely used in studies of the thermal environment.

Rohles and Milliken (1981) however used such techniques and found a number of psychological dimensions. They considered the effects of lighting, colour, and room decor on thermal comfort. Five factors were identified from responses of 432 subjects on 71 pair semantic differential scales. These were efficiency, attractiveness, thermal comfort, spatial and environmental quality. The factor 'thermal comfort' involved bipolar scales of warmth, comfort, acceptability, pleasantness, satisfaction, and good–bad temperature.

Hollies *et al.* (1979) used a similar method for rating thermal sensation felt while wearing different types of clothing. Subjects were asked to describe sensations felt over a number of conditions providing the following psychological 'dimensions' – stiff, sticky, non-absorbent, clammy, damp, clingy, rough and scratchy.

Rarely have the psychological 'models', and techniques for investigating them, been used in research into thermal sensation. There are many other techniques for investigating psychological responses that could be considered. Griffiths (1975) notes that most studies consider only the sensation of 'warmth' and says, 'this may be caused by the relative ease of having subjects operate as physical meters and the relative difficulty of measuring affective aspects of experience. If this is the case then a more sophisticated approach to subjective measurement is needed.' This clearly relates to the discussion of psychological models used for considering humans in thermal environments provided at the beginning of this chapter.

Scales of warmth sensation

Most practical studies of thermal sensation have used rating scales to quantify the psychological response and have correlated the ratings with

Table 3.2 Scales of warmth sensation

Bedford comfort scale		ASHRAE sensation scale	
Much too warm	7	Hot	7
Too warm	6	Warm	6
Comfortably warm	5	Slightly warm	5
Comfortable	4	Neutral	4
Comfortably cool	3	Slightly cool	3
Too cool	2	Cool	2
Much too cool	1	Cold	1

Sources: Bedford (1936) and ASHRAE (1966).

human thermal environmental conditions. Two scales shown in Table 3.2 have been commonly used: the 'comfort scale' of Bedford (1936) and the 'sensation scale' of ASHRAE (1966).

An often used form of the ASHRAE scale is to give the neutral sensation a value of 0 (hence hot = +3, cold = −3 etc.) − (Fanger, 1970; ISO 7730, 1994).

McIntyre (1980) suggests that the Bedford scale is less useful as it confounds warmth and comfort. However, this is not a limitation as all of the terms on the scale are affective (value judgements), providing a comparison of how the subject feels with how he or she would prefer to be. The ASHRAE scale refers only to thermal sensation. Despite the differences between the scales, McIntyre (1980) concludes from practical use of the scales, that for similar conditions similar subject ratings will be obtained. The use of thermal comfort and sensation scales is discussed further in Chapter 5, and thermal comfort is discussed in detail in Chapter 8.

MOOD, AGGRESSION, DEPRESSION AND OTHER PSYCHOLOGICAL REACTIONS

Thermal conditions will directly affect thermal sensation and comfort and can also influence the general 'psychological state' of the body, for example, a person's mood and behaviour. This will depend upon an interaction of stimuli and psychological factors. However, there is some evidence and there are many anecdotes.

Hippocrates noted the importance of the weather on health and advised physicians to study its effects. Huntington (1915) related climatic factors to national characteristics, defining an ideal climate as one that is moderately stimulating but not exhausting, changing seasons, and frequent storms to provide stimulation. Markham (1947) goes further and generally attributes the fall of the Roman Empire to their loss of interest in central heating (control of indoor climate). The search for ideal weather conditions which promote 'advanced civilization' is characteristic of the early half of the twentieth century. Hypotheses are difficult, if not impossible, to test; however even if proposals verge on the dogmatic, it is clearly considered that climatic conditions play a major role.

Lynn (1991) suggests that climatic factors influence anxiety which affects rates of suicide, psychosis, alcoholism, and caloric intake. Mills (1939) notes that storms produce irritability, restlessness and petulance in children, and quarrelsome, faultfinding and pessimistic moods in adults.

Dexter (1904) noted that abnormal winds, very low barometric pressure and excessive humidity were associated with increased use of corporal punishment in schools. Despite the apparent abundance of evidence, it is difficult to identify the mechanisms by which climate influences mood and behaviour. Mills (1939) noted that worsening of arthritis occurred in climates of increasing humidity with falling pressure. There are 'warm winds' which occur in certain regions of the world, e.g. föhn in Innsbruck, Switzerland, which have been associated with increased suicide rate, accidents, etc. However, it is not clear how the wind influences these responses; an increase in positive ions in the air has been suggested. It can be concluded that there is a great deal of evidence that weather conditions can affect the mood and behaviour of persons; however, the underlying causal mechanisms have not been shown and are not understood. Provins (1966) suggests that there is no simple direct relationship between the meteorological environment and human behaviour. A combination of person and context variables will have greater influence on behaviour than any effects of the weather.

It is clear that in this area an underlying simple deterministic model of human interaction with the thermal environment will be inadequate and laboratory experiments will be of limited value. de Dear and Brager (2002) and Nicol and Humphreys (2002) have suggested, from large global field surveys, that there is a relationship between outside temperatures and indoor conditions required for comfort. This is correlation however, not cause and effect, and there are many intervening variables. A large scale 'field experiment' is reported by Rivolier *et al.* (1988). A group of scientists studied psychological and physiological reactions to spending one winter in the Antarctic in a complex and arduous experiment using themselves as both experimenters and subjects. Tests were also conducted in a laboratory both before and after the field tests. Study of the diaries of men on previous expeditions had revealed psychological problems among members of expeditions: after a disagreement, Amundsen, the Norwegian explorer, was said not to have spoken to his leader for six months; on a ten-year expedition, Scott sent Shackleton home a year early; and there are reports of depression, disputes in multi-national parties and other inter-personal conflicts. For this reason, the study in the Antarctic involved much psychological investigation. How the psychological investigators themselves interfered with individual responses and group dynamics is not known. However, subjects gave ratings about themselves and others; peer ratings were given by the group about each of its members in turn, there were self reports and notes made by each subject about all of the other subjects in the group. In a carefully worked out plan, information was received from a number of modalities involving reporting and observational techniques.

The results showed that subjects made frequent references to boredom, weariness, homesickness, bad temper and anxiety, disturbances of mood and self-confidence. Observers noticed problems before the subjects themselves.

There were major social problems. Subjects denied that they had inter-personal problems but reported that the group was unfriendly, disharmonious and lacked cohesion. There was criticism of the organization, personal irritability, outward aggressiveness and lack of mutual support. A number of these reports were made in 'personal' diaries but not in reports to others (evidence of experimenter interference). Almost all subjects reported the superficial nature of interpersonal relationships. One subject remained isolated throughout the expedition. At some inconvenience, another man had to be evacuated for psychological reasons.

All subjects survived and in physiological terms the body (within clothing etc.) did not become excessively cold and remained in an acceptable environment. Observers had however noticed psycho-social problems from the first day of the field trial. Subjects had unsatisfactory experiences during the one-month field trial conducted in Sydney. It is likely that the wealth of data from this study will never be completely unravelled. However, what has been demonstrated is the inadequacy of a simple deterministic model of humans in thermal environments, for many practical applications. Each subject interacted with the environment (cold, isolation, social hostility, etc.), in a way involving his own personal attributes, psychological model, experiences, etc.

Despite a great deal of evidence that thermal environments can significantly influence psychological responses, the underlying mechanisms are not understood.

There have been a number of studies of thermal conditions that promote social disorder. In the UK, for example, it was noted that on the rare occasions that riots occur, they can be associated with long hot summers. It has been noted that when riots occur in India the temperature has been generally above 26 °C (79 F), although they did not occur in excessive heat. A similar finding is reported in the USA. The results of such field surveys are, however, open to other interpretations. In a laboratory experiment, Griffiths (1975) found subjects less attracted to strangers' opinions during hot conditions than when in thermal comfort. (Don't go for a job interview on a hot day!)

Baron (1972) used two temperature levels (comfortable or hot) and two levels of 'anger arousal' (manipulated by an accomplice of the experimenter posing as a subject and insulting or complimenting the other subjects) to investigate the relationship between heat and aggression. After insults or compliments had been made, subjects were asked to teach the accomplice nonsense syllables and to punish him with an electric shock for failure to learn. It was found that more angered subjects delivered longer shocks. However, shorter shocks were given in hot conditions than in the 'cool' condition. Baron and Lawton (1972) used identical conditions in a follow-up experiment but in this case half of the experimental subjects observed an aggressive 'model' teacher (gave frequent and lengthy shocks) before teaching the accomplice-subject themselves. The subjects who were not shown the model teacher behaved in a similar way to the previous experiment. However, those subjects who observed the aggressive teacher were more aggressive in the hot conditions than they were under the comfortable conditions. This may imply that under hot conditions a single violent incident may provide a model for others to become aggressive. 'If riot development at all resembles the experimental conditions (heat, anger arousal, aggression opportunity), one should expect reduced probability of mass violent disruption if no initial violent model is presented to the potential rioters.'

Behavioural responses

The simple deterministic model of human response to the environment is such that for a given set of environmental conditions human sensation or behaviour will follow. Based on this model, many studies of thermal sensation have used simple (seven point) scales of thermal sensation to determine conditions for thermal comfort or thermal sensation. A disadvantage of asking subjects to report their sensation is that it may interfere with the subjects, their feelings or activities, etc. and it may not be possible for subjects to provide reports, for example, with the very young, sick or disabled. It has been seen that a major component of human thermoregulation is behavioural – put on clothing, change posture, change activity, etc. – but there can also be other behavioural responses to thermal conditions. For these reasons, behavioural measures offer an approach to determining human response in terms of 'comfort ranges' or preferred conditions. Methods for quantifying behaviour and relating values to criteria for acceptability are not as clear. However, behavioural methods can provide little or no interference with the persons under investigation. A further development is to consider behaviour in terms of the more sophisticated psychological models, however these have been sparsely researched.

There have been a number of studies of schoolchildren to investigate 'optimum conditions' for classrooms. Humphreys (1972) used time-lapse photography to observe the clothing worn by over 400 children during 758 classroom lessons over a period of a whole summer. It had been estimated that peak temperatures in classrooms were typically above 30 °C. Teachers' comments confirmed that classrooms were often hot and that this affected children's work performance and behaviour. A pilot study demonstrated that levels of clothing worn by pupils closely correlated with classroom conditions. Observations of clothing level could then be a behavioural measure of comfort ranges without interfering with children's work. The large scale behavioural study was then complemented by a smaller scale study involving rating scales. A later study investigated teachers' views on children's behaviour. It was found that teachers consider two dimensions, the extent to which children apply themselves to the task, and the intensity of physical activity or liveliness. Teachers considered the effect of raised temperature to reduce activity and liveliness, but time of day effects were considered more significant.

Wyon and Halmberg (1972) observed the behaviour of children in a Swedish school. A perceptible response to heat index was proposed, based on clothing, posture and appearance.

Behavioural changes in individuals can be observed during hypothermia, where responses range from remaining unusually quiet to hyperactivity and delirium. When the body becomes cold the discomfort caused stimulates a behavioural response. The donning of clothing, change in posture and appearance, and level of activity are all evident. During mild hypothermia ('core' temperature 34–35 °C), Collins (1983) describes the following behavioural responses '...there are signs of muscular weakness and loss in co-ordination. Walking may begin to be affected... mental state may become perceptibly dulled... lack of response, difficulty in understanding the situation. Elderly people... often become sluggish... or immobile.'

The behavioural responses described have a direct link to thermal conditions. However, the determinants of behaviour are multi-causal and the deterministic psychological model is simplistic. Predicting behaviour in a human thermal environment is therefore complex and field studies must recognize this. Despite this, however, there have been a number of such studies, e.g. Grivel and Fraise (1978) studied the behaviour of passengers waiting for trains on the Paris Metro. These types of studies provide practical insight and should be encouraged, but, they also demonstrate the importance of the 'total' environment and the use of holistic or transactionist models. For example, on a station platform, thermal conditions and thermal sensation will be affected by crowd density (Braun and Parsons, 1991). Behaviour will be influenced by these conditions but will also be greatly affected by psychological factors of each individual providing the total interaction and subsequent behaviour. This is also true for studies of car drivers' behaviour where there is concern over the incidence of accidents in hot conditions, in the South of France for example.

Ramsey *et al.* (1983) considered the relationship between thermal conditions and worker behaviour. In an extensive study, over 14 months, three trained observers observed the behaviour of workers in two industrial plants, a metal products manufacturing plant and a foundry. The method involved identifying unsafe behaviour over a wide variety of tasks and risk levels, and using this taxonomy to classify worker behaviour as unsafe or safe. Visits were made to the factory at 'random' times and 60 thirty-second observations per day were made of workers' operations. The WBGT (wet bulb globe temperature) index was used to quantify the physical thermal environment at the workplace. Of a total of 17 841 observations, around 10 per cent were categorized as unsafe. The temperature (WBGT), workload, job risk and period (shift, time of day) were all found to be significant (statistically) resulting in unsafe behaviour ratios. There were also a number of important interactions. An attempt was made to develop an 'unsafe behaviour index' (UBI) and a U-shaped curve (see Figure 12.6) was generally found to relate to proportion of behaviours which were unsafe.

The use of the WBGT index to quantify thermal stress was convenient but it is not helpful in interpreting results in terms of human thermal environments. For example, clothing insulation levels were not considered and probably varied from about 0.5 to about 1.5 Clo. The method of defining unsafe behaviour is qualitative and provides problems in terms of identifying an appropriate psychological model of the worker. Further work is required in this area, which has great potential. Ramsey *et al.* (1983) recommend the behavioural observation technique which they say is more sensitive than traditional safety studies (e.g. invoking statistics of reported accidents). At present however, the study provides little information regarding the effects of thermal conditions on human behaviour *per se*.

Threat, expectancy and pleasure

Human responses to thermal environments will be influenced by psychological factors such as the perceived threat the climate poses, the expectation of the people exposed, and how pleasurable it will feel. If the environment does not allow people the opportunity to adapt as their thermal condition changes (e.g. move away, take off clothing – see Chapter 9), then a cool or warm environment after five minutes of exposure may be perceived as a major threat to the person if they are aware that they will be exposed to it for two or three hours and that they have no means of adjustment or escape. For people with limited adaptive opportunity (e.g. people with physical disability) an otherwise acceptable environment may be rated as unacceptable with associated adjustments to thermal sensation ratings and behaviour.

The expectations of people may influence their satisfaction with the thermal environment which may confound thermal comfort surveys. The natural bias in not wishing to report dissatisfaction may also be complicated by people expressing satisfaction because 'they don't expect any better' (or dissatisfaction because they expect much better). It is clearly difficult to distinguish this expression of satisfaction from 'true' shifts in thermal comfort requirements (e.g. across cultures) and, by definition, 'the condition of mind' is expressed. However, an important consideration will be if what would 'normally' be described as a comfortable environment is also rated as satisfactory (i.e. given better than expected), then it is possible that expectation will play a role and there has not been a shift in thermal comfort requirements. Designing for discomfort because people do not expect any better is probably not an acceptable strategy. Expectancy can be defined as a confidently predicted future state. It will depend upon the personal model of the world of an individual involving his or her psychological character-istics, experiences and culture. The response to a difference between actual conditions and what were expected will depend upon the individual, how far conditions are removed from what was expected, the predicted conse-quences, whether conditions are better or worse than expected and the opportunity to change conditions.

Thermal conditions can affect feelings of pleasure both in terms of transient phenomena (moving from cold to warm environments) and in terms of self-image. Context will be important. In the leisure industry, an otherwise unacceptable environment in terms of comfort or heat stress, for example, may not only become acceptable but also desirable, as the thermal strain enhances the self-image of the person.

DISCUSSION

Much work has been performed relating human thermal environments to psychological effects such as thermal sensation and comfort. Information obtained relates to group variation and average responses of individuals based on a simplistic-response psychological model. These data are of great use for many applications; however, they do not enhance understanding of

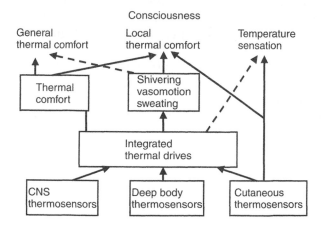

Figure 3.10 Physiological conditions for temperature sensation, thermal comfort and temperature regulation.

Source: Hensel (1981).

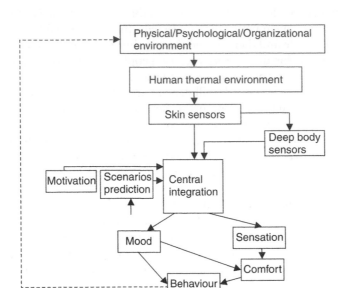

Figure 3.11 Model of human behavioural response to thermal environments.

psychological responses and are inadequate in many situations. Applied psychology has provided more sophisticated 'models of man' which can be used in practical situations. Use of the models will provide further insight into human psychological response to thermal environments. The practical point is that the most appropriate psychological model to be used will depend upon the application. It is important however to identify which model is 'assumed' and recognize that other models are available.

Hensel (1981) attempts to integrate human response to the thermal environment into a more holistic system than a simple psychological model (see Figure 3.10).

A similar approach is given in Figure 3.11; however, more detail is provided concerning the 'cognitive aspects of the model'.

This chapter has considered psychological responses to human thermal environments; methods of 'measuring' these responses are provided in Chapter 5.

4 Dehydration and water requirements

INTRODUCTION

Any reduction from 'normal' body water content (euhydration) to a water deficit (hypohydration) is termed dehydration and can lead to dramatically reduced human performance and death. In any human thermal environment there are water requirements and a programme of drinking is required to preserve balance. The evaporation of sweat from the skin cools the body and plays an active role in human thermoregulation. Together with water loss due to insensible perspiration, moisture from breathing and that in urine and faeces (as well as other avenues such as tears or loss through injury) there can be a significant water and electrolyte loss from the body. Water provides the medium in which biochemical reactions occur and for the transportation of materials. From a state of euhydration the body achieves water balance if water inputs (by drinking) equal water outputs (sweat, urine, etc.). This is a continuous and dynamic process and there is an active controlling system to stimulate water intake, retention, secretion and excretion as well as its appropriate distribution around the body.

The water required by the body is that which is lost from a state of euhydration. This appears to be a simple matter of replacement. If a person loses one litre of sweat then one litre should be replaced. While this is true, on average, it neglects the dynamic and interactive properties of the body's water regulatory system. Any effective drinking regime must therefore not only be based on how much, it must take account of the properties of the system. It is difficult to overestimate the importance of drinking for maintaining health and well-being during activity in hot environments. It has often been considered that some people (e.g. fit soldiers) can survive and perform effectively with low water rations. While some adaptation and training can occur, the overwhelming body of evidence is that this is not correct. Water is a resource required by the body and all people require it in sufficient amounts. Restricting water to people when working in hot environments is a mistake.

BODY WATER

The average 75 kg male consists of around 45 litres of body water making up 60 per cent of total body weight. The water can be considered to be divided into three compartments. One compartment, *intracellular* water, makes up 67 per cent of body water and extra-cellular water is in two compartments consisting of *intravascular* water (blood plasma – 8 per cent) and *interstitial* water (between the cells and the blood vessels – 25 per cent). The values vary, as the system is dynamic and water flows between the compartments. Older people have a lower percentage body water (around 50 per cent) and children a higher percentage (around 75 per cent). Muscle contains more water than fat, so athletes have a relatively high percentage of body water (see Tables 4.1, 4.2). Practical considerations include the ability of water to transfer from the stomach to the blood. The stomach could therefore be considered as an additional compartment.

Drinking allows water to be absorbed in the alimentary canal and into the blood (intravascular). All membranes within the body are semi-permeable and allow free flow of water across them. Arterial hydrostatic pressure (caused by the pumping of the heart) pushes water, electrolytes and nutrients through the capillary walls into the interstitial space. Diffusion causes the movement of molecules from an area of higher concentration to an area of lower concentration and osmosis is the passage of water through a membrane from a region of lower to higher solute concentration to equalize

Table 4.1 Body water distribution between the body fluid compartments in an adult male

	(%) of body weight	(%) of lean body mass	(%) of body water
Intracellular water	40	48	67
Plasma	5	6	8
Interstitium	15	18	25
Extracellular water	20	24	33
Total body water	60	72	100

Table 4.2 Water content of various body tissues for an average 75 kg man

Tissue	(%) of water	(%) of body weight	Litres of water per 75 kg	(%) of total body water
Skin	72	18	9.72	22
Organs	76	7	3.99	9
Skeleton	22	15	2.47	5
Blood	83	5	3.11	7
Adipose	10	12	0.90	2
Muscle	76	43	24.51	55

the two solute concentrations. These provide mechanisms for movement of water and materials between intracellular, interstitial and intravascular compartments. Interstitial fluid is gel-like such that water and solutes move by diffusion rather than mass movement. Excess fluids and solutes that accumulate within the interstitium will enter the lymphatic system to be returned to the intravascular compartment. Venous colloid osmotic pressure (created by plasma proteins) pulls water, electrolytes and cell waste products from the interstitial compartment into the venous capillary through the capillary walls. This profound and continuous exchange using water as the medium is affected by the state of body hydration. In low levels of hypohydration water is reduced in the extracellular spaces but at higher levels water is transferred out of the intracellular compartment. This is controlled by a system of regulation which attempts to ensure that water is available when and where needed. It can be seen therefore that water loss can affect all three compartments of the body and that simply drinking what is lost will not provide automatic rehydration, but that there will be delays as balance and distribution are restored.

Control of body water

The body reacts to a change in blood plasma volume (hyper- or hypovolaemia) and the concentrations of solutes (ions) in solution (osmolarity). If plasma volume is reduced or osmolarity is high – or both – then the hypothalamus stimulates the posterior pituitary gland to release Anti Diuretic Hormone (ADH). This causes water re-absorption in the kidneys and stimulates thirst. Water re-absorption reduces plasma osmolarity and urine flow output and increases plasma volume. Thirst causes water intake, restoring plasma volume and osmolarity levels. This system is shown in Figure 4.1.

Decreased blood volume stimulates the kidney to release renin, which leads to aldosterone release. Aldosterone is produced by the adrenal cortex and released as part of the RAA (renin, angiotensin, aldosterone) mechanism. This regulates water re-absorption in the kidney by increasing sodium uptake from the tubular fluid into the blood. Much is known about the control of fluid balance and electrolytes. For a fuller description, see Sawka (1988), Lang (1996) and Arnaud (1998).

DEHYDRATION MEASURES

Whole body blood volume and osmolarity are affected by dehydration, are the variables controlled by the body, and could be considered to be the primary indicators of hydration state. In practice, however, they are not easy to measure and it is useful to consider other measures. The dynamic nature of the water and electrolyte regulation system provides many

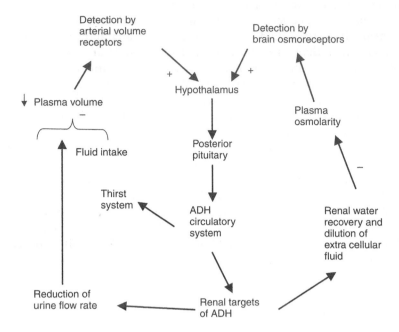

Figure 4.1 Control of water balance in the body.

measures that could be used to indicate hydration state. As with all measures they can be considered in terms of their *validity* (does the measure really relate to hydration state?), *reliability* (for the same conditions would the measure, if repeated, give the same answer?), *sensitivity* (what change in hydration state will be detected as a change in the measure?), *specificity* (if the measure changes value, is it only due to a change in hydration state or are there other causes?), and the *usability* (can it be used – is it practical to take the measure in the conditions and context of interest?).

A knowledge of what measure is appropriate and how to interpret it depends upon a knowledge of the water and electrolyte regulation system. A system of measurements may use, as its rationale, a model of this system; for example, based upon the three compartments (intracellular, intravascular, interstitial) and their interactions. Blood osmolarity and blood volume are primary indicators, however direct measures of the variables require invasive methods, medical supervision and careful analysis. Table 4.3 presents a list of possible measures and their rationale for use.

Stirling and Parsons (1998a,b) and Stirling (2000) considered measures of human hydration state for use in laboratory and field studies in hot environments. They considered the correlation between the different measures of hypohydration with particular reference to their application in preventing the dehydration of firefighters. Eight male subjects – wearing only shorts and shoes – were exposed to a hot environment ($40.3 \pm 0.1\,°C$;

Table 4.3 Measures of human hydration state

Measure	Definition	Rationale	Example of measurement method and rationale	Comment
1 Based upon analyzing urine				
Urine specific gravity – U_{sg} – ND	Ratio of mass of solution to mass of an equal volume of water	As water is retained in the body, urine concentration (mass) increases U_{sg} ↑	Hydrometer in urine – direct reading	$U_{sg} = 1.0$ for water to >1.035 for hypohydration
Urine osmolality – U_{osm} – mosmol/kg H_2O	Total concentration of discrete solute particles in urine solution	As water is retained in the body, number of particles in urine increases U_{osm} ↑	In the laboratory – osmometer using freezing point depression	U_{osm} is more accurate than U_{sg} as it is independent of solution volume
Urine temperature – U_{temp}, °C	Temperature of urine inside the body	Internal body temperature increases with hypohydration U_{temp} ↑	Urinate on sensor in a vacuum flask, initially at 37°C	Internal body temperature varies with thermal conditions, so interpretation is required
Urine volume – U_{vol} – m^3	Volume of urine over a time period	Water is retained due to hypohydration U_{vol} ↓	Measuring cylinder	24 hour measures can be inconvenient and non-specific. Normal values = 1.5 litres/24 h
Urine sodium – U_{Na} – mmol/day	Number of sodium ions in urine	Re-absorption of Na^+ causes water retention by osmosis. During hypohydration U_{Na} ↓	Flame photometry in the laboratory	Normal values around 124 mmol/day but varies from 0 to 100 mmol/day
Urine potassium – U_K – mmol/day	Number of potassium ions in urine	Re-absorption of Na^+ causes excretion of K^+ U_K ↑	Flame photometry in the laboratory	Normal values around 172 mmol/day but varies from <10 to 500 mmol/day

Table 4.3 (Continued)

Measure	Definition	Rationale	Example of measurement method and rationale	Comment
Urine pH – U_{pH} – ND	The pH of urine ($-\log[H^+]$)	Water retained $U_{pH} \downarrow$ (i.e. more acid)	Litmus paper	Values range from around 4.5–8.0
Urine colour – U_{col} – ND	Colour of urine	Increased concentration of waste products in urine due to water retention $U_{col} \uparrow$	Compare urine in a glass container in good light with colour on a urine colour chart	Indicators of U_{osm} and U_{sg} Rating scale from 'clear' to 'dark brown'
2 Based upon analysis of a 'pin prick' blood sample				
Blood haematocrit – H_{ct} – %	% of blood made up by red blood cells	A dehydrated person contains less blood plasma so $H_{ct} \uparrow$	Blood pin prick within sterile environment. Blood to microcapillary tubes, centrifuge and read value of red blood cells	H_{ct} around 40–50% for normal values
Blood haemoglobin – Hb – g/100 ml	Concentration of haemoglobin in the blood	Dehydration reduces all corpuscle size and increases % of cell occupied by haemoglobin	Blood pin prick – introduce cyanmethaeglobin reagent solution and analyse using colorimetric method	Normal values around 15 g/100 ml with 4.5% reduction with 4% hypohydration
Change in blood plasma volume – ΔPV – %	Change in volume of plasma in blood taking account of change in red cell size	Plasma volume is blood volume minus cell volume. Change in plasma volume therefore reflects both change in blood and cell volume during hypohydration	Blood pin prick Calculated from red cell volume and blood volume before and after dehydration	Use method of Dill and Costill (1974)

3 Based upon analysis of sweat

Sweat rate – SwR – litres/sec	Rate of sweat produced by the body	Sweat represents a water and electrolyte loss from the body leading to dehydration	Accurate weighing scales. Continuous weighing or before and after exposure. Account for: sweat trapped in clothing, dripping, urine, faeces, breathing	Usual practical indicator of dehydration and water requirements
Body weight loss – BW_{loss} – %	Loss in body weight over exposure	Most body weight lost as fluid	Accurate weighing scales. Continuous weighing or before and after exposure. Account for: sweat trapped in clothing, dripping, urine, faeces, breathing	Assume normal faeces 1/3 fluid by weight. As for sweat rate
Sweat sodium – Sw_{Na} – mg	Amount of sodium in sweat collected over a period of time	Re-absorption of Na^+ causes water retention, affects plasma electrolytes and hence sweat Na^+ concentration	Ventilated sweat capsules on the skin or the use of pads at sites across the body, changed at intervals. Flame photometry in the laboratory	Response non-specific to hypohydration. NaCl content of sweat falls drastically with acclimatization
Sweat potassium – Sw_K – mg	Amount of potassium in sweat collected over a period of time	Re-absorption of Na^+ causes excretion of K^+ affecting plasma electrolytes and hence sweat K^+ concentration	Ventilated sweat capsules on the skin or the use of pads at sites across the body, changed at intervals. Flame photometry in the laboratory	Response non-specific and concentration changes with acclimatization. Linked with Na^+ concentration

Table 4.3 (Continued)

Measure	Definition	Rationale	Example of measurement method and rationale	Comment
4 Based upon subjective methods				
Thirst – rating scale value	That condition of mind which elicits a desire to drink in order to alleviate a sensation of dryness in the mouth and throat	Hypohydration reduces blood volume and increases osmolality, releasing ADH and producing a thirst sensation	Scales from 'not thirsty' to 'very thirsty'	Objective measures of saliva levels using dental pads. Thirst is not specific to hypohydration and is alleviated before euhydration
Skin reaction and appearance	Appearance of skin and reaction to pressure (e.g. pinching)	In hypohydration low blood pressure, skin appears dry, slow recovery from pinching, flat blood vessels in neck, slow filling of heart valves	Observe appearance. Pinch skin and closely observe recovery rate	Severe dehydration, cold clammy skin. Qualitative general indicator of dehydration
Urine colour	*See Table 4.3:1. 'Measures of human hydration state based upon analyzing urine'*			

air temperature (equal to mean radiant temperature) 34.3 ± 0.3 per cent; relative humidity $0.41 \pm 0.1 \, \mathrm{m\,s^{-1}}$ air velocity) for 170 min, performing five sets of 20 min cycling and 10 min rest. There were two experimental sessions, separated by at least one week, in a balanced, repeated measures experimental design. In the 'drinking' session subjects replaced weight loss with cool water after each cycling period. In the 'no drink' session subjects received no fluid replacement. A battery of physiological and subjective measures (blood, urine, sweat, thirst) were taken pre- and post exercise. To ensure that subjects were in similar hydration (euhydration) and nutritional states at the start of each experimental session, diet, fluid intake and exercise, during the 15 h preceding each trial, were carefully controlled.

Measures in sweat

Subjects were weighed after each cycle bout to determine weight loss over time and over the total time. Any water consumed was also weighed (loss of weight in water bottle). This provided the amount of sweat loss due to evaporation and some dripping. To collect sweat for analysis, small pads of filter paper were placed over sites on the skin and collected (with replacements) into sealed plastic bags. They were analyzed for *sodium* and *potassium* concentration using flame photometry.

Measures in blood

Blood pinprick measures determined changes in *haemoglobin, haematocrit*, and *plasma* volume over the exercise period and between conditions. Haematocrit was measured by introducing the samples into microcapillary tubes and using a centrifuging technique. Haemoglobin was measured by introducing the samples to a cyanmethaemoglobin reagent solution and analysed using the calorimetric method. Change in plasma volume was calculated according to the method described in Dill and Costill (1974) who determined plasma volume from measurements of haemoglobin concentration and haematocrit.

Measures in urine

Urine measures were provided pre-, post- and six hours after exposure. *Urine temperature* and *urine volume* were obtained by urinating into a vacuum flask containing a sensor initially at 37 °C, and transferring to a measuring cylinder. *Urine pH* was determined using litmus paper and urine colour was judged by comparing the colour in natural daylight with that on a urine colour chart (see Table 4.4). Litmus paper and urine colour were always judged by the same person. *Urine specific gravity* was measured using a glass hydrometer and urine osmolality with an osmometer.

Table 4.4 Urine colour chart (Different patterns used here with colour descriptions. True colour chart will contain actual bands of colour for direct comparison.)

1	Clear	
2	Light yellow	
3	Yellow	
4	Dark yellow	
5	Yellow/orange	
6	Light orange	
7	Orange	
8	Brown	

Always check urine for:	*Advice and guidance*
Volume (how much)	>2 cups of urine per day
Frequency (how often)	>twice per day
Colour (how dark)	No darker than colour 3 (yellow) on the colour chart. Check regularly at similar times of day

Source: Based upon Nevola (1998).

It is essential that quality procedures are implemented in all measurements and that calibration techniques used are carefully carried out. Stirling (2000) concludes that the haematological (blood) measures did not accurately reflect hydration state and that urine measures (osmolality, specific gravity, colour and volume) were found to be practical, reliable, sensitive, and valid methods of assessing hydration state. It should be noted that the conclusions are based upon the methods and conditions of the experiment. Sawka (1988) suggests that there are many factors that influence the intravascular fluid responses to exercise and heat stress and that this explains inconsistencies in the results of different studies. He lists *premeasurement* (posture, sample arm position and stasis, ambient temperature); *stressor* (posture, exercise type, intensity and duration, thermal stress and strain) and *subject* (fitness, acclimation state,

Table 4.5 Some terms and definitions

Term	Definition
Dehydration	Dynamic loss of body water leading to a transition from euhydration to hypohydration
Diffusion	Movement of molecules from an area of high concentration to an area of low concentration
Electrolyte	Chemical which, when dissolved in water, dissociates into ions and thus can conduct an electric current
Euhydration	Normal body water content
Haemoconcentration	Net loss of plasma from intervascular space
Haemodilution	Net gain of plasma from intervascular space
Hypervolemia	Steady state blood volumes greater than normal
Hypovolemia	Steady state blood volumes less than normal
Hyperhydration	Body water excess
Hypohydration	Body water deficit
Osmolality	A measure of total solute concentration per kg of solvent
Osmolarity	Concentration of solutes (ions) in solution. The more solutes in solution, the higher the osmolarity
Palatability	How acceptable a beverage is to consume (taste, colour, smell, temperature, etc.)
Rehydration	The method of restoring a level of full hydration from a previously dehydrated state
Thirst	That condition of mind which elicits a desire to drink to alleviate a sensation of dryness in the mouth and throat

hydration level) as influencing response (dilution, concentration). He clarifies terms used and notes that, 'solute concentrations are expressed in milliosmoles per kilogram of water (osmolality) or milliosmoles per litre (osmolarity)', and that osmolality is preferred as it is not influenced by the solution temperature nor the volume that the solutes occupy within the solution. A list of terms and definitions is provided in Table 4.5. An assessment of the properties of measures and their usability is provided in Table 4.6.

Sawka (1988) reviews mechanisms and studies of hypohydration during exercise in the heat. He concludes that if sweat output exceeds water intake then hypohydration will occur with fluid loss from both intracellular and extracellular fluid compartments. During exercise hypohydration will reduce plasma volume, which will produce greater heat storage and higher body temperatures. This is due to reduced cardiac output and reduced ability to dissipate heat. He also concludes that hyperhydration does not provide an advantage but does delay dehydration (see also McArdle *et al.*, 1996).

Skin condition

Leithead and Lind (1964) note that, although hot and dry to the touch during heat stress, the skin has normal hydration and elasticity. In dehydration, however, this is reduced and the level of dehydration can be indicated in 'rough' terms by rate of recovery after pinching (see Table 4.3).

Table 4.6 Summary of the reliability, sensitivity, validity and usability of measures of hydration state

Measure		Reliability	Sensitivity	Validity	Usability
Urine colour	U_{col}	✓✓	✓✓	✓✓	✓✓
Urine specific gravity	U_{sg}	✓✓	✓✓	✓✓	✓
Urine osmolality	U_{osm}	✓✓	✓✓	✓✓	✗
Urine temperature	U_{temp}	✗	✓	✗	✗
Urine pH	U_{pH}	✗	✗	✗	✓
Urine volume	U_{vol}	✓	✓✓	✓	✓✓
Urine sodium	U_{Na}	✗	✗	✓	✗
Urine potassium	U_{K}	✗	✓	✓	✗
Urine haematocrit	H_{ct}	✗	✗	✗	✗
Urine haemoglobin	Hb	✗	✓	✗	✗
Sweat sodium	Sw_{Na}	✗	✓	✗	✗
Sweat potassium	Sw_{K}	✗	✗	✗	✗
Thirst		✗	✗	✗	✓

Source: Adapted from Stirling (2000).

Thirst

Thirst can be defined as:

> that condition of mind which elicits a desire to drink to alleviate a sensation of dryness in the mouth and throat.

The definition emphasizes that it is a condition of mind as thirst occurs, not only due to the physiological state of hypohydration, but when the mouth and throat are dry through dust or other materials, breathing in or movement of warm dry air across the face, in illness, and for social and psychological reasons. An illness that interferes with central mechanisms in the brain could produce thirst even though the mouth and throat are moist. Thirst primarily occurs through hypohydration however, and associated hypovolaemia, increased osmolality and release of ADH (see Figure 4.2).

In his review, Sawka (1988) notes that studies have consistently shown that *ad libitum* water intake results in incomplete water replacement (voluntary dehydration) during exercise in the heat. For acclimatized people this effect is reduced but voluntary dehydration still occurs.

Although objective measures of thirst can be used (sampling pads – absorbent dental cotton wool rolls – in the mouth and throat), thirst is usually measured subjectively. There are also degrees of thirst – as opposed to yes/no – so subjective scales often assume a continuum. Rolls *et al.* (1980) used a 100 mm line with the question, 'How thirsty do you feel now?' and end points, 'Not at all' and 'Very thirsty'. Stirling (2000) used a simple four-point scale – 0 = Not thirsty; 1 = Slightly thirsty; 2 = Thirsty; 3 = Very

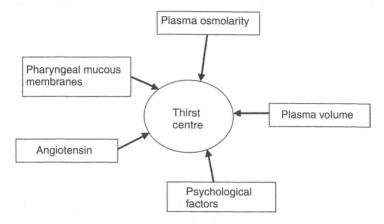

Figure 4.2 Mechanism that influences thirst.

thirsty. Thirst as a psychological phenomenon can be regarded as a desire or motivation which influences behaviour and has elicited an associated semantic structure – thirsty, dry, 'gagging', thirst quenching, etc. This structure however has not been fully explored and converted to appropriate (valid, sensitive, reliable) subjective scales of known properties.

Brunstrom and Macrae (1997a,b) and Brunstrom *et al.* (2000) have demonstrated the importance of mouth state on drinking behaviour. Cold drinks produced greater reduction in thirst ratings than did warm drinks. They propose that temperature influenced thirst reduction because of its differential effects on the post ingestive state of the mouth. They also propose that termination of drinking is governed by changes in mouth dryness due to increased saliva production during drinking. The dynamics and psychological aspects of thirst require further investigation; however, there is consistent evidence that relying on thirst alone as a measure of hypohydration will not restore water balances.

BIOELECTRICAL IMPEDANCE ANALYSIS (BIA)

The electrical resistance of the body to an alternating current varies with its water content. This is because water with dissolved electrolytes will conduct electric current (Lukaski, 1987). Bioelectrical impedance is a more comprehensive representation of the resistance which includes also the electrical capacitance and inductance of the body and will vary with the frequency of the alternating current applied. At high frequencies (>50 Hz) the impedance will reflect the total body water (TBW) as the current penetrates the cells and at lower frequencies (<5 Hz) the impedance will reflect the extracellular water (ECW) only, as the electrical current will not penetrate the cell membrane. The volume of water is given by

$$V = \rho \frac{L^2}{R},$$

where

V = volume

ρ = specific resistivity

L = length between electrodes

R = resistance.

Intracellular water (ICW) can be obtained by subtracting the volume calculated from the resistance at <5 Hz (ECW) from the volume calculated at >50 Hz (TBW). More recent methods use *Sine-Sweep* or multi-frequency techniques (bioelectrical impedance spectroscopy) where the impedance frequency response of the body is determined (Ritz, 1998).

The apparatus for measuring bioelectrical impedance is simple and portable. Subjects are in a supine position for 30 min in a fasting state and hours after an exercise. Surface electrodes are placed on the end of limbs and a current – a few microamps – is passed through the body, allowing measurement of resistance. Bioelectrical impedance can be related to body composition and devices are commercially available for general use. However, caution should be taken when interpreting results. Assumptions must be made in the application of mathematical models that relate electrical resistance to body water content. The results obtained will depend upon individual factors such as sex, body weight, age, ethnicity, and disease. Also the reliability of the technique is around 1 litre for total body water. For use in heat stress assessment where sweat loss may be 1 litre per hour, the technique may have limited application. Stirling (2000) conducted a small-scale investigation with one subject exposed to heat stress and another subject to control conditions. She found that the BIA technique was not sensitive to changes in body weight due to sweating, and that more research is required on the reliability, sensitivity and validity of the technique – as well as calibration procedures – before it could be considered for use in measuring dehydration levels for workers in hot conditions.

DEHYDRATION LIMITS

An effective management of work in hot conditions should avoid dehydration. However because of the significant effects of dehydration, where this cannot be achieved, it is necessary to set dehydration limits.

Table 4.7 Effects of body weight loss

% body weight loss (litres for a 70 kg man)	Symptoms
0.5 (0.4)	
1.0 (0.7)	Thirst threshold at rest, impaired ability to thermoregulate during exercise
2.0 (1.4)	Thirst threshold during exercise, vague discomfort, loss of appetite
3.0 (2.1)	Increasing haemoconcentration, dry mouth, reduction in urine
4.0 (2.8)	Increased effort of exercise, flushed skin, impatience, apathy, exercise capacity is reduced 20–30% below normal
5.0 (3.5)	Difficulty in concentrating, headache, impatience (*see above too*)
6.0 (4.2)	Severe impairment in exercise temperature regulation, increased heart rate, risk of heat stroke
7.0 (4.9)	Likely collapse if combined with heat and exercise
8.0 (5.4)	Dizziness, laboured breathing in exercise, mental confusion
10.0 (7.0)	Spastic muscles, inability to balance with eyes closed, general incapacity delirium and wakefulness, swollen tongue
11.0 (7.7)	Circulatory insufficiency, marked haemoconcentration and decreased blood volume, failing renal function
15.0 (10.5)	Circulatory failure and death

Source: Adapted from Nevola (1998).

Dehydration limits provide levels of dehydration beyond which it would be considered unacceptable. They will depend upon the context under consideration and could be set to maintain comfort, performance or health. They are usually presented in terms of the percentage loss in body weight and should not be ambiguous about the extent and method of water and electrolyte replacement. (Does the limit include replacement water or not?) The relationship between percentage of body weight loss and symptoms is shown in Table 4.7. Simple charts such as this provide a general indication of consequences and could be used to set limits.

ISO 7933 (1989) provides limiting values for work in hot environments based upon heat storage and dehydration. For dehydration, duration-limited exposures are calculated based upon a maximum water loss for a warning level and a danger level of around 4 per cent and 6 per cent of body weight for non-acclimatized and acclimatized people, respectively. The limits (D_{max}) for maximum water loss are shown in Table 4.8.

Table 4.8 Limits for maximum water loss

Criteria Maximum water loss		Non-acclimatized		Acclimatized	
		Warning	*Danger*	*Warning*	*Danger*
D_{max}	$W\,h\,m^{-2}$	1000	1250	1500	2000
	g	2600	3250	3900	5200

Source: ISO 7933 (1989).

The Duration Limited Exposure (DLE) is calculated from

$$DLE = 60\frac{D_{max}}{SW_p}\ \text{min},\tag{4.1}$$

where SW_p is the predicted sweat rate ($W\,m^{-2}$) and D_{max} ($W\,h\,m^{-2}$) can be found in Table 4.8. The Standard appears ambiguous in terms of replacement of drinking. A modified version of ISO 7933 (European Standard EN 12515) notes that 'the Standard assumes normal rehydration but no account is taken of amount and frequency of drinking and its effectiveness.' It is recognized that drinking will not directly replace water lost and assumes that water replaced is not available at the tissues for sweat loss on the skin. If a dehydration limit is reached or exceeded therefore, work must cease and no further exposure allowed for the rest of the day.

In a revision of the required sweat rate model that is used in ISO 7933 (1989) a new model has been proposed; i.e. the Predicted Heat Strain (PHS) index (Malchaire *et al.*, 1999, 2000). Consideration of limits for maximum dehydration and water loss suggested a 3 per cent maximum dehydration as a limit for industrial work. This was based upon Candas *et al.* (1985) who reported that a 3 per cent dehydration induces an hypertonic hypovolaemia with increased heart rate and depressed sweating sensitivity. Malchaire *et al.*, 1999 report an average rehydration rate of 60 per cent for four to eight hour shifts in German coal mines. Consideration of coal miners with sweat rates of over 2000 g over the shift indicated that 95 per cent of subjects had a rehydration rate of greater than 40 per cent. Their data can therefore be used to provide the following limits for water loss:

$$\frac{3\%}{(1-0.6)} = 7.5 \text{ per cent of the body mass for an average subject}$$

$$\frac{3\%}{(1-0.4)} = 5 \text{ per cent of the body mass for 95 per cent of the working population.}$$

Further consideration of the Predicted Heat Strain index is provided in Chapter 10.

PRACTICAL RECOMMENDATIONS FOR DRINKING

Limits for water loss and dehydration can be used to prevent unacceptable consequences when working in hot conditions and any dehydration may be considered unacceptable. The ability to maintain acceptable rehydration states is enhanced by the introduction of a management system with checks and procedures. Who is responsible? Where and how are checks recorded? Working practices etc. Working practices for hot environments are described by NIOSH (1986) and are presented in Table 10.8 (see page 290 of this book). Recommendations include ensuring water loss is replaced and electrolyte balance maintained. Goldman (1988) suggests that workers should be acclimatized, using a breaking in period of 5–7 days. Dietary salt should be supplemented during acclimatization and that drinking water should be available at all times and to be taken frequently during the working day. Athletes consider rehydration as important and recognize that it can affect performance. Drinking regimes and types of drink are the subjects of debate in the specialized area of high exertion and activity. Drinking water early and frequently is often recommended.

McIntyre (1980) provides practical information concerning anhidrotic heat exhaustion caused by water depletion which leads to heat stroke, delirium and death. A water deficit of four litres produces intense thirst and scanty urine; deficits over 5 litres lead to circulatory failure with death at around 10 litres (15 per cent body weight) deficit. As, on average, a person loses about 1.5 litres of water per day (from urine, faeces, lungs and skin), in a temperate climate a person can survive without water for about seven days. Work in the heat will dramatically reduce that and an adequate supply of drinking water is essential.

Industrial work management systems for occupational safety have become requirements and safety officers are often required to assess and record the risks in a work place. However, they have yet to be linked to recommendations based upon a model of human water balance and regulation; for example, based upon the dynamic interaction of a four compartment system (alimentary canal – including mouth and stomach, intravascular, interstitial, and intracellular).

Nevola (1998) provides a practical guide to drinking for military commanders to aid them in ensuring the optimum performance of soldiers operating in hot environments. He considers the causes of dehydration, its effects and how it can be avoided, and strategies for drinking. Advice on 'what to look for' as signs of heat exhaustion due to dehydration are presented in order of severity: from fatigue, loss of appetite, flushed skin, heat intolerance, light-headedness, small amount of dark urine, difficulty in swallowing, inelastic skin (slow recovery when pinched), sunken eyes, dim vision, painful urination, muscle spasms and, eventually, heat illness.

Of particular importance is heat stress and dehydration when wearing total (impermeable) protective clothing (Nuclear, Biological, Chemical – NBC).

Nevola (1998) provides guidance in terms of water requirements for levels of heat stress represented in Wet Bulb Globe Temperature (WBGT) index values (see Table 4.9).

The usefulness of the WBGT index in this context may not be optimum as it is strongly influenced by the natural wet bulb value which freely interacts with the environment, whereas people wearing impermeable clothing will not be represented by the index. The proposals for the avoidance of dehydration, strategies for drinking and recommendations for recovery from dehydration are presented in Tables 4.10, 4.11 and 4.12 (Nevola, 1998).

Stirling (2000) identified that fire training instructors suffered from hypohydration and that *ad libitum* drinking did not restore water balance, but a pre-determined drinking regime did. Based upon laboratory and field studies and an investigation of the requirements of the task and organization, she proposed a practical guidance document, 'Heat strain and dehydration advice for fire-fighting.' The guidance provides substantial, introductory and background material related to firefighters, including conditions when self-contained breathing apparatus and protective clothing are worn. She particularly noted the effects of dehydration at the beginning of the day after a previous evening of alcohol consumption and 'macho' attitudes which lead to over-exposure and reluctance to report symptoms. She recommends that firefighters should attempt to monitor body temperature and body weight loss through sweating. Firefighters are recommended to monitor the colour of their urine on a colour chart and to ensure that it is equivalent to colours 1, 2 or 3 on the urine chart (i.e. pale yellow or clearer – see Table 4.4).

Incorporating an effective fluid intake regime is essential to combat the effects of heat strain and dehydration and should be implemented during operational training and work (Stirling, 2000). Thirst is not a good indicator of water requirement and individuals are encouraged to drink as much as possible, without stomach discomfort, up to that which is required. Carbohydrate in solution will aid intestinal absorption, however the rate of emptying decreases above 8 per cent concentration. Where the volume of fluid in the stomach is maintained at 600 ml, most people can empty more than 1 litre per hour from the stomach (Stirling, 2000). Other recommendations agree with those of Nevola (1998) – see Table 4.10. Fluids that are advisable to drink are water, squash, juice, milk, and carbohydrate-electrolyte drinks (<8 per cent concentration).

In neutral and cold climates the average resting body will lose about 1 litre of water per day due to insensible water loss from the skin and lungs and, in warm environments, about 2 litres. Drinking is therefore required to replace this loss although dehydration is unlikely. Babies, children and others with a large surface area to mass ratio will lose a greater percentage of their body weight. In the cold, dry skin, particularly lips, is caused by wind chill and the large differences between saturated vapour pressure at the skin and that in the cold air. It is unlikely to contribute to dehydration.

Table 4.9 Heat, work and water requirements when wearing NBC (Nuclear, Biological, Chemical impermeable) protective clothing

Temperature (WBGT°C)	Work/Rest limit (min per hour)			Continuous work limit (min)			Water needed for continuous work (litres)			Estimated daily drinking water (litres)		
	v. light	light	moderate	v. light	light	moderate	v. light	light	moderate	v. light	light	moderate
33.3	NL	UW	UW	NL	79	42	1.5	2	2	15	15	15
	(NL)	(UW)	(UW)	(NL)	(92)	(43)	(1)	(2)	(2)			
31.1	NL	UW	UW	NL	95	45	1	1.5	2	15	15	15
	(NL)	(UW)	(UW)	(NL)	(125)	(47)	(1)	(1.5)	(2)			
28.9	NL	19	7	NL	133	50	1	1.5	2	7–10	10–15	15
	(NL)	(NL)	(13)	(NL)	(NL)	(52)	(1)	(1.5)	(2)			
25.5	NL	NL	16	NL	NL	57	1	1.5	2	5–7	5–7	5–10
	(NL)	(NL)	(21)	(NL)	(NL)	(62)	(0.5)	(1.0)	(1.5)			

Notes
UW = unwise; NL = no limit; ( ) = night exercise. Assumes personnel are fully hydrated, relative humidity is 50 per cent, casualties number <5 per cent, wind speed is $2 \, \text{m s}^{-1}$ (4.5 mph).

Table 4.10 Recommendations for the avoidance of dehydration

Establish a management system	Analyze the work and implement a management system and system of procedures to ensure that drinking requirements are met
Drinks and drinking	Eat a balanced diet and drink sugar and electrolyte beverages (<8% concentration). Individuals should practise and develop their own proven regime
Palatability 'If it tastes good, you'll drink it'	Ensure that the beverage is pleasant to drink. 5 °C drinks are most pleasant but 15–20 °C is acceptable. Fizzy drinks, beer and warm water are not recommended. 2 g (half a teaspoon) of salt added to a cup of beverage improves palatability when thirsty
Potability	Poor hygiene leads to bacterial growth, stomach illness, and diarrhoea. Ice made from 'local' water will still hold contaminants
Logistics	Containers may have to be carried and will need to be adequate for a fixed time. Practice may be required in container use which should allow for an acceptable drinking rate
The ideal drink	Cool, safe, ready to drink, pre-calculated volume, flavoured and lightly salted to suit taste in an easy to use container
What not to drink	Alcohol, coffee, tea, fizzy drinks and other caffeine-containing beverages. Oral rehydration solutions (for the treatment of diarrhoea), large volumes in short exercise periods, carbohydrate electrolyte drinks >8% concentration, fructose drinks during exercise, hot drinks when cool ones are available, unknown or unsafe drinks
What to drink	Plain, safe, cool or chilled water is acceptable for ≤90 min exposure. For >90 min, flavoured, with <8 per cent concentration of carbohydrate (12.5 g – two tablespoons – of sugar to one litre of water). For >240 min supplement with one teaspoon of salt per litre as well
When and how much to drink	Before, during and after heat exposure. 500 ml 2 h in advance, 300 ml 15 min prior to exposure and 200 ml during exposure. Should produce >400 ml of urine per day (of light yellow colour) to avoid decreased performance

Source: Adapted from Nevola (1998).

Cold diuresis is a response to the cold that increases urine output. This is related to changes in blood flow and distribution and is affected by posture (increases when lying down). The cold diuresis is inhibited by anti-diuretic hormone and is unlikely to contribute to dehydration.

Leithead and Lind (1964) describe water-depleted heat exhaustion as, 'water depletion due to inadequate replacement of water loses in prolonged

Table 4.11 Practical guidance for drinking

Training	All personnel should be trained in their roles and be clear about what to do for avoidance and detection of dehydration
Checks	Maintain colourless urine by checking urine colour against colour chart. Allow time for bladder emptying before starting work. Do not rely on thirst as an indicator of when to drink. Rehearse your planned drinking strategies long before trying them in real situations. Choose an appropriate drink that you like and has already been successful for you
Before heat exposure	Eat a balanced diet. Ensure that you are well hydrated with 1 litre of water 2 h before exposure and 200 ml 15–20 min prior to exposure
During heat exposure	Drink one cup of cool, slightly salted water (carbohydrate-electrolyte drink for >90 min work) every 15–20 min. Avoid alcohol and caffeine (tea, coffee, coke) and plan work – especially hard work – in the heat to allow for rests and opportunities to drink
Water rationing	Avoid if at all possible. Where essential, implement strict drinking policy with controls to ensure drinking takes place and a particular level of dehydration is maintained. Identify acceptable exposure times and recovery plan

Source: Adapted from Nevola (1998).

Table 4.12 Recommendations for recovery from dehydration

Management system	Dehydration is cumulative from day to day. Once it has been detected, a system of procedures is required for recovery and restoration to euhydration. This may involve simple procedures to eventually transportation to hospital. Systems should be implemented and people made responsible
What to drink	Water is acceptable but beverages containing sugar and salt are more effective
Severe dehydration	If conscious – provide a drink with 1 litre of water, 40 g of table sugar and 6 g of salt If unconscious – intravenous rehydration with cool IV fluid (medical supervision required)
When to drink	As soon as possible after heat exposure and every 2 h after until urine colour is light yellow or clear
How much to drink	To full recovery of body weight (>6 h for complete rehydration) and beyond (to 170% *of fluid lost*). At least 1 litre consumed after end of exposure immediately

Source: Adapted from Nevola (1998).

sweating, characterized by thirst, fatigue, giddiness, aliguria, pyrexia, and in advanced stages, by delirium and death.' Thirst is the earliest symptom, one with which the 'patient' becomes obsessed. The tongue and mouth become dry, eating and swallowing are difficult and appetite for solid food is lost. Goldman (1988) describes a wet globe thermometer with a colour scale attached to the sensor. He advises soldiers that when the pointer on the scale is, 'in the pink, drink'.

There may be confounding factors in human responses to thermal environments that influence water balance. Drugs (e.g. diuretics) can influence control of water balance. For example, so called recreational drugs may inhibit the transport of water from the stomach which could be fatal in a prolonged, frenzied dancing spree in hot conditions. Diarrhoea and vomiting are particularly important and although rarely relate directly to heat, intestinal infections with bacteria, protozoa or helminths are common causes when a person travels to a hot, unfamiliar climate. Significant and dangerous amounts of water can be lost depending upon the severity of the condition.

People who are ill, infants, enfeebled patients and those that cannot make known or effect their need for water are particularly vulnerable. An adaptive approach should be considered in risk assessment. What is the adaptive opportunity to drink? Voluntary dehydration is practised even among the fit and healthy and those without disabilities. Where there is an effort required to drink (taking off protective clothing, travelling across a hot room, water not cool or palatable) then voluntary dehydration is more likely. These obstacles can become barriers to people with disabilities and others when the opportunity to drink has effectively been removed from them. A detailed analysis of both user requirements and characteristics is required therefore in the design of any work system.

5 Measurement methods and assessment techniques

INTRODUCTION

To assess human thermal environments it may be necessary to quantify the environment (basic parameters), and its effects (physiological, psychological, etc.) and interpret the values obtained in terms of the health, comfort, and performance of those exposed. Basic environmental parameters can be quantified using measuring instruments; physiological responses can be measured using transducers connected to the body and psychological responses are quantified using subjective and behavioural measures. Assessment tools include the use of the heat balance equation and the derivation and use of thermal indices.

It is fundamental to all measuring systems that they will interfere (however slightly) with what they are intended to measure and many factors other than the parameters of interest will contribute to the measured quantity. A measuring instrument, therefore, must be designed and used such that the parameter of interest can be quantified and hence ensure that other factors are relatively unimportant for the application of interest. For example, when measuring air temperature a sensor should not be exposed to direct sunlight where it will be greatly influenced by radiation. The experimenter (or his equipment case etc.) should not stand between a radiation sensor and a radiation source, when attempting to quantify conditions in which people would normally work (in the absence of measuring equipment and experimenter). Although apparently obvious, the selection of instruments and methods of use are of great importance and errors are often made. In an attempt to quantify an environment, careful consideration should be given to what is being measured and how the instrument and its use will relate to this. Guidance will then be obtained on which instrument to use (in terms of accuracy, sensitivity and time constant, interference, etc.) and where, when and how to measure the parameter.

Air temperature (t_a)

In Chapter 1, air temperature was defined as the temperature of the air surrounding the human body which is representative of the aspect of the surroundings which determines heat flow between the body and air. Measurement of air temperature is usually made with a mercury-in-glass thermometer, a thermocouple, a platinum resistance thermometer or a thermistor. Instruments that provide continuous electrical signals are useful for recording temperature values for later use in computer analysis. It is important to establish calibration as no instrument measurement is truly linear with temperature change – this is particularly so with thermistors, where electronic equipment must be matched to specific thermistors – and it is important to ensure accurate calibration of equipment. In addition, it is useful to apply checks of equipment by, for example, placing all sensors in a stirred water bath and comparing values with that of an accurate (certified in a standards laboratory) mercury-in-glass thermometer, before and after measurement. This allows an 'absolute' check of each sensor and, importantly, a relative check between sensors and an identification of possible changes during measurement. A comparison of the properties of methods of measuring air temperature is provided in Table 5.1 and the requirements for an instrument for assessing human thermal environments are given in Table 5.2.

Table 5.1 Instrumentation for temperature measurement

Property	Thermocouple*	Thermistor	Platinum resistance thermometer	Semiconductor junction[†]	Mercury-in-glass
Long-term stability	variable	ages	stable	stable	stable
Signal for 1 °C change	10–60 μV	1% of resistance (linearized)	40 μV (at 1 mA current)	2.3 mV	
Speed of response	fast	fast	moderate	moderate	slow
Relative cost[‡]	1	4	5	2	3
Mechanical stability	robust	moderate	moderate	robust	poor
Reproducibility	moderate	good	very good	poor	very good
Linearity	moderate	linearized versions required	good	good	good
Accuracy (typical)	±2 °C	±1 °C	±0.1 °C	±1 °C	±0.1 °C (NPL calibrated)

Source: BOHS (1990) (Acknowledgement to National Physical Laboratory: Report Qu48, 1978).

Notes
* Cold junction or compensated circuit required.
† High self-heating effect.
‡ Relative cost (1: cheap; 5: expensive).

Table 5.2 Requirements for an instrument for measuring air temperature in the assessment of human thermal environments

Long term stability	Stable; the ability to recalibrate quickly an advantage
Signal for 1 °C	As high above 'noise' as possible; ability to digitize useful; visual signal on mercury-in-glass thermometer not always easy to read
Speed of response	Depends upon study; high compared with rate of change of environmental conditions
Cost	Depends upon study; in most applications moderate cost will provide satisfactory equipment
Mechanical stability	Moderate to high desirable
Reproducibility	Should be good
Linearity	Should be good, otherwise calibration required; may be possible to provide linearity within software; for example, in a spreadsheet
Accuracy	High and within ±0.3 °C. Depends upon application. There will be measurement errors in addition to sensor accuracy
Other	Thermometers should be easy to use and not interfere with the environment or other measuring instruments (e.g. anemometers); shielding from radiation is important, particularly outdoors; must be able to measure at the point of interest

The effect of radiation on the sensor is important and can be reduced by reducing the emissivity of the sensor (i.e. make it silver), shielding it (but still allowing good air circulation) and increasing air velocity across the sensor. McIntyre (1980) suggests that the equilibrium temperature (t) of a spherical sensor is given by

$$t = (1 - g)t_a + gt_r, \tag{5.1}$$

where

t_a = air temperature
t_r = mean radiant temperature
g = radiant response ratio

and

$$g = \frac{1}{(1 + 1.13v^{0.6}d^{-0.4})}, \tag{5.2}$$

where

v = air velocity (m s^{-1})
d = diameter of sensor (m).

So increasing v and decreasing d greatly reduces g and hence t tends to t_a. For low air speeds (e.g. $<0.2\,\mathrm{m\,s^{-1}}$) a mercury-in-glass thermometer can have a value of g of around 0.25 hence producing serious errors in measurement when t_r differs from t_a. The value of g may be decreased by increasing relative air movement across the thermometer bulb as in the Assman and Sling psychrometers (see below).

McIntyre (1980) concludes that the only satisfactory way of measuring air temperature when radiation is significant is to shield the sensor (in low-emissivity metal), ensure good air circulation, and draw air (rather than blow to avoid heat from the 'fan motor') across it. A wide mouthed vacuum flask would provide an effective shield for example.

Equation (5.1) above, assumes an emissivity of unity for the sensor, which is not the case for a silvered thermometer bulb for example. A more general equation is

$$t = \frac{(1-g)t_a + \varepsilon g t_r}{1 + (1-\varepsilon)g}, \tag{5.3}$$

where ε is the emissivity of the sensor. For a silvered surface the emissivity may be as low as 0.1 and for a black bulb close to 1.0. However, the relative contribution of t_a and t_r will also depend upon the difference in the values. For a measurement of air temperature, a low d and high v gives a low g value. With a low ε value this means that t tends towards t_a. Conversely, it can be seen that to measure mean radiant temperature using a globe sensor the emissivity should be high (black), the diameter of the globe large, and the air velocity low, providing a g value close to unity and a measurement of t which tends to t_r. In all cases however, the greater the difference between t_a and t_r, the greater the 'error'.

Mean radiant temperature (t_r)

In Chapter 1, we saw how mean radiant temperature can be defined in two ways. With respect to the radiant field it is the temperature of a uniform enclosure with which a 'small black sphere' at the test-point would have the same radiation exchange as it does with the real environment. The alternative definition is concerned specifically with the human body and is the same definition as above with the words 'human body' replacing 'small black sphere'. Whichever definition is used the mean radiant temperature can be derived from the temperature of a black globe: 150 mm diameter is typically used. From equations (5.2) and (5.3) above, it can be seen that the large globe size and high emissivity 'tend' globe temperature $(t = t_g)$ towards t_r. However, as t_a differs from t_r and as air velocity increases, contributing to heat transfer by convection, a correction is required to t_g to determine t_r.

Bedford and Warner (1934) consider the use of globe thermometers for measuring mean radiant temperature. They cite Leslie (1804) as demonstrating the effects of radiation using two 100 mm (4 inch) diameter spheres (one silvered, the other blackened), filled with warm water. The different rates of cooling were measured. Aitken (1887) proposed the use of the hollow black sphere and Vernon (1930) first proposed the traditional blackened copper 150 mm (6 inch) diameter globe with a thermometer at its centre, which is still used widely. These globes were easily available as they were used as ball valves in plumbing. Bedford and Warner (1934) provide an approximate correction to (150 mm) globe temperature to calculate mean radiant temperature, which is converted to SI units by McIntyre (1980) as

$$t_r = t_g + 2.44\sqrt{v}(t_g - t_a). \tag{5.4}$$

This applies only if t_r is within a few degrees of t_a. It may be noted that if v tends to zero then t_g tends to t_r. This may be achieved with a globe shielded by a polythene (transparent) globe. However, there are other factors; for a full description see McIntyre (1980). Ellis *et al.* (1972) present a number of nomograms, charts and tables for the calculation of environmental parameters, including mean radiant temperature from globe temperature, air temperature and air velocity. ISO 7726 (1998) provide the following equations. For natural convection, e.g. $v \leq 0.15$

$$t_r = \left[(t_g + 273)^4 + \frac{0.25 \times 10^8}{\varepsilon} \left(\frac{|t_g - t_a|}{d} \right)^{\frac{1}{4}} \times (t_g - t_a) \right]^{0.25} - 273 \tag{5.5}$$

and for forced convection, e.g. $v > 0.15\,\mathrm{m\,s^{-1}}$

$$t_r = \left((t_g + 273)^4 + \frac{1.1 \times 10^8 v^{0.6}}{\varepsilon d^{0.4}} \times (t_g - t_a) \right)^{0.25} - 273. \tag{5.6}$$

For the standard globe, values of $\varepsilon = 0.95$ and $d = 0.15$ m can be used.

In theory, using the correction, any globe can be used to determine mean radiant temperature. However, the correction involves an accumulation of errors from other measurement (e.g. air velocity, globe temperature, air temperature) and hence if the magnitude of correction is large, it can be inaccurate. The use of small globes, for example, may be convenient (less bulky, quicker response) but can lead to large errors in estimates of mean radiant temperature.

The material type of the black globe will not affect the reading; however, it will affect response time. The standard black copper globe requires around 20 min to reach equilibrium. This can be reduced by increasing air movement within the globe and using thermocouples (or other devices with low thermal inertia) instead of mercury-in-glass thermometers (Hellon and Crockford, 1959).

If the mean radiant temperature is defined with respect to the human body, then the black globe will provide a good approximation for most situations;

however, it may not be representative for (directional) radiation longitudinal to the body (e.g. caused by heated ceilings) where a cylindrical or ellipsoidal shaped measuring device may be more appropriate. For outdoor conditions with solar radiation the black globe will overestimate the effects of radiation on a person. The globe should have the colour of clothing or a correction should be made. Also, for this case in indoor environments (rooms), the mean radiant temperature can be calculated from the temperatures of the surrounding surfaces, angle factors and the shape, size and relative position of the surface in relation to a person.

If an assumption is made that all surfaces have high emissivity then mean radiant temperature can be estimated using:

$$t_r^4 = t_1^4 F_{p-1} + t_2^4 F_{p-2} + \cdots + t_n^4 F_{p-n}, \tag{5.7}$$

where

t_r = mean radiant temperature (K)
t_n = surface temperature (K) of surface n ($n = 1, 2, \ldots, N$)
F_{p-n} = angle factor between a person, p and surface n.

If there are only small differences between temperatures of surfaces then as an approximation (in °C):

$$t_r = t_1 F_{p-1} + t_2 F_{p-2} + \cdots + t_n F_{p-n}. \tag{5.8}$$

Angle factors can be estimated from simple diagrams as in Figure 5.1.

An equation for calculation of angle factors is given in ISO 7726 (1998) as follows:

$$\text{Angle factor} = F_{max}\left(1 - e^{-(a/c)/\tau}\right)\left(1 - e^{-(b/c)/\gamma}\right),$$

where

$$\tau = A + B(a/c)$$
$$\gamma = C + D(b/c) + E(a/c).$$

(see Table 5.3)

Plane radiant temperature and radiant temperature asymmetry

Plane radiant temperature (t_{pr}) is the uniform temperature of an enclosure where the radiance on one side of a small plane element is the same as in the non-uniform actual environment, ISO 7726 (1998). A 'net' radiometer can be used to measure this (Olesen, 1985a; ISO 7726, 1998). Mean radiant temperature provides an average value over all directions to represent the radiant environment, while plane radiant temperature provides information

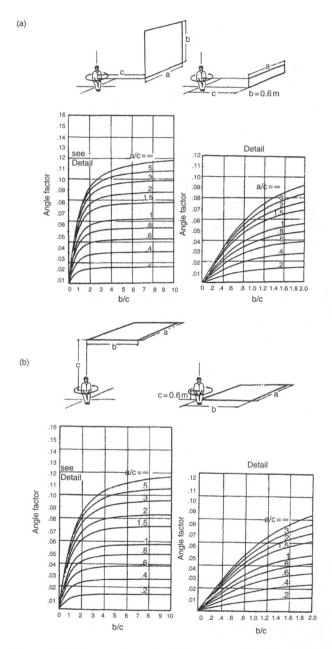

Figure 5.1 Mean values of angle factors used in the calculation of mean radiant temperature for a person, between (a) a seated person and a vertical rectangle above or below his centre; (b) a seated person and a horizontal rectangle above or below his centre; (c) a standing person and a vertical rectangle above or below his centre; (d) a standing person and a horizontal rectangle above or below his centre.

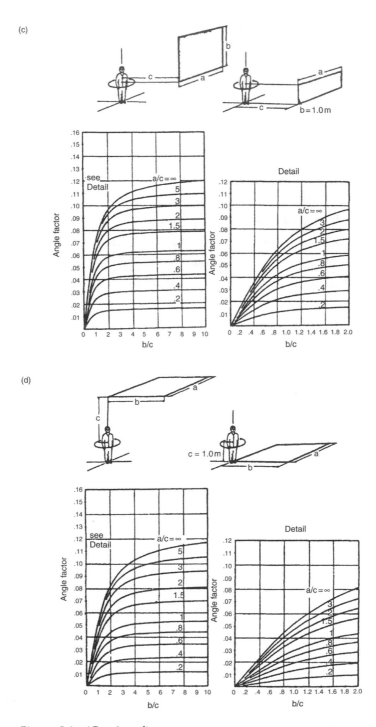

Figure 5.1 (Continued)

Table 5.3 Equations for calculations of the angle factors

	F_{max}	A	B	C	D	E
Seated person, Figure 5.1(a) Vertical surfaces: wall, window	0.118	1.216	0.169	0.717	0.087	0.052
Seated person, Figure 5.1(b) Horizontal surfaces: floor, ceiling	0.116	1.396	0.130	0.951	0.080	0.055
Standing person, Figure 5.1(c) Vertical surfaces: wall, window	0.120	1.242	0.167	0.616	0.082	0.051
Standing person, Figure 5.1(d) Horizontal surfaces: floor, ceiling	0.116	1.595	0.128	1.226	0.046	0.044

about radiation in one direction. If measures are taken in 'all' directions, it provides the variation about the mean, and can also be used to calculate mean radiant temperature. For example, ISO 7726 (1985) gives the following equation for calculating mean radiant temperature from plane radiant temperature measurements taken in the direction of six surfaces of a cube:

$$
t_r = \frac{A\left(t_{pr1} + t_{pr2}\right) + B\left(t_{pr3} + t_{pr4}\right) + C\left(t_{pr5} + t_{pr6}\right)}{2(A + B + C)}, \tag{5.9}
$$

t_{pri} = plane radiant temperature in direction i. i = direction (1) up, (2) down, (3) right, (4) left, (5) front, (6) back, and A, B, C are projected area factors given in Table 5.4.

The *radiant temperature asymmetry* is the difference between the plane radiant temperature of the two opposite sides of a small plane element. 'The concept of radiant temperature asymmetry is used when the mean radiant temperature does not completely describe the radiative environment' (ISO 7726, 1998). For example, in the case of 'local' thermal discomfort caused by a specific radiant source/sink in a given direction, radiant temperature asymmetry values will provide a more sensitive measure with which to assess the environment than mean radiant temperature.

Table 5.4 Projected area factors for calculating mean radiant temperature

	Up/down A	*Left/right* B	*Front/back* C
Standing person	0.08	0.23	0.35
Upright ellipsoid	0.08	0.28	0.28
Sphere	0.25	0.25	0.25
Seated person	0.18	0.22	0.30
Tilted ellipsoid	0.18	0.22	0.28
Sphere	0.25	0.25	0.25

Table 5.5 Instruments for measuring radiant temperatures

Instrument	Measurement
Globe thermometer	Globe temperature. Calculate t_r using v and t_a
Heated globe radiometer	Power to maintain temperature
Shielded globe thermometer*	Temperature of globe in a polyethylene evolope
Radiometer	Instrument that measures radiation
Net radiometer	Net radiation: direct, ground, sky
Pyranometer	Radiometer that measures short wave or visible radiation

Sources: * McIntyre (1980); Santee and Gonzalez (1988); NIOSH (1986).

Plane radiant temperature can also be determined from surface temperatures and shape factors using an equation similar to that for determining
mean radiant temperature, but with shape factors calculated for the case of
a small plane element parallel to a rectangular surface.

For further discussion of radiant temperature and its measurement, the
reader is referred to McIntyre (1980), ISO 7726 (1998) and ASHRAE (1997).
Requirements of instruments for the measurement of radiant temperatures
are provided below (Table 5.5) and instruments for measurement are shown
in Figure 5.2.

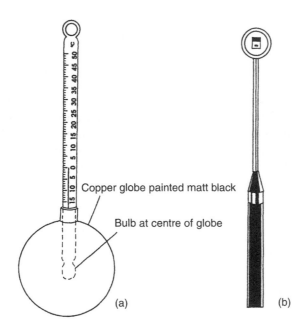

Figure 5.2 Instruments for measuring radiant temperatures: (a) the black globe
thermometer (mean radiant temperature); and (b) instrument for
measuring plane temperature and radiant temperature asymmetry.

These instruments must have a response time that is as fast as that required by application; globe temperature generally has a slow response time which can be reduced by using internal 'fanning' and the use of a sensor with low inertia. Their shapes can vary: spherical for mean radiant temperature as defined at a point; hemispherical for plane radiant temperature; and approximation to human body shape, taking account of posture, for mean radiant temperature defined for a human. The instruments' use and recording must not interfere with the work environment and must allow access to the area of interest. Electrical sensors allow recording for later analysis; but glass thermometers require reading. The instruments are usually black for mean radiant temperature defined at a point, or body and clothing colour to represent a person's mean radiant temperature.

Solar radiation

Solar radiation can be measured in terms of direction, level and spectral content. It can be measured with radiometers and pyranometers. Radiometers are instruments that measure radiation. Net radiometers are electric thermopiles protected by a transparent plastic or glass domes which screen out long wave radiation ($>3 \times 10^3$ nm). Net radiometers use clear plastic screens which allow the penetration of light from 300 to 6.0×10^4 nm. Pairs of radiometers with special screened domes are used to determine sky and ground radiation values. Pyranometers are radiometers that measure short wave or visible radiation. Coloured filters may be used to measure a narrow band of the visible spectrum hence, with a range of filters, producing spectral content of the radiation. A shadow band can be used as a first approximation (to blot out the sun) to measure diffuse radiation and then determine direct solar radiation.

Air velocity (v)

In Chapter 1, air velocity was considered to be the 'mean' air velocity over the body integrated over all directions and over an exposure time of interest. Specifications for an appropriate measuring instrument (anemometer) are given in Table 5.7. It is important to note that the body can perceive relatively low air movements ($<0.1\,\mathrm{m\,s^{-1}}$) and therefore devices involving air to move a mass (cup, fan blades, etc.) on the instrument, as used for measuring in outside conditions or in ventilation shafts for example, may be inappropriate. Important characteristics of a sensor are its sensitivity to direction of flow and to fluctuations in air velocity. A commonly used principle is to calculate the air velocity from the cooling (or heating) produced by air moving across the instrument. The Kata thermometer (Hill *et al.*, 1916), for example, is an alcohol thermometer with a large bulb which is heated and exposed to the environment. The time for the alcohol to fall (rise) over a 3 °C range, and air temperature are related to air velocity by

$$v = \left(\frac{aF}{t(t_k - t_a)} - b \right)^2, \tag{5.10}$$

where

t_k = mean temperature of Kata thermometer (°C)
t_a = air temperature (°C)
t = cooling time (seconds)
F = Kata factor of the individual thermometer (marked on the stem)— the amount of heat (m cal cm^{-2}) lost as the thermometer cools through the 3 °C range
a, b = constants depending on the type of thermometer (see Table 5.6).

Kata thermometers provide practical average values of air velocity; however, they are inadequate for quantifying fluctuations or for recording for computer analysis. Hot wire anemometers measure the cooling capacity of air moving across a 'hot' wire and relate this (using air temperature) to air velocity. They are widely used, measure fluctuations in air temperature and measures are easily recorded for later analysis. They are however directional in response and can be inaccurate in low air velocities due to natural convection of the hot wire. A similar principle to the hot wire anemometer is used by the hot-sphere anemometer, ISO 7726 (1998). Johanessen (1985) describes the development of a hot-ellipsoid anemometer involving calibration methods and spherical design features to avoid the heated element interacting with the air temperature (unheated) sensor and to produce an omni-directional instrument. The Kata thermometer provides a good general measure of air velocity; however, in fluctuating air the variation in air velocity will affect human response and should be quantified.

Table 5.6 Kata thermometer constants

Cooling range (°C)	Bulb	t_k(°C)	Constants			
			Low speeds		High speeds	
			a	b	a	b
38.0–35.0	glass	36.5	2.49	0.497	1.95	0.206
54.5–51.5	glass	53.0	2.82	0.600	2.14	0.246
38.0–35.0	silver	36.5	2.49	0.251	–	–
54.5–51.5	silver	53.0	2.82	0.310	2.31	0.046
65.5–62.5	silver	64.0	2.14	0.284	1.76	0.058

Source: McIntyre (1980).

A simple, non-trivial, method for investigating air movement around the body is to use 'visualization'. For example, a smoke puffer will blow smoke around the area of interest and show airflow, but smoke is inconvenient. Carpenter and Moulsley (1972) used neutral buoyancy soap bubbles with a helium/air mixture. Despite some change in temperature it is very useful to simply blow soap bubbles, using those which can be purchased at a children's toyshop, to demonstrate airflow, identify sources of draught and also gain some quantification of air velocity. Waving the bubble window in air to create bubbles will avoid problems with exhaled air temperature. Inexpensive gun-like devices for creating bubbles are also available as children's toys and can be useful.

Humidity

Two commonly used methods of expressing humidity are the partial vapour pressure of water vapour in air (P_a) and the relative humidity which is the ratio of P_a to the saturated water pressure at that temperature (P_{sa}), usually expressed as a percentage. The dew point, the temperature of air at which dew will begin to form if the air was slowly cooled, is also sometimes used. All three measures can be calculated from each other if air temperature is known. One of the most commonly used instruments for determining humidity is the whirling hygrometer (sling psychrometer) (Figure 5.3).

The hygrometer contains two mercury-in-glass thermometers, one whose sensor is covered in wet muslin (close meshed cotton) or silk (wet bulb) and one normal (dry bulb). The hygrometer is whirled, hence passing moving air

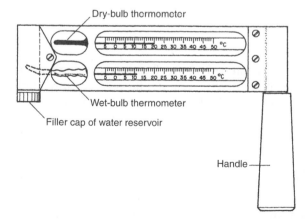

Figure 5.3 The whirling hygrometer for measuring humidity consists of two thermometers that are turned by vigorously swinging the handle and exposing the thermometers to rapid air movement. The bulb of one thermometer is covered with a silk or muslin sleeve that is kept moist and this will record wet bulb temperature. The atmospheric humidity is determined by special tables from the difference in reading between the two thermometers.

over the two sensors. This allows an effective measure of air temperature (dry bulb – t_a) and a measure of aspirated (air moved across sensor) wet bulb temperature (t_{wb}). The decrease in wet bulb temperature due to heat loss by evaporation, for a given air temperature, is related to the humidity of the environment. At 100 per cent relative humidity, there will be no detectable fall in t_{wb} which will be equal to t_a. The saturated water vapour pressure P_{sa} (mb) at 1 atmosphere, and a temperature T (°C) is given by Antoine's equation:

$$P_{sa} = \exp\left(18.956 - \frac{4030.18}{t + 235}\right) \text{ mb.} \tag{5.11}$$

Note: 1 kPa = 10 mb.

From dry bulb (air temperature t_a) and aspirated wet bulb temperature (t_{wb}) values of the partial vapour pressure (P_a), relative humidity (ϕ) and dew point (t_{dp}) can then be calculated:

$$P_a = P_{swb} - 0.667(t_a - t_{wb}),$$

$$\phi = \frac{P_a}{P_{sa}} \times 100,$$

where P_{swb} is the saturated water vapour pressure at the wet bulb temperature and dew point

$$t_{dp} = \frac{4030.18}{18.956 - \ln P_a} - 235.$$

The equations apply for temperatures above freezing. For high relative humidities, psychrometers are relatively accurate; however, at low humidities ($\phi < 20$ per cent) they are not (McIntyre, 1980). At low air temperatures inaccuracy may also occur due to errors in reading the thermometers. The whirling hygrometer is an example of a psychrometer where the wet bulb is aspirated manually. Other 'automatic' devices involve fans driven by clockwork (e.g. Assman hygrometer) and electrical fans, where air is sucked over the bulbs of the thermometers. It is important that the wet bulb is aspirated (at around 5 m s^{-1}). The aspirated wet bulb temperature should be distinguished from the unaspirated or 'natural' wet bulb temperature (t_{nwb}), which is not used in the measurement of humidity (see Chapter 10).

Dew point can be measured by cooling a shiny surface until dew condenses on it (detected optically), and measuring the temperature. Other methods of measuring humidity require calibration and include chemical sensors using lithium chloride or aluminium oxide for example, and electrical devices (so humidity can be recorded) involving a change in resistance or capacitance.

A widely used and particularly robust method is the hair hygrometer or hygrograph. The length of hair (human, horse, etc.) changes with relative humidity. A circular chart driven by clockwork is rotated and marked by a pen that is connected to the hair. Although calibration is required and accuracy is limited, particularly at extremes, a useful and continuous recording can be made.

A summary of the characteristics of measuring instruments (ISO 7726, 1985) is given in Table 5.7, and a collection of instruments for measuring basic parameters of the thermal environment is shown in Figure 5.5.

Table 5.7 Characteristics of measuring instruments

Air temperature (t_a)	
Measuring range	10–30 °C for comfort; −40–120 °C for stress
Accuracy	Required ±0.5 °C; desired ±0.2 °C for comfort
	For stress, required:
	$\quad$ −40–0 °C: $\pm(0.5 + 0.01\lvert t_a \rvert)$ °C
	$\quad$ >0–50 °C: ±0.5 °C
	$\quad$ >50–120 °C: $\pm 0.5 + 0.04\lvert t_a - 50 \rvert$ °C
	desired: *(required accuracy)*/2
	These levels are to be guaranteed at least for a deviation of $\lvert t_a - t_r \rvert = 10$ °C for comfort and 20 °C for stress
Response time (90%)	The shortest possible. Value to be specified as characteristic of the measuring appliance
Comment	The air temperature sensor shall be effectively protected from any effects of the thermal radiation coming from hot or cold walls. An indication of the mean value over a period of one minute is also desirable
Mean radiant temperature (t_r)	
Measuring range	10–40 °C for comfort: −40–150 °C for stress.
Accuracy	Required ±2 °C; desired ±0.2 °C for comfort. These values may not be achievable in some circumstances in which case the actual accuracy shall be reported
	For stress required:
	$\quad$ −40–0 °C: $\pm(5 + 0.02\lvert t_r \rvert)$ °C
	$\quad$ >0–50 °C: ±5 °C
	$\quad$ >50–150 °C: $\pm\lvert 5 + 0.08(t_r - 50) \rvert$ °C

Table 5.7 (Continued)

	desired:
	$-40-0\,°C$: $\pm 0.5 + 0.01|t_r|\,°C$
	$>0-50\,°C$: $\pm 0.5\,°C$
	$>50-150\,°C$: $\pm |0.5 + 0.04(t_r - 50)|\,°C$
Response time	The shortest possible. Value to be specified as characteristic of the measuring appliance
Comment	When the measurement is carried out with a black sphere, the inaccuracy relating to the mean radiant temperature can be as high as $\pm 5\,°C$ for comfort and $\pm 20\,°C$ for stress, according to the environment and the inaccuracies in measurement of air temperature, air velocity and globe temperature
Air velocity (v_a) Measuring range	$0.05-1.0\,\mathrm{m\,s^{-1}}$ for comfort: $0.2-10\,\mathrm{m\,s^{-1}}$ for stress
Accuracy	Required for comfort: $\pm|0.05 + 0.05v_a|\mathrm{m\,s^{-1}}$
	desired: $\pm|0.02 + 0.07v_a|\mathrm{m\,s^{-1}}$
	Required for stress: $\pm|0.1 + 0.05v_a|\mathrm{m\,s^{-1}}$
	desired: $\pm|0.05 + 0.05v_a|\mathrm{m\,s^{-1}}$
	These levels shall be guaranteed whatever the direction of flow within a solid angle $\omega = 3\pi s_r$
Response time (90%)	Required: $1.0\,s$ and desired $0.5\,s$ for comfort. For stress, the shortest possible. Value to be specified as characteristic of the measuring appliance
Comment	Except in the case of a unidirectional air current the air velocity sensor shall measure the effective velocity whatever the direction of the air. An indication of the mean value for a period of three minutes is also desirable. The degree of turbulence is an important parameter in the study of comfort problems; it is recommended that it be expressed as standard deviation of the velocity. In a cold environment it is recommended that comfort instrumentation be used for both comfort and stress analysis
Absolute humidity $(P_a -$ as partial pressure of water vapour)	
Measuring range	$0.5-2.5\,kPa$ for comfort; $0.5-6\,kPa$ for stress.
Accuracy	$\pm 0.15\,kPa$. This level should be guaranteed even for air and wall temperatures equal to or greater than $30\,°C$ and for a difference $|t_r - t_a|$ of at least $10\,°C$
Response time (90%)	The shortest possible. Value to be specified as characteristic of the measuring appliance

Source: ISO 7726 (1985).

MEASURING KITS AND COMPOSITE INSTRUMENTS

There are collections of instruments and charts (kits) and instruments that measure a number of the relevant parameters (composite instruments) which can be used to measure human thermal environments. Some of these are described below. The usual integrated set of instruments for measuring climate is the Stevenson's box. However it should be noted that such an enclosed box will generally be inadequate for detailed assessment of human thermal environments.

'Admiralty box'

During the Second World War, a committee within the UK, chaired by Thomas Bedford, considered the problem of environmental warmth and its measurement with particular reference to problems of human performance below decks in HM ships. The result of this work was the production of a report (published by Bedford in 1940) which recommended the introduction of a measurement 'kit' into all ships. This 'Admiralty Box' was a simple self-contained measuring system and included a whirling hygrometer, Kata thermometer, globe thermometer and a series of charts for obtaining basic parameter values from instrument values (as well as stop watch, vacuum flask for hot water, cloth for drying Kata thermometer, etc.). These 'kits' are still used in industry today and provide a robust, practical method for measuring and assessing an environment. A version is shown in Figure 5.4.

Ellis *et al.* (1972) converted the charts to SI units. A number of variations of the kits are also available, for example, replacing the more cumbersome Kata thermometer with electrical instruments, but generally this kit provides an inexpensive, reliable and reasonably accurate method for assessing the environment.

Figure 5.4 Commonly used kit for measuring human thermal environments.

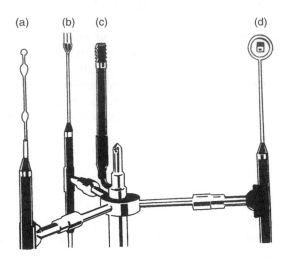

Figure 5.5 Instruments for measuring environmental parameters (a) air velocity; (b) shielded air temperature; (c) humidity; (d) plane radiant temperature.

Indoor climate analyser

This collection of instruments provides a relatively expensive but accurate method for measuring individual basic parameters (and also plane radiant temperature); see Figure 5.5. Air temperature is measured by a platinum resistance thermometer that is shielded from radiation – a contact thermometer is also available. A plane radiant temperature sensor measures in two directions simultaneously and hence allows calculation of radiant temperature asymmetry and mean radiant temperature (if orientated in a number of directions). The humidity sensor is an optical dew point sensor and air velocity is measured with an omni-directional constant temperature anemometer. Electronic analysis of the data allows the basic parameter values to be directly displayed (Olesen, 1985a).

Other instruments

There are a number of other kits and composite instruments; for example, mounts containing air temperature, aspirated (with a fan) wet bulb and globe temperature sensors allow estimates to be made of basic parameters. Some instruments measure the environment and integrate it into values of thermal indices. For example, the WBGT (wet bulb globe temperature) meter (Olesen, 1985b), a heated ellipse used to calculate the predicted mean vote (PMV) thermal comfort index (Olesen *et al.*, 1982) and the Botsball (Botsford, 1971).

More recent and important developments include the use of data logger recording devices which can be programmed. Values from a number of transducers can be recorded over time (including those involved in physiological measures) and date, and later transferred to a digital computer (laptop, palmtop, specialist processor, etc.). Analysis software can then produce values of basic parameters, calculations of thermal indices and responses of thermal models, calculation of mean skin temperature and so on. Access to standard spreadsheet software allows calculations to be performed as well as almost automatic production of a printed report on the thermal environment.

Direct measures of heat transfer

Direct measures of heat transfer can be made from the human body using heat flow discs and plate calorimeters (Clark and Edholm, 1985). Heat flow discs measure the temperature difference between the surfaces of a disc placed on the body. The heat flow can be calculated from the temperature difference and the thermal properties of the disc. Surface plate calorimeters measure the power required to maintain the plate at skin temperature. This is then a measure of the dry heat loss (convection and radiation) from the skin covered by the calorimeter.

Zhou *et al.* (2002a,b) describe a garment for measuring sensible heat transfer between the human body and its environments and the derivation of heat transfer coefficients on clothed people. Sensors mounted in pockets of 'normal' office clothing measured dry heat transfer and radiation sensors mounted on the surface of clothing measured radiant heat transfer allowing the calculation of convective heat transfer. It is concluded that the garment offers a convenient method for the measurement of heat transfer from human subjects in office climates.

Thermal manikins

Thermal manikins are heated to represent the human body and the power to maintain that temperature is used to estimate the heat transfer between a person and the environment. If the manikin is used in nude state under controlled conditions and then in a clothed state in the same conditions, by subtraction, the thermal properties of the clothing can be derived. Holmér (1999) includes evaluation of clothing and evaluation of environments (HVAC systems – buildings, vehicles, incubators) as two important applications of manikins. Whole-body manikins have up to 30 segments to allow measurement of whole-body and local heat transfer. Individual body part manikins (e.g. hands and feet) are used to measure the properties of gloves and boots. Moving and sweating manikins extend the breadth of measurements that can be made.

Madsen (1999) describes a heated, breathing thermal manikin – close winding of nickel wire on top of 16 parts of a body shell provides both a system of heating and temperature measurement that can be used to control independently each of the manikin segments.

MEASUREMENT OF PHYSIOLOGICAL RESPONSE

Skin temperature

Skin temperature can be measured by placing sensors on the skin (thermocouples, thermistors, etc.) or by measuring thermal radiation which has the advantage of non-contact.

Contact techniques must overcome the problem of interference, for example due to insulation and pressure and also must account for the heat transfer between sensor and the environment – air, radiation, etc. Skin temperature devices include harnesses, which expose the sensor to the skin with good contact but do not insulate it, so-called air and vapour permeable tape, which covers the sensor and metal-based tape, allowing conduction between skin and sensor. All methods have errors and corrections to readings may be necessary. Further details are provided by Michel and Vogt (1972), McIntyre (1980) and Clark and Edholm (1985).

Clark and Edholm (1985) provide a discussion, and a number of examples, of the measurement of skin temperature (and the surface temperature of clothing), using infrared thermography. A single 'point' measurement can be made as well as a colour thermal image using scanning techniques in 'thermovision'. It is important to recognize that a specialist instrument is required with restricted range and high sensitivity for measuring skin temperature. In addition, careful consideration must be given to emissivity values (often assumed to be 1.0) and calibration techniques.

Mean skin temperature (t_{sk}) is the mean value of the skin over the whole body and can be measured with an infrared scanning radiometer (Clark and Edholm, 1985). More traditional methods estimate t_{sk} from weighted average measurements of temperature at specific points on the skin. The weighting used is usually related to body surface area (e.g. trunk has a high weighting, hand a low one, etc.), but other methods could consider distribution of thermoreceptors or sweat glands over the body (see Chapter 2). Many different methods have been proposed and used. The general principle derives from statistical sampling methods. The greater the variation in temperature, the more points needed to have a representative sample. For the human body therefore, fewer points are required for assessment in the heat where skin temperature is homogenous (due to vasodilation) and more points are required in the cold where skin temperature is heterogeneous (due to

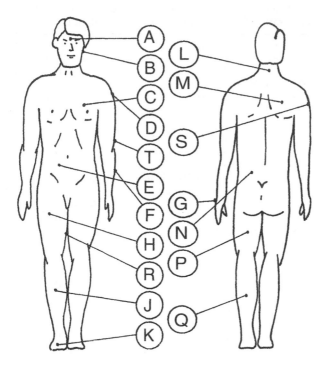

A   forehead
B   left side of the face
C   left upper chest
D   left front shoulder
E   right abdomen
F   left outer mid lower arm
G   left hand
H   right anterior thigh

J   right shin
K   right instep
L   neck
M   right scapula
N   left paravertebral
P   left posterior thigh
Q   left calf
R   right mid inner thigh

Figure 5.6 Measurement sites used, over a number of methods, in the estimate of mean skin temperature over the body, from a weighted average of skin temperatures at the individual skin sites (see also Table 5.8).

vasoconstriction). A summary of methods is provided in Figure 5.6 and Table 5.8.

A number of studies have compared the use of fewer points (which is more convenient) to results obtained using a large number of points, in an attempt to compare methods. The results however, will depend upon conditions and differ between studies. A practical consideration is the time required to prepare a person for measurement and the interference caused by the associated wires and equipment. For this reason telemetry systems are sometimes used. The use of skin temperature for assessing human

Table 5.8 Weighting coefficients used in formulae for calculating estimates of mean skin temperature from skin temperatures at individual sites

	1	2	3	4	5	6	7	8	9	10	11
A	0.07	0.07								0.07	1/14
B			0.10	0.149	0.14						
C		0.0875	0.125	0.186	0.19	0.30	0.50			0.175	1/14
D			0.07	0.107		0.30					
E	0.35	0.0875									1/14
F	0.14	0.14	0.07		0.11		0.14				
G	0.05	0.05	0.06		0.05				0.16	0.05	1/14
H	0.19	0.095	0.125	0.186	0.32	0.20				0.19	1/14
J	0.13	0.065	0.15			0.20	0.36		0.28		1/14
K	0.07	0.07	0.05								1/14
L									0.28		1/14
M		0.0875							0.28	0.175	1/14
N		0.0875	0.125	0.186	0.19						1/14
P		0.095									1/14
Q		0.065								0.20	1/14
R			0.125	0.186				1.00			
S										0.07	1/14
T										0.07	1/14

Notes
1 – Hardy/DuBois, 7pt; 2 – Hardy/DuBois, 12pt; 3 – QREC, 10pt; 4 – Teichner, 6pt; 5 – Palmes/Park, 6pt; 6 – Ramanathan, 4pt; 7 – Burton, 3pt; 8 – medial thigh, 1pt; 9 – ISO, 4pt; 10 – ISO, 8pt; 11 – ISO, 14pt.

thermal environments will be discussed in later chapters, particularly Chapters 8, 9, 10 and 11. In general, mean and local skin temperature can be related to thermal discomfort and local skin temperatures (e.g. fingers, toes) can be used as 'safety limits' for exposure to cold environments. Hand skin temperature is related to manual performance in cold environments.

Measurement of internal body temperature

There are numerous methods for measuring internal body (core) temperature. These include measurements in all accessible places (oriphae) and other derived methods involving heat flow for example. Instruments used generally involve mercury-in-glass thermometers, thermistors and thermocouples, but other equipment such as radio pills, heat-flow meters and infrared thermometers are also used. It is important to note that all of these measurements will differ in temperature and each will have its own characteristics and method of interpretation. In addition, some will be more convenient to measure, some more acceptable to the subject, and all will have practical considerations in measurement and require experience in their use. These are discussed below (from top to tail!).

Tympanic temperature

A small sensor touching the tympanum will reflect the temperature of blood in the internal carotid artery that also supplies the hypothalamus, the main centre for thermoregulation. After careful inspection of the external auditory meatus (ear canal), contact is made between the sensor and the tympanum. This can prove very painful (and cause fainting) to the subject and great care must be taken. This method is not generally found acceptable to subjects either in laboratory or field investigations. If contact is maintained and the sensor is insulated from outside conditions, then tympanic temperature provides a reliable estimate of changes in brain temperature with time. Use of infrared (non-contact) sensors makes the measurement of tympanic temperature acceptable. Infrared thermometers for measuring tympanic temperature are readily available (Figure 5.7) and may become widely accepted (Mekjavic *et al.*, 1992). Caution must be taken however to ensure that these are used effectively and in appropriate operating conditions. Shiraki *et al.* (1988) have uniquely measured both brain hypothalamus temperatures and tympanic temperatures in an unanaesthetized human patient. They concluded that tympanic temperature did not reflect brain temperature. However, Cabanac (1992) identifies a number of experimental artefacts in this work and suggests that tympanic temperature is a good estimate of brain temperature.

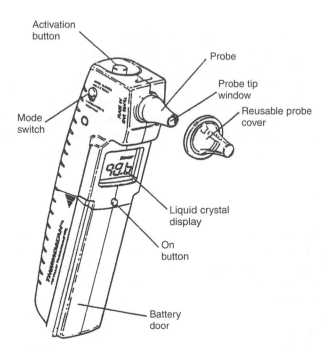

Figure 5.7 Infra-red thermometer to measure tympanic temperature.

Aural temperature

Based on a similar principle, but a more acceptable and widely used method than tympanic temperature, aural temperature is measured in the external auditory meatus (ear canal). The small sensor is placed close to the tympanum. Greater precaution is required to insulate the sensor from the external environment to avoid conduction of heat down wires and penetration of external environmental conditions. Insulation can be provided by 'fitted' plastic ear plugs, foam ear plugs (which roll up, are placed in the ear canal and expand to take its shape) and additional tape and ear muffs or other harness devices. They can provide good insulation and help keep the sensor in position. Adequate insulation is difficult to achieve in cold conditions. One active method involves a 'zero-gradient' thermometer that consists of earmuffs containing a heated element, which maintains a zero-gradient with a sensor inside the ear canal. This method works very effectively but requires an additional power supply. Experience with the use of aural temperature can provide a useful estimate of the change in brain temperature with time. Acceptability will depend upon fitting and sensors can cause discomfort over long periods of time. For any measure of ear canal temperature, especially when isolation is involved, there will be an interference with acoustic performance that may be unacceptable for some investigations.

Forehead temperature

The measurement of skin temperature on the forehead can give an estimate of 'brain' temperature, particularly when the person is hot (in fever for example). Thermistors and thermocouples will provide measures; however, an often used method of measurement on babies is to stretch some calibrated thermosensitive material across the baby's forehead and numbers on the material will change colour according to temperature. This method will be affected by external conditions, in some applications, because vasoconstriction in the forehead is limited, it will provide a useful and acceptable first approximation.

Oral temperature

This method is widely used in medical applications. The sensor (mercury-in-glass maximum thermometer, thermistor or thermocouple) is placed under the tongue, hence near the lingual artery. It is convenient that, with guidance, subjects place a mercury-in-glass thermometer under the tongue themselves. External conditions may have great influence and the mouth should be held closed for a few minutes (i.e. >4), the thermometer removed and the reading taken. In very hot (high air and/or radiant temperatures) conditions, it may be necessary to take the reading with the sensor still in the mouth, or take other precautions, use of a beaker of water for example (to prevent the mercury rising up the stem when exposed to the hot environment). Continuous recording can be made using a dental mount holding a thermometer under the tongue. Breathing and saliva will influence the measure which causes difficulties during physical activity, eating, talking, etc. Oral temperature is sometimes underestimated in its usefulness; it can be acceptable and used effectively if careful attention is given to the measurement method.

Oesophageal temperature

A sensor placed in the oesophagus at the level of the heart will reflect internal blood temperature, for example of the aorta going to the brain and the rest of the body. A small sensor (coated with analgesic gel to reduce discomfort) is introduced through the nose and inserted to the required position (25 per cent of a person's height gives a rough guide to catheter length). This method is used in laboratory investigations; however it may not be accepted by subjects, or experimenter – it is uncomfortable and can be painful and even dangerous. Once positioned, however, there are few problems and a reliable measure is taken, although swallowed drink, food and saliva will influence the readings.

Subclavian temperature

A thermometer placed in the armpit is affected by the subclavian artery and can provide a rough estimate of internal body temperature. Errors can however be very large, particularly in cold environments. When subjects are vasodilated it may provide a reasonable approximation to internal body temperature where the armpit is closed by the arm being held tightly against the body. This method should be used as a last resort.

Intra-abdominal temperature

The temperature in the abdomen is representative of the central trunk temperature. It is usually measured using a radio pill which is swallowed and provides an FM signal related to temperature as it passes through the whole of the alimentary canal (after which it can be recovered by the subject). The pills are wrapped in a protective coating (rubber condoms are sometimes used) and the whole package is relatively large and sometimes difficult to swallow, so it can be considered unacceptable. However, using a similar pill, intra-abdominal pressure has been measured in the assessment of lifting using many nurses (Stubbs *et al.*, 1983; Mairiaux *et al.*, 1983). Careful consideration should be given to calibration and the ability of the system to receive the signal. This measure is not directly affected by environmental conditions, however the passage of food, drink, saliva, etc. will affect the temperature of the alimentary canal, particularly near the mouth.

The system has been used for long term recording of internal body temperature, for example astronauts on space missions.

Rectal temperature

A temperature sensor is inserted (>100 mm) by the subject into the rectum and provides a measure representative of the temperature of a large mass of deep body tissue, independent of ambient conditions. Once located the sensor is found comfortable and generally acceptable. A 'swelling' in the cable inserted just inside the anus, and tape, will hold the sensor in position during exercise. In some subjects there is general dislike for this measure (mainly the concept) and particularly with connotations regarding the spread of viral diseases. This measure gives a value of 'average' internal body temperature; however, it may not be representative of brain temperature and particularly changes in temperature. It will also be affected by 'cold' blood or 'hot' blood from the legs during vasoconstriction or exercise respectively.

Vaginal temperature

Vaginal temperature in females can provide a measure of deep body temperature and is similar in measurement to rectal temperature. It can be used in clinical applications, however as a measure of internal body temperature it is generally unacceptable and offers no advantage over rectal temperature.

Urine temperature

The temperature of urine inside the body is representative of deep body temperature and can be measured with negligible heat loss after excretion from the body. A mercury-in-glass thermometer, a thermistor or thermocouple can be placed in the flow of urine and a measure taken, e.g. using a thermistor placed in a funnel over a beaker (Collins, 1983). Although a sensor placed in a vacuum flask (rinsed with water at around 37 °C) into which urine is excreted, will provide a measure. This method provides an acceptable but 'one off' method of average internal body temperature. It can be used as a comparative check or supportive method at the end of a study, or when other methods are not acceptable, e.g. in studies of the discomfort of the aged.

Transcutaneous deep-body temperature

A method for measuring deep-body temperature on the skin surface is to heat the skin until there is no thermal gradient between internal body temperature and skin temperature. An insulated disc with a heating element and two thermistors vertically separated can achieve this. Fox *et al.* (1973) describe a measuring probe which is attached on the skin surface (e.g. the chest) and heated until the sensors reach equilibrium. This provides a measure of deep body temperature that is acceptable to subjects and responds to changing internal body temperature (Clark and Edholm, 1985). However, it does affect the area of skin used and the device requires a power supply.

Preferred measure of internal body temperature

The preferred measure of internal body temperature will depend upon the practical application. This will involve a number of factors including usefulness in interpreting physiological stress as well as acceptability to subjects and other health, safety and ethical considerations. A combination of measures may be useful, for example, measure continuously in both ears (related to changes in brain temperature), rectum (reliable whole-body measure) and oral (with a mercury-in-glass thermometer). Use of the clinical thermometer will be useful to clinicians and gives a measure independently of electrical systems. Whatever the method used, ethical considerations are of paramount importance. These will include ensuring subjects know that they can withdraw from investigation at any time without recrimination. Professional ethics dictate that the experimenter must create conditions such that this is true (including avoidance of psychological pressure) and that health and safety is of primary importance. Further information is provided in the Helsinki agreement (WMA, 1985) but is implicit in the choice of measurement method.

For a further discussion of measurement, interpretation and limiting values of body 'core' temperature the reader is referred to ISO 9886 (1992) and Fox (1974).

Measurement of thermal effects on heart rate

Heart rate (measured for example) in beats per minute (bpm) provides a general index to stress on the body (or anticipated exertion, pleasure, etc.) which can be caused by activity, static exertion, thermal strain, psychological responses and so on. The electrical activity of the heart, as represented by an electrocardiogram (ECG), describes the rhythmic cycle of contractions etc. of components of the heart, making up the beat (auricles and ventricles) and can also be affected by stress on the body. For example, in hypothermia or with abnormal heart function, non-typical ECG patterns can be seen. ECG requires more sophisticated equipment than simple measurements of heart rate.

To determine the thermal effects on heart rate one can divide overall heart rate into a number of (non-independent) components.

$$HR = HR_0 + \Delta HR_m + \Delta HR_s + \Delta HR_t + \Delta HR_n + \Delta HR_e, \qquad (5.12)$$

where

HR = overall heart rate
HR_0 = heart rate at rest in thermal neutrality
ΔHR_m = component due to activity
ΔHR_s = component due to static exertion
ΔHR_t = component due to thermal strain
ΔHR_n = component due to psychological reactions
ΔHR_e = residual component—circadian rhythm, breathing, etc.

If HR_0 is measured and other components can be assumed to be negligible ΔHR_t can be estimated from

$$\Delta HR_t = HR_r - HR_0,$$

where HR_r is the heart rate after recovery from activity in the heat but still containing the thermal component. This recovery time up to the break is, on average, four minutes (ISO 9886, 1992), but it may be longer, for high activity levels. The 'recovery' curve should be measured to identify this point (see also Vogt *et al.*, 1983). The ΔHR_t component provides a measure of thermal strain and can be related to deep body temperature. A similar principle can be used to determine ΔHR_m from HR and hence estimate metabolic heat production (see Chapter 6).

Heart rate can be measured using a simple pulse detected with the hand placed on the wrist or neck, for example. ECG measures give heart rate but are over-elaborate if this is the only parameter of interest. Infrared techniques, electrical sensors and temperature sensors relating to changing blood flow in the finger or earlobe, for example, can also be used.

As well as portable pulse reading devices attached by wires to sensors and carried on belts, there are a number of simple reading devices available. One such device comprises electrodes strapped around the chest, which send radio signals to a watch worn on the wrist. These devices are reliable and data can be stored for later analysis, allowing recording at a pre-programmed rate, of heart rate over 24-hour periods if required.

If a number of devices are used, careful consideration should be given to range of reception (maybe approximately 1 m) and interference between devices. Methods of determining heart rate should be considered and ideally compared with ECG traces over a number of conditions.

Measurement of body mass loss

Body mass loss is related to thermal strain, mainly due to sweat loss. It is also affected by evaporative loss due to breathing and differences in mass between expired CO_2 and inspired O_2.

A simple method of measurement is to weigh a subject nude before exposure (N_B), clothed before exposure (C_B), clothed after exposure (C_A) and nude after exposure (N_A). The following can then be determined (assuming all mass loss is sweat).

$N_B - N_A$	Total mass (sweat) loss from the body
$C_B - N_B$	Dry weight of clothing
$C_A - N_A$	Wet weight of clothing
$(C_A - N_A) - (C_B - N_B)$	Sweat trapped in clothing
$(N_B - N_A) - [(C_A - N_A) - (C_B - N_B)]$	Sweat evaporated

Dripping of sweat can also be important and in a laboratory can be measured by collecting it in a pan of oil placed beneath the subject. Naturally, all inputs (food, drink, etc.) and outputs (urine, stools, etc.) should be accounted for. Scales used should ideally have an accuracy at around 5 g and not less than 50 g although it depends upon application. Caution should be taken with some scales with digital readings, which falsely imply great accuracy. Simple devices such as the use of small sheets of absorbent paper inside clothing have been used, e.g. to assess comfort of seating (Grandjean, 1988) and total sweat loss estimated.

Continuous measurement of mass loss can be made with highly sensitive balances, however, they are used mainly in the laboratory. Vapour pressure sensors on the skin (within clothing) will also provide continuous estimate of mass loss with time (Candas and Hoeft, 1988) as can sweating capsules which are ventilated devices placed on specific areas of the skin and can provide measures of local sweating.

Personal monitoring systems

Although originally considered to be relatively expensive specialist equipment, the wide availability of (now inexpensive) physiological measuring equipment has led to the use of personal physiological monitoring systems. Where workers are exposed to extreme hot or cold environments, especially when wearing protective clothing and equipment, a method of ensuring safety may be for them to wear personal monitoring systems, including heart rate, internal body temperature, and maybe, skin temperature sensors (particularly on hands and feet in the cold). Personal monitoring systems have been used by physiologists in field surveys, however, they have the potential to be used in normal working practices. Issues remain as to which equipment to use for what purpose, who should use it, specification of the instrumentation (accuracy, sensitivity, reliability, etc.) and how to interpret the measures (e.g. limiting values – when do the levels become unacceptable?). Although personal monitoring systems are a practical option, care should be taken to ensure that they are implemented correctly into safe working practices.

Specification of physiological monitoring instruments

Just as environmental measuring equipment has a specified performance appropriate for its application (Table 5.7), so should instruments for physiological measurement. This has not been obvious however, and instruments for measuring tympanic temperature using infra red radiation, heart rate monitors, skin and body temperature thermometers have been made widely available without reference to application requirements. If the measuring instrument does not meet with required specifications or is operated outside of its limits then the interpretation can be misleading and the consequences dangerous.

A proposed British standard is in early stages of development to provide specification of instruments for measuring deep body temperature, skin temperature, heart rate and body mass loss. Specification is in terms of range, accuracy, frequency of measurement, precision, resolution, response time, stability and calibration frequency. Guidance is also provided on system integration and use and case study examples are provided. Such a standard is essential for the appropriate selection of physiological monitoring equipment. It is also essential that if personal monitoring systems are introduced, a full implementation of a quality safety system is performed.

MEASUREMENT OF PSYCHOLOGICAL RESPONSES

One of the best ways of determining whether a group of people are comfortable is to ask them. Thermal sensation, comfort, pleasure, pain, as well as behavioural responses are all psychological phenomena. There have been many useful studies to 'correlate' physical conditions and physiological responses with psychological responses. However, no model provides a more accurate prediction than measuring psychological responses directly.

Methods for measuring psychological responses range from psychophysical techniques (method of limits, method of magnitude estimation, multidimensional scaling, etc. – see Guildford, 1954) and simple (seven point) scales often used in laboratory experiments, to the integration of techniques into questionnaires for practical surveys as well as behavioural measures. Consideration of the psychological model of a person which is assumed, is important (e.g. man as a simple 'comfort' meter etc. – see Chapter 3).

Most practical studies to date have used simple (seven point) scales to measure thermal sensation and comfort. Although not useful for explaining why persons perceive an environment as uncomfortable, they provide very important and practical information. Simple practical methods for measuring psychological response are discussed below.

Subjective measures

There are a number of subjective scales which have been used in the assessment of thermal environments; the most common of these are the seven-point scales of Bedford (1936) and ASHRAE (1966) (see Table 3.2).

The form and method of administering the scales is important. For example, a continuous form of the scale would be to draw a line through all points. This would allow subjects to choose points between ratings (e.g. between cool and cold, a rating of 1.6 on the ASHRAE scale for example). In analysis of results this would enable parametric statistics to be used. However, maybe the investigator does not consider data 'strong enough' for this and is prepared only to use ordinal data (ranks) and non-parametric statistics. These and other points are of importance and for further information the reader is referred to a text on the design and analysis of surveys – e.g. Moser and Kalton (1971) – and on the use of subjective assessment methods – e.g. Sinclair (1990).

The 'psychological' interaction when the scale is administered may also influence results. Subjects are usually given the scale and asked to tick the place which represents 'how *they* feel *now*', for example. It is important to avoid ambiguity which may lead to a person providing his or her own interpretation, e.g. what the environment is *generally* like, or how *other* people may perceive it, etc. Other issues include range effects – the range provided, e.g. hot to cold, influences the judgement – and leading questions 'you are uncomfortable aren't you?'. Sinclair (1990) identifies the following important issues to be considered when constructing questionnaires: question specificity, language, clarity, leading questions, prestige bias, embarrassing questions, hypothetical questions and impersonal questions. Other issues include whether knowledge of results is given – for example, if responses are requested over time, is the subject informed of previous ratings he made? – and whether the ratings are given in the presence of others.

Investigations involving subjective measures therefore must be carefully planned. It should be emphasized that although there are many pitfalls, most can be relatively easily overcome and the use of simple subjective methods allows easy collection of important data, which can prove invaluable in the measurement of psychological responses.

ISO 10551 (1995) presents the principles and methodology for the construction and use of scales for assessing the environment. Scales are divided into two types: personal and environmental. Those related to the personal thermal state may be perceptual – how do you feel now? (e.g. hot) – affective – how do you find it? (e.g. comfortable) – and preference – how would you prefer to be? (e.g. warmer). Those related to the environment fall into two types: acceptance – is the environment acceptable?; and tolerance – is the environment tolerable? An interesting point for an International Standard is translation between languages, since in French, for example, one cannot use together 'warm', 'hot' and 'very hot'. The fundamental

principles and psychological phenomena however apply over all national-
ities although language and cultural difference – in some cultures subjects
may be reluctant to express dissatisfaction – will be important.

Although subjective measurement techniques can be useful for measuring
extreme environments, they should not be used as a primary measure of
heat stress, cold stress and in health and safety. In these conditions, the
ability of a person to make a 'rational' subjective judgement may be
impaired. While a subject must always be allowed to withdraw from investigation, he does not have the overriding right to remain in it. It is the
investigator's judgement as to whether he should remain exposed (based on
physiological responses etc.) even if the subject is willing (enthusiastic) to
do so. A classic case is response to heat, where subjects may be highly
motivated to remain in the environment and strongly express this opinion.
The case where the subject is normally higher in the organizational hierarchy
than the experimenter can cause dangerous consequences.

The selection of subjective scales will depend upon the population under
investigation and an initial investigation may be necessary to identify meaningful dimensions. For example, in the investigation of the thermal comfort
of clothing (Hollies *et al.*, 1979) the seven-point thermal sensation scales are
in general use, however, scales of stickiness, wetness, etc. are used for
specific applications. The construction and use of two simple questionnaires
used in a thermal survey (Parsons, 1990; BOHS, 1990), are provided in
Figures 5.8 and 5.9.

Form 1 (Figure 5.8) was used in a moderate office environment where
employees had been complaining about general working conditions. The
assessment of the thermal environment formed part of an overall environmental ergonomics study (lighting, air quality, noise, etc.). Details about
individuals' characteristics, location, etc. were collected separately. The thermal environment was assessed by the method described in ISO 7730 (1994)
and the form was handed to workers for completion at their workplace.

Question 1 determines the workers' sensation vote on the ASHRAE scale.
Note that this can be compared directly with the measured PMV thermal
comfort index, as described in ISO 7730 (1994). Question 2 then provides an
evaluative judgement. So while question 1 determines the subject's sensation
(e.g. warm), question 2 compares this sensation with how the subject would
like to be. Questions 3 and 4 provide information about how workers
generally find their thermal environment. This is useful where it is not
practical to survey the environment for long periods, although other methods
such as the use of logs or diaries could also be considered. Questions 5 and 6
are catch-all questions about worker satisfaction and any other comments.
Answers to these two questions will provide information about whether more
detailed investigation is required. Answers will also indicate factors that are
obvious to the workers but not obvious to the investigator.

Form 2 (Figure 5.9) was used in a cold environment. The suitability of
clothing and working practices in cold environments was investigated as

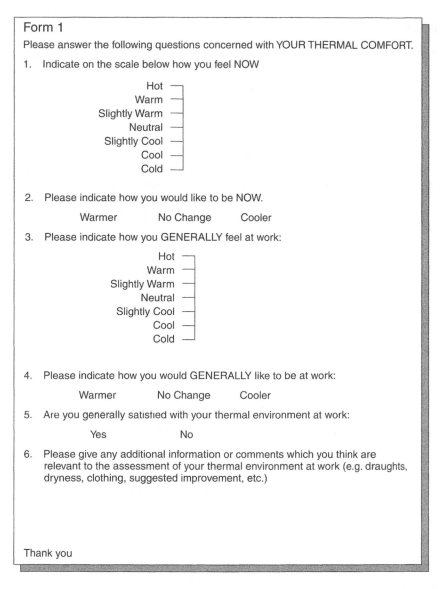

Figure 5.8 Form 1: single subjective form used in the investigation of thermal comfort.

part of an ergonomics assessment of the working environment. It may be noted that physiological measures (including mean skin and core temperatures) were also measured in this study.

Question 1 determines the worker's sensation vote overall and for six areas of the body. The nine-point scale allows the extremes of 'very hot' and

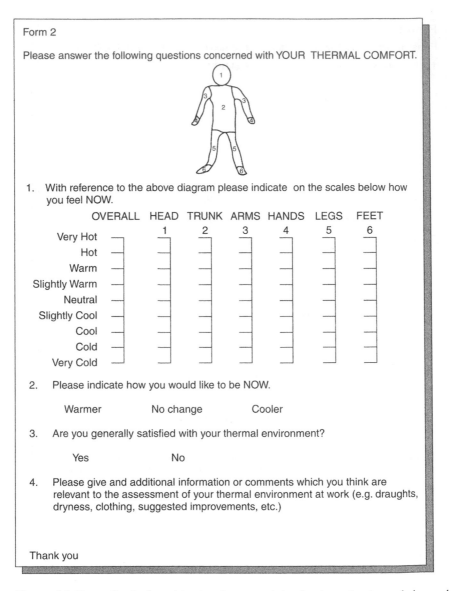

Figure 5.9 Form 2: single subjective form used in the investigation of thermal comfort overall and in specific areas of the body.

'very cold' to be recorded. It is particularly important in cold environments to record sensation votes at the extremities (hands and feet). Question 2 then provides an overall evaluative judgement; allowing comparison of the worker's sensation with how he would like to be. Questions 3 and 4 serve the same purpose as questions 5 and 6 on form 1.

These two forms provide useful examples of easily administered measures of psychological response which provide valuable information and which can allow the 'follow up' of points which subjects raise in more detail. An example of a questionnaire used for the assessment of vehicle environments is similar in format to that for buildings and is shown in Figure 5.10.

THERMAL COMFORT ASSESMENT

Date:
Subject:
Time: _____
Session:

Neutral, Pre , 0, 5, 10, 15, 20, 25, 30, Post

1. THERMAL ENVIRONMENT
  Please rate on these scales how YOU feel NOW

	overall	Head	Trunk Front	Trunk Rear	Arms	Upper Legs Front	Upper Legs Rear	Lower Legs Front	Lower Legs Rear	Feet
		1	2	3	4	5	6	7	8	9

7 Extremely hot
6 Very Hot
5 Hot
4 Warm
3 Slightly warm
2 Neutral
1 Slightly cool

	overall	Head	Trunk Front	Trunk Rear	Arms	Upper Legs Front	Upper Legs Rear	Lower Legs Front	Lower Legs Rear	Feet
		1	2	3	4	5	6	7	8	9

4 Very Uncomfortable
3 Uncomfortadle
2 Slightly Uncomfortable
1 Not Uncomfortable

	overall	Head	Trunk Front	Trunk Rear	Arms	Upper Legs Front	Upper Legs Rear	Lower Legs Front	Lower Legs Rear	Feet
		1	2	3	4	5	6	7	8	9

4 Very Sticky
3 Sticky
2 Slightly Sticky
1 Not Sticky

2. Please rate on the scale how YOU would like to be NOW:

Much warmer   Warmer   Slightly warmer No change Slightly cooler Cooler   Much cooler

3. Please rate on the scale how YOU feel NOW in this thermal environment:

Very pleasant   Pleasant   Slightly   Neither pleasant   Slightly   Unpleasant   Very
                            pleasant   nor unpleasant   unpleasant   unpleasant

4. Please indicate how acceptable YOU find this thermal environment NOW:

☐ Acceptable   ☐ Not Acceptable

5. Please indicate how satisfied YOU are with this thermal environment NOW:

☐ Satisfied   ☐ Not Satisfied

Figure 5.10 Questionnaire used for the assessment of vehicle environments.

Behavioural and observational measures

Thermal environments can affect the behaviour of individuals (move about, curl up, put on or take off clothing, become aggressive or quiet, change thermostat setting, etc.). This behaviour can be observed and aspects of it quantified and hence measured. A number of studies have observed the behaviour of householders in controlling internal temperatures by measuring temperatures in the homes (e.g. Weston, 1951; Humphreys, 1978). The causes are not always clear. They may be related to increased expense of fuel to the preference for wearing lighter clothing for example. More direct observations of behaviour have been made on schoolchildren using time-lapse photography and two-way mirrors. For these methods observer interference is of great importance and should be carefully considered (Humphreys, 1972; Wyon and Holmberg, 1972). Other behavioural measures could include the occupational density of a room (where there is choice) or a measure of accidents or critical incidents.

Drury (1990) provides a description of direct observational methods that he says have a high degree of face validity but a low degree of experimental control. For example, even if behavioural measures are correlated with thermal conditions, it cannot be concluded that thermal conditions are the (sole) cause. If the establishment of causality is not important however, then observation of behaviour provides a useful measure. Drury suggests that observation data are either of events or states. For example, an *event* could be the observed occurrence that a person removed an item of clothing and is recorded with thermal conditions and time of the event. Analysis of event data can provide sequence of events, duration of events, spatial movements or frequency of events. *States* relate to system conditions, being constant between events, and for example may relate to the activity and position of all persons in the room, number of people who have removed a garment, etc. System states can be recorded at discrete time intervals. The method of recording ranges from 'pen and paper' to electronic keyboard techniques and the use of video. Observers will have to be trained and it may be useful to use more than one observer for each situation. Analysis and interpretation of results must also be carefully planned to reduce observer bias. Observation of a system will always lead to an abstraction of what is happening and any quantification will be a 'subset' of all events and states. For further information the reader is referred to Wilson and Corlett (1995). Behavioural measures can provide a method of observing both quantitative and qualitative 'measures' with little interference with what is being observed.

THE THERMAL INDEX – AN ASSESSMENT TECHNIQUE

A useful tool for describing, designing and assessing thermal environments is the thermal index. The principle is that factors that influence human response to thermal environments are integrated to provide a single index value. The aim is that the single value varies as human response varies and can be used to predict the effects of the environment. A thermal comfort index for example, would provide a single number that is related to the thermal comfort of the occupants of an environment. It may be that two different thermal environments (i.e. with different combinations of various factors such as air temperature, air velocity, humidity and activity of the occupants) have the same thermal comfort index value. Although they are different environments, for an ideal index, identical index values would produce identical thermal comfort responses of the occupants. Hence environments can be designed and compared using the comfort index.

A useful idea is that of the *standard environment*. Here the thermal index is the temperature of a standard environment that would provide the 'equivalent effect' on a subject as would the actual environment. Methods of determining equivalent effect have been developed. One of the first indices using this approach was the *Effective Temperature (ET) index* (Houghton and Yagloglou, 1923). The ET index was in effect the temperature of a standard environment – air temperature equal to radiant temperature, still air, 100 per cent relative humidity for the activity and clothing of interest – which would provide the same sensation of warmth or cold felt by the human body as would the actual environment under consideration.

Rational indices

Rational thermal indices use heat transfer equations (and sometimes mathematical representations of the human thermoregulatory system) to 'predict' human response to thermal environments. In hot environments the heat balance equation can be rearranged to provide the required evaporation rate (E_{req}) for heat balance ($S = 0$) to be achieved, e.g.

$$E_{req} = (M - W) + C + R, \tag{5.13}$$

where thermal conductivity is ignored and W is the amount of metabolic energy that produces physical work. Because sweating is the body's major method of control against heat stress, E_{req} provides a good heat stress index. A useful index related to this is to determine how wet the skin is; this is termed skin wettedness (w) where:

$$w = \frac{E}{E_{max}} = \frac{\text{actual evaporative rate}}{\text{maximum evaporation possible in those conditions}}.$$

In cold environments the clothing insulation required (IREQ) for heat balance can be a useful cold stress index based upon heat transfer equations (Holmér, 1984).

Heat balance is not a sufficient condition for thermal comfort. In warm environments sweating (or skin wettedness) must be within limits for thermal comfort and in cold environments skin temperature must be within limits for thermal comfort. Rational predictions of the body's physiological state can be used with empirical equations which relate skin temperature, sweat rate and skin wettedness to comfort.

Empirical indices

Empirical indices are derived using samples of human subjects who are exposed to a range of thermal conditions of interest. Their responses form a database or empirical model (e.g. curve-fit to data) that can be used to predict thermal responses. The effective temperature (ET) index described above is an example of an empirical index where human subjects give ratings of warmth over a range of conditions. Results were related to the equivalent sensation in a standard environment and the index value was the temperature of that standard environment.

Direct indices

Direct indices are measurements taken on a simple instrument that responds to similar environmental components to those to which humans respond. For example a wet, black globe with a thermometer placed at its centre will respond to air temperature, radiant temperature, air velocity and humidity. The temperature of the globe will therefore provide a simple thermal index, which with experience of use can provide a method of assessment of hot environments. Other instruments of this type include the temperature of a heated ellipse and the integrated value of wet bulb temperature, air temperature and black globe temperature giving the WBGT index.

An engineering approach employing what is known as the dry resultant temperature (CIBSE, 1986) is to use the equilibrium temperature of a 100 mm diameter globe thermometer placed in the environment. The temperature of the globe approximates to an average of air temperature and mean radiant temperature. The index needs to be corrected for air movement greater than $0.1 \, \mathrm{m \, s^{-1}}$, and assumes that relative humidity lies in the range of 40–60 per cent.

A major proportion of work into human response to thermal environments has been to establish the 'definitive' or most useful thermal index. There have been numerous reviews (see McIntyre (1980) and Goldman (1988)). The notion of the definitive index is fundamentally flawed as all factors that influence human response will never be known. An index can then be viewed as a simple thermal model. The utility of the thermal index as a single number is however great. It provides a simple practical method for allowing the effective design, construction and evaluation of human thermal environments.

6 Metabolic heat production

INTRODUCTION

An estimate of metabolic heat production in the body is fundamental to the assessment of human thermal environments. If no heat were produced, as in death, the body would achieve thermal equilibrium with surrounding conditions. Heat is produced in the cells of the living body and most of the investigation into human response to thermal environments can be regarded as the study of the distribution and dispersion of this heat. It is often considered that heat is a by-product from cells; however this is misleading, especially in homeotherms where the heat production is an important integral part of the living system.

WHERE DOES THE HEAT COME FROM?

Energy is originally derived from the electro-magnetic radiation, i.e. light and other radiation, emitted from the sun which plants, using carbon dioxide and water, convert by the process of photosynthesis into adenosine triphosphate (ATP – providing potential energy to the cell which can be quickly released), glucose (energy store), and oxygen. Humans, like plants, require ATP to supply energy to the cell, for use in membrane transport, chemical reactions and mechanical work, but they make it from glucose which they obtain from food – other animals and plants – and oxygen with the release of carbon dioxide and water, hence completing the ecological cycle. The above description is a simplification and there are other more minor avenues of energy production. The process is not completely understood. For a fuller description refer to Guyton (1969), Astrand and Rodahl (1986) and Pfeiffer (1969).

Air comprises 78.08 per cent nitrogen, 20.94 per cent oxygen, 0.03 per cent carbon dioxide, 0.95 per cent other gases as well as varying levels of water vapour. The human extracts oxygen from air in the lungs and ingests food as a combination of carbohydrates, fats, and proteins. Carbohydrates are converted in the gut and liver to glucose before they reach the cell. The proteins are converted into amino acids and fats into fatty acids. All (and oxygen) are transported to the cell via the bloodstream, where they cross the cell membrane into the cytoplasm of the cell. In the cell, enzymes convert glucose into pyruvic acid – glycolysis, anaerobic with the release of a small amount of energy – and fatty acids and most amino acids into acetoacetic acid. These chemicals are converted in the mitochondria, 'power houses of the cell' into the compound acetyl Co-A where, with oxygen, it is acted upon by a series of enzymes to produce ATP, carbon dioxide, and water (Krebs cycle). Approximately 95 per cent of energy is produced in the mitochondria and 5 per cent anaerobically outside the mitochondria. ATP therefore provides the potential energy for the cell that is immediately available when required. The ATP is steadily regenerated by 'burning' carbohydrates, fats and proteins in oxygen. This process is shown in Figure 6.1.

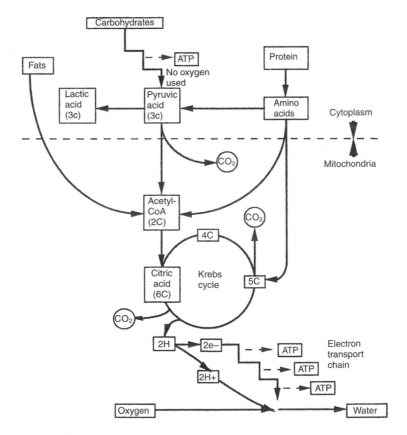

Figure 6.1 Organic compounds (fats, carbohydrates, proteins) containing carbon and hydrogen may be totally combusted to CO_2 and H_2O by oxygen, i.e. by the process of catalytic oxidation with the aid of enzymes in the living cell. For some of the steps the number of carbon atoms involved are given.

Source: Astrand and Rodahl (1986).

Essentially the energy is stored in the additional phosphate bond made by converting ADP (adenosine diphosphate) to ATP, and released where ATP is converted to ADP. Provision is made for the non-availability of resources (oxygen and food); the body stores of glucose (glycogen), fat, and even protein can be used and the system can function anaerobically (without oxygen) for a short period. The above process is usually correctly represented as the method for providing energy for all functions, however when viewed in these terms it is very inefficient. Most energy released in the process is released as heat which is distributed, mainly by blood, around the body. The integration of the heat produced provides the total metabolic heat production of the body, H. The total energy produced is termed the metabolic rate, M, and the total work is termed the external work (W) performed by the body. Metabolic heat production can then be given by

$$H = M - W. \tag{6.1}$$

The production of energy is a complicated process; however the above simple description provides an indication of how the metabolic heat production (H) (a basic parameter in the assessment of human thermal environments) may be estimated. For example it could be measured directly by observing total heat loss from the body in thermal equilibrium (e.g. calorimetry). It may be estimated from oxygen consumption, carbon dioxide production, and the energy released when food is burnt in oxygen (indirect calorimetry) to give metabolic rate (M) and a calculation of external work performed (W) from the work task. There are many methods for estimating metabolic heat production and a number are described below.

Units

To attempt to reduce the individual variability in estimates of metabolic heat production for a specific activity, the value is usually related to surface area of the body or body mass. Thus units of $W\,m^{-2}$ or kcal/min/kg are used. Frequently, values of $1.84\,m^2$ are assumed for the surface area and 65–70 kg for the mass of a man or 55 kg for a woman.

The body surface area is included in the calculation as this is directly related to the heat exchange between the body and the environment; however body weight (mass) is used for many load moving activities (e.g. walking upstairs) which will contribute to the heat produced. In many tables that provide estimates of metabolic heat production for different activities, it is important to remember that the values assume an average man and may require correction for specific populations or individuals. A unit sometimes used is the Met where 1 Met is $50\,kcal/m^2/h = 58.15\,W\,m^{-2}$ and is said to be the metabolic rate of a seated person at rest. The surface area of the body can also be estimated from the DuBois surface area (A_D) by

$$A_D = 0.202 \ W^{0.425} \ H^{0.725}, \tag{6.2}$$

(DuBois and DuBois, 1916) where

W = weight (kg)
H = height (m).

This value has been found to provide an underestimate of body surface area; however this is not a problem in heat transfer analysis as any systematic error is cancelled out (McIntyre, 1980). The division of heat production by surface area allows comparison between individuals of different sizes, only over a restricted range of sizes however as the relationship does not hold for 'extremes'.

ESTIMATION OF METABOLIC HEAT PRODUCTION

Calorimetry

A method of determining metabolic heat production is to directly measure the heat produced by the person in a whole-body calorimeter. There are a number of devices ranging from controlled rooms (e.g. Dubois, 1937) to water controlled suits. Because of variations in body temperature, exposure time must be long, however if careful measurement and calculation of all heat pathways are made, the average heat production by the body can be determined. If the calorific value of food eaten is compared with heat output (taking account of all possible avenues of heat loss), on average they will be found to be equivalent. This finding may not be surprising; however when first performed it demonstrated that living things were subjected to the laws of physics.

Murgatroyd *et al.* (1993) state that direct calorimetry measures energy expenditure rate at which heat is lost from the body to the environment. It is usually measured using a controlled chamber. Non-evaporative heat loss (radiation, convection, conduction) is measured passively from the temperature gradient across the walls of a poorly insulated chamber (gradient layer chamber method) or, 'actively as the rate at which the heat must be extracted from a chamber to avoid heat loss through well insulated walls (heat sink calorimetry)' (Murgatroyd *et al.*, 1993). Evaporative heat loss is derived from the increase of water vapour in the air. Total heat loss is then the sum of non-evaporative and evaporative heat loss.

Although modern calorimeters provide efficient mechanisms for determining metabolic heat production, they are not useful outside the laboratory. For an estimate of metabolic heat production in an applied setting, a more practical method is required.

Indirect calorimetry

An estimate of metabolic rate, M, can be made by measuring how much oxygen has been used to 'burn' food. This is measured by collecting expired air from a subject while he is performing a task of interest. The calorific value of the food (and oxygen used) will then provide an estimate of energy produced. The calorific value of food will depend upon the proportion of carbohydrates, fat and protein in the food. A measure of carbon dioxide produced will give an indication and is represented by the respiratory quotient (RQ) where

$$RQ = \frac{\text{volume of } CO_2 \text{ given off}}{\text{volume of } O_2 \text{ consumed}}.$$

RQ values vary from about 1.0 for pure carbohydrates to 0.71 for fat (see Table 6.1).

For a 'normal' mixed diet an RQ value of around 0.85 would be typical giving an energy equivalent, EE, in Wh per litre of O_2, of 5.68. If RQ is known then EE can be calculated from ISO 8996 (1990).

$$EE = (0.23RQ + 0.77) \times 5.88. \tag{6.3}$$

Metabolic rate, M, in $W\,m^{-2}$, is then

$$M = EE \times V_{O_2} \times \frac{1}{A_D}, \tag{6.4}$$

where V_{O_2} is oxygen consumed in litres per hour at standard temperature and pressure, dry (STPD). The above equations are relatively simple; however the calculation is complicated because values refer to those at STPD and measurements are made under different conditions – for a full description of calculations involved (see ISO 8996, 1990).

McIntyre (1980), from Weir (1949) and Liddell (1963) presents a simple equation based on oxygen utilized (fraction of oxygen in inspired air

Table 6.1 Values related to energy production from burning food in oxygen

	Fat	Carbohydrate (monosaccharide)	Protein	Alcohol
Respiratory quotient	0.710	1.000	0.835	0.667
Oxygen consumption (litre g^{-1})	2.010	0.746	0.952	1.461
Energy equivalent of Oxygen (kJ litre^{-1})	19.61	21.12	19.48	20.33
Energy density (kJ g^{-1})	39.4	15.76	18.55	29.68

Source: Adapted from Murgatroyd *et al.* (1993).

(0.2093) minus fraction of oxygen in expired air (O_e)) and ventilation rate (V in $1\ s^{-1}$):

$$M = \frac{20600\ V(0.2093 - O_e)}{A_D}.$$ (6.5)

There are a number of methods for the collection of expired air: usually a Douglas bag or a spirometer is used but also other portable devices, e.g. K–M (Kofranyi–Michaelis) meter (Eley *et al.*, 1978), Oxylog recording systems (Humphrey and Wolff, 1977) etc. These devices can involve mouthpieces, valves, corrugated tubing, nose clips, and so on, although full face masks can also be used. The principle aim is not to interfere with the task being 'measured' and to avoid problems with leaks, experimenter variability and calibration, especially of gas analysis meters. There are many practical procedures and precautions that need to be considered since if care is not taken serious errors can be made. For a fuller description the reader is referred to Bonjer *et al.* (1981) and Murgatroyd *et al.* (1993).

The method of indirect calorimetry provides an estimate of metabolic rate, M. To obtain an estimate of metabolic heat production, external work (W) must be subtracted. In practice, W ranges from 0 to 20 per cent of M and is difficult to measure. Because of inaccuracies in estimating both W and M, W is often assumed to be zero.

Murgatroyd *et al.* (1993) describe indirect calorimetry as a method of measuring the rate at which heat is produced in the body. If there is no net heat storage or heat loss from the body, direct (heat output) and indirect (heat production) calorimetry will give the same overall result. There will be phase differences over time however, depending upon the activity of the person of interest and the thermal inertia of the body.

Indirect calorimetry can be carried out using whole-body chambers (analysis of well mixed in-going and out-going air), tents, hoods, and bags or tanks (attached to appropriate mouth pieces and nose-clips) and masks. All methods have advantages and disadvantages and calibration and practical techniques must be adhered to rigorously avoid errors. Some methods use breathing as the forcing mechanism for collection of expired air, others use a continuous stream of forced (mechanical) ventilation. Gas analysis can be continuous (breath by breath) or based upon a value of expired air collected over a period of time.

Chemical methods of gas analysis are being replaced by paramagnetic oxygen or infrared CO_2 analyses. Douglas bags are often used in laboratories and can be used in field work where, for ambulatory purposes, the subject can carry the bag on the back (or followed with a trolley). Douglas bags have been made of canvas over a rubber lining, PVC and aluminium-lined plastic. Although impermeable, leakage of CO_2 by diffusion is a problem and the contents of the bags should be as full as possible and analysed as soon as possible after collection (within 20 min, maybe longer, depending upon the properties of the bag).

Murgatroyd *et al.* (1993) describe other portable and ambulatory methods. The Kofranyi–Michaelis (K–M) or Max Planck respirators are portable devices that measure expired air values through low air resistance valves. Gas analysis is performed on expired air collected in a small rubber bag linked to the system. An oxylog system measures cumulative oxygen consumption every minute and an Integrated Motor Pneumotachygraph (IMP) can be used for measuring energy production in athletes. The COSMED K2 system provides data on expired air volumes and O_2 concentration using polarographic O_2 electrodes. Clinical systems include various types of ventilated hood or tent systems. The Deltatrac Metabolic Monitor system uses the method of indirect calorimetry involving a plastic tent, face canopy or hood which may be connected to a mechanical ventilator and can be used for measurements in adults and children.

For use in exercise tests, the Beckman metabolic measurement count (MMC) is useful for performing a number of forms of indirect calorimetry with instruments mounted on a trolley.

Indirect calorimetry has the advantage over direct calorimetry in that *in vivo* carbohydrate and fat oxidation processes can be studied as well as those of protein (estimated from urine nitrogen content) and alcohol (from its rate of disappearance – as measured by breath for example). Hydrogen and methane exchanges may also be measured providing insight into gut fermentation processes. A practical adaptation of the whole-body chamber method would be the measurement of CO_2 (production) of people occupying rooms. This would provide a method of estimating average metabolic heat production of people in a room and will contribute to environmental design by providing data for ventilation and heat load calculation. Ventilation rates (and CO_2 build-up) are traditionally used in measures of air quality assessment and, assuming good sampling methods aided by mixing of air, measures of room O_2, CO_2 and air exchange rates could be used to estimate metabolic heat production. This would be a practical, non-invasive and much needed method for field studies but its effectiveness has yet to be demonstrated.

Collection and analysis of expired air

Methods of indirect calorimetry are divided into two approaches: the partial method and the integral method. The partial method is more commonly used and involves the collection of expired air while the subject is performing a light to moderate task (assumes aerobic). The integral method requires collection while the subject is performing a heavy task and during recovery (to account for anaerobic activity and hence O_2 debt). The partial method is considered below.

To estimate metabolic rate using the partial method the following are required.

1   Personal subject data: sex, weight, height, age.
2   Duration of measurement, T (hours).
3   Measured atmospheric pressure, p (kPa).
4   Volume of expired air, V_{ex} (litres).
5   Temperature of expired air recorded, t (°C).
6   Fraction of O_2 in expired air, F_{O_2} (ND).
7   Fraction of CO_2 in expired air, F_{CO_2} (ND).

Expired air will probably have been recorded under climatic chamber, laboratory or working conditions and measured and analysed under laboratory conditions. The recorded values must be converted to their equivalent values at standard conditions for temperature 0 °C, barometric pressure 101.3 kPa, and dry (STPD) by multiplying V_{ex} by a factor f where

$$f = \frac{273(p - pH_2O)}{(273 + t)101.3},$$
(6.6)

where

pH_2O = saturated water vapour pressure at expired air temperature,

$$= 0.1 \ \exp\left(\frac{18.956 - 4030.18}{t_{ex} + 235}\right) \text{kPa.}$$

Volume flow at STPD ($1\,h^{-1}$):

$$\dot{V}_{ex} = \frac{V_{ex}|_{STPD}}{T}.$$
(6.7)

Oxygen consumption (litres of O_2 per hour):

$$\dot{V}_{O_2} = \dot{V}_{ex} \times (0.209 - F_{O_2}).$$
(6.8)

Carbon dioxide production (litres of CO_2 per hour):

$$\dot{V}_{CO_2} = \dot{V}_{ex} \times (F_{CO_2} - 0.0003).$$
(6.9)

Contraction of expired volume (for $RQ \neq 1.0$):

$$\dot{V}_{O_2} = \dot{V}_{ex}[0.265(1 - F_{O_2} - F_{CO_2}) - F_{O_2}]$$
(6.10)

$$\dot{V}_{CO_2} = \dot{V}_{ex}\left[F_{CO_2} - 0.380 \times 10^{-3} \times (1 - F_{O_2} - F_{CO_2})\right]$$
(6.11)

$$RQ = \frac{\dot{V}_{CO_2}}{\dot{V}_{O_2}}$$
(6.12)

$$EE = (0.23RQ + 0.77) \times 5.88.$$

Metabolic rate:

$$M = EE \times \dot{V}_{O_2} \times \frac{1}{A_D},$$

where A_D is calculated from the subject height and weight using equation (6.2).

The doubly-labelled water (DLW) method

The doubly-labelled water (DLW) method involves the subject drinking an accurately weighed oral loading dose of the isotope $^2H_2$ ^{18}O. Deuterium (^2H) labels the body's water pool and its rate of disappearance from the body (k_2) is related to water turnover (r_{H_2O}). ^{18}O labels both the body's water and bicarbonate pools and its rate of disappearance (k_{18}) provides a measure of combined water and bicarbonate turnover ($r_{H_2O} + r_{CO_2}$). The subject's carbon dioxide production (bicarbonate turnover) is the difference between the two rate constants ($k_{18} - k_2$). Isotope enrichment can be measured in any biological fluid (urine, saliva, blood) but is usually measured in urine (or saliva in young infants).

CO$_2$ production is related to energy expenditure using indirect calorimetry calculations. As accurate measurements are required, they are made over at least two half-lives of the isotopes. The test ranges from a minimum of six days in children to twelve days in adults and longer in the elderly. As an environmental design and assessment method therefore, the method is limited. However, it could be used to complement diary (activity analysis) and indirect calorimetry measures. Issues of measurement include dosing level, fractionation (partial retention of the heavier isotope), sequestration and exchange (isotopes may be incorporated into body solids), and an assumption about RQ in order to ascribe an energy value to the CO$_2$ production rate. As the method provides an energy expenditure over days, it is more useful for determining food and energy requirements than for an estimate of heat production for use in assessing the human thermal environment. Further details of the method can be found in Murgatroyd *et al.* (1993), Black *et al.* (1996) and IDECG (1990).

External work (W)

When a person performs a task, some energy will be used to perform external work, W. For well-defined tasks, e.g. walking, cycling, W can be estimated from calculations. Work done is calculated by multiplying the force by the distance moved in the direction of the force. When walking uphill (or on a sloped treadmill) this will be the force (mass of body times acceleration due to gravity, $g = 9.81\,\mathrm{m\,s^{-2}}$) multiplied by the distance moved upwards (i.e. potential energy = mgh) which will depend upon the actual distance travelled (length and rate of treadmill) and the angle of the slope. Other tasks are more difficult to measure.

The mechanical efficiency of the body doing work (η) is given by

$$\eta = \frac{W}{M}.\tag{6.13}$$

Maximum values of η are around 20–25 per cent (for cycling on a bicycle ergometer) and are close to zero for many tasks. When a person does work 'on' the external world this is *positive work*, but if the 'world does work on the person' – for example when walking downhill or 'pedalling' an electrically driven bicycle ergometer (Nadel *et al.*, 1972) – the body still performs work but it is of a different nature and the external work is called *negative work*. In this latter case, however, heat production is the metabolic rate minus the work performed by the body in performing the task (and not the work performed on the body). For example, the work involved in descending a step is approximately one third of the work involved in ascending (Atha, 1974; Larson, 1974). External work, W, is therefore that part of the total energy produced by the body which is not given off as heat. It can be considered in terms of the activity of (muscle) cells in the body, but can be estimated from a calculation of the positive work performed by the body on the world.

The metabolic heat production is then calculated by subtracting W from metabolic rate, M – measured by indirect calorimetry for example.

THE USE OF TABLES AND DATABASES

In practice it is often inconvenient to measure or estimate metabolic heat production from calorimetry or indirect calorimetry. A commonly used method is to obtain an estimate by matching a description of the task with that for which previous measurements or estimates have been made (e.g. Durnin and Passmore, 1967). There are a number of methods for doing this and some are described below (see ISO 8996, 1990).

General description of work

A simple method for estimating metabolic heat production is to divide the activities into rough categories. Table 6.2 presents a table providing estimates of metabolic rate based on a general description of work.

General description by occupation

A second method for estimating metabolic rate is by occupation. Since actual tasks vary within any occupation, direct measures of energy

Table 6.2 Estimates of metabolic rate based on a general description of work

General description	Metabolic rate ($W m^{-2}$)
Resting	65
Low	100
Moderate	165
High	230
Very high	290

Table 6.3 Estimates of metabolic rate based on a general description of occupation

Occupation	Estimate of metabolic rate ($W m^{-2}$)
Bricklayer	110–160
Painter	100–130
Foundry worker	140–240
Tractor driver	85–110
Secretary	70–85

expenditure can provide a range of values for a given occupation. Typical values have been tabulated in Table 6.3.

Summation method

This method attempts to analyse the metabolic heat production into separate components. It involves an evaluation of basal metabolic rate (usually an estimate of heat production of a completely resting subject) to which is added estimates of metabolic rates for each component activity. For example, suppose a man was performing two-arm work standing stooped. The metabolic rate:

$$M = B_M + M_P + M_W + M_M,$$

where

M = estimate of metabolic rate
B_M = estimate of basal metabolic rate (e.g. $45 \, W m^{-2}$)
M_P = estimate of additional metabolic rate due to posture
   (e.g. standing stooped – $30 \, W m^{-2}$)
M_W = estimate of additional metabolic rate due to type of work
   (e.g. average two-arm work – $85 \, W m^{-2}$)
M_M = estimate of additional metabolic rate due to whole-body
   movement (e.g. no movement – $0 \, W m^{-2}$).

So, for our example, M (in $W m^{-2}$) is given by

$$M = 45 + 30 + 85 + 0 = 160.$$

Table 6.4 Estimates of metabolic rate for basic activities

Basic activity	Estimate of metabolic rate $(\mathrm{W\,m^{-2}})$
Lying	45
Sitting	58
Standing	65
Walking on level even path at $2\,\mathrm{km\,h^{-1}}$	110
Walking on level even path at $5\,\mathrm{km\,h^{-1}}$	200
Going upstairs (0.172 m/step), 80 stairs per minute	440
Transporting a 10 kg load on the level at $4\,\mathrm{km\,h^{-1}}$	185

It will be seen from this type of treatment that the various components in this calculation might be refined and related more specifically to rates of work, durations of dominant posture and descriptions of loads moved. Much of this is available in reference tables and subdivides into calculations of the energy involved in basic activities and component occupational tasks. These tables are derived initially from indirect calorimetry and empirical methods.

Estimation for basic activities

There are certain common basic activities. These include standing, sitting, lying and walking. Many 'jobs' simply involve combinations of tasks related to these basic activities, e.g. walking with a load, walking upstairs, etc. An example of estimates of metabolic rate for basic activities is provided in Table 6.4.

Estimation for component occupational tables

This method involves an estimate of metabolic rate from a description of component tasks involved in specific professions or occupations. An example from the building industry is provided in Table 6.5.

Table 6.5 Estimates of metabolic rate for occupational components

Task	Estimate of metabolic rate $(\mathrm{W\,m^{-2}})$
Bricklaying	
Solid brick (3.8 kg)	150
Hollow brick (15.3 kg)	125
Building a dwelling	
Mixing cement	155
Pouring concrete for foundations	275
Loading a wheelbarrow with stones and mortar	275

EMPIRICAL MODELS

There have been a number of studies where metabolic rates have been measured by indirect calorimetry and regression techniques involving task components, walking speed, etc. have been used to calibrate metabolic rate; however, care must be taken as they are only valid under conditions for which they were calibrated.

An example of a model is provided by Randle *et al.* (1989), where oxygen uptake V_{O_2} (related to metabolic rate) for intermittent arm load carriage can be predicted from:

$$V_{O_2} = 36.3 - 1.74W - 1.76D - 7.17F + 0.027W^2$$
$$+0.041WD + 0.196WF + 0.783DF,$$

where

V_{O_2} = oxygen uptake (ml kg min^{-1})

W = weight of load (range: $10-30$ kg)

D = distance of load carriage (range : $9-15$ m)

F = frequency of carriage (range : $2.5-3.5$ per min).

The model was derived using fifteen young male subjects who carried loads on a level treadmill. The load was carried for a predetermined period, placed on a shelf, then the subject continued to walk, pick up another load, and the cycle started again. The task was selected to represent industrial tasks.

Towle *et al.* (1989) describe a knowledge-based system which includes models for estimating metabolic rate from load carriage for body load (Givoni and Goldman, 1971; Pandolf *et al.*, 1977; Legg and Pateman, 1984) and for arm load work (Morrissey and Liou, 1984; Randle, 1987). The equations for each of these 'models' are provided in Table 6.6.

Heart rate method

Heart rate can be considered to be made up of a number of components (resting, psychological, thermal, activity, etc. – see Chapter 5), and for values of between approximately 120–160 bpm there may be a linear relationship between metabolic rate and increase in heart rate above a resting level, i.e.

$$HR = HR_0 + RM(M - BM),$$

Table 6.6 Regression equations for calculating estimates of metabolic rate

---

Load carriage in the arms
Givoni and Goldman (1971)

$$M = [0.015L^2v^2] + n(W + L)[2.3 + 0.32(v - 2.5)^{1.65} + G(0.2 + 0.07(v - 2.5))]$$

Garg *et al.* (1978)

$$M = 60[0.024W + 10^{-2}[68 + 2.54WV^2 + 4.08LV^2 + 11.4L + 0.379(L + W)GV]]$$

Morrissey and Liou (1984)

$$m = -75.14 + 3.11W + V^2(2.72L + 87.75) + 13.36(W + L)(L/W)^2$$

Morrissey and Liou (1984)

$$m = 312.94 + V[2.39(W + L) - 481.62] + V^2(218.3 + 0.36Z) + 17.35(W + L)(L/W)^2$$

Notes
M = metabolic rate (kcal h^{-1}); m = metabolic rate (W); v = walking speed (km h^{-1}); V = walking speed (m s^{-1}); L = load weight (kg); W = body weight (kg); G = treadmill gradient (per cent); n = terrain factor (treadmill = 1); Z = boxwidth (cm).

where

HR = heart rate
HR_0 = resting heart rate
M = metabolic rate
BM = basal metabolic rate
RM = increase in heart rate per unit of metabolic rate.

The values for the above equation are determined experimentally (e.g. for RM) and can be used to estimate metabolic rate from heart rate. Although average values can be used, an RM value for each subject should be obtained, in effect providing a calibration curve for the subject in the laboratory which can then be conveniently used in application. This can be obtained by exposing the subject to a range of workloads in laboratory conditions and measuring heart rate and energy production by the method of indirect calorimetry. When the relationship has been established, heart rate can be measured in field studies and metabolic rate estimated.

SUBJECTIVE METHODS

An estimate of metabolic rate can be achieved by asking subjects how hard they think they are working. In effect this is similar to asking subjects to use tables of metabolic rate themselves according to their perceived exertion. The most widely used scale is the rating of perceived exertion (RPE) scale of Borg (1998) – see Table 6.7.

These scale values can be related to metabolic rate estimates by matching the subjective ratings with general descriptions in tables. They also provide

Table 6.7 Rate of perceived exertion (RPE) scale

Score	Subjective rating
6	
7	very, very light
8	
9	very light
10	
11	fairly light
12	
13	somewhat hard
14	
15	hard
16	
17	very hard
18	
19	very, very hard
20	

Source: Borg (1982).

an approximation to one tenth of the heart rate of the working subjects. The use of so many points on the scale may not be effective as the variation in response between and within individuals may cover a number of the points.

Subjective methods offer great potential as they are convenient to use; however, they should be carefully developed with an understanding of the user population. The fitness and disposition of the subjects will be important. For a given metabolic rate one subject may find the task hard and another fairly light. Some subjects may be reluctant to say that they feel the task hard. The Borg scale uses the terms light and hard; these may be ambiguous, since for example, hard may be confused with complexity and difficulty of the task. Further guidance on the use of scales is given in Chapter 5.

The effects of temperature and thermoregulation

If the temperature of the body is increased the rate of chemical reactions (in cells) increases by around 13 per cent for each 1 °C rise in temperature. Muscle stiffness is reduced and blood and synovial fluid becomes less viscous enabling faster rates of body activity.

When the body cools, for a resting man there is a lower critical temperature below which metabolic heat production starts to increase by non-shivering thermogenesis or shivering. In man the ability to produce heat by the utilization of brown fat appears to be restricted to babies. Shivering can increase metabolic rate by up to 4–5 times that of the non-shivering subject. The effects of behavioural thermoregulation should not be underestimated in hot and cold environments. In hot environments people may adapt by slowing down to maintain comfort or avoid heat strain even in what appears to be paced tasks. Metabolic rate may be increased by adaptive mechanisms such as opening windows or moving from areas of discomfort. In cold environments there is good incentive to move around and increase metabolic heat and the donning and doffing of clothes and the effects of heavy clothing on day to day activities will affect heat production.

Combination of activities

In all persons metabolic rate will vary with time due to a change in activity. Through thermoregulation the body maintains a dynamic equilibrium with the thermal environment. In some industrial work for example, metabolic rate may vary widely ranging from rest to heavy work. Any estimate of metabolic rate will be an average value over a time of interest. One method of integrating metabolic rate values into a single representative value, recommended by ISO 8996 (1990), is to take the time weighted averages over values representative of each activity, i.e.:

$$M = \frac{t_1 M_1 + t_2 M_2 + t_3 M_3 + \cdots + t_n M_n}{T},$$

where

M = overall average metabolic rate
t_i = time on activity i
M_i = metabolic rate representative for activity i
T = total time over all activities.

A method of combining activities is important for setting control values for thermal stress in industry, in terms of thermal indices for example. A time-weighted method is generally convenient but may not provide a good indication of physiological strain. For example, it implies that over a total period of one hour a 'work hard for 5 min, rest for 5 min, repeated six times' regimen is equivalent to a 'work hard for 30 min, rest for 30 min' regimen. Morris and Graveling (1986) considered this potential problem for assessing work in British Coal mines using ISO Standards – see Chapter 13. They considered both the composition and pattern of physical workload on human heat tolerance. They found that the average physiological responses to intermittent dynamic workload did not differ from those to continuous dynamic work of the same intensity. There were, however, wide individual differences and the use of the time-weighted average method could cause problems where peak loads may present a health risk. They also concluded that heat stress criteria derived originally for continuous workloads, were likely to be equally appropriate for intermittent combined workloads.

Most, if not all, work involves combined activities, so care should be taken when selecting a metabolic rate for a job. For example, in an office where a person is working at a desk, an estimate of 58 W m^{-2} for metabolic heat production may underestimate the overall activity where the person may (i.e. has the right to) move around. The variation in activity must be considered in any practical assessment of work places and conversely can be used in the design of work/rest regimes for jobs.

Task and activity analysis methods

The variation of tasks and activities performed by persons throughout the day provides a difficulty in estimating metabolic heat production values essential for the design and assessment of thermal environments. For an accurate representation of values a system of estimation is required. This can be provided by task analysis methods where tasks are described in terms of human requirements (cognitive and physical) and in sophisticated form can lead to work system analysis and design (see Stammers *et al.*, 1990). Work study methods can also provide an indication of energy required, and hence produced, over a period of work. Activity analysis involves dividing tasks up into appropriate physical activities and recording when the activities are performed and for how long. Observers should be trained, consistent and able to use 'common sense'. An estimate of metabolic heat production for each activity can provide time variation information and total metabolic heat production. This method is not only used for estimating metabolic heat production. A further analysis of these data can also provide specification for clothing systems (see Chapter 7) and estimation of food (calorific) and water requirements for example (Edholm, 1981; Weiner and Lourie, 1981; Weiner, 1982).

A method of improving the accuracy of estimation is to develop charts and tables for specific jobs. The heat production for each activity and combined activities can be determined through indirect calorimetry (in the laboratory, for example) and the values used in calculations from the recordings of activities performed.

Accuracy of methods

For practical applications it is debatable whether an accurate estimate of metabolic heat production can be made. Whole-body calorimetry is impractical and indirect calorimetry, as well as being inconvenient, involves many potential errors. Parsons and Hamley (1989) suggest problems with indirect calorimetry as: inconvenience and unacceptability to the subjects; interference with activity and subject (e.g. hyperventilation); inter- and intra-subject and experimenter variation for the same activity; and problems with calibration, leaks, etc. The use of tables is convenient; however, it is difficult to provide a value for all activities. Methods which involve detailed descriptions of tasks are probably no better, though no worse, than those which use a general description of activity. Parsons and Hamley (1989) suggest an accuracy of around 50 per cent of the 'actual' value which is probably a realistic assessment.

A practical problem with the use of extensive tables (or databases) of metabolic heat production estimates, is difficulty in matching data about a task with those in the database – a similar problem was found for databases of clothing (McCullough and Jones, 1984 – see Chapter 7). Parker and Parsons (1990) describe a computer database system which greatly aids in the effectiveness of finding appropriate data.

The principles of measurement and advantages and disadvantages of techniques for estimating metabolic rate are presented in Table 6.8.

SPECIAL POPULATIONS – CHILDREN, PEOPLE WITH PHYSICAL DISABILITIES AND PEOPLE WEARING PROTECTIVE CLOTHING AND EQUIPMENT

Metabolic rates for children may differ from adults as they can be more active and their bodies have a higher surface area to mass ratio and hence higher heat exchange rate with the environment than adults. This is particularly relevant in hospitals, for determination of conditions for the newborn and for the design of babies' incubators (see Clark and Edholm, 1985). The wide range and frequently changing levels (including crawling and toddling) of activities of children in homes and nurseries makes for challenging environmental design. The changing metabolic rate of children may be regulated to maintain an acceptable thermal condition for the child, however there will be a greater sensitivity to changes in the six basic factors (t_a, t_r, v, rh, Clo, activity) than that of adults. Adaptive opportunity will be particularly important (small children and babies will not be able to take off blankets or clothing or put them on if too hot or cold). Environmental and organizational design should provide opportunities for behavioural thermoregulation and parents and teachers may have to provide adaptive advice and control. There are few studies of metabolic heat production in children and their activities. As a first approximation the data presented in ISO 8996 (1990) for adults may be used with appropriate correction for reduced mass and surface areas of the body. This relationship may be inaccurate at extremes (e.g. babies) however. Black *et al.* (1996) describe average levels of 'free living energy expenditure' from affluent societies (e.g. well nourished) to determine the influence of body weight, height, age and sex. A survey of 574 doubly-labelled water measurements provided Total Energy Exposures (TEE) and basal metabolic rates (directly recorded from measurements during sleep) for people aged 2–95 years, including people with physical disabilities and during different stages of pregnancy. It was found that height and age were significant predictors of energy expenditure (free living – i.e. tasks and activities not controlled) and that females expended

Table 6.8 Advantages and disadvantages of methods of measuring metabolic heat production

Method	Principle	Advantages	Disadvantages
Whole-body direct calorimetry	Measures rate at which heat is lost from a person in a room where all avenues of heat exchange (non-evaporative and evaporative) can be measured	Accurate Precise Fast-responding Direct energy measurement Good environment for strictly controlled studies	Must account for heat transfer in food, drink, excreta, lighting, TV etc. which may be 10 per cent of total Expensive Complex mechanical engineering Provides no information about substrates from which energy is derived Not easy to combine with invasive measurements Artificial environment
Whole-body indirect calorimetry	Measures rate at which heat is produced in the body by a person in a room, from analysis of CO_2 and O_2 content and rate of, in-going and out-going air in the room	Simple in principle Accurate and precise Much world-wide expertise Fast responding Provides information about substrates from which energy is derived Good environment for strictly controlled studies	Requires careful design, engineering and computing Requires best possible instruments Expensive Not easy to combine with invasive measurements Artificial environment
Douglas bag method	Measures rate at which heat is produced in the body by collection of a person's expired air into an impermeable bag and measurement of its O_2 and CO_2 content and volume over a fixed time	Relatively simple and robust Yields reliable results Relatively inexpensive	Diffusion of CO_2 and chances of leakage of expired air Prompt analysis of expired air required after collection

Table 6.8 (Continued)

Method	Principle	Advantages	Disadvantages
			Suitable for short periods of activity only Cumbersome, socially undesirable Inconvenient, uncomfortable for subject May interfere with normal activity of subject
K–M respirometer	Portable system which measures rate at which heat is produced in the body from expired air volume through a low resistance valve meter and mouthpiece and analysis of expired gas in a small rubber collection sample bag	Smaller, more compact and lighter than Douglas bag apparatus Useful for light to moderate physical activity Relatively simple and robust Yields reliable results	Suitable only for short periods of light activity Discomfort to subject on prolonged use due to mouthpiece and nose clip Requires separate analysis of gases by chemical methods or gas analysers Prompt analysis of expired air samples required after collection
Oxylog system	Portable digital system recording cumulative oxygen consumption (using polargraphic cells) and inspiratory volume (meter and mouthpiece)	Smaller, more compact and lighter than Douglas bag Can be used for light, moderate and even strenuous activity Does not require separate gas analysers; provides instantaneous gas analysis Provides minute-by-minute oxygen consumption data during the activity	Suitable only for short periods of activity Prolonged use leads to discomfort with mask or mouthpiece and nose clip Not simple to operate, needs frequent calibration and checking of functioning of oxygen electrodes Initial investment considerable

Deltatrac metabolic monitor	Clinical method using a mechanically ventilated tent (face canopy or hood) and analysis of O_2 consumption and CO_2 production	Accurate and reliable data Gas analysers and computer incorporated Values for RQ available Automatic with good display systems User-friendly	Needs periodic checks by alcohol/butane burns Rather expensive
Activity diaries and time budget determinations	Recording (in real time or from video recording by an observer or subject) of activity performed over time for estimate of variation in metabolic rate and cumulative or average (time-weighted) energy production	Inexpensive for large population groups (retrospective) Provide good data on patterns of activity	Subject co-operation essential (diary) Expensive, large manpower needed (time-and-motion) Observers must be trained and may require previous activity and task analysis May influence habitual activity pattern (diary, time-and-motion) Requires measures of energy cost of each activity for computation of energy expenditure Validity of energy expenditure data variable
Heart rate monitoring techniques	Measure heart rate over time and relate metabolic rate from previously determined individual 'calibration' curve	Can be used in free living individuals performing habitual activity Simple, relatively inexpensive Socially acceptable and non-intrusive Can be used on children Provides useful activity data even without conversion to energy expenditure	Needs individual assessment of HR to V_{O_2} relationship in each subject for a wide range of activities HR not only affected by metabolic rate Beyond an acceptable range HR may not reflect changes in O_2 uptake

Table 6.8 (Continued)

Method	Principle	Advantages	Disadvantages
			In groups proximity of instruments is important as they may interfere
			Need to ensure that electrodes are in contact all the time
			Extremes of ambient temperature and humidity as well as sweating may alter relationships
Doubly-labelled water method	Subject drinks $^{2}H^{18}O$ isotopes and rate of disappearance in the body fluids (e.g. urine) is related to CO_2 production	Measures free-living TEE	Very high capital and recurrent cost
		Accurate	Technically challenging
		Non-invasive, non-intrusive	Time-consuming analysis
		Simple for subject	May be invalid in exceptional circumstances
		Usable in uncooperative subjects	Elite technique
		Representative of habitual TEE integrated over many days	No fine detail of expenditure
		Estimates energy cost of activity	Measurement period >1 week

Sources: Adapted from Murgatroyd et al. (1993) and ISO 8996 (1990).

11 per cent less energy than males after adjustment for weight, height and age. Black *et al.* (1996) provide extensive tables of energy expenditure over a range of populations. People with physical disabilities may have altered metabolic rates due to their disability, its treatment and control, and the consequences of the disability. Black *et al.* (1996) quantify basal metabolic rates of people with cerebral palsy and show some differences when compared with those of other people without physical disabilities. Changes in muscle tone or nervous tremor, and those with altered thermoregulatory capacity, as well as physical asymmetry and reduced surface area, should be considered. Those that are immobile will have a reduced activity range and hence metabolic heat production, however operation of a wheelchair or the greater effort to complete a task should be taken into account. Those with mental disability will have additional requirements as behavioural control may be required from a carer. Drugs taken by people with disabilities and those who are ill may affect metabolic heat production. For

Table 6.9 Some substances associated with hyperthermia (i.e. heating of the body)

Drug	Action on thermoregulation
Alcohol	Inhibition of central nervous function. Dehydration peripheral vasodilation. Impairment of behavioural thermoregulation and judgement
Antidepressants e.g. tricyclics	Hyperthermia with high doses especially in combination with other agents, e.g. amphetamines. Counteract reserpine-induced hypothermia
Hypnotics e.g. barbituates	Central nervous depressant. Body temperature increases in hot environments. Effects augmented by alcohol
Psychotropics e.g. phenothiazanes	Hyperthermia in high ambient temperature. Central effect on thermoregulation with possible peripheral actions
Cannabis	Hyperthermia in hot environments
Morphine	Hyperthermia usually with low doses
Amphetamines	Central nervous stimulant. Vasoconstriction Increased peripheral heat production
Anaethetics	Central nervous depression of thermoregulatory centres. Hyperthermia in hot environments 'Malignant hyperthermia' a rare complication of e.g. halothane anaethesia involving muscle contracture in susceptible subjects
Cocaine	Overdosage may result in heat-stroke
Anticholinergics e.g. atropine	'Atropine fever'. Effects on thermoregulatory centres Inhibition of sweating
Organophosphates e.g. pesticides	Potential heat stroke via alteration of 'set-point' (see Table 6.10.)

example, BOHS (1996) notes that antithyroids (e.g. carbimazole) reduce metabolic heat production and hypoglycaemics (e.g. biguanides) reduce metabolic heat production during treatment of diabetes mellitus (see Tables 6.9 and 6.10).

If protective clothing and equipment is worn there is not only a restriction to heat transfer between the body and the environment but also there may be a significant increase in metabolic heat production. This will depend upon activity level. For a resting person, heavy boots and clothing will have

Table 6.10 Some substances associated with hypothermia (i.e. cooling of the body)

Drug	Action on thermoregulation
Alcohol	Central nervous inhibition. Hypoglycaemia. Peripheral vasodilation. Impairment of behavioural thermoregulation and judgement in low temperatures. May alter 'set-point'
Antidepressants	Rapid depression of body temperature. Reduced awareness of cold. Potentiate hypothermic effect of barbituates and alcohol
Tranquillizers e.g. benzodiazepines	Can induce hypothermia, especially in high doses
Hypnotics	Lower body temperature in cold environments Peripheral vasodilation. Augmented by alcohol. Hypothermia may occur with non-barbituate hypnotics (nitrazepam) especially in the elderly
Psychotropics e.g. phenothiazanes	Powerful hypothermic action in cold with long-term central action on body temperature. May have peripheral action
Cannabis	Fall in body temperature in ambient temperatures within or below thermoneutral. Hypothermia with overdosage of cannabinoids
Morphine	Hypothermia with high dosage and during withdrawal of drugs
Anaethetics	Central nervous depression. Hypothermia may develop in ambient temperatures below 24 °C
Hypoglycaemics e.g. biguanides	Reduced metabolic heat production during treatment of diabetes mellitus
Antithyroids e.g. carbimazole	Reduced metabolic heat production
Sympathetic and ganglion-blocking agents e.g. reserpine	Reduced vasoconstriction in cold causing fall in body temperature
Organophosphates e.g. pesticides	Anticholinesterase action on autonomic nervous system resulting in reduced body temperature. May alter 'set-point' Can also lead to heat stroke (see Table 6.9)

little effect on heat production, but in walking up a hill or stepping they will have a significant effect. BS 7963 (2000) provides estimates of increases in metabolic rate due to wearing PPE (see Table 6.11).

Table 6.11 Estimated increases in metabolic rate due to wearing PPE

PPE item	Increase in metabolic rate due to wearing PPE ($W m^{-2}$)				
	Resting	Low metabolic rate	Moderate metabolic rate	High metabolic rate	Very high metabolic rate
Safety shoes/short boots	0	5	10	15	20
Safety boots (long)	0	10	20	30	40
Respirator (low/moderate performance, e.g. P1, P2, see note 2)	5	10	20	30	40
Respirator (high performance, e.g. P3, see note 2)	5	20	40	60	80
Self-contained breathing apparatus	10	30	60	95	125
Light, water vapour permeable chemical coverall (e.g. disposable)	5	10	20	30	40
Chemical protective water vapour impermeable ensemble (e.g. polyvinyl chloride [PVC]) with hood, gloves and boots	10	25	50	80	100
Highly insulating, water vapour semi-permeable ensemble (e.g. firefighters' gear consisting of helmet, tunic, over trousers, gloves and boots)	15	35	75	115	155

Notes
1 Values have been rounded to the nearest $5 W m^{-2}$.
2 Respirator classifications P1, P2 and P3 are defined in BS EN 143 (1991).
3 Very high metabolic rates probably cannot be maintained when wearing some forms of PPE.
4 It is not appropriate to add the increase in metabolic rate due to wearing footwear when tasks are stationary or sedentary.
5 The values in this table have been extrapolated from experimental data. The values are approximate and more accurate measurements might be obtained in practice using the method given in BS EN 28996. This more accurate method of measuring metabolic rate requires no correction for the PPE worn.

7 The thermal properties of clothing

INTRODUCTION

At a recent meeting to develop an International Standard for assessing the reaction of human skin to hot surfaces, the Japanese delegate reminded experts that there were many people in the world who wore no clothes, and we should remember that if we are considering an International Standard, exposed skin in people is not always confined to a small percentage of the surface area of the body e.g. hands, face, etc. This point is important and raises interesting issues about our 'natural state'. However, it is probably the case that for most people, most of their skin is covered by clothing for most of their lives.

Clothing provides a thermal resistance between the human body and its environment. A functional role of clothing is therefore to maintain the body in an acceptable thermal state, in a variety of environments. While there are many other functional roles of clothing, it should be remembered however that 'fashion' is of great importance. In a 'pilot' study of four outdoor suits (light trousers and jackets made of similar materials with different properties) a laboratory test led to comments and some amusement about the four different colours of the suits. In a field trial where subjects walked through an area populated by people they knew, however, comments increased to objections about wearing 'the bright green' suit. Investigations (especially subjective ones) into thermal properties would therefore have been confounded by the appearance of the suits, if they had been investigated in the main study. A person's perception of how they appear to others in clothing is of great importance; soldiers for example may not wish to wear 'stylish pink' into battle even if it was shown to give optimum performance in other respects. Image is also important and should complement and be an integral part of an organization's mission. Nurses' uniforms should portray a caring, efficient confident impression. The public should identify and respect firefighters or policemen and so on. For a discussion of clothing in a wider context (see Renbourn, 1972). The important contribution which clothing makes to human thermal responses is discussed below.

The thermal behaviour of clothing on an (active) person is complex and dynamic, not fully understood, and is difficult to quantify. While much is not known, much of what is known is mainly derived from theoretical and empirical research. Factors affecting the thermal behaviour of clothing will include the dry thermal insulation, transfer of moisture and vapour through clothing (e.g. sweat, rain), heat exchange with clothing (conduction, convection, radiation, evaporation and condensation), compression (e.g. caused by high wind), pumping effects (e.g. caused by body movement), air penetration (e.g. through fabrics, vents and openings), subject posture, and so on.

An approach to assessing the thermal properties of clothing is to identify simple thermal models of clothing behaviour and attempt to estimate values of quantities required for the thermal models. For example, a simple value for dry thermal insulation or vapour permeation characteristics.

A SIMPLE CLOTHING MODEL

The dry thermal insulation value of clothing materials and clothing ensembles is of fundamental importance and has been extensively investigated. A simple thermal model is of a heated body with a layer of insulation (Figure 7.1). For the body to maintain equilibrium heat flows to the skin, determining skin temperature, through the insulation to the clothing surface, determining clothing temperature, and to the outside environment. If the body were not continuously heated (i.e. death), heat would flow out of the body until equilibrium was reached when body temperature, skin temperature and clothing temperature were at environmental temperature. For a continuously heated body (by metabolic heat production), a dynamic

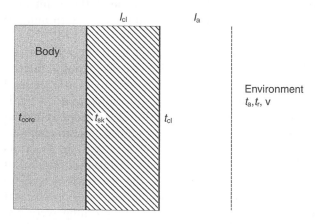

Figure 7.1 A simple thermal model of a heated body with a layer of clothing insulation.

equilibrium is maintained where (generally) body temperature is greater than skin temperature which is greater than temperature of the clothing surface which in turn is greater than environmental temperature.

That temperature of clothing is higher than environmental temperature emphasizes that the environment also provides insulation (the boundary or air layer). The properties of this 'layer' are very important to heat exchange and can be affected by the external environment. The basic or intrinsic clothing insulation (I_{cl}) will, according to the model, be independent of the external environmental conditions. The thermal properties of each of the components of the model presented in Figure 7.1 are discussed below.

Heat transfer from the body to the skin

Metabolic heat production occurs in all parts of the body and the thermoregulatory system regulates how much is transferred to the skin. This will involve heat transfer through tissues and will be greatly determined by the degree of vasodilation. Burton and Edholm (1955) provide a detailed description. In terms of the model shown in Figure 7.1, it is important to know that this will affect skin condition (temperature and sweat – see later) which is important for considering heat transfer through clothing.

Intrinsic clothing insulation, I_{cl}

Intrinsic (or basic) clothing insulation is a property of the clothing itself (and not the external environment or body condition) and represents the resistance to heat transfer between the skin and the clothing surface. Rate of heat transfer through the clothing is by conduction, which depends on surface area (m^2), temperature gradient (°C) between skin and clothing surface and the thermal conductivity ($W\,m^{-2}/°C$) of the clothing. Intrinsic clothing insulation is the reciprocal of clothing conductivity with units of $m^2\,°C\,W^{-1}$.

Gagge *et al.* (1941) first proposed the *Clo unit*, to replace the rather physical unit of $m^2\,°C\,W^{-1}$ with something easily visualized and related to clothing worn on the human body. One Clo was said to be the thermal insulation required to keep a sedentary person comfortable at 21 °C. It is said to have an average value of $0.155\,m^2\,°C\,W^{-1}$ and is representative of the insulation of a typical business suit. It is important to note that the m^2 term in this unit refers to surface area of the body. A neck tie worn on its own for example may be given an estimated thermal insulation value of 0.1 Clo, while a suit made of the same material as the necktie may have an estimated insulation value of 0.8 Clo. The important point therefore is that the Clo value gives an estimate of insulation as if any clothing were distributed evenly over the whole body. This can cause confusion for those used to considering the thermal insulation of materials. A unit of the thermal insulation of material is the Tog (Pierce and Rees, 1946). One Tog is given a value of $0.1 \, \text{m}^2 \, {}^\circ\text{C} \, \text{W}^{-1}$ where the m^2 term refers to the area of the material tested.

Thermal resistance of the environment – the air 'layer' (I_a)

If the environment had perfect conductivity i.e. no resistance, the surface temperature of the clothing would 'fall' to that of the environment. This is not the case however. The environment provides significant thermal resistance depending upon how environmental conditions affect heat transfer – mainly by convection (C) and radiation (R). The thermal resistance (insulation) of the environment for a nude body is:

$$I_a = \frac{1}{h},\tag{7.1}$$

where

$h = h_r + h_c$
h_r = radiative heat transfer coefficient
h_c = convective heat transfer coefficient.

For a clothed body, the surface area for heat transfer is increased by an amount depending upon the thickness of the clothing layer. This is taken into consideration using the term f_{cl}, which is the ratio of the clothed surface area of the body to the nude surface area of the body, i.e.

$$I_a(\text{clothed}) = \frac{1}{f_{cl}h} = \frac{I_a}{f_{cl}}.$$

A difficulty is in determining a value for f_{cl}. On an active person only an approximation can be made, and even on a copper manikin it is difficult to measure (McCullough and Jones, 1984). More sophisticated systems include photographic techniques and computer aided anthropometric scanners (e.g. Jones *et al.*, 1989). An approximation can be based on the intrinsic clothing insulation value. For example, McCullough and Jones (1984) gives the following equation as a rough estimate for indoor ensembles:

$$f_{cl} = 1.0 + 0.31 \, I_{cl},\tag{7.2}$$

where I_{cl} is in Clo. It should be emphasized that this equation is an approximation and other values of the coefficient for I_{cl} have been used, e.g. 0.20 by Fanger (1970).

'Total' insulation (I_t) and effective insulation (I_{cle})

The necessary principle and quantities for the model shown in Figure 7.1 are provided above. There are however other values which are mentioned in the scientific literature and it is useful to discuss these briefly here. Total clothing insulation is the combined insulation provided by clothing and surrounding air layer. Parsons (1988) gives the following equation:

$$I_t = I_{cl} + I_a \text{ (clothed)},$$

$$\text{but} \quad I_a \text{ (clothed)} = \frac{I_a(\text{nude})}{f_{cl}}.$$

So the terms I_a (clothed) and I_a (nude) are redundant and I_a can always be regarded as the insulation of the environment on a nude person. f_{cl} provides the correction for clothing.

To determine I_{cl}, by using a thermal manikin for example, it is possible to measure I_t using a clothed manikin, and I_a using a nude manikin. However determination of f_{cl} is less accurate. From above:

$$I_{cl} = I_t - \frac{I_a}{f_{cl}}. \tag{7.3}$$

A more convenient term for measurement is therefore effective clothing insulation (I_{cle}) where

$$I_{cle} = I_t - I_a.$$

Although more convenient to determine, I_{cle} is not consistent with the model provided in Figure 7.1 and can lead to confusion.

The Burton thermal efficiency factor (F_{cl})

Oohori *et al.* (1984) provide the following equation for the dry heat loss from the skin:

$$D_{ry} = F_{cl}f_{cl}h(T_{sk} - T_o),$$

where

$$F_{cl} = \text{Burton thermal efficiency factor} = \frac{I_a}{(I_{cl} + I_a)},$$

$$\text{where in this case } I_a = I_a(\text{clothed}) = \frac{1}{f_{cl}h} = \frac{I_a(\text{nude})}{f_{cl}}.$$

Efficiency factors (clothing indices, etc.) are often useful for simplifying calculations, providing terms which have physical meaning and linking with experimental method. They can however cause confusion and the essential concepts for the model provided in Figure 7.1, are I_{cl} and I_a. The essential point about this model is that I_{cl} is intrinsic to the clothing and is not affected by external conditions. The point that the model may be inadequate in many conditions will be discussed later. Methods of determining I_{cl} values are discussed below.

Method of determining the dry thermal insulation of clothing

The thermal insulation of clothing materials can be measured on standardized equipment which usually involves placing a sample of material on the equipment and, by measuring heat flows or temperature, the thermal insulation can be calculated. Such equipment would include standardized heated flat plates and cylinders. More sophisticated methods involve heated manikins with a temperature distribution across the body similar to that of a human subject – see McCullough and Jones (1984), Kerslake (1972), Olesen *et al.* (1982), Wyon (1989) and Holmér (1999) for reviews.

The role of manikins in clothing development is becoming more clearly defined and they are becoming an essential part of clothing design and evaluation. Although still specialized and expensive, many countries of the world now have manikin test centres. Dry thermal insulation values have been determined for many types of clothing using thermal manikins, and databases (tables) of insulation values have been provided. ISO 9920 (1995) provides such a database as do McCullough and Jones (1984). Olesen and Dukes–Dubos (1988) provide examples for whole clothing ensembles (Table 7.1) and for individual clothing garments (Table 7.2).

Values of I_{cl} are obtained directly from Table 7.1. However, it is likely in practice that the clothing ensembles in the database (tables) will not be identical to the ensemble for which an I_{cl} value is required. In this case estimates of I_{cl} values can be made from I_{clu} values which are 'effective' insulation values for garments; i.e. insulation values calculated without taking account of an increase in surface area due to the garment. Olesen and Dukes–Dobos (1988) found that simply adding Clo values for garments gives a realistic estimate of ensemble I_{cl} values. This was different from other methods where simple regression equations were used to estimate the insulation of clothing ensembles from garment values (e.g. Sprague and Munson, 1974).

ISO 9920 (1995) provides tables of insulation values for clothing ensembles and garments. Garments can be identified in terms of garment type (e.g. shirt), style (e.g. long sleeved, short sleeved, etc.) and material type (e.g. 100 per cent cotton, cotton polyester). Parker and Parsons (1990) describe a computer database of the ISO clothing ensembles and a method of matching required clothing with those for which I_{cl} values are required. This is the most effective way to use such a database. McCullough and Jones (1984) report on a user trial of their database 'manual' and found that even clothing science students were unable to use it effectively. ISO 9920 (1995)

Table 7.1 Work clothing ensembles: dry thermal insulation values

Work clothing	I_{cl}	
	Clo	$m^2 \, ^{\circ}C \, W^{-1}$
Underpants, boiler suit, socks, shoes	0.70	0.110
Underpants, shirt, trousers, socks, shoes	0.75	0.115
Underpants, shirt, boiler suit, socks, shoes	0.80	0.125
Underpants, shirt, trousers, jacket, socks, shoes	0.85	0.135
Underpants, shirt, trousers, smock, socks, shoes	0.90	0.140
Underwear with short sleeves and legs, shirt, trousers, jacket, socks, shoes	1.00	0.155
Underwear with short sleeves and legs, shirt, trousers, socks, shoes	1.10	0.170
Underwear with long legs and sleeves, thermojacket, socks, shoes	1.20	0.185
Underwear with short sleeves and legs, shirt, trousers, jacket, thermojacket, socks, shoes	1.25	0.190
Underwear with short sleeves and legs, boiler suit, thermojacket + trousers, socks, shoes	1.40	0.220
Underwear with short sleeves and legs, shirt, trousers, jacket, thermojacket and trousers, socks, shoes	1.55	0.225
Underwear with short sleeves and legs, shirt, trousers, jacket, heavy quilted outer jacket and overalls, socks, shoes	1.85	0.285
Underwear with short sleeves and legs, shirt, trousers, jacket, heavy quilted outer jacket and overalls, socks, shoes, cap, gloves	2.00	0.310
Underwear with long sleeves and legs, thermojacket + trousers, outer thermojacket + trousers, socks, shoes	2.20	0.340
Underwear with long sleeves and legs, thermojacket + trousers, parka with heavy quilting, overalls with heavy quilting, socks, shoes, cap, gloves	2.55	0.395

Table 7.2 Individual clothing garments: dry thermal insulation values

Garment description	Thermal insulation Clo (I_{clu})
Underwear	
Panties	0.03
Underpants with long legs	0.10
Singlet	0.04
T-shirt	0.09
Shirt with long sleeves	0.12
Panties and bra	0.03
Shirts/blouses	
Short sleeves	0.15
Lightweight, long sleeves	0.20
Normal, long sleeves	0.25
Flannel shirt, long sleeves	0.30
Lightweight blouse, long sleeves	0.15
Trousers	
Shorts	0.06
Lightweight	0.20
Normal	0.25
Flannel	0.28
Dresses/skirts	
Light skirt (summer)	0.15
Heavy dress (winter)	0.25
Light dress, short sleeves	0.20
Winter dress, long sleeves	0.40
Boiler suit	0.55
Sweaters	
Sleeveless vest	0.12
Thin sweater	0.20
Sweater	0.28
Thick sweater	0.35
Jackets	
Light summer jacket	0.25
Jacket	0.35
Smock	0.30
High insulative, fibre-pelt	
Boiler suit	0.90
Trousers	0.35
Jacket	0.40
Vest	0.20
Outdoor clothing	
Coat	0.60
Down jacket	0.55
Parka	0.70
Fibre-pelt overalls	0.55

Table 7.2 (Continued)

Garment description	Thermal insulation Clo (I_{clu})
Sundries	
Socks	0.02
Thick ankle socks	0.05
Thick long socks	0.10
Nylon stockings	0.03
Shoes (thin soled)	0.02
Shoes (thick soled)	0.04
Boots	0.10
Gloves	0.05

provides the following equations for estimating the I_{clu} for garments and hence I_{cl} values for clothing ensemble:

$$I_{clu} = 0.095 \times 10^{-2} A_{cov} \ \text{m}^2 \, {}^\circ\text{C W}^{-1},$$

$$I_{cl} = \sum_{i} I_{clu,i},$$

(7.4)

where

I_{cl} = intrinsic clothing insulation for clothing ensemble

$I_{clu,i}$ = effective insulation for garment i

A_{cov} = body surface area covered by the garment (per cent).

The simple model presented in Figure 7.1 allows ensemble estimates to be made for parameters of the model (I_{cl} and I_a). Figure 7.2 shows the estimate of ranges of I_a values according to Burton and Edholm (1955).

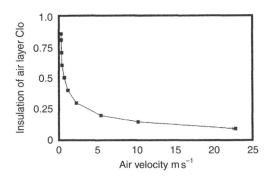

Figure 7.2 Insulation of the 'air layer' around the body as affected by air velocity. The effect is independent of air temperature within 0.05 Clo.

Source: Burton and Edholm (1955).

The model in Figure 7.1 provides a representation of clothing in the stationary, comfortable, or cold human body in many conditions. It provides only an approximation however to many other circumstances and will have limitations when applied to active, sweating people.

THE TWO-PARAMETER MODEL

An important limitation of the model shown in Figure 7.1 is that it does not consider wet clothing. Moisture can transfer heat between the body and the environment. This is particularly important when the skin sweats. A simple two-parameter model of clothing would be to consider 'dry' heat transfer (Figure 7.1) and moisture transfer (Figure 7.3) as separate and independent mechanisms, which combine to provide the total effect.

Intrinsic resistance of clothing to vapour transfer (I_{ecl})

The model in Figure 7.3 is analogous to that of Figure 7.1. In Figure 7.1, however, the temperature gradient between the skin and the outside surface of the clothing provides the driving potential for heat loss. In Figure 7.3, it is the vapour pressure difference between skin and environment that provides the driving potential. Liquid (sweat) on the skin evaporates at the skin surface and is transported through the clothing to the environment. The resistance to this vapour transfer is termed the intrinsic evaporative resistance I_{ecl}. The units of I_{ecl} are therefore $m^2\,kPa\,W^{-1}$. The partial vapour

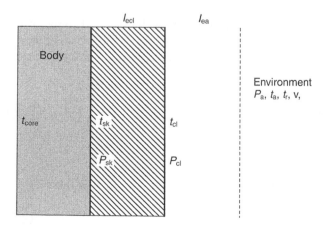

Figure 7.3 A simple model of a heated body showing resistance to moisture transfer through a layer of clothing insulation.

pressure at the skin is assumed to be the saturated vapour pressure at skin temperature. From Antoine's equation:

$$P_{sk,s} = \exp\left(18.956 - \frac{4030}{t_{sk} + 235}\right) \text{mb}. \tag{7.5}$$

The partial vapour pressure in the air (P_a) is related to relative humidity (ϕ) by

$$\phi = \frac{P_a}{P_{sa}},$$

where P_{sa} is the saturated vapour pressure in air at air temperature and can be obtained by substituting into Equation 7.5 P_{sa} and t_a for $P_{sk,s}$ and t_{sk} respectively.

Resistance of the environment to vapour transfer from clothing (I_{ea})

When vapour reaches the clothing surface it transfers to the environment at a rate depending upon the evaporative heat transfer coefficient h_e. The resistance to vapour transfer is therefore $I_{ea} = 1/h_e$.

It has been traditional to determine h_e by relating it to the convective heat transfer coefficient by the Lewis number, defined as the ratio of mass transfer coefficient by evaporation to heat transfer coefficient by convection (no radiation). At sea level for air:

$$LR = \frac{h_e}{h_c} = 16.5 \, \text{K kPa}^{-1} = 1.65 \, \text{K mb}^{-1} = 2.2 \, \text{K Torr}^{-1}.$$

The Lewis relationship (LR) is not affected by the size and shape of the body or by air speed or temperature. It is affected by the physical properties of the gases involved (in the case of air and water vapour) and by atmospheric pressure.

$$LR = \frac{1.65}{P_{atm}} \, \text{K mb}^{-1}, \tag{7.6}$$

where
P_{atm} = atmospheric pressure in atmospheres.
The maximum evaporation from the skin surface to air is therefore:

$$E_{max} = h_e\left(P_{sk,s} - P_a\right), \tag{7.7}$$

where $h_e = 16.5 h_c \, \text{m}^2 \, \text{kPa W}^{-1}$.

In terms of the model in Figure 7.3,

$$h_e = \frac{1}{I_{ea}} \quad \text{or} \quad I_{ea} = \frac{1}{h_e} = \frac{1}{16.5\, h_c},$$

where I_{ea} is the vapour resistance of the air layer. The maximum heat loss from the skin, through clothing, to the environment is therefore,

$$E_{max} = \frac{1}{I_{ea} + I_{ecl}} (P_{sk,s} - P_a).$$

For the case where the skin is not completely wet the skin wettedness w is used to calculate evaporative heat loss (E) where:

$$w = \frac{E}{E_{max}},$$

i.e.

$$E = \frac{w}{I_{ea} + I_{ecl}} (P_{sk,s} - P_a).$$

With respect to vapour transfer through clothing therefore, vapour transfer resistance of the air layer:

$$I_{ea} = \frac{1}{h_e} = \frac{1}{16.5\, h_c}.$$

Intrinsic vapour transfer resistance of clothing (I_{ecl}) is therefore required. An important point is that the human skin is always 'wet' to some extent. There is always a flow of vapour from wet tissues below the skin to drier air above (diffusion through skin – insensible heat loss). The vapour permeation properties of clothing are therefore of great importance.

The traditional approach to the consideration is presented above, however to be consistent with the dry heat transfer models, consideration should be given to the increase in surface area due to clothing and hence the f_{cl} value.
That is:

$$I_{et} = I_{ecl} + \frac{I_{ea}}{f_{cl}}, \tag{7.8}$$

To continue the parallel further it may be useful to define a unit for vapour permeation properties similar to the Clo value for dry insulation. This could be

$$1 \text{ Unit} = 0.0155 \, \text{m}^2 \, \text{kPa} \, \text{W}^{-1},$$

for typical clothing. The greater the value, the greater the resistance to vapour transfer and hence the more impermeable the clothing.

Vapour permeability indices

To provide easy to understand values of clothing permeability, or to aid in simplifying heat transfer equations, a number of vapour permeability indices for clothing have been produced. Although all are essentially based on the two-parameter model, they have been developed independently and only recently has the relationship between them been determined (Oohori *et al.*, 1988). As with the indices for dry clothing insulation it is debatable whether the apparent advantage of simplification, provided by the indices, outweighs the ability to directly relate terms to the three clothing model parameters (Figure 7.3) and to refer directly to the fundamental units of physics.

Woodcock 'moisture permeability index' (i_m)

A property of clothing is that it impedes evaporative heat transfer more than it does sensible heat transfer (Kerslake, 1972). Woodcock (1962) proposed the permeability index (i_m), which compares this property with that of air (h_e/h_c). The moisture permeability index of a material is defined by ASHRAE (1997) as the ratio of the actual evaporative heat flow capability between the skin and the environment to the sensible heat flow capability as compared to the Lewis Ratio. That is, i_m is $(1/I_{et})/(1/I_t)$ divided by the Lewis ratio $(1/I_{ea})/(1/I_a)$, no radiation.

So,

$$i_m = \frac{I_t/I_{et}}{LR} = \frac{I_t}{LR\, I_{et}}, \tag{7.9}$$

i_m ranges from 0 for a material impermeable to water vapour to 1 for air. For a nude subject, if radiation exchange is present, the index is:

$$i_m = \frac{h_e/h_0}{h_e/h_c} = \frac{h_c}{h_0} = \frac{h_c}{h_r + h_c} \quad \text{i.e.} \quad <1.$$

The i_m value does not provide a value intrinsic to clothing as it is affected by external environmental conditions. A value of 0.5 is a typical value for a nude subject, 0.4 for normal clothing and 0.2 for impermeable type clothing.

Authors have presented the Lewis relation and i_m in different ways. Goldman (1988) describes i_m as a theoretical index which is the ratio of the maximum evaporative cooling, at a given ambient vapour pressure, from a 100 per cent wetted surface through a fabric, to the maximum evaporative cooling of a psychrometric wet bulb thermometer at the same vapour pressure.

Permeation efficiency factor (F_{pcl})

Nishi and Gagge (1970) proposed a permeation efficiency factor (F_{pcl})

$$F_{pcl} = \frac{I_{ea}}{I_{ea} + I_{ecl}}, \tag{7.10}$$

where

I_{ea} = the resistance of air to the transfer of water vapour
I_{ecl} = the resistance of clothing to the transfer of water vapour,

i.e. analogous to Burton F_{cl} for 'dry' heat exchange.
From the above equation:

$$F_{pcl} = \frac{1}{1 + I_{ecl}/I_{ea}}.$$

But from the Lewis relation:

$$I_{ea} = \frac{1}{h_e} = \frac{1}{16.5\,h_c},$$

so

$$\frac{1}{I_{ea}} = 16.5\,h_c.$$

Also for clothing we could consider a Lewis number, k, where

$$k = \frac{h_{ecl}}{h_{cl}} = \frac{I_{cl}}{I_{ecl}},$$

as

$$I_{cl} = \frac{1}{h_{cl}} \quad \text{and} \quad I_{ecl} = \frac{1}{h_{ecl}},$$

so

$$I_{ecl} = \frac{I_{cl}}{k},$$

therefore

$$F_{pcl} = \frac{1}{1 + \left(\frac{16.5\,h_c}{k} \times I_{cl}\right)}. \tag{7.11}$$

Note that if equation (7.12) were used

$$I_{et} = I_{ecl} + \frac{I_{ea}}{f_{cl}}.$$
(7.12)

Then

$$F_{pcl} = \frac{I_{ea}/f_{cl}}{I_{ea}/f_{cl} + I_{ecl}} = \frac{I_{ea}}{I_{ea} + f_{cl}I_{ecl}} = \frac{1}{1 + f_{cl}I_{ecl}/I_{a}}.$$

So, from the above analysis:

$$F_{pcl} = \frac{1}{1 + f_{cl}\frac{16.5h_c}{k}I_{cl}}.$$
(7.13)

which is in line with the version provided by ASHRAE (1997) – Table 7.3.

Nishi and Gagge (1970) conducted empirical experiments involving light clothing and naphthalene sublimation and vapour transfer. Relating it to water vapour transfer, they found:

$$F_{pcl} = \frac{1}{1 + 0.143\, h_c I_{Clo}},$$

where

I_{clo} = Effective clothing insulation in Clo.

Further work by Lotens and Linde (1983) and Oohori *et al.* (1984) provided the current form of the equation:

$$F_{pcl} = \frac{1}{1 + 0.344\, h_c I_{Clo}}.$$

It should be noted that I_{clo} is effective clothing insulation and therefore does not correct for the increase in surface area due to clothing.

Oohori *et al.* (1984) describe the relationship between three permeation ratios i_a, i_{cl}, and i_m. They are defined as the ratios of the Lewis number of each layer to that of an equivalent non-radiative air layer. These are shown in Table 7.3.

Oohori *et al.* (1984) show how the indices can contribute to the human heat balance $(S = 0)$ equation:

$$S = M_{sk} - 16.5wf_{cl}h_c\left(P^*_{sk} - P_{dp}\right)F_{pcl} - f_{cl}h(t_{sk} - t_o)F_{cl},$$

i.e. the equation based on the two-parameter model for clothing. ASHRAE (1997) provide a summary of parameters used to describe clothing. Relationships between clothing parameters and their use in heat loss equations are provided by Oohori *et al.* (1988) (see Table 7.3).

Table 7.3 The relationship between the vapour permeation ratios for clothing

	Dry (A)	Evaporative (B)	Lewis No. (B/A) K Torr^{-1}	Permeation ratio (B/2.2A)
Air	$h = h_r + h_c$	h_e	h_e/h	$i_a = h_c/h$
Clothing	h_{cl}	h_{cle}	$\frac{h_{cle}}{h_{cl}}$	$i_{cl} = \frac{h_{cle}}{2.2\, h_{cl}}$
Combined clothing and air	hF_{cl}	$h_c F_{pcl}$	$\frac{h_c F_{pcl}}{hF_{cl}}$	$i_m = \frac{h_c F_{pcl}}{2.2\, hF_{cl}}$

Also,

$$\frac{1}{i_m} = \frac{F_{cl}}{i_a} + \frac{(1 - F_{cl})}{i_{cl}}, \quad \text{(ND)}$$

and

$$i_m = F_{pcl} i_a + \frac{(1 - F_{pcl})}{i_{cl}}, \tag{7.14}$$

where

    h = combined heat transfer coefficient
    h_c = convective heat transfer coefficient
    h_r = radiative heat transfer coefficient
    h_e = evaporative heat transfer coefficient
    i_a = vapour permeation ratio of the air layer
    i_{cl} = vapour permeation ratio of the clothing
    i_m = Woodcock moisture permeability index
    h_{cl} = dry heat transfer coefficient of clothing
    h_{cle} = evaporative heat transfer coefficient of clothing
    F_{cl} = Burton thermal efficiency factor
    F_{pcl} = permeation efficiency factor.

The usefulness of clothing indices can be seen in the above equation. However, index values for clothing systems are not widely available and the index values do not directly demonstrate the mechanisms of the model. In addition, these values depend upon environmental conditions unlike the intrinsic values, I_{cl} and I_{ecl}. Intrinsic clothing insulation values for dry heat transfer (I_{cl}) are now available for a wide variety of garments and clothing ensembles (McCullough and Jones, 1984; ISO 9920, 1995). Values of I_{ecl} are required and some exist (McCullough *et al.*, 1989 – Table 7.4). It could now be argued that the index values have 'outlived their usefulness' and what are required for application are databases of intrinsic values for clothing.

Table 7.4 Index values for range of clothing ensembles

Ensemble	Total evaporative resistance – $R_{e,t}$(kPa m²/W)	Moisture permeability index – i_m	Evaporative resistance of clothing* – $R_{e,cl}$(kPa m² W^{-1})	Moisture permeability index for clothing – icl
Men's business suit	0.044	0.37	0.033	0.32
Women's business suit	0.039	0.40	0.028	0.35
Men's summer casual	0.027	0.43	0.015	0.36
Jeans and shirt	0.031	0.40	0.020	0.32
Summer shorts and shirt	0.023	0.42	0.010	0.34
Women's casual	0.026	0.45	0.014	0.41
Women's shorts and sleeveless top	0.022	0.40	0.009	0.27
Athletic sweat suit	0.029	0.45	0.017	0.41
Sleepwear and robe	0.035	0.41	0.024	0.37
Overalls and shirt	0.035	0.40	0.024	0.35
Insulated coverall and long underwear	0.048	0.39	0.037	0.35
Work shirt and trousers	0.037	0.40	0.025	0.34
Cleanroom coverall	0.039	0.38	0.028	0.32
Wool coverall	0.042	0.38	0.031	0.33
Firestop cotton coverall	0.038	0.40	0.027	0.35
Modacrylic coverall	0.038	0.41	0.027	0.36
Tyvak coverall	0.045	0.33	0.034	0.26
Gortex two-piece suit	0.044	0.38	0.033	0.33
Novex coverall	0.039	0.40	0.028	0.35
PVC/polyester xnit acid suit	0.105	0.15	0.094	0.11
PVC/vinyl acid suit	0.126	0.13	0.115	0.09
Neoprene nylon suit	0.120	0.14	0.109	0.10

Source: McCullough *et al.* (1989).

Note
* Calculated based on an air layer resistance of 0.014 kPa m² W^{-1}.

Modification of the two-parameter model – wicking

For most practical applications the simple dry insulation 'model' is used to quantify clothing insulation (e.g. Fanger, 1970). For more specialist evaluations of clothing the two-parameter model is needed especially for the assessment of hot environments and all environments involving activity where sweating and hence vapour permeation properties will be of great importance.

For a more detailed representation of clothing, models are required which involve other important factors of clothing thermal behaviour. A simple modification to the two-parameter model is to consider wicking of liquid through clothing materials with evaporation occurring within clothing. Kerslake (1972) provides a model (Figure 7.4) in which the latent heat of vaporization is removed from within the clothing and not at the skin.

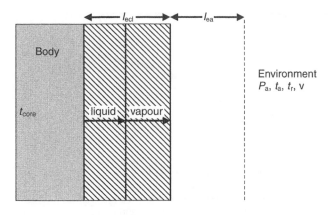

Figure 7.4 Evaporation within clothing.
Source: Model from Kerslake (1972).

McIntyre (1980) summarizes Kerslake's work. The result is a loss of efficiency of sweating. In a simple case, where the sweat is wicked through to the outer layer of clothing, this efficiency (η) becomes,

$$\eta = \frac{1}{1 + 0.155 \, hI_{cl}},$$ (7.15)

The efficiency term means that if the total evaporation rate is E_{sw} then only ηE_{sw} is removed from the body. Evaporation does not always occur at the clothing surface however and, in addition, condensation can supply heat to the clothing.

Ventilation in clothing

An extension of the two-parameter model is to include clothing ventilation. As well as dry heat and water vapour transfer through clothing depending upon the clothing resistance (I_{cl}, I_{ecl}), water vapour and heat can be lost through gaps in the clothing as well as direct penetration of air through clothing. In a cool climate therefore, clothing insulation may be much less effective at preserving body heat than would be expected if there was an assumption of no clothing ventilation. In a hot climate, evaporative heat loss may be greater than expected. Methods of 'accounting' for the effects of clothing ventilation have ranged from general guidance such as: for active people clothing insulation can be reduced by 50 per cent in the cold (ISO TR 11079, 1993) to empirical equations and measurement methods. Crockford *et al.* (1972) proposed a tracer gas method for determining the ventilation

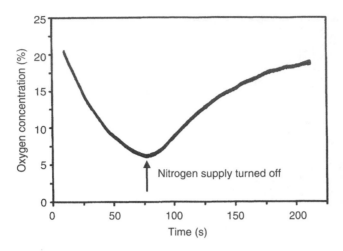

Figure 7.5 Typical pattern traced from oxygen analyser.

properties of clothing. The technique is derived from that used to investigate the ventilation of rooms in occupational hygiene and air quality studies. The principle is that if you introduce a tracer gas into an environment, the rate at which the environment returns to 'normal' (e.g. decay in tracer gas) or a measurement of how much gas is required to maintain a level, is related to the ventilation of the room. From the volume of the room an absolute value of air (and hence heat) exchange can be determined. Crockford *et al.* (1972) used a tracer gas technique (first carbon dioxide then nitrogen with detection of return of oxygen to 'normal' levels) by introducing the gas through pipes beneath the clothing on active human subjects and monitoring the return of oxygen concentration (e.g. from 20 per cent to less than 10 per cent and back to 18 per cent when the tracer gas was switched off). They used this to investigate the ventilation properties of weatherproof clothing in fishermen. The fishermen's smock was found to be an excellent example of how clothing can be adapted to adjust heat flow and ventilation was an integral part of that system. However if the fishermen fell into the sea it restricted chances of survival. Crockford and Rosenbloom (1974) extended the method to include a measurement of clothing volume and this was further developed by Sullivan *et al.* (1987). This technique involves a completely sealed (difficult to obtain in practice) oversuit where the air is extracted until it touches the outer surface of the clothing and then further until all air is extracted (determined by pressure). The ventilation rate (min^{-1}) can be determined from a curve fitting technique involving the exponential recovery (see Figure 7.5) of oxygen level (Crockford *et al.*, 1972; Birnbaum and Crockford, 1978; Bouskill, 1999). The absolute value (litre/min) is the ventilation index (VI) where

$$VI = \text{volume} \times \text{ventilation rate}$$
$$\text{litre/min}^{-1} = \text{litre} \times \text{min}^{-1}.$$

(Birnbaum and Crockford, 1978). There are many practical issues such as gas distribution through clothing, equipment calibration, non-leaking suits for measuring volume, effective sampling and so on. Angel (1995) further developed the method and Bouskill *et al.* (1998a) and Bouskill (1999) continued with its development and provided ventilation index values for a range of clothing (see Table 7.5).

Lotens and Havenith (1988), Havenith *et al.* (1990) and Lotens (1993) further developed the tracer gas technique into a system that involved continuous flow of tracer gas (Argon, N_2O) and mass spectrometer detection techniques. The improved gas distribution and detection provided a method that could assess clothing in around 3 min and did not require measures of volume. This compares favourably to the more cumbersome methods of Crockford *et al.* (1972). Further developments have included the work of Reischl *et al.* (1987) which involved locally administered tracer gas and specially constructed sensors. Bouskill *et al.* (2002) and Havenith and Zhang (2002) have developed further the effects of wind, human movement, and

Table 7.5 Mean ventilation index values for a range of clothing ensembles and conditions

Study	Clothing	Conditions	Ventilation Index VI (1 min^{-1})
1	Foul weather suit	Slow pacing, still air	45.4
		Jog-trotting, still air	99.2
		Slow pacing, wind @ 10 knots	108.7
		Jog-trotting, wind @ 10 knots	113.8
2	T shirt and work shirt	Standing manikin	26.1
		Standing person	18.6
		Walking person, 1.5 mph	68.8
		Walking person, 2.5 mph	127.3
3	Work pants, polo shirt and sweater	Sitting, $v = 0.7\,\text{m s}^{-1}$	97.1 ± 44.1
		Standing, $v = 0.7\,\text{m s}^{-1}$	105.1 ± 36.0
		Walking, $0.3\,\text{m s}^{-1}$, still air	101.3 ± 12.4

Source: A review by Bouskill (1999).

Study 1 Birnbaum and Crockford (1978).
2 Reischel *et al.* (1987).
3 Havenith *et al.* (1990).

multiple clothing layers. A report of the meeting of the Clothing Science Group concerning ventilation in clothing is provided by Lumley *et al.* (1991). Ventilation properties of clothing are provided in Table 7.5.

Clothing ventilation – empirical methods

There have been a number of studies of the ventilation properties of clothing based upon comparisons between clothing properties worn on stationary manikins (and humans), and the clothing properties provided to manikins (humans) when walking and in wind. McCullough and Hong (1992) investigate the changes in insulation values of typical indoor clothing between a stationary and walking manikin (no wind). They relate changes in insulation value to a number of properties of clothing and provide the following equation for absolute change in intrinsic clothing insulation, I_{cl}.

$$I_{cl} = (0.504 \times \text{ICLSTAND}) + (0.00281 \times \text{WALKSPEED}) - 0.240,$$
(7.16)

where

> ICLSTAND = Intrinsic insulation measured on a standing manikin $\text{m}^2 \,^{\circ}\text{C}\,\text{W}^{-1}$
> WALKSPEED = Steps per minute (range 30–90 steps/min 1.23–3.70 Km/h).

Parsons *et al.* (1998, 1999), Havenith *et al.* (1998, 1999, 2000) and Holmér *et al.* (1998, 1999) provide the following equation based upon a programme of research to develop the Predicted Heat Strain (PHS) model (Malchaire *et al.*, 2001 – see Chapters 10 and 14).

Dynamic insulation of clothing is determined by correcting the total (including air layer) clothing insulation using empirical equations:

$$I_{\text{tot st}} = I_{\text{cl st}} + \frac{I_{\text{a st}}}{f_{cl}},$$
(7.17)

where the increase in surface area due to clothing:

$$f_{cl} = 1 + 1.97\, I_{\text{cl st}} \quad \left(I_{\text{cl st}} \text{ in } \text{m}^2\,\text{K}\,\text{W}^{-1}\right)$$
(7.18)

$$I_{\text{tot dyn}} = C_{\text{orr,tot}} \times I_{\text{tot st}}$$
$$I_{\text{a dyn}} = C_{\text{orr},I_a} \times I_{\text{a st}}$$
$$I_{\text{cl dyn}} = I_{\text{tot dyn}} - \frac{I_{\text{a dyn}}}{f_{cl}},$$
(7.19)

where

$$C_{orr,tot} = C_{orr,cl} = e^{\left(0.043 - 0.398\,\text{Var} + 0.066\,\text{Var}^2 - 0.378\,\text{Walksp} + 0.094\,\text{Walksp}^2\right)}$$

$$(7.20)$$

for $I_{cl} \geq 0.6$ Clo.

For nude person or adjacent air layer:

$$C_{orr,tot} = C_{orr}, I_a = e^{\left(-0.472\,\text{Var} + 0.047\,\text{Var}^2 - 0.342\,\text{Walksp} + 0.117\,\text{Walksp}^2\right)}$$

$$(7.21)$$

for 0 Clo $\leq I_{cl} \leq 0.6$ Clo

$$C_{orr,tot} = (0.6 - I_{cl})C_{orr}, Ia + I_{cl} \times C_{orr,cl}. \tag{7.22}$$

With Var limited to $3\,\text{m s}^{-1}$ and Walksp limited to $1.5\,\text{m s}^{-1}$.
When walking speed is undefined or the person is stationary:

$$\text{Walksp} = 0.0052(m - 58) \text{ with Walksp} \leq 0.7\,\text{m s}^{-1}. \tag{7.23}$$

The evaporative resistance of clothing is derived using the clothing permeability index i_m where i_{mst} is i_m for static conditions and i_{mdyn} is i_{mst} corrected for the influence of air and body movement.

$$i_{mdyn} = i_{mst} \times C_{orr,E},$$

where $C_{orr,E} = 2.6 \times C_{orr,tot}{}^2 - 6.5 \times C_{orr,tot} + 4.9.$ (7.24)

If $i_{mdyn} > 0.9$ then $i_{mdyn} = 0.9$.
Dynamic evaporative resistance:

$$R_{tdyn} = I_{tot\ dyn}/i_{mdyn}/16.7. \tag{7.25}$$

Required sweat rate:

$$SW_{req} = \frac{E_{req}}{r_{req}}$$

and required skin wettedness:

$$w_{req} = \frac{E_{eq}}{E_{max}},$$
$$\text{where } E_{max} = \frac{(P_{sk,s} - P_a)}{R_{tdyn}}.$$

$$(7.26)$$

More complex clothing models

Factors that have not been included in the models described above can have significant effects on the thermal properties of clothing. Kerslake (1972) considered some of the practical effects on the thermal properties. He noted that insulation is provided by fabrics themselves and the layers of air trapped between the skin and clothing and the clothing layers. The insulation of fabrics is mainly due to the air trapped in and between them. On a clothed person heat will be exchanged by penetration of air (or expulsion due to human movement) through vents and openings and directly through layers of the fabric. A simple model of this is provided in Figure 7.6.

The interactions between wind penetration, pumping, clothing ventilation, and thermal insulation of fabrics are not easily modelled and depend greatly on clothing design, thermal condition of the body and a person's activity. Lotens (1988, 1990) describes a four-layer model of clothing consisting of underclothing, trapped air, outer clothing and adjacent air layer, and including ventilation through apertures. Lotens states that the model was evaluated for the effects of (transient) moisture absorption, condensation, semipermeability, heat radiation and ventilation. The model is linked with a model of the human body and thermoregulation to provide a 'whole-body' model of the physiological response of clothed humans (Lotens, 1993). A similar approach has been adopted by Thellier *et al.* (1992). A model of human thermoregulation with clothing has been incorporated into a computer system for environmental design in buildings.

More complex models of clothing therefore can have a number of characteristics in addition to the simple two-parameter approach. No fully comprehensive model exists and the possibility for this is restricted. As inputs to such a model, a detailed description of the human skin

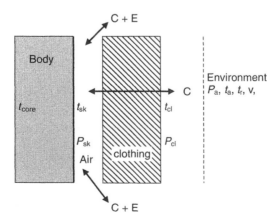

Figure 7.6 Air trapped between clothing and skin.

(e.g. involving a thermal model of the human body), the nature and properties of the clothing, of the environment and the dynamic behaviour of the person would all be required. Even for relatively 'steady state' conditions knowledge does not exist for a fully comprehensive model. For transient conditions where less is known about the effects of clothing, and where thermal inertia (mass) of the clothing will be an important consideration, few models are available (e.g. Jones *et al.*, 1990, 1994). Linking transient and steady state responses in a truly comprehensive dynamic model of clothing (interfaced with body and environment) has not yet been achieved. Empirical studies of clothing for example, involving human subjects, will always improve upon, and complement, information about clothing behaviour obtained from simple models. This raises the important practical question of how one would design and assess clothing for a specific practical application. An important point has been that the practitioner need not be interested in I_{cl} and I_{ecl} values, detailed equations and so on. The point is, does the clothing 'work' in practice?

DETERMINATION OF THE THERMAL PROPERTIES OF CLOTHING

Tests are available for determining specific properties of clothing, using heated flat plates or cylinders for example to determine I_{cl}, I_{ecl} values (e.g. Umbach, 1984; McCullough, 1990; ISO 11092, 1993). A comprehensive program to design and evaluate clothing will involve a number of tests and trials. Laboratories specializing in clothing science have developed integrated methods for assessing clothing.

A number of authors have proposed the 'clothing triangle' as a method for developing and evaluating clothing (Figure 7.7). The wide base of the triangle represents the wide range of simple tests which are performed on fabrics using simple heat transfer apparatus, i.e. they are easy to conduct and repeatable but unrealistic as they do not use human subjects. The narrow peak of the triangle represents field evaluation trials using humans wearing the clothing. The methods require relatively large resources and are difficult to control, but they are realistic. Goldman (1988) presents five levels of clothing evaluation. Apparatus methods include level 1, the physical analysis of materials and level 2, the biophysical analysis of clothing ensembles (e.g. using manikins) and predictive modelling. Levels 3, 4 and 5 involve human subjects and involve controlled climatic chamber tests, controlled field trials, and field evaluations respectively. As the costs of the tests increase with increasing level, it is important to use information from lower levels in the planning of tests at a higher level.

Umbach (1988) summarizes the work of the Hohenstein Institute in Germany in a five-level system shown in Figure 7.7. Some of the methods are described below.

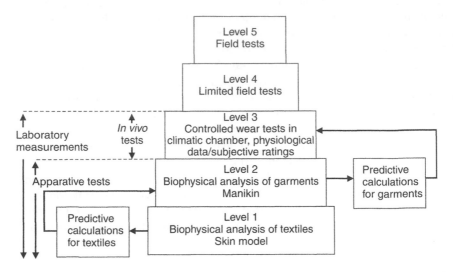

Figure 7.7 A five-level system for the analysis of the physiological properties of textiles and garments.

Source: Umbach (1988).

Level 1: Biophysical analysis – textiles

Level 1 involves testing the thermal insulation (R_{ct}) and water vapour resistance ($R_{e,t}$) of the fabric layers for both steady-state and transient conditions. The amount of water vapour absorbed by a sample of material during an exposure time of one hour on the skin model (standardized wet flat plate – Figure 7.8) is given the symbol F_i (per cent). Values obtained depend on fabric thickness. To allow comparison between fabrics an i_m value is used which compares the ratio of thermal with water vapour resistance of a fabric to the ratio found with an air layer of the same thickness as the fabric, i.e.

$$i_{mt} = 0.6 \frac{R_{ct}}{R_{e,t}},$$

where the coefficient 0.6 stems from the ratio of thermal to water vapour resistance of an air layer with the same thickness as the fabric. This makes i_{mt} a dimensionless number from 0 (impermeable) to 1.

Umbach (1988) states that F_i and i_{mt} can be used to judge a fabric's physiological quality (high i_{mt} and F_i preferable). Using essentially the same tests, values for a fabric can be determined for different conditions of interest, for example, to determine the effect of fabric compression or when the fabrics become wet. Dynamic changes with pressure and time can then

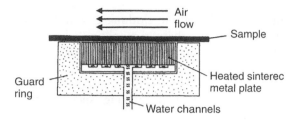

Figure 7.8 The principle of the skin model.
Source: Umbach (1988).

be measured; these will be important for example in the avoidance of post-exercise chill of subjects. For the assessment of a fabric's buffering capacity against water (simulating the effects of a moving, sweating body) the dynamic (transient) characteristics can be recorded using a modified form of the apparatus, shown in Figure 7.9. The moisture regulation index (K_d), temperature regulation index (β_T), buffering index (K_f), moisture transport index (F_i), and moisture regain (Δ_G) can all be determined from the time history of values measured during the test.

Further tests determine the fabrics buffering capacity against liquid sweat (as opposed to water vapour) providing the K_f value. In practical use 'normal' and transient heat conditions occur in combination. Both properties are therefore important. Empirical studies, under typical conditions (for underwear), provided a predicting equation for a comfort rating W_{CT} on the scale:

1 very good
2 good
3 satisfactory
4 sufficient
5 defective, and
6 unsatisfactory.

$$W_{CT} = -5.64 i_{mt} - 0.375 F_i - 1.587 K_d - 4.512 \beta_t - 4.532 K_f + 11.553.$$

Umbach (1988) presents data which suggest that the W_{CT} value predicted for a fabric is in very good agreement with comfort sensation ratings when the fabric is made into clothing and worn by human subjects. If this is the case then great expense could be saved in conducting user trials. The formulation of the equation suggests however, that this is fabric specific and will apply only over a limited range of practical conditions. It may however be useful in a preliminary selection of textiles that could be suitable for an application and warrant investigation at a higher level in the system (Figure 7.7).

Level 2: Manikin tests

Level 2 involves an investigation of the whole clothing ensemble (underwear, outer fabrics, air layers, etc.) using a thermal manikin ('Charlie' – see Figure 15.3). Three thermal resistance values R_c are determined:

1   with the manikin standing still;
2   for the manikin moving with a defined walking speed but with openings in clothing sealed with tape; and
3   for the manikin walking with openings not sealed.

From these values, both thermal resistance and ventilation rate of clothing can be found. The water vapour resistance, R_e, cannot be found directly as the manikin cannot sweat. This is derived from a model based on values derived in level 1. The effect of wind, compressing a garment, is determined by blowing air at the clothed manikin. Umbach (1988) found that this effect could reduce thermal insulation by up to 50 per cent mainly due to compression of air layers within clothing. This is also similar to the figure suggested by Kerslake (1972).

Thermal manikins are now widely and routinely used in the testing, certification, and development of clothing ensembles and garments. Holmér (1999) reviews the use of manikins in research and standards. Thermal manikins provide a realistic simulation of whole-body and local heat exchange, allow measurement of 3D heat exchange, integrate dry heat losses in a realistic manner and they provide a quick accurate and repeatable objective method for measurement of clothing thermal insulation. Holmér (1999) also cites International and European Standards that describe test methods that use thermal manikins. These include heated body parts (EN 345 – Safety boots, EN 397 – Safety helmets, EN 511 – Protective gloves against cold) to whole-body manikins (ASTM F1291 – Standard method for measuring the thermal insulation of clothing using a heated thermal manikin, EN 342 – Protective clothing against cold). Manikins were traditionally made of copper, but more recently aluminium has been used (e.g. to test survival suits (see Smallhorn, 1988)). Madsen (1999) describes a female manikin for thermal comfort and air quality assessment as well as clothing insulation (see Figure 7.9). Meinander (1999) describes the sweating thermal manikin (Coppelius) which can be used to measure heat and water vapour transmission through clothing systems. Further consideration of thermal manikins is provided in Chapter 15 – Thermal models and computer aided design and in Nilsson and Holmér (2000).

Higher level tests

Mecheels and Umbach (1977) provide an empirically derived model for values obtained at level 1 and 2 of the system (see Figure 7.7) with those

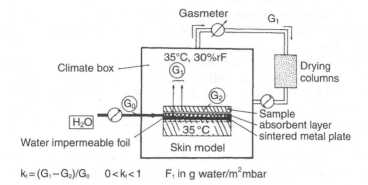

$k_f = (G_1 - G_2)/G_0 \quad 0 < k_i < 1 \quad F_1$ in g water/m^{2}mbar

Figure 7.9 The measurement of a fabric's buffering capacity against liquid sweat. Source: Umbach (1988).

likely to be obtained at level 3 which involves controlled tests in a climatic chamber. In the climatic chamber trials, subjects perform a defined programme of physical work (walking or cycling) under fixed climate conditions. Rectal temperature, skin temperature, heart rate, sweat production, metabolic rate, and temperature and humidity (next to the person's skin) are measured. Subjects' comfort sensation ratings are also taken. It was found that physical measures could be used to predict a subject's sensation. A model that predicts physiological responses of the clothed body can then be used to predict conditions for comfort. Umbach (1988) provides a psychrometric chart on which can be plotted a comfort range (range of utility) in terms of air temperature and humidity. Predictive formulae for the wear comfort of clothing ensembles are given as follows:

1   For warm conditions, i.e. heat load

$$W_C = 1.52t_{re} + 0.3t_s + 3.1K_f + 0.02HR - 67.85.$$

2   For cold conditions, i.e. heat loss

$$W_C = -5.61\Delta t_{re} + 0.6\Delta t_s + 1.51,$$

where

W_c = subjective comfort vote from 1 (very good) to 6 (completely unsatisfactory)
t_{re} = rectal temperature
t_s = mean skin temperature
K_f = skin wettedness (comfort factor)
HR = heart rate
Δt_{re} = decrease in rectal temperature
Δt_s = decrease in mean skin temperature

All of the above can be calculated from the predictive model. The work presented also provides an example of an integrated and comprehensive approach to determining the thermal properties of clothing that is used in practical application.

The use of multiple linear regression equations cannot provide a rational underlying causal model of wearer comfort, but they can be useful over the range of conditions for which they were determined. In addition although the work is comprehensive it has not been validated by other laboratories and may well apply to the specific equipment and laboratory used. It will be apparent from previous discussions that the methods described above extend beyond the simple two-parameter model of clothing. However, it would be useful if the values representing the properties of clothing (R_{ct} and R_{et}) were related to intrinsic insulation (I_{cl} and I_{ecl}) values taking account of the air layer around clothing (I_a and I_{ea}).

User tests and trials

User performance tests and trials of clothing can be relatively expensive to conduct, however they provide realism. A significant deficiency of any system not involving human subjects would be in the interaction of clothing insulation with human activity. People do not only walk, stand or lie down; they bend, run, and sit at different rates and angles. There are numerous factors where humans differ from manikins, so using human subjects to evaluate clothing will reduce control but it is the only way to provide a realistic and comprehensive evaluation.

The degree of control varies according to the conditions: in climatic chambers human subjects can be used 'as manikins' to provide measurements under controlled conditions; and user trials can involve human subjects wearing the clothing from day-to-day under normal operating conditions.

Performance tests

An example of user performance tests to determine the thermal properties of clothing is provided by Parsons (1988). An initial test involves eight male subjects standing stationary in a climatic chamber at air and mean radiant temperature of 5 °C, 50 per cent *rh* and still air. Two sessions are required (same time of day on separate days): one where the subject is clothed, the other minimally clothed (shorts and shoes only). The difference, in mean, mean skin temperature (over subjects), after one hour of exposure, between the two conditions is given a clothing index value $d_{t_{sk}}$. Using a model of human thermoregulation (Nishi and Gagge, 1977) an I_{cl} value can be calculated, i.e. one which would give a similar fall in mean skin temperature over time to that of the mean fall for human subjects. This method was extended (Parsons, 1991) to include the effects of pumping, whereby subjects were exposed to a further one hour session in which they moved their arms and legs alternatively and slowly in wide circles, to provide maximum pumping. The resultant insulation (in Clo) of the clothing can then be determined from:

$$I_{res} = I_{cl} - \frac{A\Delta I_{cl}(33 - t_a)}{28} \quad \text{for} \quad t_a > 5\,^\circ\text{C}, \tag{7.27}$$

where

I_{cl} = intrinsic clothing insulation
ΔI_{cl} = maximum fall in clothing insulation due to pumping at $5\,^\circ\text{C}$
t_a = air temperature
A = activity factor, 0 (for standing subject) to 1 (for maximum pumping).

The method appears to work well, but it is possible that results are confounded by some (thin) subjects who may shiver at $5\,^\circ\text{C}$ and it is important to provide good control over the environment. An advantage of this and similar methods is that the data are realistic – it clearly demonstrates wide individual differences and can quantify these. Further development is required; however, it is possible that such methods may provide preferred alternatives to the use of manikins. Hollies (1971) described standardized tests of the assessment of clothing, in which human (female) subjects provide subjective judgements of clothing while in a thermal chamber which provides transient changes in air temperature and humidity. The system has been used for laboratory testing of consumer clothing. User performance tests vary in level of control and realism; however none consider clothing under actual use.

The classic measures taken in user evaluations of clothing involve physiological, subjective and physical measures. These are provided in Table 7.6.

A number of user performance tests have been used to determine the thermal properties of clothing by fixing environmental conditions and at known physiological states (e.g. onset of sweating, heat balance). For example studies of evaporative heat loss in human subjects can be carried out in thermal chambers with air and wall (radiant) temperatures held at $36\,^\circ\text{C}$ (approximate skin temperature). The radiative and convective heat loss from the body can be assumed to be zero (no temperature difference to drive heat transfer) and the heat balance equation used to study evaporation of sweat. Kenney *et al.* (1988, 1993) describe a method for measuring total clothing insulation (I_T) and resistance to water vapour permeation (R_e) on exercising clothed subjects. The upper limit of the prescriptive zone (ULPZ, Lind (1963) – see Chapter 10) is identified as the point where heat gain is exactly matched by heat loss and above that point rectal temperature begins

Table 7.6 Measurements useful in the user assessment and evaluation of clothing

Measurement	Examples of method	Possible interpretation
Internal body (core) temperature	Rectal, aural, oral, urine, etc.	Indication of body heat stored or lost. Preferred indicator for hot or cold environments. Can the clothing maintain heat balance?
Mean skin temperature	Over the body. Weighted method More points needed in cold	Related to cold discomfort. Can link with core temperature to indicate heat storage or loss
Extremity skin temperature	Hand, big toe, ear, face	Safety limits especially in cold. Effects on manual performance
Weight of subject and clothing	Balance with accuracy within 50 g and prefer 10 g. Account for any intake of food and drink and excretions. Measure immediately before and after exposure	Weigh subject before nude (W_{bn}), before clothed (W_{bc}), after clothed (W_{ac}), and after nude (W_{an}). Then $W_{bc} - W_{bn}$: weight of clothing can be related to I_{cl}. $T = W_{bn} - W_{an}$: weight loss from body (sweat + respiration) Relate to dehydration and heat strain. $e = W_{ac} - W_{bc}$: relate to amount evaporated (cooling to avoid heat strain) ($W_{ac} - W_{an}$) − ($W_{bc} - W_{bn}$): moisture in clothing. Related to comfort and heat stress when sweating e/T: index of clothing efficiency during sweating
Subjective scales	Many scales and methods. Questionnaires. Votes during exposure. Thermal sensation, stickiness comfort, satisfaction over body areas (feet, hands) useful in the cold	Most effective methods for thermal comfort and satisfaction. Use with other methods (to complement objective and behavioural methods). Not for extreme environments. Simple but important principles of design, presentation and analysis. Careful control over method of administration required

to rise (also indicated by a rise in heart rate which just precedes the inflection point in the rectal temperature curve). Subjects walk continuously on a motor-driven treadmill (at 30 per cent, $VO_{2 \text{ max}}$) for up to 2.5 h. Air velocity is 0.3 m s^{-1}. Thirty per cent of maximum capacity is typically given by industry as the highest work rate a worker can sustain over an 8 h working day. Each subject undergoes two separate tests in the same clothing (randomized/balanced order over subjects).

Test 1 T_{db} is held constant and after 30 min, Pa is increased incrementally in steps of 1 torr every 5 min. This test determines the 'critical water vapour pressure' or P_{crit} (ULPZ when rectal temperature just begins to show disproportionate rise) at the fixed T_{db}.

Test 2   Pa is held constant at low relative humidity and after 30 min, T_{db} is systematically raised 1 °C every 5 min providing 'critical air temperature' or T_{crit} (ULPZ when rectal temperature just begins to show disproportionate rise).

At the critical points, determined independently from different conditions but wearing identical clothing, heat gain equals heat loss with no heat storage. So, from the heat balance equation (see Chapter 1)

$$(M - W) - (C_{res} + E_{res}) - (R + C) - E = 0.$$

At the critical point for Test 1:

$$(M_1 - W_1) - (C_{res_1} + E_{res_1}) - (R_1 + C_1) - E_1 = 0.$$

At the critical point for Test 2:

$$(M_2 - W_2) - (C_{res_2} + E_{res_2}) - (R_2 + C_2) - E_2 = 0,$$

where

$$(R_i + C_i) = (T_{db_i} - T_{sk_i})/I_T \text{ W m}^{-2}$$
$$E_i = (P_{s,sk_i} - T_{dp_i})/R_e \text{ W m}^{-2} \quad i = 1 \text{ or } 2.$$

Hence two simultaneous equations with two unknowns, I_T and R_e. Values for the heat balance equation are calculated for each test with the following equations:

$$M = 352(0.23 \, RQ + 0.77)\frac{VO_2}{A_D} \text{ W m}^{-2}$$
$$W = 0.163 \, M_b \times V_w \times \frac{F_g}{A_D} \quad \text{W m}^{-2}$$
$$C_{rcs} = 0.0012 \, M \, (34 - T_{db}) \quad \text{W m}^{-2}$$
$$E_{res} = 0.0173 \, M \, (5.87 - P_{dp}) \quad \text{W m}^{-2}$$

where

RQ respiratory quotient (ND)
VO_2 oxygen uptake (litre min^{-1})
A_D Dubois surface area (m^2)
M_b body mass (kg)
V_w walking velocity (m min^{-1})
F_g fractional grade of the treadmill
T_{db} dry bulb temperature at critical point (T_{crit} in Test 2)
P_{dp} dew point pressure at critical point (P_{crit} in Test 1).

Kenney *et al.* (1993) suggest that when the vapour permeability of the garments is low, heat storage should be accounted for in the heat balance equation. They also note that the largest source of experimental error is measurement of metabolic rate.

Performance trials

The principle of the user performance trial is to investigate clothing while in actual use, hence providing 'practical' information. This will involve identifying a sample of users of the clothing and 'observing' the properties over a period of time representing realistic conditions. Questionnaire techniques and possibly some physiological measures can be used.

For example, in a final evaluation of a clothing ensemble, the clothing will be given to a sample (or all) of users for a period of time (e.g. days, weeks or months). The users may be observed at frequent intervals, interviewed or asked to complete questionnaires. If experimenter presence is not possible or desirable, then the subjects may be asked to complete a diary of their experiences. A more convenient method may be to ask the subjects to complete a questionnaire at the end of the trial period.

Examples of user trials would include giving a sample of consumers a new type of shirt, or sports clothing. Trials are also conducted for military use e.g. tests of new sleeping bags or protective clothing are performed in both laboratory and field trials.

The measurement techniques used will depend upon the clothing application. An evaluation of boots for example may include a measure of the frequency and severity of blisters as well as questions concerning sweating or cold feet. A comparison of different clothing ensembles is often required and the measures in Table 7.6 provide useful techniques.

A user trial, although confounded with many factors that cannot be controlled, should be conducted using correct technique of experimental design in order to maximize useful information obtained. A discussion of field evaluation methods for military clothing is provided by Behmann (1988). It becomes apparent when considering field trials for the thermal properties of clothing that any evaluation will be context dependent. Isolating only thermal properties will be difficult and they will interact with many other factors, for example ventilation properties, bulk and fit (allowing ease of movement). When considering the thermal properties of clothing in user trials therefore it will be necessary to consider the wider context.

The design of functional clothing

The development of the 'ergonomics systems' methodology (e.g. Singleton, 1974) led to a change of philosophy in designing and evaluating 'components' of systems. Clothing can be seen therefore in terms of a dynamic component of an overall man-machine-environment-organization system. Despite the general acceptance of this philosophy, clothing design and assessment has been slow to use it in practice. For a more complete description of systems ergonomics and practical ergonomics methodology the reader is referred to Singleton (1974) and Wilson and Corlett (1995) but a brief description of how it might relate to the thermal properties of clothing is provided in Figure 7.10.

Rosenblad-Wallin and Kärholm (1987) describe a technique called 'environmental mapping' for the design and development of functional clothing; that is matching the clothes to the requirements of the job, including thermal requirements for the clothing. They use task and activity analysis techniques involving interviews and 'on the job' observation to identify the requirements for the clothing. A rating of the clothing requirement is then obtained in terms of level of activity (related to metabolic rate), thermal environment (climate) and a pollution factor. The following scales are used:

Climate	Pollution	Activity
1 = very warm	1 = low degree of pollution	1 = low activity
2 = warm indoor	2 = moderate degree of pollution	2 = moderate activity
3 = normal indoor	3 = high degree of pollution	3 = high activity
4 = cool climate		
5 = indoor–outdoor changes		
6 = cold/outdoor climate		

The principle is then that the three ratings provide information to the designer; the clothing is then produced and evaluated. In practice, the three ratings provide simple guidance and much more information and expertize is required to produce the final design. The overall method therefore needs

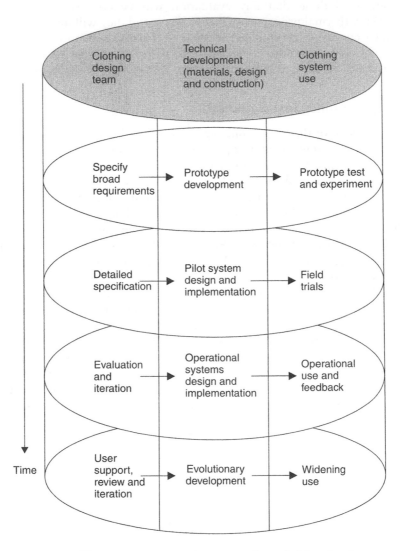

Figure 7.10 The systems ergonomics approach to clothing development.

much development. However, this 'user-orientated product development' approach has been used successfully in the design of clothing for fishermen, foundry workers, nurses, agricultural workers, for military boots and in other applications.

Ilmarinen *et al.* (1990) used a simple systems approach to the design, development and evaluation of 'meat cutters' clothing. An interesting point is that a team of investigators is required for this type of study involving physiologists, designers and ergonomists. The tasks and problems of the

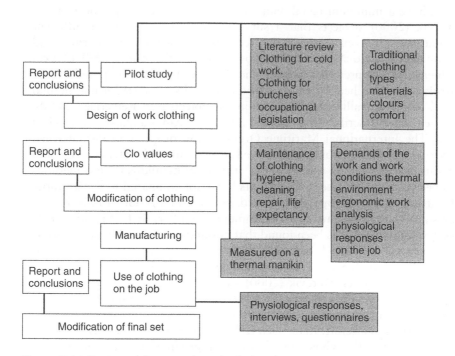

Figure 7.11 Design of functional work clothes for meat cutters.
Source: Ilmarinen *et al.* (1990).

meat cutter were studied in depth and clothing requirements identified. This involved literature reviews, questionnaires and interviews, work analysis and physiological measurements. Particular attention was paid to the activity of the meat cutters and the thermal and general working conditions in slaughterhouses. The overall systems approach to the design, development, and implementation of the clothing is shown in Figure 7.11, but even with this relatively comprehensive study, only the simple model of dry thermal insulation (I_{cl}) was used to specify clothing requirements for thermal environments.

More complex thermal models and their component values (I_{cl}, I_{ecl}, etc.) and general methodologies, provide further practical information. However there is some way to go before all 'ingredients' are linked to provide a comprehensive thermal design and evaluation methodology for clothing. Cost and convenience will always play a role; however, the benefits of optimum design of clothing should not be underestimated.

Survival suits

Immersion in cold water can greatly increase heat loss from a person to the environment and can rapidly lead to hypothermia and death. Survival suits are often required to maintain a person in acceptable thermal condition (e.g. above a minimum rectal temperature) for a specified period of time. Fœrevik (2000) suggests that a survival suit, 'should provide sufficient insulation to prevent the lethal effects of accidental hypothermia and attenuate the cold-shock response by reducing the rate of fall of skin temperature and prevent leakage'. The thermal insulation required can be calculated from a heat balance equation and suits are tested using thermal manikins (e.g. Smallhorn, 1988) or human subjects (e.g. Reinertsen, 2000). Påsche (2000) describes International and European Standards for survival suits. The International Maritime Organization propose an insulated 6 h suit (so named because it must be tested over 6 h in cold water (<2 °C)) and an uninsulated suit (1 h suit). The use of thermal manikins for test methods has been slow to be accepted as results have not compared well with those of human subjects. Proposed standards include Pr EN ISO 15027 Part 1: Constant wear suits; Part 2: Abandonment suits; Part 3: Test requirements and test methods. It has been found that cooling of 'victims' bodies in some accidents has been much quicker than expected from tests due to low air temperatures, high wind speeds, waves and water spray which are not included in the tests. Fœrevik (2000) suggests that the design of survival suits is a systems issue and should consider operational requirements. Pilots, for example, may have to continually wear survival suits and may suffer hyperthermia in aircraft. Reinertsen (2000) suggests that not only considering insulation of the whole-body but also improving insulation in selected areas of the body would increase chance of survival.

An interesting issue is whether the survival suit should allow the wearer to generate metabolic heat by movement. In cold water it is generally considered that heat loss due to movement (destruction of the boundary layer) is greater than metabolic heat gain. If survival bags provided significant insulation then additional metabolic heat may enhance survival. However, this needs to be tested and a window of application to be established.

Active and smart clothing

Active clothing systems provide additional power for the body to maintain acceptable thermal state by enhancing cooling power (e.g. ice vests) when the body is too hot or enhancing heating power (e.g. heated gloves or vests) when the body is too cold. In some situations active clothing systems are essential for survival. For example, for workers in hot conditions where personal protective clothing and equipment is worn (e.g. to protect from fire or nuclear, biological or chemical hazards) and work must be carried out over a period of time. Evaporation of sweat through impermeable clothing may be impossible and the worker will not survive without assistance. This can be demonstrated in heat balance calculation and the amount of heat required to be removed by an active system can be calculated. Conversely in cold environments the amount of heat input (or reduced loss) required can be calculated. In any risk assessment, of prime interest will be whether the heat or cold stress can be reduced and whether the workers need to be exposed to the heat or cold. Hence consideration should be given to work organization and design. Where workers must be exposed, then active clothing systems can reduce the thermal strain.

Selection of the optimum active clothing system will depend upon the requirements of the task. There is little point in wearing an ice vest when running up a hill, as the metabolic cost of carrying the ice will outweigh the benefit of cooling. For a sedentary task in hot conditions an ice vest may be useful. There are principles that can be used to select active clothing. The thermal audit (heat balance analysis) should always be carried out to identify requirements. Active clothing systems should complement the body's natural system of thermoregulation. Wearing an ice jacket or a neck scarf through which cool water is pumped through tubes (e.g. Bouskill and Parsons, 1996) may remove significant heat. However, in hot conditions it will restrict the evaporation of sweat. Any system must therefore ensure that it removes (or preserves in the cold) at least as much heat as the device prevents removing and then any benefit is in addition to that. Cool vests add insulation and prevent evaporation of sweat, systems add weight and increase metabolic rate and so on. A system of cool compressed air flowing across sweating skin will remove sweat by evaporation with little restriction. Where systems can be placed near workers (so compressed air does not have to be carried) and suits are provided, an appropriate microclimate can be created where workers may work indefinitely. If the body is considered as a system consisting of a pump (heart) and pipes (blood vessels) most of which are insulated within the body, then any effective cooling system should be placed where heat exchange can take place. Where large blood vessels are near the skin surface (neck, wrists, under arms) good heat exchange is required so that liquid cooling systems with pipes containing fluid should run close to the skin (e.g. square pipes with flat surfaces). Care must be taken when enhancing heating or cooling to the body not to 'trick' the body into restricting thermoregulation. Heating limbs with electrically heated material may stimulate general vasodilation or even sweating when the body should be preserving heat. Cooling hands or the neck may cause vasoconstriction and cessation of sweating and so on.

There are three main types of microclimate cooling systems: ice vests; air-cooling; and liquid cooling. All have advantages and disadvantages and are chosen depending upon application. Air cooling systems that enhance evaporation of sweat complement the thermoregulation system. Ice vests can be worn for fixed periods where weight is not an issue. Ice vests have used frozen water or dry ice systems. Liquid cooling systems include whole body underwear that contain tubes through which cool water or other fluid is pumped. Such systems have been worn by pilots, space men and women and people who wear furry suits to entertain children in amusement parks.

Factors that must be considered are power supply requirements, weight, contact with the skin and comfort. In cold conditions thermal insulation is the main method of preserving heat. However, enhanced heat input can be provided by 'vests with heated water or other material', warm air pumped into clothing and warm water pumped through tubed underwear or through diving suits. Where hot or cold materials are used it is important not to promote local contact injuries. In some extreme conditions, for example in increased air pressure, the range of temperature for comfort and survival may be narrow and careful control is required. Breathing warm or cool air will have some effect on heat loss or heat gain, especially where pressurized. Again, careful analysis and design are required. Electrically heated clothing is available, particularly gloves and boots, but also liners in garments that can allow reduced clothing bulk to perform tasks and keep hands and feet warm. Chemical polymers can also supply the same effect, as can electrically heated polymers. Portable battery packs are now powerful and small and, coupled with careful design, may be able to provide significant systems for keeping people warm in the future.

Smart clothing is where sensors in the clothing detect the thermal state of the body and respond to preserve thermal comfort or acceptable thermal strain. Those involved in sports clothing and fashion, massive popular markets, take a keen interest in developments. Phase-change materials can be micro-encapsulated into materials for mass production. They change from liquid to solid or solid to liquid at selected temperatures (e.g. 33 °C). A vest made of such a material may 'detect' that the skin temperature has risen above 33 °C and the material in the capsules may change from solid to liquid, taking the latent heat of fusion from the skin and cooling it down. When the skin cools below 33 °C, the liquid changes to solid, releasing the heat. While the system 'works' the laws of thermodynamics apply and the heat requirements of the whole body may be an order of magnitude greater than what is supplied. There may however be application for local cooling or heating of hands and feet. Careful heat transfer analysis will be required.

Other smart clothing include the use of micro-pumps distributed throughout clothing and nano-technology. 'Memory materials', change shape at a specific temperature (which they 'remember'). For example, they may be flat below 36 °C and bulky and enlarged (enclosing air) above 36 °C. If encompassed into a garment liner, when the person becomes hot (skin temperature above 36 °C) a large air layer will be produced for the evaporation of sweat to cool the person. Biomimetics is the study of animals and plants to gain ideas from biological 'design'. The study of the thermal behaviour of penguin feathers, how polar bears survive the cold, and why fish do not freeze all provide ideas for new materials.

Technology has provided opportunities through active and smart clothing to extend the environmental conditions within which people would consider acceptable. Each new development should however be considered in terms of the principles of the human heat balance equation and thermal physiology.

CLOTHING SPECIFICATION AND THE WINDOW
OF APPLICATION

The analysis of the human thermal environment can provide a specification of clothing in terms of its thermal properties. How much thermal insulation (I_{cl}) is required to maintain thermal comfort and acceptable thermal strain can be calculated for the range of thermal conditions as can the vapour permeation and ventilation properties of clothing (see Chapter 1 for the heat balance equation and Chapter 15 for thermal models). Clothing can therefore be specified in terms of its wearer performance. If this were common practice then it would provide guidance on selection of appropriate clothing. Clothing is rarely specified in terms of its thermal performance however. Protective clothing and equipment do have specification requirements, however the performance is mainly in terms of fabric and material properties. Although these are important (for protection against hazards) they provide no guidance of the range of thermal conditions within which people can safely wear the clothing. They also do not provide detail of the contribution a garment or piece of equipment will make to thermal strain on the wearer of a full personal protective equipment (PPE) ensemble. Parsons (1995) and Bethea and Parsons (1998a) have described the concept of the 'window of application' for clothing. This is analogous to the tolerance envelope for aircraft; a multi-factor envelope of conditions within which the aircraft will fly and outside of it, which it will not. For clothing, the six basic parameters and exposure time can be combined to describe the range of conditions according to maximum acceptable strain (e.g. internal body temperature between 36 °C and 38.5 °C) or thermal comfort (based upon skin temperature and sweat rates). Such windows of applications can be derived from thermal models. They may also be confirmed by selected experiments with human subjects, where subjects wear whole ensembles (and active and smart clothing systems), completing tasks of interest and physiological responses are monitored.

8 Thermal comfort

INTRODUCTION

Thermal comfort is often defined as 'that condition of mind which expresses satisfaction with the thermal environment' (ASHRAE, 1966; ISO 7730, 1984). It is convenient to have such a generally agreed definition as it saves laborious and insoluble arguments over semantics and emphasizes that comfort is a psychological phenomenon, not directly related to physical environment or physiological state. An understanding of why a person reports thermal comfort (or discomfort) or related feelings of warmth, freshness, pleasure and so on, is complex and not known. That thermal environments affect such feelings is easily demonstrated as are the consequences of not achieving thermal comfort, when humans will complain, health and productivity can be affected, morale can fall, and workers may refuse to work in an environment. For this reason, for almost all of the twentieth century, and often before, there has been an active interest in research into the conditions that produce thermal comfort. This interest and associated debate has continued with vigour into the twenty-first century. Emphasis has not been on understanding why people report comfort or discomfort, but on what conditions will produce thermal comfort and acceptable thermal environments. That is for a group of people, what human thermal environment (in terms of the interaction of the six basic parameters) will produce comfort and what will be the effect, in terms of thermal comfort and satisfaction, of deviating from these conditions?

Thermal sensation and thermal comfort are bipolar phenomena; i.e. they range from uncomfortably cold to uncomfortably warm or hot with comfort or neutral sensations being somewhere around the middle of these. In steady state, thermal comfort can be regarded as a lack of discomfort as, although positive feelings such as thermal pleasure are experienced, they are transient in nature (e.g. experienced when a cold person moves to a 'warm' environment) and are not experienced in steady-state conditions. Acceptable conditions are therefore often described in terms of the average subjective expressions of a group of subjects on a scale of comfort or thermal sensation (see Table 3.2).

For example, an acceptable environment could be described as any environment where the average rating of a large group is between −1 (slightly cool) and +1 (slightly warm) on the scale (+3 hot; +2 warm; +1 slightly warm; 0 neutral; −1 slightly cool; −2 cool; −3 cold). An optimum environment would be for an average rating of 0 (neutral).

Although there have been many studies into thermal comfort over the twentieth and into the twenty-first century, a full historical review will not be provided here (see for example, Chrenko, 1974; McIntyre, 1980). Some important early studies will be reviewed, however, as fundamental methods and principles, and some results, are still applied today. Emphasis will be given to work conducted over the last 40 years where major developments have taken place. Discussion of the subject will be divided into whole-body comfort, which is concerned with overall feelings of comfort and discomfort and local thermal comfort which is concerned with comfort or discomfort to specific areas of the body (head, hands, feet, etc.).

WHOLE-BODY THERMAL COMFORT

Historical perspective

The provision of thermal comfort is fundamental to human existence and hence has always been a consideration even in the absence of controlled research and publication of results. People must always have adjusted their clothing, activity and regulated their position with regard to heat sources (artificial or natural) and sought or constructed shelter from heat and cold to meet with their requirements. The number of scientific studies of thermal comfort increased with the requirements of the industrial revolution. Chrenko (1974) provides a brief historical review of early British work and cites Michael Faraday (1835) providing evidence to a House of Commons committee on the inadequacy of using air temperature alone in determining optimum conditions. Other researchers also identify that it is a combination of environmental parameters that affect thermal comfort and have produced a number of instruments which would improve on using the 'normal' thermometer alone. Heberden (1826) observed the rate of cooling of a thermometer heated above 100 °F and Leslie (1804) proposed a large bulbed thermometer for measuring air velocity. Aitken (1887) used a black globe thermometer '... which would show the effects of radiation as tempered by the wind', and the General Board of Health Commissioners (1857) recommended that the walls of a room be at least as high in temperature as the general temperature of the room (Chrenko, 1974). It can be seen therefore that even as early as the nineteenth century the general principles behind the establishment of comfort conditions were known.

When considering thermal comfort studies world wide, it is important to understand the perspective of the study. In hot climates, consideration is generally given to how to cool indoor environments to provide thermal comfort conditions, for example, by the use of air-conditioning systems or increased air movement. In cooler climates, consideration is given to how to heat environments to provide comfort. In the USA much research has been co-ordinated by ASHVE (American Society of Heating and Ventilation Engineers – later ASHRAE (American Society of Heating, Refrigerating and Air-Conditioning Engineers)) into the relative importance of air temperature and humidity which is of great interest, especially in the hot, humid climates experienced in some parts of the USA. In Europe, humidity has not been considered in such detail nor air-conditioning requirements. In the UK consideration has been given to environmental warmth and freshness.

Much of the early work in the twentieth century has been to determine thermal comfort indices, i.e. to develop a system of integrating relevant (six basic) parameters to provide a single number that indicates thermal comfort. Sir Leonard Hill *et al.* (1916) developed a Kata thermometer, a heated alcohol in glass thermometer, where its rate of cooling was said to be related to the effects on the body. However, the instrument was found to be too sensitive to air velocity to be useful as a thermal comfort index (so much so that it is still used today as an anemometer – see Chapter 5).

Effective temperature

A comprehensive series of studies was conducted on behalf of ASHVE in their Pittsburgh laboratory, USA, which led to the influential Effective Temperature (ET) index (Houghton and Yagloglou, 1923, 1924; Yagloglou and Miller, 1925). Incredibly, the studies contained a fundamental experimental error. Three subjects were used and each walked between two chambers and compared different combinations of air temperature and humidity in terms of subjective impressions of warmth. However, subjects gave their immediate impressions that would be largely determined by the effects of transient absorption and evaporation of moisture from skin and clothing. The data may therefore be useful for studies of transient effects but will overestimate the effects of humidity when considering steady-state conditions, which was the aim of the study. The extensive number of comparative judgements made, allowed the relative importance to subjective warmth of air temperature (dry bulb), humidity (plotted as aspirated wet bulb temperature) and air velocity to be plotted on charts for determining the ET index.

The ET index uses the concept of the temperature of a standard environment as the index value (a technique widely used in later studies). ET is the temperature of a standard environment that contains still, saturated air that would provide the same sensation of warmth as in the actual environment. The person in the standard environment would have the same clothing and activity as in the actual environment. Two charts were produced: one

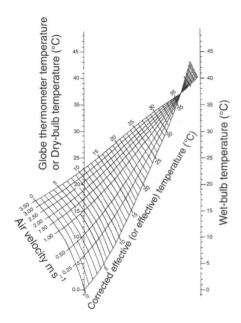

Figure 8.1 Chart showing basic scale of corrected effective (or effective) temperature.

for persons stripped to the waist (Basic Effective Temperature (BET) Figure 8.1) and another for persons 'normally clad' (Normal Effective Temperature (NET) Figure 8.2).

A number of methods of correction of the index were proposed to allow radiation to be taken into account. The generally accepted method was to use the 150 mm diameter globe temperature measurement on the scale in place of dry bulb temperature (Vernon and Warner, 1932) and the index is then termed Corrected Effective Temperature (CET). Both ET and CET have been influential and widely used throughout the world initially as a comfort index. They are still used but not generally recommended now, although they have been found useful for assessing hot environments.

Resultant temperature

Missenard (1935, 1948) in France proposed a number of methods for determining a thermal index that he called the resultant temperature. He was one of the many researchers who have used wet and dry globe thermometers of appropriate sizes to mimic the response of the human body. He later attempted to overcome the problems of the ET index by defining a transitory form and a steady state form of resultant temperature with supporting nomograms for calculation – see McIntyre (1980) and Missenard (1948, 1959) – and provided definitions in terms of standard environments.

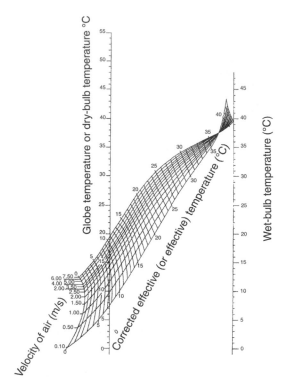

Figure 8.2 Chart showing normal scale of corrected effective (or effective) temperature.

The Chartered Institute of Building Services Engineers (CIBSE, 1986) in the UK provide recommendations for indoor environments which they refer to as resultant temperature; i.e. the temperature of a 100 mm diameter black globe. Further discussion on the use of globes to provide thermal indices is provided in Chapters 10 and 15.

Equivalent temperature

Dufton (1929, 1936) developed a heated black copper cylinder to mimic the thermal behaviour of the human body (eupatheoscope – see Chapter 15), and he termed its temperature 'equivalent temperature'. Bedford (1936) reported on a large study of (mainly female) factory workers '... to investigate the relation between the physical environment and personal feelings of warmth of people engaged in light industrial work'. Using his thermal comfort scale (Table 3.2) he correlated subjective judgements with a number of thermal index values (Table 8.1).

Table 8.1 Correlation between warmth votes of workers and values of environmental parameters and indices

Parameter or index	Correlation coefficient	Range	Standard deviation
Equivalent temperature			
(Mk 1 eupatheoscope) (°C)	0.52	8–24	2.4
150 mm globe temperature (°C)	0.51	12–24	2.2
Effective temperature (°C)	0.48	11–21	1.6
Air temperature (°C)	0.48	12–24	2.0
Mean radiant temperature (°C)	0.47	12–27	2.6
Kata cooling power (mcal cm^{-2} s^{-1})	−0.43	4–9	0.8
Air speed (m s^{-1})		0.05–0.5	0.07

Source: The survey of factory workers by Bedford (1936).

Using these data he also derived an equation for his equivalent temperature index:

$$t_{eq} = 0.522\,t_a + 0.478\,t_r - 0.21\sqrt{v}(37.8 - t_a), \tag{8.1}$$

where temperatures are in °C and air velocity in m s^{-1}. Further analysis
involving the four basic environmental parameters provides

$$S = 11.16 - 0.0556\,t_a - 0.0538\,t_w - 0.0372\,f + 0.00144\sqrt{v}(100 - t_a),$$

where, in the original units

S = value assigned to the sensation of warmth
    (comfortable = 4)
t_a = air temperature (°F)
t_w = mean temperature of the surroundings (walls, etc.) (°F)
f = partial pressure of water vapour in the air (mm Hg)
v = air velocity (Ft/min).

(Chrenko, 1974).

The term *equivalent temperature* was adapted for a thermal comfort
index for vehicles (Wyon *et al.*, 1985). This considers the distribution of
conditions across vehicle occupants and can involve heated, thermal manikins. An extensive programme of research was conducted to evaluate and
develop this index (Holmér *et al.*, 1999) and details are provided later in this
chapter.

Kansas laboratory trials

Nevins *et al.* (1966) and Rohles and Nevins (1971) report the history of the use and eventual replacement of ET by ASHVE/ASHRAE and on extensive climatic chamber trials conducted at Kansas State University in the 1960s to provide recommendations for comfort conditions for ASHRAE standards.

After the establishment of ET by Houghton and Yagloglou (1923), they determined a 'comfort zone' and comfort line in terms of ET values using 130 subjects under laboratory conditions. The comfort zone was defined as including those ET values over which 50 per cent of people are comfortable. On this basis, the comfort zone was found to be 62–69 °F (16.7–20.6 °C) ET with a comfort line of 64 °F (17.8 °C) ET (ASHVE 1924). This 'comfort chart' was modified according to the laboratory results of Yagloglou and Drinker (1929) to provide a summer comfort zone (66–75 °F (18.9–23.9 °C) ET) and comfort line (71 °F (21.6 °C) ET), including 100 per cent votes indicating comfort, and a winter comfort zone (63–71 °F (17.2–21.6 °C) ET) with a comfort line of 66 °F ((18.9 °C) ET). These conditions were recommended by ASHRAE until 1961 with a few additional notes. For example, a note that the conditions apply for buildings whose occupants experienced three hours or more exposure and that they apply to northern USA cities such as Pittsburgh. For other areas an increase of 1 °F (0.56 °C) ET was recommended for each 5° reduction in Northern latitudes.

There was an increasing amount of practical data that suggested that ET was adequate when considering subjects in a dynamic state, but for steady state conditions too much emphasis was placed on relative humidity (e.g. Glickman *et al.*, 1950). ASHRAE moved their laboratory to Cleveland where Koch *et al.* (1960) conducted experiments to determine comfort conditions and the laboratory later moved to Kansas where Nevins *et al.* (1966) and Rohles and Nevins (1971) conducted further extensive laboratory trials.

Nevins *et al.* (1966) used subjective ratings of 360 male and 360 female college-aged students to re-evaluate the conditions for comfort as affected by air temperature (dbt = dry bulb temperature) and relative humidity. Subjects were exposed in groups of ten (5 male, 5 female) seated (playing cards, reading), in light clothing (estimate 0.52 Clo), for three hours. Seventy-two conditions were investigated involving combinations of 9 air temperatures (66, 68, 70, 72, 74, 76, 78, 80, 82 °F (18.9 to 27.8 °C)) and 8 relative humidities (15, 25, 35, 45, 55, 65, 75, 85 per cent). Rohles and Nevins (1971) later conducted a similar, more extensive, study involving 800 male and 800 female students and extended the range of air temperature from 60 °F to 98 °F (15.5–36.6 °C) with clothing now estimated as 0.6 Clo (i.e. the Kansas State University (KSU) uniform). They concluded that, over all subjects, air temperatures for these conditions rated as comfortable covered the range of

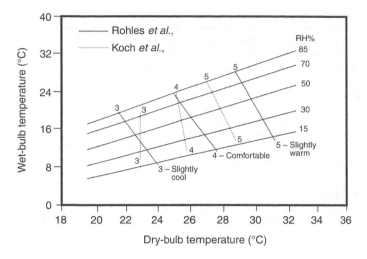

Figure 8.3 A comparison of lines of thermal sensation obtained from laboratory studies of large groups of subjects.

Source: Rohles and Nevins (1971).

62–98 °F (16.7–36.6 °C). It took 1.5 h for males to adapt completely to the conditions, whereas females adapted more quickly and sensitivity to changes in air temperature was 7–9 times that of sensitivity to changes in relative humidity. The results are summarized in Figure 8.3.

Fanger (1970)

The most significant landmark in thermal comfort research and practice was the publication of the book *Thermal Comfort* by Fanger (1970), which outlines the conditions necessary for thermal comfort and methods and principles for evaluating and analysing thermal environments with respect to thermal comfort. Fanger considered that existing knowledge of thermal comfort was inadequate and unsuitable for practical application, and his book is based upon research undertaken at the Technical University of Denmark and at Kansas State University, USA. The methods that he developed are now the most influential and widely used throughout the world. The reason for this success has been the consideration of the 'user requirements'. He had the vision to recognize that it is the combined thermal effect of all (six basic parameters) physical factors which determines human thermal comfort, and that a practical method was required which could predict conditions for 'average thermal comfort' and consequences, in terms of thermal discomfort, e.g. percent of people dissatisfied, of exposure to conditions away from those for 'average thermal comfort'.

Table 8.2 Terms used in the Predicted Mean Vote (PMV)

H = Internal heat production in the human body
E_d = Heat loss by water vapour diffusion through skin
E_{sw} = Heat loss by evaporation of sweat from skin surface
E_{re} = Latent respiration heat loss
L = Dry respiration heat loss
K = Heat transfer from skin to outer surface of clothing
R = Heat transfer by radiation from clothing surface
C = Heat transfer by convection from clothing surface

Source: Comfort equation of Fanger (1970).

The comfort equation

Fanger (1970) defines three conditions for a person to be in (whole-body) thermal comfort:

1   the body is in heat balance;
2   sweat rate is within comfort limits; and
3   mean skin temperature is within comfort limits.

A fourth condition is the absence of local thermal discomfort and this will be discussed later. The objective was to produce a comfort equation requiring input of only the six basic parameters and based on the above three conditions, to calculate conditions for thermal comfort. This was achieved using a rational analysis of heat transfer between the clothed body and the environment and experimental research. Fanger's original work was not in SI units so the SI version presented below is taken from Olesen (1982), ASHRAE (1993) and ISO 7730 (1994).

Heat balance

Fanger's conceptual heat balance equation is:

$$H - E_d - E_{sw} - E_{re} - L = K = R + C, \tag{8.2}$$

where a description of terms used is given in Table 8.2.

Heat is generated in the body and lost at the skin and from the lungs. It is transferred through clothing where it is lost to the environment. Logical considerations, reasonable assumptions, and a literature review provide equations for each of the terms such that they can be calculated from the six basic parameters (t_a, t_r, rh, v, Clo, Met) (see Table 8.3).

Sweat rate and skin temperature for comfort

Heat balance is a necessary but not a sufficient condition for comfort. The body can be in heat balance but uncomfortably hot due to sweating or uncomfortably

Table 8.3 Equations for components of the heat balance equation used by Fanger (1970) in determining the *PMV* thermal comfort equations (see Table 8.2 for terms)

$$H = M - W$$

$$E_d = 3.05 \times 10^{-3}(256\,t_s - 3373 - P_a)$$

$$t_s = t_{sk,req} \quad \text{i.e. for thermal comfort}$$

$$= 35.7 - 0.0275\,H$$

$$E_{sw} = E_{rsw,req} \quad \text{i.e. for thermal comfort}$$

$$= 0.42(M - W - 58.15)$$

$$L = 0.0014\,M(34 - t_a)$$

$$E_{re} = 1.72 \times 10^{-5}M(5867 - P_a)$$

$$K = \frac{(t_s - t_{cl})}{0.155\,I_{cl}}$$

$$R = 3.96 \times 10^{-8}f_{cl}\left[(t_{cl} + 273)^4 - (t_r + 273)^4\right]$$

$$C = f_{cl}h_c(t_{cl} - t_a)$$

Note
Units for all components: $W\,m^{-2}$.

cold due to vasoconstriction and low skin temperatures. Skin temperatures and sweat rates required for comfort $t_{sk,req}$ and $E_{rsw,req}$, depend upon activity level. Rohles and Nevins (1971) provide the following equations, which are shown in graphical form in Figures 8.4 and 8.5.

$$t_{sk,req} = 35.7 - 0.0275(M - W)\,°C \tag{8.3}$$
$$E_{rsw,req} = 0.42(M - W - 58.15)\,W\,m^{-2}. \tag{8.4}$$

By substituting $t_{sk,req}$ and $E_{rsw,req}$ terms into the heat balance equation, the method of combination of the six basic parameters which produce thermal comfort are expressed in the comfort equation:

$$M - W \qquad\qquad\qquad\qquad\qquad\quad H$$
$$-3.05[5.73 - 0.007(M - W) - P_a] \qquad E_d \quad (\text{for } t_{sk} = t_{sk,req})$$
$$-0.42[(M - W) - 58.15] \qquad\qquad E_{sw} \quad (\text{for } E_{sw} = E_{rsw,req})$$
$$-0.0173\,M(5.87 - P_a) \qquad\qquad\qquad E_{re}$$
$$-0.0014\,M(34 - t_a) \qquad\qquad\qquad\quad L$$
$$= 3.96 \times 10^{-8}f_{cl}\left[(t_{cl} + 273)^4 - (t_r + 273)^4\right] \quad R$$
$$+ f_{cl}\,h_c(t_{cl} - t_a) \qquad\qquad\qquad\qquad C$$

where

$$t_{cl} = 35.7 - 0.0275(M - W) - 0.155\,I_{cl}[(M - W)$$
$$- 3.05(5.73 - 0.007(M - W) - P_a) - 0.42[(M - W) - 58.15]$$
$$- 0.0173\,M(5.87 - P_a) - 0.0014\,M(34 - t_a)] \tag{8.5}$$
$$h_c = \max(2.38(t_{cl} - t_a)^{0.25}, 12.1\sqrt{v})$$
$$f_{cl} = 1.0 + 0.2\,I_{cl} \quad \text{for } I_{cl} \le 0.5$$
$$\phantom{f_{cl}} = 1.05 + 0.1\,I_{cl} \quad \text{for } I_{cl} > 0.5.$$

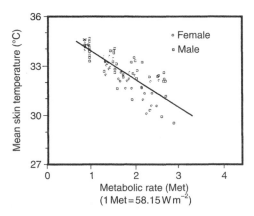

Figure 8.4 Mean skin temperature as a function of the activity level for persons in thermal comfort.

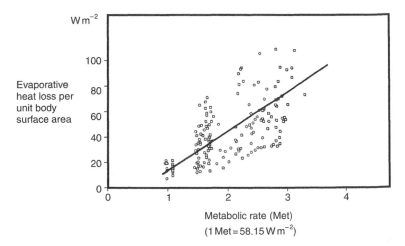

Figure 8.5 Evaporative heat loss as a function of the activity level for persons in thermal comfort.

Predicted mean vote (*PMV*) and predicted percentage dissatisfied (*PPD*)

To provide a method for evaluating and analysing thermal environments, Fanger made the proposal that the degree of discomfort will depend on the thermal load (*L*). This he defined as 'the difference between the internal heat production and the heat loss to the actual environment for a man hypothetically kept at the comfort values of the mean skin temperature and the sweat secretion at the actual activity level'. In comfort conditions the thermal load will be zero. For deviations from comfort the thermal sensation experienced will be a function of the thermal load and the activity level. For sedentary activity, Nevins *et al.* (1966) and Fanger (1970) provide data and McNall *et al.* (1968) provide data for four activity levels (from 1396 subjects exposed for 3 h in 0.6 Clo KSU uniform). This provided an equation for the predicted mean vote (*PMV*) of a large group of subjects if they had rated their thermal sensation in that environment on the following scale:

hot	+3
warm	+2
slightly warm	+1
neutral	0
slightly cool	−1
cool	−2
cold	−3

$$PMV = (0.303e^{-0.036M} + 0.028) \times [(M - W)$$
$$- 3.05 \times 10^{-3}\{5733 - 6.99(M - W) - P_a\}$$
$$- 0.42\{(M - W) - 58.15\} - 1.7 \times 10^{-5}M(5867 - P_a)$$
$$- 0.0014\,M(34 - t_a) - 3.96 \times 10^{-8}f_{cl}\{(t_{cl} + 273)^4$$
$$- (t_r + 273)^4\} - f_{cl}h_c(t_{cl} - t_a)], \tag{8.6}$$

where

$$t_{cl} = 35.7 - 0.028(M - W) - 0.155\,I_{cl}[3.96 \times 10^{-8}f_{cl}$$
$$\times \{(t_{cl} + 273)^4 - (t_r + 273)^4\} + f_{cl}h_c(t_{cl} - t_a)]. \tag{8.7}$$

Fanger (1970) presents tables showing *PMV* values for 3500 combinations of the variables. These are now unnecessary as the calculations of *PMV* can easily be made on a personal computer (see ISO 7730, 1994; Parsons, 1993).

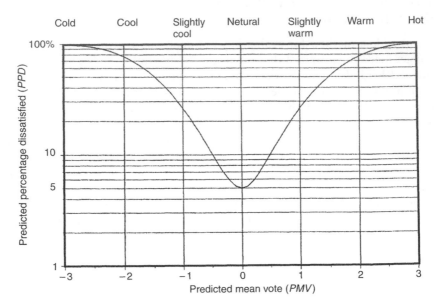

Figure 8.6 The predicted percentage of dissatisfied (*PPD*) persons as a function of the Predicted Mean Vote (*PMV*) index.

The predicted percentage of dissatisfied (*PPD*) provides practical information concerning the number of potential complainers. The data of Nevins *et al.* (1966), Rohles (1970) and Fanger (1970) provided a relationship between the percentage of dissatisfied and the mean comfort vote.

$$PPD = 100 - 95 \exp -\left(0.03353 PMV^4 - 0.2179 PMV^2\right) \qquad (8.8)$$

(ISO 7730, 1994) (see Figure 8.6).

Fanger (1970) describes a method for the use of the *PMV* and *PPD* in practical applications and includes examples involving an analysis of a large room and the use of a thermal non-uniformity index, lowest possible percentage of dissatisfied (LPPD), indicating the 'best' that can be achieved by changing the average *PMV* value in the room only – by changing average air temperature, for example – and hence indicating where specific areas of the room require attention.

Data to allow the determination of some *PMV* and *PPD* values are provided in Tables 8.4, 8.5 and 8.6.

The comfort studies of A. P. Gagge *et al.* – ET*, PMV*, SET, the two-node model and enthalpy

In parallel with the work of Fanger (1970), workers at the J. B. Pierce Foundation laboratory, USA, inspired by A. P. Gagge, conducted research

Table 8.4 Examples of estimates of clothing insulation values (I_{cl}) for use in the
PMV thermal equation of Fanger (1970)

Clothing ensemble	Clo	$m^2 K W^{-1}$
Naked	0	0
Shorts	0.1	0.016
Typical tropical clothing outfit Briefs (underpants), shorts, open neck shirt with short sleeves, light socks and sandals	0.3	0.047
Light summer clothing Briefs, long lightweight trousers, open neck shirt with short sleeves, light socks and shoes	0.5	0.078
Working clothes Underwear, cotton working shirt with long sleeves, working trousers, woollen socks and shoes	0.8	0.124
Typical indoor winter clothing combination Underwear, shirt with long sleeves, trousers, sweater with long sleeves, heavy socks and shoes	1.0	0.155
Heavy traditional European business suit Cotton underwear with long legs and sleeves, shirt, suit comprising trousers, jacket and waistcoat (US vest), woollen socks and heavy shoes	1.5	0.233

Table 8.5 Examples of estimates of metabolic rates (M) for use in the PMV
thermal comfort equation of Fanger (1970)

Activity	Met	$W m^{-2}$
Lying down	0.8	47
Seated quietly	1.0	58
Sedentary activity (office, home, laboratory, school)	1.2	70
Standing, relaxed	1.2	70
Light activity, standing (shopping, laboratory, light industry)	1.6	93
Medium activity, standing (shop assistant, domestic work, machine work)	2.0	116
High activity (heavy machine work, garage work)	3.0	175

into determining improved rational indices of thermal comfort. A number
of indices were developed, all using the method of relating actual conditions
to the air temperature of a standard environment which would give equivalent effect, e.g. comfort (see Table 8.7).

The Effective Temperature index (ET) – Houghton and Yagloglou (1923) –
was defined as the temperature of a standard environment ($rh = 100\% \ t_a = t_r$,

Table 8.6 Predicted mean vote (*PMV*) values from Fanger (1970). Assume $rh = 50\%$; still air; and $t_a = t_r PMV$: +3, hot; +2, slightly warm; +1, warm; 0, neutral; −1, slightly cool; −2, cool; −3, cold

$t_a = t_r$	I_{cl}, Clo						
	0.1	0.3	0.5	0.8	1.0	1.5	2.0
$M = 1$ Met; var = $0.1\,\mathrm{m\,s^{-1}}$							
10						−2.2	−1.4
12						−1.8	−1.0
14					−2.5	−1.4	−0.7
16				−2.5	−1.9	−1.0	−0.3
18				−1.9	−1.4	−0.5	0.0
20			−2.3	−1.3	−0.9	−0.1	0.4
22		−2.3	−1.5	−0.7	−0.3	0.4	0.8
24	−2.3	−1.4	−0.8	−0.1	0.2	0.8	1.1
26	−1.2	−0.5	0.0	0.6	0.8	1.2	1.5
28	−0.1	0.4	0.8	1.2	1.4	1.7	1.9
30	1.0	1.3	1.6	1.8	1.9	2.1	2.3
32	2.0	2.2	2.3	2.4	2.5	2.6	2.6
$M = 1.2$ Met; var = $0.1\,\mathrm{m\,s^{-1}}$							
10						−2.7	−1.6	−0.9
12					−2.8	−2.2	−1.2	−0.6
14					−2.3	−1.8	−0.9	−0.3
16				−2.8	−1.8	−1.3	−0.5	0.0
18			−2.9	−2.1	−1.2	−0.8	−0.1	0.3
20			−2.2	−1.5	−0.7	−0.4	0.2	0.6
22		−2.3	−1.4	−0.8	−0.2	0.1	0.6	0.9
24	−1.4	−0.7	−0.2	0.3	0.6	1.0	1.3
26	−0.5	0.1	0.4	0.8	1.0	1.4	1.6
28	0.4	0.8	1.1	1.3	1.5	1.7	1.9
30	1.3	1.5	1.7	1.8	1.9	2.1	2.2
32	2.0	2.1	2.2	2.3	2.3	2.4	2.4
$M = 1.6$ Met; var = $0.1\,\mathrm{m\,s^{-1}}$							
10					−2.0	−1.5	−0.7	−0.2
12				−2.6	−1.6	−1.2	−0.4	0.0
14			−2.9	−2.1	−1.3	−0.9	−0.2	0.3
16			−2.4	−1.7	−0.9	−0.5	0.1	0.5
18	−2.8	−1.8	−1.2	−0.5	−0.2	0.4	0.7
20	−2.1	−1.3	−0.7	−0.1	0.2	0.6	0.9
22	−1.4	−0.7	−0.2	0.3	0.5	0.9	1.2
24	−0.7	−0.2	0.2	0.7	0.8	1.2	1.4
26	0.0	0.4	0.7	1.1	1.2	1.5	1.6
28	0.7	1.0	1.2	1.5	1.6	1.8	1.9
30	1.4	1.6	1.7	1.9	1.9	2.0	2.1
32	2.1	2.2	2.2	2.3	2.3	2.3	2.4

$v < 0.15\,\mathrm{m\,s^{-1}}$) which would provide equivalent warmth to that of the actual environment. Data that achieved the determination of equivalent warmth were derived empirically. Gagge *et al.* (1971) proposed a *new* Effective Temperature (*ET**) which was the temperature of a standard environment (*rh* = 50%,

Table 8.7 The principle of using a standard environment to define a thermal index. For the hypothetical case shown, the index value is the air temperature of the standard environment that would produce equivalent effect to that of the actual environment under consideration

		Human thermal environment			
		Standard	Examples		
			Actual$_1$	Actual$_2$	Actual$_3$
Air temperature	(t_a)	index	20	30	40
Mean radiant temperature	(t_r)	$t_r = t_a$	30	30	20
Air velocity	(v)	0.15	0.15	0.2	0.5
Relative humidity	(ϕ)	50%	40	50	30
Clothing	(I_{cl})	0.6	1.0	0.5	0.8
Activity	(Met)	1.0	1.0	1.6	1.2
Index value: $t_a(°C)$			27	31	32

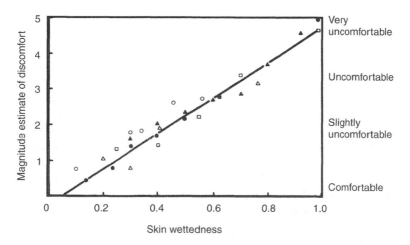

Figure 8.7 Skin wettedness is a good predictor of warm discomfort. The measurements were obtained over a wide range of temperatures and humidities so that a given skin wettedness covers a range of absolute sweat rates.
Source: Gonzales and Gagge (1973).

$t_a = t_r$, $v < 0.15 \, \mathrm{m\,s^{-1}}$) that would give equivalent effect, but based on the premise that a person experiencing an equivalent effect would have the same skin wettedness (w), same mean skin temperature (t_{sk}), and same thermal heat loss at the skin (H_{sk}). Evidence for this had been provided by Gagge *et al.* (1967), Berglund and Cunningham (1986) and Gonzalez and Gagge (1973) (see Figures 8.7 and 8.8).

The ET^* index is applicable to persons with sedentary activity and light clothing (for both actual and standard environment), is related specifically to

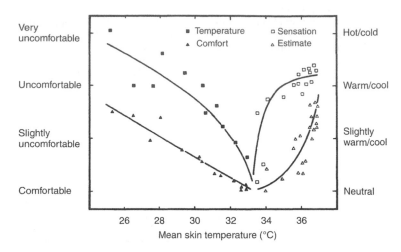

Figure 8.8 Comfort and sensation as a function of mean skin temperature. In the warm, sweating limits the rise in skin temperature, making the increase of sensation appear very steep.

Source: Gagge *et al.* (1967).

thermal comfort and has been used by ASHRAE (1997). An extension of the index to include a range of activity and clothing values provides the *Standard Effective Temperature* (SET) thermal index (Gagge *et al.*, 1972). The SET is defined as the temperature of an isothermal environment with air temperature equal to mean radiant temperature, 50 per cent relative humidity, and still air ($v < 0.15\,\mathrm{m\,s^{-1}}$) in which a person with a standard level of clothing insulation would have the same heat loss at the same mean skin temperature and the same skin wettedness as he does in the actual environment and clothing insulation under consideration (see Table 8.8). The ET^* is therefore equivalent to the SET for sedentary activities (and light clothing).

The activity is the same in the standard environment as it is under actual conditions. To allow SET to provide consistent thermal sensation predictions over a wide range of conditions (hot to cold), the clothing insulation in the standard environment depends upon activity level (Nishi and Gagge, 1977).

$$I_{cls} = \frac{1.33}{(Met - Wk + 0.74)} - 0.095, \tag{8.9}$$

where

I_{cls} = clothing insulation in the standard environment (Clo)
Met = metabolic rate
Wk = mechanical work.

Table 8.8 Standard environments used to define thermal indices: Effective Temperature (*ET*); New Effective Temperature (*ET**); and Standard Effective Temperature (*SET*)

	ET	ET*	SET	Example actual
Air temperature (t_a)	ET			30
Mean radiant temperature (t_r)	$t_r = t_a$	$t_r = t_a$	$t_r = t_a$	40
Air velocity (v)	0.15	0.15	0.15	0.25
Relative humidity (ϕ)	100	50	50	30
Clothing (I_{cl})	actual	0.8	f(Met)	0.8
Activity (Met)	actual	1.0	actual	1.0
Equivalence:	From charts of experimental data	H_{sk} 53.7 t_{sk} 35.6 w 0.4	53.7 35.6 0.4	53.7 35.6 0.4
Index value for actual environment (°C)	23	33	33	

Note
The *ET** index applies only to sedentary activity and light clothing such as found in office work, it is a subset of the *SET* index. The standard clothing insulation value for SET is a function of the activity level.

For example, for a metabolic rate of 1 Met, clothing in the standard thermal environment is 0.67 Clo. However at 1.1 Met, the standard clothing is 0.6 Clo and so on (see Table 8.9).

Where the environment does not allow steady-state conditions to be reached, the state of the body after one hour (after starting from thermo-neutral conditions) is used. The SET is the most comprehensive thermal index, integrating all six basic parameters, and applicable over hot, moderate,

Table 8.9 Clothing insulation for the standard environment used in the definition of standard effective temperature (SET)

Activity (Mets)	Standard clothing (Clo)	Physiological state of neutrality	
		Mean body temp. (°C)	Skin wettedness
0.8	0.7	36.26	0.06
1.1	0.6	36.35	0.07
2.0	0.5	36.56	0.14
2.9	0.4	36.71	0.21
3.8	0.3	36.88	0.28

Source: Nishi and Gagge (1977).

Table 8.10 Relationship between standard effective temperature (SET) index
levels and thermal sensation

SET (°C)	Sensation	Physiological state of sedentary person
>37.5	Very hot, very uncomfortable	Failure of regulation
34.5–37.5	Hot, very unacceptable	Profuse sweating
30.0–34.5	Warm, uncomfortable, unacceptable	Sweating
25.6–30.0	Slightly warm, slightly unacceptable	Slight sweating, vasodilation
22.2–25.6	Comfortable and acceptable	Neutrality
17.5–22.2	Slightly cool, slightly unacceptable	Vasoconstriction
14.5–17.5	Cool and unacceptable	Slow body cooling
10.0–14.5	Cold, very unacceptable	Shivering

and cold thermal conditions. The relationship between SET and thermal
sensation effects on the body is shown in Table 8.10.

Unfortunately, the SET index was 'ahead of its time' and was cumbersome to determine involving the use of computers (which were not then
widely available) or tables (Markus and Morris, 1980).

Two steps in calculation were involved. First the mean skin temperature (t_{sk}),
skin wettedness (w) and heat loss at the skin (H_{sk}) are determined for a person
under the 'actual' conditions of interest. Then the temperature of the standard
environment which provides the same t_{sk}, w, H_{sk} (i.e. equivalent effect) is
determined (see Table 8.8). These calculations are performed using a two-node
model of human thermoregulation (Gagge *et al.*, 1973). The availability of
personal computers and appropriate software makes this a simple process and
the SET a comprehensive index with great practical value. It also however
demonstrates the practical value of models such as the two-node model and
changes the way practitioners design for thermal comfort (see Chapter 15).

PMV, PMV*, ET* and enthalpy

A series of studies have been conducted at the Pierce laboratories to improve
the human heat balance equation and rationalize indices for determining
clothing insulation, particularly vapour permeation (Oohori *et al.*, 1984),
and to attempt to describe the heat transfer from a clothed human in terms of
classical theory. Fobelets and Gagge (1988) posed the question, 'can the basic
theory of energy transfer from a wet surface through convection be applied to
the energy balance of a partially wet skin covered with clothing?'.

Gagge *et al.* (1986) argue that the – widely used and adopted by ISO 7730
(1984) – PMV index of Fanger (1970) underestimates the effects of
humidity, when predicting thermal comfort response, and proposed a new
index, *PMV**, which is identical to the PMV but with the value for *ET**
used in the *PMV* equation to replace operative temperature (t_o). Fanger

(1970) however argues that the *PMV* defines conditions for thermal comfort, based on his work and the work of Rohles and Nevins (1971) for example, and that the effects of humidity are relatively unimportant 'near to' comfort conditions. It is likely that both arguments apply and that *PMV** will provide improvement, but since the arguments have proven rather subtle for those involved in practical application, the *PMV** has not been widely used.

Fobelets and Gagge (1988) develop the arguments of Gagge *et al.* (1986) to provide a new perspective on thermal indices. They argue that heat and mass transfer from clothed man can be regarded as being between two extreme situations: one where man is analogous to a psychrometric wet bulb; and the other where he is analogous to a dry bulb. For any wet surface the dry bulb temperature (DBT) and wet bulb temperature (WBT) measure the potential of the environment to exchange dry heat and total heat (respectively) through convection. The Lewis relation links the coefficients of heat and mass transfer and enthalpy describes heat transfer per unit mass. Fobelets and Gagge (1988) argue that by analogy with heat transfer from a wet surface, operative temperature t_o, and ET^* measure the dry heat and total heat exchange by convection and radiation with a human whose skin is partially wet and protected by clothing. This provides a classical description and substantiates the use of ET^* as a universally applicable thermal comfort index. There is however some debate about the usefulness of the concept of effective enthalpy as enthalpy is primarily concerned with transfer from an exposed wet surface.

THERMAL MODELS AND THERMAL COMFORT

The 'search' for the definitive thermal comfort index has continued for over one hundred years and the development of the rationally-based SET and of simple instruments, such as the 'comfytest', a heated ellipsoid for determining the *PMV* – see Chapter 5 – provide comprehensive practical indices which are appropriate for most situations. The development and wide availability of personal computers however has changed the nature of the task for those who design environments. Thermal models, e.g. the two-node model – see Chapter 15 – can be used not only to calculate indices but also to predict the dynamic physiological response of an average person. It could therefore be argued that it is the response of the model itself and not a single index value that is of most value in environmental design and assessment. This will lead to computer aided design software. The development of the personal computer has therefore not changed the principles underlying thermal comfort but provided much greater scope for translating these principles into practice. The use of thermal models does raise many other issues however and these are discussed in Chapter 15. Although this area is still at an early stage in development, models and computer systems already exist which could be used in the evaluation and design of steady state and transient conditions (Parsons, 1989; Jones and Ogawa, 1992; Candas *et al.*, 1998; Fiala, 1998; Zhou, 2001).

Local thermal discomfort

Thermal discomfort may be caused by the body (as a whole) being too warm or too cool but also by a part of the body being too warm or too cool. For whole-body comfort, Fanger (1970) proposes that three conditions must be met: heat balance; sweat rate within comfort limits; and mean skin temperature within comfort limits. A fourth condition for thermal comfort is therefore that there must be no local thermal discomfort. Local thermal discomfort is usually considered in terms of draught, but can also be caused by temperature differences across the body, contact with cold or hot surfaces, and high radiant temperature asymmetry. These issues are discussed in detail by McIntyre (1980), Olesen (1985a), Berglund (1988) and ISO 7730 (1994).

Draughts

ISO 7730 (1994) defines a draught as unwanted local cooling of the body caused by air movement. The perception of a draught is dependent on air velocity and degree of disturbance, air temperature, area of the body exposed and the thermal state of the person. A warm person may perceive air movement as a pleasant breeze whereas the same air movement may be perceived as an unwelcome draught by a cool person. This is of great importance and 'a sound practical rule ... should be to get the ambient temperature to an optimum level' (McIntyre, 1980). A number of laboratory studies have been conducted into draughts, at feet and ankles, for example (Houghton *et al.*, 1938; Nevins, 1971; Fanger and Pedersen, 1977; McIntyre, 1980; Christensen *et al.*, 1984; Fanger and Christensen, 1986; Berglund and Fobelets, 1987; Toftum and Nielsen, 1996a,b; Zhou, 1999; Toftum, 2002; Griefahn *et al.*, 2001). A draught rating index is proposed in ISO 7730 (1994) where

$$DR = (34 - t_a)(v - 0.05)^{0.62}(0.37vT_u + 3.14),$$

where

DR = draught rating, i.e. the percentage of people dissatisfied due to draught
t_a = local air temperature (°C)
v = local mean air velocity (m s^{-1})
T_u = local air turbulence (per cent) defined as the ratio of the standard deviation of the local air velocity to the local mean air velocity.

The air velocity should be measured with an omnidirectional anenometer having a 0.2 s time constant (ASHRAE, 1997). 'The model...is based on studies comprising 150 subjects exposed to air temperatures of 20–26 °C, mean air velocities of $0.05-0.4$ m s^{-1} and turbulence intensities of 0–70 per cent. The model applies to people at light, mainly sedentary, activity with a thermal sensation for the whole body close to neutral.' The sensation of draught is lower for active people and for people warmer than neutral, although local evaporation of sweat will confound the problem. It is possible that people who are already cold will feel draught more severely; however, this has yet to be demonstrated. Where turbulence intensity is not known, a figure of 40 per cent can be taken.

Asymmetric thermal radiation

All practical thermal environments have an asymmetric radiation field to some degree. If the asymmetry is sufficiently large then it can cause discomfort; for example, to persons exposed to the open door of a furnace, direct sunlight, heated ceilings or large cold windows or walls. The effect of the radiation asymmetry will be reduced if the environment is otherwise thermally neutral (McIntyre, 1980). Plane radiant temperature (t_{pr}) describes the radiation in one direction and the radiant temperature asymmetry (Δt_{pr}) is the difference between the plane radiant temperature of the two opposite sides of a small plane element. A method for measuring and calculating radiant temperature in a room is provided by Olesen (1985a) (see Chapter 5).

A relationship between radiant temperature asymmetry and thermal discomfort is described by Olesen (1985a) for seated subjects dressed in

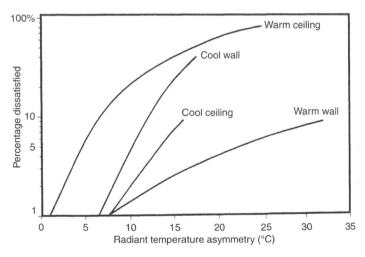

Figure 8.9 Percentage of people dissatisfied as a function of degree and type of asymmetric radiation.

Source: ASHRAE (1989a).

standard clothing (= 0.6 Clo), and exposed to either an overhead warm surface or a vertical cold surface, but in an overall thermally neutral climate. The percentage of dissatisfied subjects is presented in Figure 8.9.

It is emphasized that the percentage of people feeling local discomfort and overall discomfort (*PPD*) should not be added to obtain total dissatisfied.

It is generally concluded that for people in light clothing and sedentary activity, the radiant temperature asymmetry from cold vertical surfaces should be less than 10 °C (in relation to a small vertical plane 0.6 m above the floor) and from a warm (heated) ceiling should be less than 5 °C (in relation to a small horizontal plane 0.6 m above the floor) (Olesen, 1985a; ISO 7730, 1994; ASHRAE, 1997).

Vertical air temperature differences

It is typical for air temperature to vary throughout a space and, because warm air rises, for temperatures to be higher at the head than at the feet of a person. This is particularly prevalent in relatively small spaces such as in vehicles. Chrenko (1974) suggests that 'such conditions tend to produce a stuffy feeling in the head, while the feet are cold'. Chrenko (1974) cites Reid (1844) who suggested that coaches should have floor heating and a report to the General Board of Health on the heating and ventilation of dwellings (1857) suggested that '... in a comfortable and healthy apartment the floor should be at the highest temperature in the room'.

Olesen (1985a) summarizes a number of studies of thermal comfort and vertical air temperature difference. He also notes that if the gradient is sufficiently large, local warm discomfort can occur at the head and/or cold discomfort can occur at the feet, although the body as a whole is thermally neutral. Olesen *et al.* (1979) exposed seated subjects, in overall thermal neutrality, to vertical air temperature differences between head and ankle.

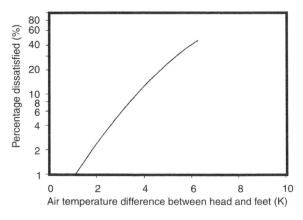

Figure 8.10 Local discomfort caused by vertical air temperature difference. Applies when the temperature increases upwards.

Figure 8.10 shows the percentage of dissatisfied in terms of vertical air temperature difference for people in light sedentary activities. ISO 7730 (1994) and ASHRAE (1992) both recommend that the 'vertical air temperature difference between 1.1 m and 0.1 m above floor (head and ankle level) should be less than 3 °C'.

Recent studies concerning thermal comfort requirements for environments involving chilled ceiling and displacement ventilation (cool air to the floor which rises by natural convection) have suggested that, although this may provide general guidance, vertical temperature differences of greater than 3 °C can provide comfort (Wyon and Sandberg, 1996; Loveday *et al.*, 2002).

Local discomfort caused by surface contact

The effects of skin contact with solid surfaces are discussed in detail in Chapter 13, where mechanisms, sensations and skin damage are considered. Discomfort caused by contact will be greatly reduced by clothing (especially shoes) which will provide thermal insulation and will depend upon surface material type, condition and temperature. Contact between cold floors and bare feet has been studied by Nevins *et al.* (1964a,b), Nevins and Feyerherm (1967) and Olesen (1977) and recommendations are reported in Olesen (1985) (see Table 8.11).

ISO 7730 (1994) makes the following recommendations for light, mainly sedentary, activities:

> The surface temperature of the floor shall normally be between 19 and 26 °C, but floor heating systems may be designed for 29 °C.
>
> ISO 7730 (1994)

Table 8.11 Temperatures of floors to provide comfort for people with bare feet

Flooring material	Optimum floor temperature (°C)		Recommended floor temperature range
	Floor contact time		
	1 min	10 min	
Textiles (mats)	21	24.5	21–28
Cork	24	26	23–28
Pinewood floor	25	26	22.5–28
Oakwood floor	26	26	24.5–28
PVC sheet with felt underlay on concrete	28	27	25.5–28
Hard linoleum on wood	28	26	24–28
5 mm tessellated floor on gas concrete	29	27	26–28.5
Concrete floor	28.5	27	26–28.5
Marble	30	29	28–29.5

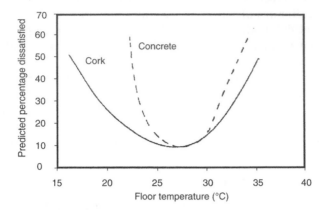

Figure 8.11 The predicted percentage dissatisfied *(PPD)* for people standing with bare feet on cork or concrete floors.

Source: After Olesen (1977).

Recommendations can also be provided in terms of use of an artificial foot. DIN 52614, for example, uses a 150 mm diameter water-filled cylinder with a rubber base as an artificial foot and provides recommendations in terms of integrated heat flow after one and ten minutes.

Olesen (1977) produced predicted percentage dissatisfied *(PPD)* values for people in thermal neutrality standing on floor surfaces of different construction and temperature – see Figure 8.11. ISO 7730 (1994) (from ISO/TS 13732–2) provides a more general relationship (Figure 8.12).

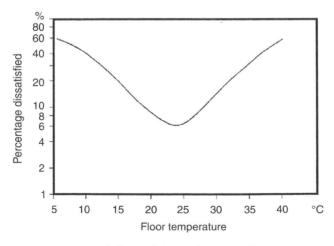

Figure 8.12 Local thermal discomfort caused by warm or cold floors.

DO REQUIREMENTS FOR COMFORT UNIVERSALLY APPLY?

The above discussion outlines the principles and recommendations for determining human thermal comfort in terms of the six basic parameters. However, the implication is that they apply in all situations. For example, do old male Chinese or Japanese people on night work in rooms with walls painted red have the same requirements for comfort as young female American College students during the day in rooms with walls painted blue? Do the sick or disabled in rooms in winter have similar requirements to the able, fit and healthy in rooms in summer?

Fanger (1970) considers this point in detail and concludes that, if one specifies comfort requirements in terms of the six basic parameters, there are other influential factors, but they have no effect of sufficient magnitude to be of practical engineering significance. This has been controversial and has led to many studies, especially of the thermal comfort requirements of those of different national–geographic locations and race. The results are summarized below.

National–geographic location

It is natural to expect that people who live in parts of the world with warm and hot climates might require different conditions for thermal comfort than those who live in parts of the world with cool and cold climates. Despite a great deal of research this has not been shown to be the case. There are two main reasons: first we consider the conditions for thermal comfort and not therefore heat or cold stress where acclimatization may be important and differences may be found; and, second, the thermal conditions of the person are defined in terms of the six basic parameters. These include clothing and activity factors that will vary with cultural and climatic differences. Fanger (1970) considered whether the comfort equation (used in the calculation of the *PMV*) was applicable for populations from different parts of the world. He performed experiments similar to those of Nevins *et al.* (1966) and found no significant difference between conditions preferred by Danish and American subjects. He concludes that the comfort equation can be applied over temperate regions of the world and that previously reported differences between the USA and Europe are probably due to differences in clothing worn, a parameter contained in the comfort equation.

For those 'acclimatized' to more tropical climates Fanger (1970) cites Ellis (1953) who, in an investigation of Europeans and Asians in Singapore, found an optimal temperature of 27 °C (for $rh = 80\%$, $v = 0.4\,\mathrm{m\,s^{-1}}$, light tropical clothing (=0.4 Clo)) where Fanger's comfort equation would have predicted a value of 27.4 °C. Fanger (1972) summarizes other results – see Table 8.12 – and concludes that the *PMV* comfort equation can be used effectively for conditions in the tropics.

Table 8.12 Comparison of preferred ambient temperatures determined by studies conducted in tropical climates with those predicted by the PMV comfort equation

Location	Study	Relative humidity (%)	Preferred ambient temperature (°C)			
			Rel. air velocity (m s⁻¹)	Field study	PMV eqn.	Diff.
Calcutta	Rao (1952)	70	0.1	26	26	0
Singapore	Ellis (1953)	80	0.4	27	27.5	−0.5
Nigeria	Ambler (1955)	70	0.1	26	26	0
Singapore	Webb (1959)	80	0.4	28.5	27.5	+1
North Australia	Wyndham (1963)	80	0.1	26	26	0
New Guinea	Ballantyne et al. (1967)	80	0.1	26	26	0

Source: Fanger (1972).

There have been a number of attempts to validate the *PMV/PPD* index, in different parts of the world. Tanabe *et al.* (1987) obtained thermal sensation votes from 172 Japanese college-age subjects exposed for three hours in climatic chamber trials. Subjects were sedentary and wore light clothing (= 0.6 Clo). The experiment was similar to those conducted by Nevins *et al.* (1966) on American students and by Fanger (1970) on Danish students. A mean 'neutral' temperature of 26.3 °C was found and results corresponded well with both *PMV* and *PPD* predictions for these Japanese subjects.

In an extensive field study in the United Kingdom, Fishman and Pimbert (1978) recorded the comfort votes of 26 office staff, every working hour, over one year. This 'Watson House survey' was carefully controlled and carried out and involved detailed measurements of workplace environments and comfort votes using the Bedford scale. They found that the 'neutral temperature' was 22 °C irrespective of whether subjects were male or female (a small difference), it was winter or summer, or morning or afternoon. Subjects generally kept themselves comfortable by adjusting clothing.

Using an average estimated metabolic rate of 80 W m⁻² and clothing insulation of 0.8 Clo, Fishman and Pimbert (1978) showed a remarkable agreement between the Watson House survey data and the results predicted by the *PMV* and *PPD*. The important point about such field trials however, is that people can adjust their activity and clothing to achieve comfort (Humphreys, 1976).

An interesting point about the *PMV* is why should it be so successful. Its logical derivation provides only comfort conditions and predictions away from comfort are only tentatively based. In studies of seated subjects the effects of the seat are rarely taken into account yet the index still makes reasonable predictions. The success of the index is in part to do with the latitude afforded in the estimate of metabolic rate and clothing insulation levels. Any inaccuracy of the index for example can easily be explained in terms of poor estimates of these parameters. Another important reason for its success in practical application is that it gives users what they want; i.e. a consistent prediction of average response with an indication of variation in terms of those likely to be dissatisfied and complain.

The effects of age

It is generally considered that older people prefer higher air temperatures for comfort than do younger, more active people. Collins and Hoinville (1980) compared the comfort votes of 16 elderly and 16 young adult subjects. They found that, when the data were 'corrected' for the effects of clothing to 1.0 Clo, both groups of subjects would prefer an air temperature of 21.1 °C on average. It was concluded that the vulnerability of elderly people in their homes is due to a lifestyle involving low activity and increased risk because of poor thermoregulatory responses and blunted perception of temperature changes. Conditions for comfort however did not vary from those of young adults.

Rohles (1969) found a similar result with a questionnaire survey of 64 adults of over 70 years of age when compared with the results of laboratory studies of younger subjects. Langkilde (1977) reviewed the literature and also determined the preferred air temperatures of 16 subjects over eighty years of age in a climatic chamber. It was concluded that age has no significant effect on the ambient temperatures preferred by man (or woman).

Cena and Spotila (1984, 1986) investigated 209 elderly persons in North America (mean age 74.2 years). They concluded that average responses were similar to those predicted by the *PMV* studies of Fanger (1970) and Nevins *et al.* (1966). In the population investigated, no effects were found of sex, age, health, smoking, alcohol usage, income and house value on activity level, clothing insulation, oral temperature, whole-body, hand or feet sensation, indoor air temperature, equivalent temperature and *PMV*.

Fanger (1970) reviews the literature and finds that neutral temperatures of elderly and college-age students are similar (see Table 8.13).

He suggests that the reason why elderly people prefer higher temperatures is because they have lower metabolic rate. As this is one of the six basic parameters, Fanger concludes that the *PMV* thermal comfort index is appropriate for predicting the comfort responses of elderly people. This as well as lifestyle and clothing factors appears to be the conclusion of most studies into the effects of age. Further work is still required on the responses of children where behavioural measures may be particularly appropriate.

The effects of gender

The results presented in Table 8.13 show no significant differences between preferred temperatures of male and female subjects. There is evidence that females are more sensitive to deviations from the optimum (Fanger, 1970).

Table 8.13 Comparison of neutral temperatures provided by groups of different gender, age and geographical location

	Neutral Temp (°C)	Diff. (°C)	PMV scale diff.	significant (5% level)
College age, Danish	25.71			
College age, American	25.55	0.16	−0.1	no
College age, Danish	25.71			
Elderly, Danish	25.71	0.00	0.0	no
College age, Danish females	25.50			
College age, Danish males	26.07	−0.57	0.2	no
Elderly, Danish females	25.22			
Elderly, Danish males	26.50	−1.28	0.4	no
College age, American females	25.91			
College age, American males	25.09	0.82	−0.3	yes

Source: Fanger (1970).

The effects of the menstrual cycle on female responses were also found to be insignificant despite the associated changes in internal body temperature over the month.

It is sometimes noted that males and females have different thermal comfort responses and this is naturally confounded with the clothing that they wear. A laboratory study by Breslin (1995; see also Parsons, 2002) compared the responses of 16 male and 16 female subjects over a three hour period in a thermal chamber. Three conditions were investigated, predicted to be 'slightly warm' to 'warm' ($PMV = 1.4$; air temperature $t_a = 29\,°C$), 'neutral' to 'slightly cool' ($PMV = -0.4$; $t_a = 23\,°C$) and 'cool' ($PMV = -2.0$; $t_a = 18.5\,°C$). For all conditions subjects sat relaxed in an office chair watching television (non related video). Males and females wore identical light clothing (shirt, long trousers) and their own underwear, socks and shoes (an estimated clothing insulation value of 0.8 Clo, including the insulation of the chair). Radiant temperature was equal to air temperature, there was no forced air movement and relative humidity was 50 per cent. Subjective impressions were provided every 15 min on the ASHRAE/ISO seven-point thermal sensation scale (hot, warm, slightly warm, neutral, slightly cool, cool, cold) overall, and for areas of the body. Thermal comfort and other subjective measures were also taken. A similar experiment on a different set of eight male and eight female subjects, was conducted by Webb and Parsons (1997). Similar conditions were employed but the subjects also wore a sweatshirt. This raised the estimated clothing insulation value to 1.0 Clo and hence modified the PMV values (+1.5 – slightly warm to warm; 0 – neutral and −1.5 – slightly cool to cool). The results are shown in Figure 8.13.

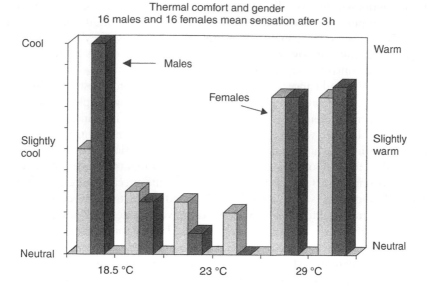

Figure 8.13 Thermal comfort responses of male and female subjects.
Sources: After Breslin (1995), Webb and Parsons (1997) and Parsons (2002).

Comparison of the *AMV*s and *PMV*s showed that the *PMV* values (BS EN ISO 7730, 1995) predicted *AMV*s within 0.5 of a scale value for neutral and warm conditions. However in slightly cool to cool conditions ($PMV = -1.5$) both male and female subjects were between neutral and slightly cool and hence warmer than predicted. For cool conditions ($PMV = -2.0$) females tended to be significantly cooler ($p < 0.05$) than males and their responses were close to the *PMV* values. Local sensation responses from the study of Breslin (1995) suggested that this was due to females reporting significantly cooler hands than males. Overall it appears that there are few gender differences for neutral and warm conditions, but that females tend to be cooler than males in cool conditions; this was found in young college age students by Breslin (1995).

The effects of acclimation state on thermal comfort requirements

When human subjects are systematically exposed to hot environments, over a number of days, their physiological responses to heat change and, in particular, they sweat earlier and more in response to a given heat stimulus. This is termed 'acclimatization' and if conducted in a laboratory, 'acclimation' (Parsons, 1993). It is often thought that people living in hot environments would have different comfort requirements from those living in more temperate climates. Either (for a given set of conditions, clothing and activity) they require higher air temperatures because they have adapted to the conditions or lower air temperatures as they compensate for (or demonstrate control over) the hot, external conditions.

Fanger (1972) and Fanger et al. (1977) found, in laboratory tests in a climatic chamber, that temperature preferences were similar among people, likely to be acclimatized in different ways and to different degrees. These included winter swimmers, cold store workers, people from the tropics, and control groups. In similar experiments, no significant effects of acclimatization were found by de Dear *et al.* (1991) who conducted laboratory experiments on heat acclimated students in Singapore. Gonzalez (1979) considered the effects of natural acclimatization during a heat wave in the USA. He found that active subjects preferred higher temperatures after the heat wave but there was no effect on resting subjects. De Dear (1998) reviews evidence of physiological adaptation as an explanation for a shift in preferred temperatures inside of buildings with outside climate. He concludes that subjective discomfort and acceptability are not affected by the acclimatization state of resting or lightly active building occupants.

Brierley (1996; see also Parsons, 2002) investigated whether subjects who had undergone an acclimation programme required different comfort conditions than those before they were acclimated. This was part of a larger project which also involved acclimation and heat familiarization programmes for athletes going to the Atlanta Olympic Games in 1996. Six male college age subjects were exposed for three hours in a thermal chamber in conditions identical to those used by Webb and Parsons (1997) (1.0 Clo etc. – see above). On Day One, conditions were neutral ($PMV = 0$, 23°C, 50% rh) and on Day Two, they were between slightly warm and warm ($PMV = 1.5$, 29°C, 50% rh). Following a five-day break, the subjects were exposed over four days to an acclimation programme in 45°C and 40 per cent rh, for two hours per day. The subjects performed stepping exercises and cycled on a cycle ergometer. On completion of the acclimation period the thermal comfort sessions were repeated in reverse order to limit the loss of acclimation.

All subjects showed acclimation responses (increased sweat rate, reduced heart rate, small changes in 'core' temperature and greater tolerance to heat stress). Figures 8.14a and 8.14b show the mean thermal sensation responses of subjects pre- and post-acclimation. It can be seen that there were only small changes in thermal comfort responses for both conditions. It was concluded that these changes were unlikely to be of practical significance in terms of thermal comfort requirements.

Other factors

Havenith *et al.* (2002) consider personal factors related to clothing and metabolic rate. The effects on thermal comfort conditions of body build, ethnic differences, circadian rhythm, colour of surroundings, food, crowding,

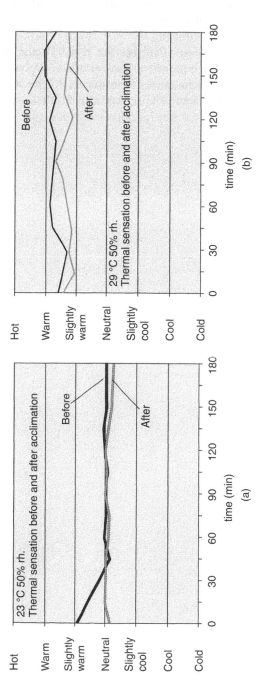

Figure 8.14 Thermal sensation response to a moderately warm environment (a) 23 °C, 50 per cent *rh* (before and after acclimation) (b) 29 °C, 50 per cent *rh* (before and after acclimation) (Brierley, 1996; Parsons, 2002).

and air pressure are all considered in detail by Fanger (1970). He concludes that none of the above factors have an influence which is not accounted for in the six basic parameters (and hence the *PMV* equation) or which is of practical engineering significance. He concludes that the *PMV* index therefore has universal practical application.

Whether this index is the best available (or the *ET** for example) or whether there are always more practical factors which are of importance and hence a judgement cannot be made, is immaterial. The body of scientific data has yet to provide evidence that conditions for thermal comfort, at least to a first approximation, cannot be universally defined.

9 Thermal comfort for special populations, special environments and adaptive modelling

INTRODUCTION

Although principles for the assessment of human thermal environments generally apply, there are specific populations and environments with particular characteristics where thermal comfort (and heat and cold stress) assessment may require special consideration. These include people with disabilities, babies, children, the sick, pregnant women and people from cultures different from those considered in accepted standards, as well as vehicle environments and particular types of environmental design such as the use of displacement ventilation and chilled ceilings. Consideration of special populations and special environments is provided in this chapter as well as the adaptive approach to thermoregulation, which is relevant in general thermal comfort assessment, but particularly relevant when assessing special populations and environments.

PEOPLE WITH DISABILITIES

Until recently, there was little information concerning whether existing thermal comfort indices, methods and standards are appropriate for determining the thermal comfort requirements for people with physical disabilities and if not, what those requirements are. Knowledge of those requirements will not only enhance understanding of thermal comfort but it will provide guidance for those who wish to design for thermal comfort. This will provide a contribution to the requirements of the Disability Discrimination Act (1995) which are to make goods, facilities and services more accessible to disabled people. Reasons why thermal comfort requirements may differ between people with and without physical disabilities include the effects of the disability on a person's thermoregulatory perception and response. For example, a disability may affect perception of thermal sensation, vaso-control of body skin temperature, ability to cool down through sweating and so on. A physical disability may also require drugs for treatment that may affect thermoregulatory response and the disability may require technical aids such as wheelchairs that will affect thermal comfort (Humphreys *et al.*, 1998). The nature of the disability will determine requirements, as different disabilities will affect thermoregulatory responses in different ways.

Yoshida *et al.* (1993) report a joint Hungarian and Japanese study where 15 people with physical disabilities were exposed to a variety of thermal conditions in a thermal chamber. It was concluded that there were differences in thermo-physiological responses between the disabled group and a control group. Risks of overheating due to restricted sweating responses and overcooling due to disorders of the peripheral blood flow were reported. Other relevant studies have been conducted into computer modelling of human responses where it is possible to change the nature of the thermo-regulatory controlling system, for example, to match that expected in the disabled person (Yoshida *et al.*, 1993). Giorgi *et al.* (1996) conducted a survey of studies of responses of disabled persons to thermal environments on behalf of ASHRAE. The survey, although extensive, made little progress. They found that confounding factors such as types and numbers of subjects, physiological parameters, effects of drugs and missing data, made the task of drawing conclusions complex. They found a wealth of data on disease and thermoregulatory deficits but very few articles on thermal comfort and disability. They found no general conclusions and suggested further work.

The International Standards Organisation (ISO) had identified a need for a standard on thermal comfort requirements for people with disability (also the aged) through their system of international voting. ISO/TC 159 SC5 WG1, 'Ergonomics of the thermal Environment', was the working group that was given the task of producing a standard specifically in that area. After some years, with a Japanese lead and supported by other countries, a draft standard has been produced (ISO CD 14415, 1999). It is expected that the draft standard will be greatly improved with further data and that it will eventually lead to a British, European and International Standard that will provide definitive guidance for the area (see Parsons, 2000 and Chapter 14).

Yoshida *et al.* (2000) conducted a questionnaire survey in five special schools in Yokohama, Japan, of teachers of children who suffer from combined physical and mental disabilities similar to severe cerebral palsy. Teachers observed that the children had major thermoregulatory disorders not usually found in cerebral palsy adults. Among 182 children, 67 suffered from disorder of body temperature control and heat retention. Higher disorder rates were found in younger children, explained by the combination of the original disability and incomplete physiological systems that are still growing. All classrooms were 'comfortably air-conditioned for healthy adults', yet hypothermia and heat retention problems were reported. It was concluded that teachers and families of young, severely disabled children should maintain their children's body temperature and that further research is required into how to design appropriate thermal environments for groups of such children housed together.

The Loughborough studies into thermal comfort requirements for people with physical disabilities

An extensive research programme was carried out over four years in the Human Thermal Environments Laboratory, Loughborough University, UK, to determine thermal comfort conditions for people with physical disabilities (Parsons and Webb, 1999). The Programme was in three parts. *Part One* involved field studies and provided information concerning thermal comfort experiences and requirements for people with physical disabilities in a range of 'real' environments and contexts. Two questionnaire studies were carried out. One obtained responses from 391 people with physical disabilities and the other was completed by 38 carers and professionals who had responsibility for caring for people with physical disabilities (see Hill *et al.*, 2000). *Part Two* consisted of a series of laboratory trials set in a climatic chamber laid out to be similar to a domestic living room. It included a total of 531 subject trials (96 of three hours duration and 435 of two hours duration). These covered subjects with a wide range of physical disabilities exposed to environments of 29 °C, 23 °C and 18.5 °C, corresponding to a predicted mean sensation of 'slightly warm to warm', 'neutral', and 'slightly cool to cool' respectively for people without physical disabilities. *Part Three* integrated the results of the laboratory experiments into a software tool that could be used to determine the thermal comfort requirements for people with physical disabilities (Webb *et al.*, 2000).

In an exploratory experiment, Webb and Parsons (1997) compared the responses of two groups of sixteen subjects (8 male, 8 female) with and without physical disabilities in three, three-hour experimental sessions ($PMV = 1.5$; 0; -1.5: see Chapter 8 in section on gender) in a simulated living room. Subjective ratings of thermal sensation, comfort, preference, stickiness, dryness and draught were recorded every 15 min throughout the experiment. The physical disabilities of subjects included: cerebral palsy, spina bifida, stroke, Friedrich's ataxia, blindness, paralysis, heart condition, encephalitis, Guillain–Barré syndrome, missing limbs and metal in legs.

The results are presented in Figures 9.1 and 9.2 and showed no significant differences ($p > 0.05$) between mean responses of subjects with physical disabilities and those without. There were also no interactions with gender. However, variation in responses was greater in people with physical disabilities for slightly cool to cool and neutral conditions but was less for slightly warm to warm conditions. Although no differences were found on average it was concluded that thermal comfort requirements for people with physical disabilities should be considered on an individual basis.

Loughborough field studies

As a result of observation and open ended questionnaire sessions on six people with physical disabilities, issues and areas of concern were identified.

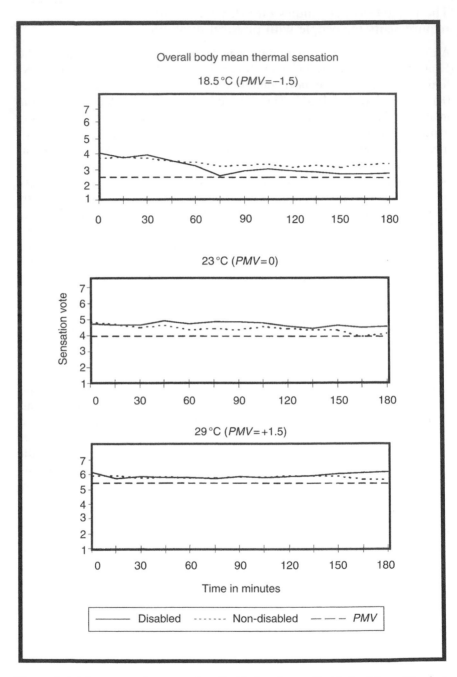

Figure 9.1 Mean sensation votes for disabled and non-disabled subjects (7 – hot; 6 – warm; 5 – slightly warm; 4 – neutral; 3 – slightly cool; 2 – cool; 1 – cold).

Source: Webb and Parsons (1997).

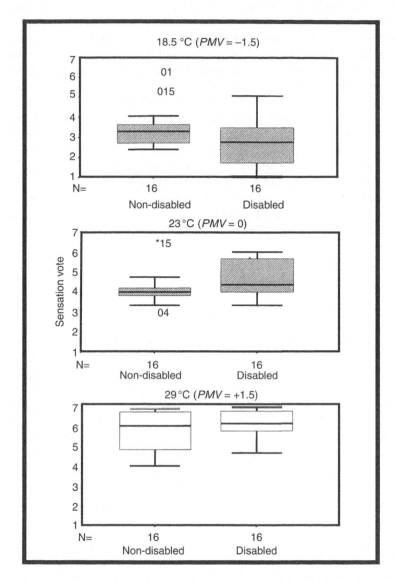

Figure 9.2 Median and range of sensation votes for disabled and non-disabled
subjects (7 – hot; 6 – warm; 5 – slightly warm; 4 – neutral;
3 – slightly cool; 2 – cool; 1 – cold).

Source: Webb and Parsons (1997).

This led to the production of a main questionnaire and survey where a total
of 391 people completed a 'stand alone' questionnaire concerning their
thermal requirements and comfort experiences as well as possible effects

of technical aids and drugs. People with physical disabilities were selected from ten day centres and residential homes, subjects who had taken part in the laboratory trials (see below) (94 responses), and a postal questionnaire of those at day centres, residential homes and holiday centres. Subjects who required assistance to complete forms were assisted by the researchers either in their own homes or at day centres. Visits to day centres and residential and people's own homes allowed data to be collected from those who required assistance. The results showed that subjects with a range of disabilities had responded. These included multiple sclerosis, stroke, rheumatoid arthritis, osteoarthritis, spina bifida, polio, paraplegia, muscular dystrophy and spinal injury. There were 391 respondents of whom 203 were male, 187 were female, and one did not state. The average age of onset of the disability was 26 years with a standard deviation of 20 years. One hundred and two people lived alone, 249 lived with their families, 26 lived in residential accommodation, and 11 in other situations and 3 gave no answer. Of the people who acquired their disability after they were 18 years of age 53 per cent wanted their environment to be warmer than before being disabled, 13 per cent cooler and 33 per cent the same as before they were disabled. Forty-three per cent felt they wore more clothing than before being disabled, 11 per cent less clothing, and 46 per cent about the same. Seventy-four per cent of respondents said that being too warm or too cool affected their ability to carry out tasks and activities. Thirty-seven per cent of respondents stated that they limited the time they spend in public places because they were too warm or too cool, 49 per cent did not limit their time, and 14 per cent did not answer. Different public places elicited different responses to how people found the environment. Except for supermarkets and restaurants 50 per cent of people or less found public facilities 'just right' thermally. Twelve to twenty two per cent of people found the following too cool (in ascending order): cinema, friend's car, local bus, local shops, public swimming pool, supermarket, friends' homes and church. In contrast 11 per cent to 35 per cent of people found the following too warm (in ascending order): hotel, restaurants, friend's home, friend's car, supermarket, shopping centres and hospitals.

In a further study, the thermal comfort requirements of people with physical disabilities were obtained from a questionnaire of the experiences and perspectives of carers. A 'stand alone' questionnaire was designed and piloted. It included questions on environmental conditions, problems and complaints, human responses and issues concerning specific disabilities. Carers were selected from those available at day centres, residential homes and holiday centres for people with physical disabilities. All were completed without assistance and some were posted to holiday centres and some were left when visits were made to day centres and residential homes. The results show that of the 38 carers, 9 were male, 29 were female, 4 were disabled and 34 were not disabled. Twenty nine of the 38 carers thought that people with physical disabilities had different temperature needs. Of the complaints received the most common request was for people to ask to be warmer although the staff tended to want it to be cooler. People with arthritis and stroke tend to wear more clothing when compared to the non-disabled.

Loughborough laboratory studies

The subjective responses of 145 subjects with physical disabilities were obtained in a thermal chamber for slightly warm to warm, neutral and slightly cool to cool conditions (i.e. 435 subject sessions).

The mean and variation in responses and comparison with *PMV* values were determined for subjects with cerebral palsy, spinal injury, spinal degeneration, spina bifida, hemiplegia, polio, osteoarthritis, rheumatoid arthritis, head injury and multiple sclerosis.

It was found that some groups were comfortable in what would be predicted for people without physical disabilities to be 'slightly warm to warm', 'slightly warm' and 'neutral' environments. No group preferred the predicted 'slightly cool to cool' environment although there are individual differences. It is particularly important to consider individual factors and characteristics. For the majority of people in each disability group a preferred environment could be identified, they were as follows: 23 °C, $PMV = 0$; 27 °C, $PMV = +1$ slightly warm for spina bifida, spinal injury, hemiplegia and osteo-arthritis and 29 °C, $PMV = +1.5$, slightly warm to warm for people with rheumatoid arthritis. People with polio were satisfied in the range from 23 °C, $PMV = 0$, neutral to 29 °C, $PMV = +1.5$, slightly warm to warm.

A software tool was designed that can be used to provide guidance on the thermal comfort requirements of people with different types of physical disability. General guidance is provided in terms of a summary of findings as well as mean and variation in responses and the possibility to interrogate individual responses (Webb *et al.*, 2000).

The laboratory results show that BS EN ISO 7730 (1995) can be used without modification as a first approximation to the average responses of people with and without physical disabilities. The variation in responses is different from those of people without physical disabilities. They also show the importance of considering individual characteristics of people with physical disabilities especially when they vary from those that would produce thermal neutrality.

Physical disabilities and the adaptive model

The most influential and established model of thermal comfort throughout the world is undoubtedly that based upon the work of Fanger (1970) which is presented in the British, European and International Standard, BS EN ISO 7730 (1995). Recent developments in thermal comfort research seek to improve on that model by recognizing that true thermal comfort response involves human behaviour. In essence if people are uncomfortable they behave in a way which allows them to become comfortable. It is unlikely therefore that if people are 'hot' for example, that they will remain hot if they can take steps to achieve comfort. A prediction of 'hot' therefore may be misleading. This change in philosophy would then lead to the establishment of ranges of conditions within which people will be able to adapt to maintain comfort (by changing postures, moving around, etc.). Attempts to use this adaptive approach in a general way have been made by Humphreys and Nicol (1995), de Dear *et al.* (1997) and others. Parsons (2002) and Raja and Nicol (1997) have taken a more systematic approach at the level of human behaviour, attempting to identify ranges of adaptive response. Of particular importance is the use of the adaptive approach for people with physical disabilities. The present research has led to hypotheses that suggest that thermal comfort requirements for people with physical disabilities should take account of the restricted ability of people to adapt to the thermal environment. An example would be the perceived threat to people with physical disabilities in a hot or cold environment where those without physical disabilities could change posture or move away but those with physical disabilities could not because they have restricted adaptive opportunity.

BABIES, CHILDREN, THE SICK AND THE PREGNANT

Most studies of thermal comfort have investigated healthy adults and there have been few studies that have varied from this. Specific populations across the human lifespan are significant in number and it is useful to consider their requirements for thermal comfort.

Babies are a specific population with developing systems of thermoregulation and specific size, shape and temperature profile. The relatively large head, large surface area to mass ratio, and high blood flow and limited sweating capacity are conducive to heat loss and gain. A newly born baby may be wet and require attention to maintain appropriate temperatures. What is the meaning of thermal comfort in babies? Health and survival are of major concern but how does a baby perceive discomfort and respond to heat and cold and how can we measure it? Behavioural measures such as crying or moving may be appropriate. In young children behavioural measures may be revealing including mood changes, distraction and irritability. Clothing will be important and require careful consideration, where clothing is determined, not by the child, but by an adult. Clark and Edholm (1985) consider incubation temperatures for naked infants. The thermoneutral temperature was around 32 °C. Infrared thermometry showed that, as temperatures were cooled to 27 °C, skin temperatures changed, with structures under the skin (e.g. blood vessels) becoming visible. Because of the large surface area for heat exchange and limited thermoregulatory control it is likely that thermo-neutral temperatures occur within a narrower range of conditions than those of adults'. Whether this has similar meaning as that of comfort in adults and how the six basic parameters interact to produce comfort in babies is open for study and debate.

As people grow from children to adults, thermal comfort requirements will be similar to that of adults and any specific requirements will probably be due to lifestyle and behaviour. Wyon (1970) studied requirements for school children in classrooms where he used a two-way mirror to observe changes in clothing, attention and posture. Meese and Schiefer (1984) considered thermal comfort requirements for school children in South Africa. There is a need for research into the metabolic rate of children; however, there is no reason to believe that the comfort equation and the *PMV/PPD* index proposed by Fanger (1970) will not apply.

Thermal comfort requirements for the sick will depend upon the nature of the sickness. However, it is probably reasonable to assume that any additional thermal strain on the body is undesirable. That is, in conditions where there is no requirement to provide a thermal stimulus for treatment ('sweat it out', use of cool water to lower body temperature, etc.) or where the condition itself calls for specific conditions (high metabolic rate, effect of drugs). For subjects seated or lying, the metabolic heat production will be low and this should be considered as well as the light clothing often worn when sick. A confounding factor may be that those tending to the sick will be active and clothed and may become uncomfortable in what they perceive as warm environments that are actually thermo-neutral for the sick. Sweating will be a thermoregulatory strain on the body and may cause skin irritation and dehydration. Vasoconstriction due to cold will reduce blood flow to the extremities which may influence healing. Particular attention should be given to local discomfort (e.g. draughts) where the sick have a limited ability to move and adapt to conditions.

There has been no research into thermal comfort requirements for pregnant women. Logical considerations would suggest that hormonal changes may have influence and that, as gestation progresses, the added weight due to baby size and fluid retention will increase metabolic rate and reduce adaptive opportunity. Conditions that provide comfort may therefore be affected and sensitivity to conditions away from comfort may increase.

For babies, children, the sick and pregnant there is a need for systematic research into the interaction of the six basic parameters and thermal comfort.

BEHAVIOURAL THERMOREGULATION, THERMAL COMFORT AND THE ADAPTIVE MODEL

People adapt to preserve comfort. When there is a 'heat wave' in the United Kingdom, enquiries are often received (from the press, public, etc.) on how to 'keep cool' in the hot conditions. Advice in the context of the six basic parameters could be to reduce clothing, increase air movement using fans, stay in the shade and be inactive. In extreme cases take a cool bath. It became clear however, that there are simpler solutions based upon a paradigm shift in thermal comfort from that of traditional laboratory and field research. This led to my (author) lecture to the Department of Human Sciences at Loughborough, 'If you can't stand the heat, go to the supermarket.' Supermarkets are cool places and provide the opportunity to avoid becoming uncomfortably hot. People are not passive receptors of discomfort.

The most effective form of thermoregulation to ensure survival, comfort and performance is classically known as behavioural thermoregulation. Moving to more desirable thermal conditions, adjusting clothing, seeking shelter, opening windows, changing posture, cuddling, lighting of fires, switching on air conditioners or fans, and more are all examples of behavioural thermoregulation. The profound change in the six basic parameters due to behavioural thermoregulation demonstrates the potential of behavioural thermoregulation and emphasizes the continuous dynamic interaction between people and their thermal environments.

Although it has always been accepted that people are not passive, little account has been taken of human behaviour in design and assessment for thermal comfort. Attempts to influence accepted methods for the design and assessment of thermal comfort have used the term 'adaptive modelling' and proposals to use adaptive models of thermal comfort have been discussed for over thirty years. The term 'adaptive' is unfortunate because it implies longer term, even evolutionary adaptation, however it is now widely used in the context of thermal comfort and although requires more stringent definition, it does not seem to cause confusion. Early 'adaptive modellers' include Auliciems (1981) from Australia and Humphreys (1976) from the United Kingdom. The drive for adaptive modelling has continued with researchers from Australia, the United Kingdom, and the United States of America involving theoretical issues and worldwide field studies. The debate is on-going, however the recently constructed ASHRAE, 1998 global database of thermal comfort field experiments and associated adaptive model (de Dear, 1998; de Dear and Brager, 1998, 2002) and the interest of the International Standards Organization (ISO) ensured that adaptive models are being given serious consideration (see also Nicol and Parsons, 2002). From a scientific perspective it is unfortunate that research has been confounded with a number of issues that often appear as 'crusades' (e.g. belief that people across the world have different thermal comfort requirements and that existing standards do not apply. Hence an attempt to find evidence to support the belief. We must prove this so that energy can be saved or costs reduced). The most important point about behavioural or adaptive modelling is the paradigm shift and the new opportunities that it affords to develop our knowledge of thermoregulation. That the current standards (ISO 7730, ASHRAE 55) may be inadequate, that adaptive approaches are more energy efficient, and that simplified adaptive models are convenient to use are all important but do not enhance our understanding of thermal comfort or the application of that understanding.

Adaptive modelling and ASHRAE

The American Society of Heating, Refrigerating and Air-Conditioning Engineers (ASHRAE) has sponsored a series of field studies to evaluate thermal comfort models in support of its thermal comfort standard – ASHRAE 55 (Schiller *et al.*, 1988; de Dear and Fountain, 1994; Donnini *et al.*, 1996; de Dear, 1998; de Dear and Brager, 1998). Early studies compared the effectiveness of the *PMV* index (Fanger, 1970) with the *ET** index (Gagge *et al.*, 1971) which led to the ASHRAE thermal comfort software WinComf (Fountain and Huizenga, 1996). Later studies led to the development of a global database of thermal comfort field experiments (de Dear, 1998) and an adaptive model of thermal comfort and preference (de Dear and Brager, 1998).

The global database was collected from field researchers around the world who supplied raw data in terms of physical measures, clothing, activity and associated subjective thermal comfort responses. Around 21 000 sets of raw data were collected from 160 buildings and used to evaluate existing thermal comfort standards and models. A system of quality control was implemented which divided data into three classes. Class I was the highest quality where all measurements met recommendations of standards. Class II collected all required data but not necessarily to the standards (e.g. measurement heights not included) and Class III where only simple measurements (e.g. temperature, humidity) were made. The complete database and supporting material are available on the world wide web and have generated a number of re-analyses (e.g. Humphreys and Nicol, 2002).

The ASHRAE database allowed the production of a proposed adaptive model for thermal comfort and preference (de Dear and Brager, 1998) which has been considered for inclusion in future versions of standards ASHRAE 55 and ISO 7730. What are becoming accepted as adaptive models are regression equations between outdoor conditions and indoor 'comfort' temperatures. This is unfortunate as they lack the sophistication and contribution to understanding of thermal comfort that a truly behavioural or adaptive paradigm could offer. Two models are provided (see Figure 9.3): one for buildings with centralized Heating Ventilation and Air-Conditioning (HVAC) systems, for which the *PMV* index makes accurate predictions (Figure 9.3a); and one for buildings with natural ventilation where the *PMV* predicts that people would be warmer (in high air temperatures) or cooler (in low air temperatures) than they actually are (Figure 9.3b).

A challenge for the regression models is to determine what is representative of outdoor conditions? In early analyses de Dear and Brager used New Effective Temperature (*ET**) which is derived from measured data and assumptions. Later they changed to mean outdoor air temperature and indoor globe temperature (de Dear and Brager, 2002).

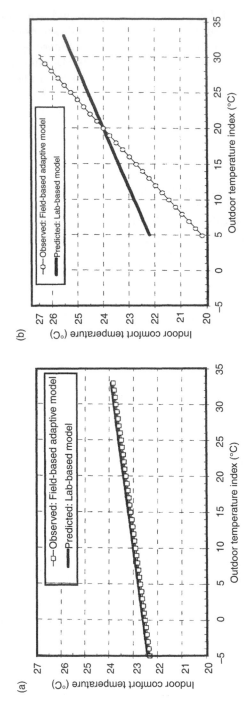

Figure 9.3 Adaptive methods derived from the ASHRAE database: (a) Air-conditioned buildings; (b) Naturally ventilated buildings.
Source: de Dear and Brager (1998).

British studies and adaptive models

British studies of the adaptive approach have come in two phases. The early work of Humphreys and Nicol in the 1970s concluded that different groups of people across the world were comfortable at different temperatures. This led to a regression approach and an adaptive model relating an exponentially averaged mean outdoor temperature to indoor comfort temperature. The second phase of the work began around 1994 (Oseland and Humphreys, 1994 – BRE Proceedings) and has continued into the twenty-first century with further field studies (UK, Europe, Pakistan) and re-analysis of the ASHRAE database.

Nicol and Humphreys (1972) noted that people from warm climates seem to prefer higher temperatures for thermal comfort than those from England, and discussed thermal comfort as part of a self-regulating system involving adaptive mechanisms including adjustment of clothing, metabolic rate and the thermal environment. Humphreys (1976, 1978) presented data from field studies and derived a regression equation between monthly mean outdoor temperature (based upon the average of the daily maximum and minimum for the relevant month) and temperature for thermal comfort. Further field studies (e.g. Nicol and Raja, 1997; Nicol and McCartney, 1997; McCartney and Nicol, 2002) have led to new versions of an adaptive model relating an exponentially averaged running mean outside temperature to temperature required for comfort (T_C). Exponential averaging is used as it allows previous experience to be included so the running mean temperature on a particular day (T_{RM}^n) is calculated from the running mean temperature on the previous day (T_{RM}^{n-1}), and the daily mean temperature on the previous day (T_{DM}^{n-1}). McCartney and Nicol (2002) used the following algorithm to control buildings in the United Kingdom and Sweden.

$$T_C = 0.302\,T_{RM} + 19.39 \quad \text{for} \quad T_{RM} > 10\,^\circ\text{C} \tag{9.1}$$
$$T_C = 22.88\,^\circ\text{C} \quad\quad\quad\quad \text{for} \quad T_{RM} \leq 10\,^\circ\text{C}, \tag{9.2}$$

where

$$T_{RM}^n = 0.8\,T_{RM}^{n-1} + 0.2\,T_{DM}^{n-1}. \tag{9.3}$$

In a preliminary analysis they concluded that using the adaptive algorithm, comfort could be maintained at lower energy costs than when using traditional control methods.

Oseland (1997) conducted three field studies and compared actual thermal comfort votes with those predicted using thermal models and standards. Surveys compared thermal comfort votes in UK houses in Winter and Summer; naturally ventilated and air-conditioned offices, and those of the same subjects wearing the same clothing conducting the same activities when in their homes, their office, and a climate chamber. He concluded that there were differences in comfort votes between homes, types of office and the climate chamber. The poor estimation of metabolic rate and clothing insulation accounted for most of the discrepancy. Prescribing the temperature ranges actually observed to provide comfort is a more relevant practical approach to achieving comfort. Oseland *et al.* (1998) consider building design and management for thermal comfort and provide guidance on adaptive options to maintain comfort. This attempts to quantify the effects of adaptive behaviour on required 'optimum' temperatures. Regression equations relate outside conditions to indoor temperatures for comfort. An assumption is therefore made concerning human adaptation and behaviour; empirically quantified but not explained (e.g. indoor clothing is reduced when outdoor temperatures rise). The adaptive approach explains the variation in thermal comfort temperatures.

De Dear and Brager (1998) consider three categories of thermal adaptation: behavioural, physiological and psychological. Behavioural adjustments include personal (remove clothing), technological (turn on fan), or cultural (siesta in midday heat). Physiological adaptation includes acclimatization which will provide greater tolerance to heat (Gonzalez, 1979) but has been found not to affect comfort requirements (Brierley, 1996; Parsons, 2002). Psychological adaptation '... refers to altered perception of, and reaction to, sensory information due to past experience and expectations' (de Dear and Brager, 1998).

Behavioural adjustment is identified as being most effective for people to maintain their thermal comfort (Williams, 1996). The extent to which the environment allows such behaviour is termed as the adaptive opportunity (Baker and Standeven, 1996). Physiological acclimation is not likely to be a factor in moderate conditions. Psychological adaptations are identified as playing a significant role in explaining the differences in responses in air-conditioned versus naturally ventilated buildings. De Dear and Brager (1998) cite the work of Pacink (1990) who notes the importance of perception of control in terms of available control (adaptive opportunity), exercised control (behavioural adjustment) and perceived control (expectation). Expectation is considered a likely explanation for differences between thermal comfort predictions (*PMV*) and responses in non-air-conditioned buildings. Fanger and Toftum (2002) identified this as an explanation and introduced an expectancy factor e ranging from $e = 0.5$ (low expectancy) to $e = 1.0$ (high expectancy). To obtain the true prediction, taking account of expectation, *PMV* is multiplied by e. Although this correction appears rather arbitrary, in the absence of further information it provides a practical adjustment. Whether an adjustment is needed however, and whether it depends upon building type remains a matter for debate.

Equivalent Clothing Index (I_{EQUIV})

In any environment there will be opportunity to adapt to maintain thermal comfort and these adaptive adjustments (behaviours) can take many forms and can occur in combination (e.g. adjust clothing, change activity, open a window, change posture). Each of the actions and their combination will have an effect on human heat exchange. It is therefore possible to represent these effects in terms of the equivalent effect of changing one of the parameters in the heat balance equation. This could be any of the six basic parameters, however a convenient approach would be to relate the total effect of all adaptive behaviour to the equivalent effect of adjusting clothing. An equivalent clothing index (I_{EQUIV}) can be described as follows:

> The Equivalent Clothing Index (I_{EQUIV}) is the clothing insulation that would give equivalent thermal comfort to people with no adaptation as the thermal comfort of people who adapt to their thermal conditions.

A group of people initially wearing 1.0 Clo who change clothing, change activity, change posture and open a window to maintain a neutral thermal sensation may have an equivalent clothing index value of 0.2 Clo, where 0.2 Clo would represent the clothing insulation required to maintain a neutral thermal sensation in the original conditions. The total of all adaptive behaviour therefore summates to a reduction of 0.8 Clo. The equivalent clothing index value can then be substituted into the *PMV* equation, instead of the clothing insulation value, to give a *PMV* that takes account of adaptive behaviour. The equivalent clothing index can then be used to attempt to quantify the relationship between outside conditions and indoor comfort temperatures.

By definition, people continually adapt to maintain thermal comfort. In air-conditioned buildings the requirement for behavioural adjustment is low and probably directly related to clothing adjustment. A re-analysis of de Dear and Brager (1998) provides the following curve for air-conditioned buildings (Figure 9.4).

Assume a neutral or comfortable indoor condition, $t_a = t_r$, $rh = 50$ per cent, sedentary activity, $v = 0.15 \, \mathrm{m \, s^{-1}}$.

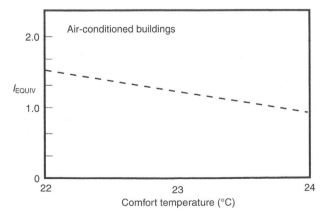

Figure 9.4 I_{EQUIV} for air-conditioned buildings.

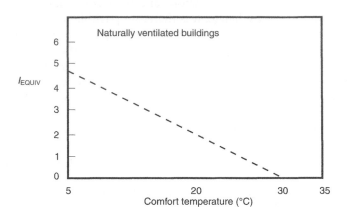

Figure 9.5 I_{EQUIV} for naturally ventilated buildings.

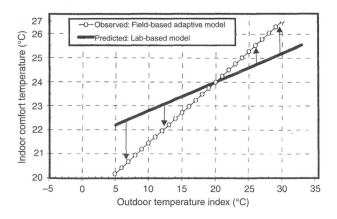

Figure 9.6 Indoor comfort temperature for naturally ventilated buildings, predicted using I_{EQUIV}.

For naturally ventilated buildings the following applies (Figures 9.5 and 9.6).

The arrows show the adjustments to predictions taking account of adaptation. The predicted responses now match the observed responses. If I_{EQUIV} is then used in a more dynamic *PMV* model, the supposed discrepancy between the 'adaptive model' and static model (*PMV*) no longer exists. This assumes that if any discrepancies occur, they are accounted for using behavioural adjustment.

Table 9.1 shows a possible relationship between adaptive opportunity and I_{EQUIV}.

$$I_{EQUIV} = I_{start} - \left(I_{ADJ} \times I_{start}\right) \text{ in the heat} \tag{9.4}$$
$$I_{EQUIV} = I_{start} + \left(I_{ADJ} \times I_{start}\right) \text{ in the cold.} \tag{9.5}$$

For cases where no adaptation is possible then clothing, posture, activity, physical environment (e.g. windows) cannot be adjusted. Therefore, the

Table 9.1 I_{ADJ} values for a range of adaptive opportunities

Adaptive opportunity	I_{ADJ}
Minimum	0
Low	0.25
Medium	0.5
High	0.75
Maximum	1.0

equivalent clothing is the actual clothing worn at the beginning of the exposure period ($I_{EQUIV} = I_{START}$). For maximum adaptation, it will be possible to take off all clothing as well as making other adaptive adjustments (open windows, reduce activity). For these conditions it may be useful to adjust parameters separately (e.g. metabolic rate) in the rational index (e.g. *PMV*) as well as using $I_{EQUIV} = 0$. In these circumstances, thermal comfort is an unusual concept and, in practice, adaptive effects greater than taking off all clothing would be unusual. It may therefore be reasonable to assume $I_{EQUIV} = 0$, a minimum practical value. A similar argument applies for I_{EQUIV} in cold conditions where doubling of clothing insulation is a realistic maximum with more detailed analysis required for special cases. It should be noted that I_{ADJ} may be different for hot conditions than for cold conditions as different adaptation may be required; for example, if opening windows is the only adaptive opportunity, then I_{ADJ} may be 'medium' in the hot but 'minimum' in the cold etc. In practical terms there will be an I_{EQUIV} required to maintain comfort. If this is less than the maximum adjustment possible then comfort will be achieved.

I_{EQUIV} and the comfort temperature range

Although a rational index such as the PMV provides a versatile tool with which to determine thermal comfort conditions in terms of the six basic parameters (variables), there is a requirement to provide temperature ranges within which people can maintain comfort. Suppose people in a building wear 1.0 Clo, perform sedentary activity in conditions with no radiant load ($t_a = t_r$), and vapour pressure of 1.0 kPa in still air. They can reduce their clothing but cannot increase it; they can move around and open windows. We could assess the adaptive opportunity as high in the heat and low in the cold. A *PMV* of 0 (neutral) is then obtained at 24 °C for $I_{EQUIV} = I_{START} = 1.0$ Clo; 28.5 °C for $I_{EQUIV} = 0.25$ Clo and 22.8 °C from $I_{EQUIV} = 1.25$ Clo. This provides a comfort temperature range of 22.8–28.5 °C. If a range between +1 (slightly warm) and −1 (slightly cool) is regarded as acceptable then an acceptable range would be 18.2–30.5 °C.

The *PMV* is an example of a rational index and its validity does not affect the principles of the above. This behavioural adjustment method is a practical way forward. It is preferable to 'expectancy' adjustments where statements such as, 'It is what they are used to and do not expect better', border on the unethical and, in any case, are flawed as if people claim comfort at 30 °C, how do we know that they would also not claim comfort at $PMV = 0$ (e.g. 24 °C).

Thermal comfort in special environments

People interact in a dynamic way with dynamic environments defined by the six basic parameters. In that context the term special environment has no meaning. All conditions are particular cases of the Human Thermal Environment and the same general principles apply. It is useful however to consider such special environments as they often occur, they vary from the norm and there may be special issues to address. They may be regarded as categories of application. It should be noted that general principles of the science of Human Thermal Environments will still apply. Do we really require a special thermal comfort index for vehicles for example? Any universal science of human thermal environments should of course include all thermal conditions. All human thermal environments will affect individuals in total (including noise, light, pressure, etc.) environments and in a particular context.

Displacement ventilation and chilled ceiling environments

If cool fresh air (e.g. 19 °C) is continuously introduced into a room at floor level, equipment and people will heat the air and it will rise in convective plumes to the ceiling. If it is then removed at the ceiling the system is called displacement ventilation (Figure 9.7). Traditional mixed ventilation systems introduce air and mix it with 'older' air. People in rooms therefore breathe and re-breathe air and also breathe air that other people have breathed. Any pollution of the air will be included in the mixture. Displacement ventilation systems, if correctly implemented, provide a continuous stream of fresh air. This is an energy efficient system that has potential to improve the quality of the environment. To enhance heat extraction, a chilled ceiling, often chilled by cold water in pipes, is used. Any other system such as evaporation from the upper surface, or rapid circulation of cool air across a thin partition could be used, depending upon practicality. It is the low surface temperature and radiant effect that is required. The surface should be at a temperature above the dew point for the room to avoid office rain. This is because displacement ventilation alone may not be sufficient to remove the required heat load. The nature of displacement ventilation and chilled ceiling environments creates moving air at low temperatures around the feet, thermal gradients and asymmetric radiation conditions which may require special consideration for comfort. There have been a number of studies of thermal comfort in such environments, mostly conducted in climatic chambers but some in buildings (Wyon, 1996; Wyon and Sandberg, 1996; Olesen *et al.*, 1979; Loveday *et al.*, 1998; Akimoto *et al.*, 1999).

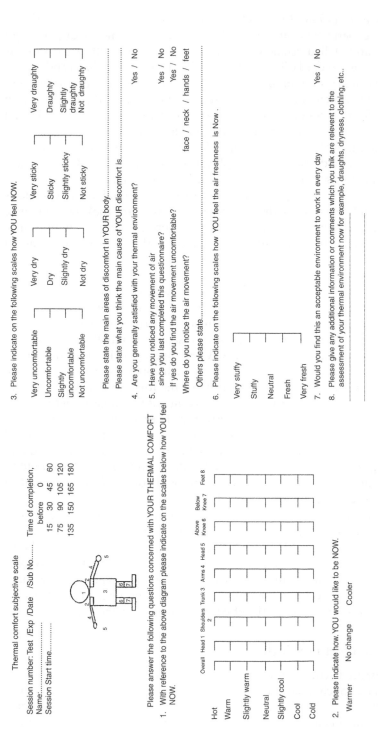

Figure 9.7 Subjective scales used to measure thermal comfort in displacement ventilation and chilled ceiling environments. Source: Loveday *et al.* (1998).

Loveday *et al.* (1998, 2000, 2001, 2002) report the results of a series of laboratory experiments into thermal comfort conditions in displacement ventilation and chilled ceiling environments. A total of 200 seated subjects were dressed in standard office clothing (0.75 Clo including chair) and carried out VDU work in groups of four (2 males, 2 females) over three hour sessions.

Twelve experimental conditions were studied. Eight of these were set up as follows: four values of chilled ceiling surface temperature (14, 16, 18 and 21 °C) at two levels of relative humidity ('medium' and 'low') corresponding to approximately 50 per cent *rh* and 25 per cent *rh* respectively; for these eight conditions, a displacement supply air temperature of 19 °C (the typical design value for sedentary occupants) at 3.9 air changes per hour was maintained (fresh air requirement for subjects). A total of 128 subjects (64 males, 64 females) in the age range 21–60 years took part in these experiments. A further four conditions were set up as follows: four air change rates (2.5, 3.9, 6.0 and 8.0 *ac/h*) at 19 °C supply with a ceiling temperature of 18 °C and 'medium' relative humidity; 64 subjects (32 males, 32 females) took part in these experiments. In all tests, subjects completed a questionnaire (Figure 9.7) at 15 min intervals throughout the three-hour period, the data from the last 30 min being used in the analysis (steady state conditions).

For investigations concerning the effects of vertical radiant asymmetry on thermal comfort, eight subjects were tested individually. Here, ceiling temperatures of 22, 18, 14 and 12 °C were investigated over a three-hour period, the predicted mean vote (*PMV*) for each condition being maintained at 'neutral' by making small adjustments to the four wall surface temperatures. In this way, any departure from neutrality of the subject's actual mean vote (*AMV*) could be attributed to vertical radiant asymmetry alone.

The overriding results were remarkable in that for the wide range of conditions studied, almost no thermal discomfort was experienced and that the existing standard EN ISO 7730 (PMV/PPD from Fanger, 1970) could be used without modification to give accurate predictions of thermal comfort votes of subjects (see Figures 9.8–9.11).

It is interesting that the adaptive approach and this controlled, closed environment approach to thermal comfort, both claim energy efficiency. An important finding in field studies of displacement ventilation and chilled ceiling systems in offices is that careful consideration should be given to installation, monitoring and maintenance. Instances where people have closed off (covered over) floor vents, added fans and opened windows are common and have destroyed the effect and undermine the application.

Thermal comfort in vehicles

When compared with buildings, human thermal environments in vehicles typically have much greater variation in space and time. They can often be greatly influenced by outdoor conditions in particular solar radiation through glazing in closed cabins (O'Neill and Whyte, 1985). The assessment

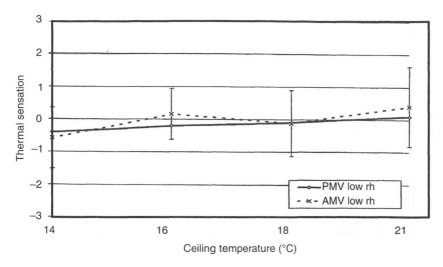

Figure 9.8 Mean thermal sensation for a range of ceiling temperatures, and low humidity, in displacement ventilation and chilled ceiling environments (−3–cold; −2–cool; −1–slightly cool; 0–neutral; 1–slightly warm; 2–warm; 3–hot).

Source: Loveday *et al.* (1998).

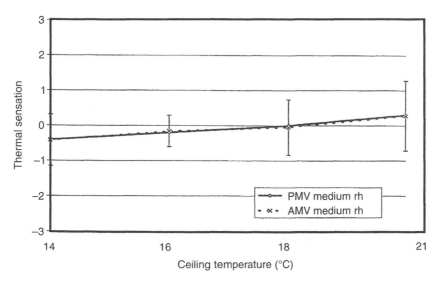

Figure 9.9 Mean thermal sensation for a range of ceiling temperatures, and medium humidity, in displacement ventilation and chilled ceiling environments (−3–cold; −2–cool; −1–slightly cool; 0–neutral; 1–slightly warm; 2–warm; 3–hot).

Source: Loveday *et al.* (1998).

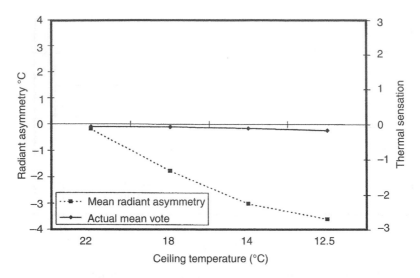

Figure 9.10 Mean thermal sensation for a range of radiant asymmetries in displacement ventilation and chilled ceiling environments (−3 – cold; −2 – cool; −1 – slightly cool; 0 – neutral; 1 – slightly warm; 2 – warm; 3 – hot).

Source: Loveday *et al.* (1998).

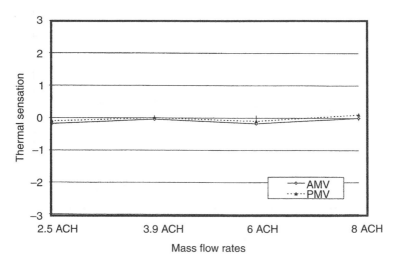

Figure 9.11 Mean thermal sensation for a range of mass flow rates in displacement ventilation and chilled ceiling environments (−3 – cold; −2 – cool; −1 – slightly cool; 0 – neutral; 1 – slightly warm; 2 – warm; 3 – hot).

Source: Loveday *et al.* (1998).

methodology will depend upon the type of vehicle. For example in large closed vehicle environments such as can be found in ships, railway carriages and buses, environments can be assessed as for rooms. For environments not similar to rooms in buildings, special conditions may apply.

Rohles and Wallis (1979) conducted thermal comfort trials in vehicle simulators and concluded that in cars, cool air should be blown towards the chest area to provide comfort. Wyon *et al.* (1985) conducted field trials in cars using a heated thermal manikin (VOLTMAN) and modified the original Equivalent Temperature Index (Bedford, 1936) into an Equivalent Temperature Index for vehicles. Madsen and Olesen (1986) also use thermal manikins in vehicle assessment European research (EQUIV) evaluated the Equivalent Temperature Index. Nilsson *et al.* (1999) cite SAE (1993), defining the Equivalent Temperature as, 'The uniform temperature of the imaginary enclosure with air velocity equal to zero in which a person will exchange the same dry heat by radiation and convection as in the actual non-uniform environment.' Nilsson *et al.* (1999) extend the above definition (whole body Equivalent Temperature) to define Segmental Equivalent Temperature (for a body part), Directional Equivalent Temperature and Omnidirectional Equivalent Temperature. Methods of measurement include basic climatic parameters (t_a, t_r, v), heated sensors, a heated ellipsoid sensor, a flat surface sensor, a local discomfort metre and a thermal manikin. The relationship between Equivalent Temperature and thermal comfort has yet to be derived. The emphasis being on standardizing measurement so that values can be compared and used in testing.

Hodder and Parsons (2001a) present a series of laboratory and field studies into thermal comfort in vehicles with particular reference to the effects of solar radiation. This was part of a European programme to investigate vehicle glazing. A specially constructed solar radiation simulation chamber in the Human Thermal Environments Laboratory at Loughborough was used to investigate the responses of human subjects sitting in a car seat and dressed in standard clothing (white shirt with sleeves rolled up to the elbows and light trousers – 0.7 Clo including the seat). In a series of experiments they investigated the effects of radiation level (0, 200, 400, 600 W m^{-2}), radiation type and glazing type (to investigate the effects of radiation spectrum) and the contribution of re-radiation from a dashboard to thermal comfort when combined with direct simulated solar radiation. Subjects were exposed to radiation for 30 min in a standard test protocol (trunk and legs exposed to direct radiation, head shielded) and reported thermal sensation, discomfort, stickiness, preference, pleasantness and acceptability using a subjective form administered every five minutes. The results are shown in Figures 9.12–9.15.

In a further experiment, Ohnaka *et al.* (2002) considered the relationship between head only radiation exposure, trunk only radiation exposure and trunk and head together. The results showed that radiation level, directly from the source (simulated solar lights) had dominant influence on thermal

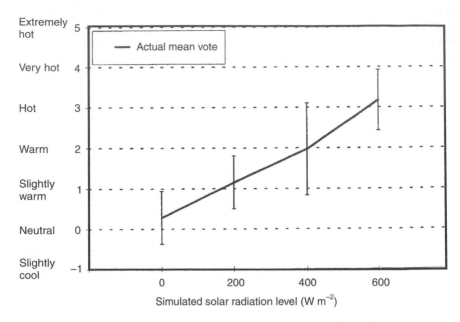

Figure 9.12 The effect of simulated solar radiation level on thermal sensation.

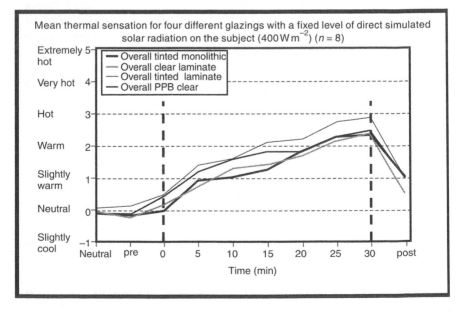

Figure 9.13 The effect of spectral content of simulated solar radiation on thermal sensation over a 30 min exposure.

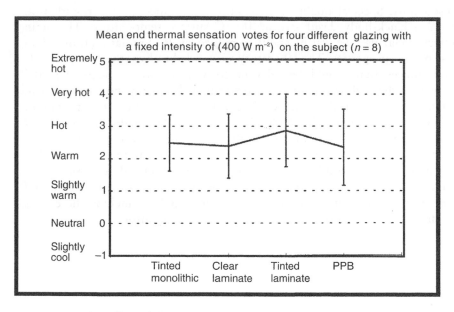

Figure 9.14 The effect of glazing type on thermal sensation at the end of a 30 min exposure to $400 \, W \, m^{-2}$ simulated solar radiation.

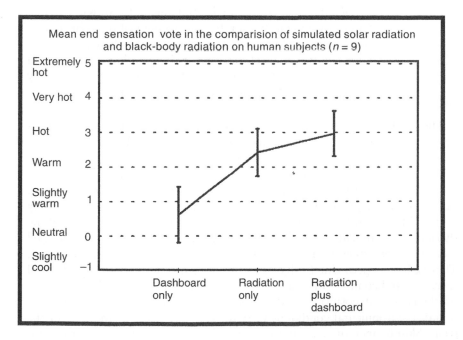

Figure 9.15 Comparison between dashboard radiation and direct simulated solar radiation on thermal sensation.

discomfort. Spectral quality was of little importance as was re-radiation from a black-body dashboard at 100 °C surface temperature. The main effect of glazing type was its ability to reduce levels of radiation rather than its spectral properties. The laboratory studies led to hypotheses for models and indices that would be relevant to vehicle environments. For the extended thermal sensation scale (extremely hot, very hot, hot, warm, slightly warm, neutral, slightly cool), a modified *PMV* index was proposed, where an increase in radiation level by 200 W m^{-2} provides an increase of one scale value (e.g. neutral to slightly warm or 400 W m^{-2} from slightly warm to hot etc.). Further research is required to ensure that the extension of the scale corresponds to the behaviour of the ISO/ASHRAE scale (hot to cold) and that the model applies for people starting with thermal sensations away from neutral. The modified *PMV* to include the effects of direct solar radiation is given in the following equation.

$$PMV_{\text{solar}} = PMV + \frac{RAD}{200},$$

where

PMV_{solar} = Predicted Mean Vote (from +5, extremely hot; through 0, neutral; to −5, extremely cold) taking account of the effects of solar radiation,

PMV = Predicted Mean Vote (ISO 7730; Fanger, 1970) with correction to clothing insulation to take account of the vehicle seat, and t_r measured inside of the vehicle in the shade,

RAD = Direct radiation level received at the position of the person (estimate from 1000 W m^{-2} = maximum; 600 W m^{-2} = very high; 200 W m^{-2} = low; and 0 W m^{-2} = minimum).

The model, along with a number of others, was evaluated in field trials conducted in Spain (Hodder and Parsons, 2001a,b). The location was selected as it provided a consistent high temperature and clear sky European environment and allowed controlled experiments to be carried out in air-conditioned cars for extreme conditions. Two Rover Freelander cars were driven on a straight motorway between Seville and Cadiz over a ten-day period. The cars were extensively instrumented to measure all aspects of the thermal environment received by the vehicle occupants. Each car contained a driver, experimenter and an experimental subject in front and back seat. That is, four experimental subjects (male) gave subjective ratings of thermal sensation and comfort for a range of conditions. These included glazing type on the cars, seat position, car direction and time of day (morning or after-noon). Environmental conditions were measured in the cars throughout the trials. Skin temperatures of subjects were also recorded. The extensive data-base of responses showed that subjective measures were highly correlated

Table 9.2 Correlation of subjective and objective measures of thermal comfort in
vehicle field trials

Variable	Correlation coefficient	Rank	n
Thermal sensation vote	1	1	640
Thermal comfort vote	0.9	2	640
Stickiness vote	0.8	3	620
Thermal preference vote	0.8	4	640
Pleasantness vote	0.8	5	640
Dashboard surface temperature	0.5	6	640
Globe temperature dash	0.5	7	640
Air temperature dashboard	0.4	8	620
Mean radiant temperature dash board	0.4	9	605
Glazing internal temperature	0.4	10	625
Weighted tr dash and centre	0.4	11	635
Weighted ta of dash, subject and centre	0.4	12	640
PMV with solar on dash over 200	0.4	13	622
Radiation at dashboard	0.4	14	640
Weighted ta at subject knees & feet	0.4	15	640
Mean skin temperature	0.4	16	640
Mean temperature for subject	0.3	17	640
Weight ta central and local	0.3	18	625
PMV + calculated MRT	0.3	19	620
Weighted ta at subjects' shoulder & knees	0.3	20	640
Weighted ta at subject's low 3 sites	0.3	21	619
Forehead temperature	0.3	22	620
PMV + solar combined	0.3	23	640
PMV + solar legs	0.3	24	640
Globe temperature centre	0.3	25	640
Teq Bedford	0.3	26	640
PMV centre + solar (chest)	0.3	27	640
PMV using centre shaded measures	0.3	28	640
External air temperature rear	0.3	29	640
Mean radiant temperature centre	0.3	30	640
Air temperature centre	0.3	31	640

Source: Hodder and Parsons (2001a).

with each other (e.g. thermal sensation, thermal comfort etc.). The modified
PMV (PMV_{solar}) model performed well in comparison with other measurements and indices (e.g. WBGT, Globe temperature, air temperature, etc.).
The overall performance (correlation between predictive model and thermal
sensation) of models is shown in Table 9.2.

The modified PMV index (PMV_{solar}) requires further investigation, however it provides a practical index for use in vehicles that links with existing
methods and seems to have potential.

Other issues which make vehicles special environments include crowding
(e.g. on a commuter train) where heat transfer rates are restricted (especially
evaporation of sweat) by other people. Braun and Parsons (1991) suggested

a crowding correction factor based upon laboratory experiments of the physiological effects of crowd density. Psychological factors will also apply. The thermal properties of the chair in terms of its material construction, have been extensively studied by Fung and Parsons (1996) and Wyon *et al.* (1985) has described a ventilated seat for use by lorry drivers. This leads to the issue of personal control in vehicles where each occupant can control their own environmental conditions. Use of directed vents is common in aircraft and further possibilities may occur with the use of heated seats. Brooks and Parsons (1999) found that people could maintain comfort over a range of temperatures with a controllable heated seat. ISO 14405 parts 1–3 are currently under development and will provide guidance on assessing vehicle environments, the Equivalent Temperature index, the use of thermal manikins in cars and subjective assessment methods (see Chapter 14).

Other special environments

There are many other special environments from space stations to submarines and it is important to remember that in assessing all of these, fundamental principles will apply. For example, the six basic parameters should be considered, humans are homeotherms, heat balance equations can be constructed and the thermal audit can be carried out. For people immersed in water, survival times can be calculated from heat balance equations and comfort may be related to similar physiological parameters (restricted sweating but skin temperatures close to water temperature) as for environments in air. For people around swimming pools minimal clothing and wet skin can be included in standard assessments. It may be, however, that comfort criteria and the meaning of comfort vary in some special environments. During light activity a feeling of pleasure may occur related to self-image and sense of achievement (as well as hormonal changes – endorphins) which may confound the more passive concept of comfort. In warm environments or during activity the sense of pleasure from high air velocities and evaporation of sweat from skin may also dominate comfort responses. Cabanac (1981) suggests that one should differentiate between pleasant states which are dynamic, based upon correcting stimuli (the body striving for neutrality) and thermal neutrality where stimuli are indifferent.

In special environments where air pressure varies from normal, heat transfer will be affected. In hyperbaric environments the ability to evaporate sweat will be greatly reduced and the convective heat transfer will increase. This will affect respiratory heat transfer and heat transfer from the skin. Parsons (1992c) demonstrated this for application in compressed air tunnelling (see also O'Brian, 1996; O'Brian *et al.*, 1997). The restriction to heat transfer by evaporation is proportionately greater than the increase in convective heat transfer. In hypobaric environments (on top of mountains, in aircraft) evaporative heat loss is increased and convective heat transfer decreased. Nishi and Gagge (1977) describe an Effective Temperature scale for use in hypo- and hyperbaric environments. One effect of increased pressure is that it narrows the range of conditions (e.g. temperatures) over which comfort can be provided. In diving, for example, the range of temperatures from comfort to heat or cold stress is small and careful control of conditions is required. Using the heat balance analysis of Nishi and Gagge (1977) thermal comfort conditions in aircraft (maintained at 8000 feet (2500 m)) can be derived. Of interest is the effect on whole body comfort conditions but also on local discomfort. Do draught rating equations apply? If displacement ventilation systems were used on aircraft, would problems occur?

10 Heat stress

INTRODUCTION

Where human thermal environments (in terms of air temperature, radiant temperature, humidity, air velocity, clothing, and activity) provide a tendency for body heat storage, the body's thermoregulation system responds to attempt to increase heat loss. This response can be powerful and effective but it can also incur a strain on the body (and sufficient heat may not be removed), which can become unacceptable and eventually lead to heat illness and death.

History has provided many examples of heat illness and death caused by heat stress. Leithead and Lind (1964) provide a brief review. Cases occur in both outdoor and indoor environments and are distributed worldwide. There are examples from military training and action, industrial activity and normal civilian activity including tourism. A key factor is how accustomed those exposed are to the heat stress. Where people used to more temperate climates are exposed to hot environments there can be major problems if those exposed are neither behaviourally nor psychologically acclimatized to the heat. This is demonstrated by the high incidence of heat illness in European and North American cities during heat waves. Incidents have also been reported in steel mills and at the Boulder Dam site in the USA, in miners in India and South Africa, many activities in North Africa, Arabia and India, including the Crusades, military campaigns in both the First and Second World Wars, workers in the oil industry, pilgrims to Mecca and workers on ships in the Red Sea, Persian Gulf and the Gulf of Aden. Cases have also been reported on troop carrying and prison ships during the Second World War where liners not equipped for tropical conditions were used (e.g. the passenger liner, Queen Mary).

Leithead and Lind (1964) also cite the cases where, in one night, 123 of 186 British soldiers died of heat stroke in the Black Hole of Calcutta and also the 194 of 281 civilians who died while imprisoned for one night in the Sudan. In the First World War, in Mesopotamia, 426 heat stroke deaths were counted in one month, descriptions being recorded of how men began to die in delirium or convulsions with dry skin and body temperature from

41.5 to 46.1 °C. They also cite the case of 125, mainly military, recruits in the Southern USA who died in intensive training camps.

There is a general figure of around 2 per 1000 workers expected to be at risk. Failure to appreciate the dangers is often a contributory factor. Leithead and Lind (1964) for example, describe the incident of a young electrician who worked on a ship in the Red Sea. Approximately one hour after entering a space to effect a repair, the electrician collapsed in convulsions and coma with a body temperature of 42.2 °C. There have been fatalities in the baking industry where men have been trapped while maintaining conveyor belts in production line ovens for producing bread and working practices were either inadequate or ignored.

More recent surveys have shown similar findings. In a BBC television programme (also in The Listener, 1990) entitled 'Dressed to Kill', a number of examples of deaths are presented, during military training. The recent trend for large numbers to travel by air to hot countries for vacations has also produced casualties and the hypothesis of global warming has led to speculation of climate changes.

Although heat stress causes many casualties, the mechanisms are well understood. Consideration of human thermal environments in terms of the six basic parameters, including clothing properties and activity causing metabolic heat production, is important. Working practices for hot environments (e.g. NIOSH, 1986) are well established, including the use of appropriate heat stress indices, acclimatization programmes, and the importance of water replacement. Where practices have been implemented, numbers of casualties have greatly reduced (e.g. Wyndham, 1974). There seems, however, to be a tendency to forget the requirement for working practices, despite a satisfactory knowledge, lessons seem to have to be re-learnt. Perception of risk is often lower than actual risk and risk assessment should take account of the ability to be removed from an exposure (Parsons, 2000). A hot process (baking, car manufacturing production line, nuclear power station) may require long cooling times before maintenance and repair can be carried out. Delay is expensive and incentive to work in dangerously hot conditions is great. Appropriate working practices and procedures are essential for safety.

PHYSIOLOGICAL RESPONSES TO HEAT

In heat stress, the body temperature may rise and receptors sensitive to change in temperature in the skin, muscle, stomach and other areas of the central nervous system, as well as in the anterior hypothalamus itself, all send signals via the central nervous system to the anterior hypothalamus. The ratio of sodium to calcium ions is also monitored. Where temperatures are above 'set point' levels, blood circulation is controlled in specific areas of the body through the sympathetic nervous system which dilates the cutaneous vascular bed and hence increases skin blood flow and invokes the sweating mechanism if necessary. This provides greater potential for heat to flow from the body and hence maintain body temperature. Because the heart cannot supply blood to all of the body's organ systems the autonomic nervous and endocrine system control allocation of blood to competing organs.

During exercise there is an initial sympathetic vasoconstriction so that blood may flow to active muscles. If heat is required to be dissipated there is an increased cutaneous blood flow. During continuous work, in the heat, central nervous blood volume decreases as the cutaneous vessels dilate. The stroke volume falls and the heart rate must increase to maintain cardiac output. The effective circulatory volume also decreases as water is lost through sweating (Hales and Richards, 1987).

Sweat glands are stimulated by cholinergic sympathetic nerves and secrete sweat onto the surface of the skin. Sweat rates of 1 litre per hour are common and for each litre evaporated 675 W of heat are lost per hour (NIOSH, 1986). However, large sweat losses reduce body water content and hence thermoregulatory effectiveness. During sweating salt is lost at about 4 g per litre (1 g per litre in acclimatized persons). As a normal diet provides 8–14 g per day, a normal diet is often sufficient. Salt tablets can irritate the stomach and heavier use of salt at meals is preferred, but salt supplementation will probably not normally be required (NIOSH, 1986). Potassium is also lost in sweat and a high salt intake may increase potassium loss. In most cases however potassium will be replaced by a normal diet (especially fruit and vegetables).

The overall physiological response for continued heat storage is therefore vasodilation to increase skin temperature and then sweating leading to profuse sweating (including ineffective dripping of some sweat losing insignificant heat but important water). As 'core' temperature continues to rise and the skin is completely wet, hidromeiosis (a reduction in sweating) may occur due to swelling and blocking of sweat glands in the wet humid conditions (Kerslake, 1972). This is often confused with so-called sweat gland fatigue. The decrease in sweating promotes a further, often rapid, increase in 'core' temperature to beyond 38–39 °C where collapse may occur to above 41 °C (rectal temperature), where heat stroke may occur. There will be mental confusion, behavioural changes, failure in central nervous thermoregulation and sweating and death with eventual denaturing of body protein. NIOSH (1986) consider age, gender, body fat, drugs (including alcohol) and other non-thermal disorders as important individual factors. There is a large individual variability in the mechanisms of response, which are not fully understood. Physical fitness however, has been shown to be of great importance (Havenith, 1997). The mnemonic, 'SHAFTS' can be used to advise people how to increase tolerance to heat. This is: Sensible (i.e. appropriate behaviour); Hydrated; Acclimatized; Fit; Thin; and Sober (including avoidance of alcohol and other drugs).

Heat disorders

Leithead and Lind (1964) conclude that heat disorders occur for one or more of three reasons:

1   the existence of factors such as dehydration or lack of acclimatization;
2   the lack of proper appreciation of the dangers of heat, either on the part of the supervising authority or of the individuals at risk; and
3   accidental or unforeseeable circumstances leading to exposure to very high heat stress.

They conclude that many deaths can be attributed to neglect and lack of consideration and that even when disorders do occur, much can be done if all the requirements for the correct and prompt remedial treatment are available. In climates such as those found in Singapore, military personnel are exposed to hot, humid conditions and must carry out essential tasks in protective clothing. Although heat stroke will occur, severe consequences have been avoided by organizational methods including extensive training of personnel and efficient back-up systems to transport casualties rapidly to hospital.

There are a number of classification systems for heat disorders. The mechanisms are summarized in Figure 10.1 (Belding, 1970) and a description is provided in Table 10.1 (Goldman, 1988).

There are a number of other complaints related to heat exposure. For example, in mildly sunburnt skin, sweat can be trapped and accumulated under the dead surface layer and cause discomfort as well as reduce evaporative efficiency. In industries where chemicals and particulates are present in the air, they may interact with sweat on the skin surface to cause complaints. The interaction between chemical substances in the air and a sweating person has yet to be fully explored but can be significant. For example, studies were conducted into the effects of poisonous gas (on volunteer subjects) in hot conditions, in Australia during the Second World War. It was found that areas where most sweating occurred on the body were those areas most affected. Chemical and biological hazards may therefore have greater effect in hot conditions than in cold conditions. Protective clothing and equipment may promote sweating and increase the risk they are protecting against.

HEAT STRESS INDICES

Introduction

A heat stress index is a single number that integrates the effects of the basic parameters in any human thermal environment such that its value will vary with the thermal strain experienced by the person exposed to a hot

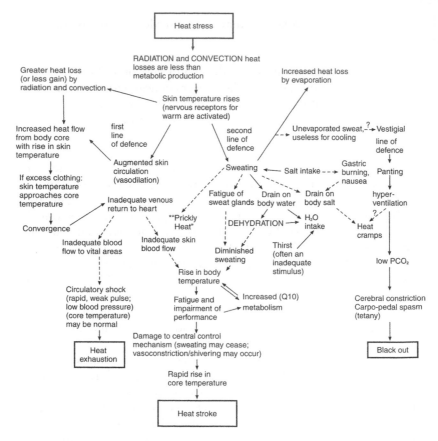

Figure 10.1 A diagrammatic representation of the effects of heat stress on the body. Source: Adapted from Belding (1970).

environment. This index value (measured or calculated) can then be used in design or in working practices to establish safe limits for work. Much research has gone into determining the definitive heat stress index and there is much discussion about which is the 'best'. For example, Goldman (1988) presents 32 heat stress indices and there are probably at least double that number used throughout the world. Many indices do not consider all six basic parameters although all have to take them into consideration in application. The use of indices will depend upon individual contexts, hence the production of so many, their incorporation into working practices being with experience of use. Some indices in use, for example, are inadequate theoretically yet can be justified for specific applications based on experience in a particular industry. Kerslake (1972) notes that 'it is perhaps self evident that the way in which the environmental factors should be

Table 10.1 Classification, medical aspects and prevention of heat illness

Category and clinical features	Predisposing factors	Underlying physiological disturbance	Treatment	Prevention
1. Temperature regulation				
Heatstroke: (1) Hot dry skin usually red, mottled or cyanotic; (2) t_{re}, 40.5°C (104°F) and over; (3) confusion, loss of consciousness, convulsions, t_{re} continues to rise; fatal if treatment delayed	(1) Sustained exertion in heat by unacclimatized workers; (2) Lack of physical fitness and obesity; (3) Recent alcohol intake; (4) Dehydration; (5) Individual susceptibility; and (6) Chronic cardiovascular disease	Failure of the central drive for sweating (cause unknown) leading to loss of evaporative cooling and an uncontrolled accelerating rise in t_{re}, there may be partial rather than complete failure of sweating	Immediate and rapid cooling by immersion in chilled water with massage or by wrapping in wet sheet with vigorous fanning with cool dry air, avoid overcooling, treat shock if present	Medical screening of workers, selection based on health and physical fitness, acclimatization for 5–7 days by graded work and heat exposure, monitoring workers during sustained work in severe heat
2. Circulatory hypostasis				
heat syncope Fainting while standing erect and immobile in heat	Lack of acclimatization	Pooling of blood in dilated vessels of skin and lower parts of body	Remove to cooler area, rest recumbent position, recovery prompt and complete	Acclimatization, intermittent activity to assist venous return to the heart
3. Water and/or salt depletion				
(a) Heat exhaustion (1) Fatigue, nausea, headache, giddiness;	(1) Sustained exertion in heat; (2) Lack of	(1) Dehydration from deficiency of water;	Remove to cooler environment, rest	Acclimatize workers using a breaking-in

Table 10.1 (Continued)

Category and clinical features	Predisposing factors	Underlying physiological disturbance	Treatment	Prevention
(2) Skin clammy and moist; complexion pale, muddy, or hectic flush; (3) May faint on standing with rapid thready pulse and low blood pressure; (4) Oral temperature normal or low but rectal temperature usually elevated (37.5–38.5 °C) (99.5–101.3 °F); water restriction type; urine volume small, highly concentrated; salt restriction type; urine less concentrated, chlorides less than 3 g/l	acclimatization; and (3) Failure to replace water lost in sweat	(2) Depletion of circulating blood volume; (3) Circulatory strain from competing demands for blood flow to skin and to active muscles	recumbent position, administer fluids by mouth, keep at rest until urine volume indicates that water balances have been restored	schedule for 5–7 days, supplement dietary salt only during acclimatization, ample drinking water to be available at all times and to be taken frequently during work day
(b) Heat cramps Painful spasms of muscles used during work (arms, legs, or abdominal); onset during or after work hours	(1) heavy sweating during hot work; (2) Drinking large volumes of water without replacing salt loss	Loss of body salt in sweat, water intake dilutes electrolytes, water enters muscles, causing spasm	Salted liquids by mouth, or more prompt relief by I-V infusion	Adequate salt intake with meals; in unacclimatized workers supplement salt intake at meals

4. Skin eruptions				
(a) Heat rash (miliaria rubra, 'prickly heat')				
Profuse tiny raised red vesicles (blister-like) on affected areas, pricking sensations during heat exposure	Unrelieved exposure to humid heat with skin continuously wet with unevaporated sweat	Plugging of sweat gland ducts with retention of sweat and inflammatory reaction	Mild drying lotions, skin cleanliness to prevent infection	Cool sleeping quarters to allow skin to dry between heat exposures
(b) Anhidrotic heat exhaustion (miliaria profunda)				
Extensive areas of skin which do not sweat on heat exposure, but present gooseflesh appearance, which subsides with cool environments; associated with incapacitation in heat	Weeks or months of constant exposure to climatic heat with previous history of extensive heat rash and sunburn	Skin trauma (heat rash; sunburn) causes sweat retention deep in skin, reduced evaporative cooling causes heat intolerance	No effective treatment available for anhidrotic areas of skin, recovery of sweating occurs gradually in return to cooler climate	Treat heat rash and avoid further skin trauma by sunburn, periodic relief from sustained heat
5. *Behavioural disorders*				
(a) Heat fatigue – transient impaired performance of skilled sensorimotor, mental, or vigilance tasks, in heat	Performance decrement greater in unacclimatized and unskilled worker	Discomfort and physiologic strain	Not indicated unless accompanied by other heat illness	Acclimatization and training for work in the heat

Table 10.1 (Continued)

Category and clinical features	Predisposing factors	Underlying physiological disturbance	Treatment	Prevention
(b) Heat fatigue – chronic Reduced performance capacity, lowering of self-imposed standards of social behaviour (e.g. alcoholic over-indulgence), inability to concentrate, etc.	Workers at risk come from temperate climates, for long residence in tropical latitudes	Psychosocial stresses probably as important as heat stress, may involve hormonal imbalance but no positive evidence	Medical treatment for serious cases, speedy relief of symptoms on returning home	Orientation on life in hot regions (customs, climate, living conditions, etc.)

Source: Goldman (1988).

combined must depend on the properties of the subject exposed to them, but none of the heat stress indices in current use make formal allowance for this'. The surge in standardization (e.g. ISO 7933, 1989; ISO 7243, 1989) has led to pressure to adopt similar indices worldwide. It will be necessary however to gain experience with the use of any new index.

Most heat stress indices consider, directly or indirectly, that the main strain on the body is due to sweating. For example, the more sweating required to maintain heat balance and internal body temperature, the greater the strain on the body. For an index of heat stress to summarize the human thermal environment and to predict heat strain, therefore, a mechanism is required to estimate the capacity of a sweating person to lose heat in the hot environment.

An index related to evaporation of sweat to the environment is useful where persons maintain internal body temperature essentially by sweating. These conditions are generally said to be in the prescriptive zone (WHO, 1969). Hence deep body temperature remains relatively constant while heart rate, and particularly sweat rate, rise with heat stress. At the upper limit of the prescriptive zone (ULPZ), thermoregulation is insufficient to maintain heat balance and the body temperature rises. This is termed the environmentally driven zone (WHO, 1969). In this zone heat storage is related to internal body temperature rise and can be used as an index to determine allowable exposure times, e.g. based on a predicted safety limit for 'core' temperature of 38 °C (see Figure 10.2).

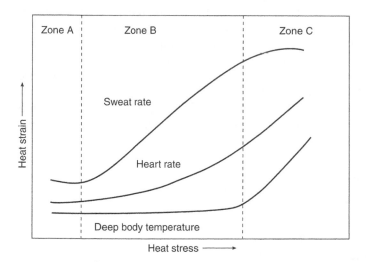

Figure 10.2 The variation of three measures of heat strain with increasing heat stress. In Zone B, the prescriptive zone (PZ), the deep body temperature is held constant by the increasing sweat rate. In Zone C, the environmentally driven zone (EDZ), sweat rate can no longer increase and the body temperature rises. The transition is termed as the upper limit of the prescriptive zone (ULPZ).

Source: WHO (1969).

Heat stress indices can be conveniently divided into rational, empirical and direct indices. Rational indices are based upon calculations involving the heat balance equation; empirical indices are based upon establishing equations from the physiological responses of human subjects (e.g. sweat loss), and direct indices are based upon the measurement (usually temperature) of instruments used to 'simulate' the response of the human body. The most influential and widely used heat stress indices are described below.

Rational indices

The heat stress index (HSI)

The heat stress index (Belding and Hatch, 1955) is based on a comparison of evaporation required to maintain heat balance (E_{req}) with the maximum evaporation that could be achieved in the environment, E_{max}

$$HSI = \frac{E_{req}}{E_{max}} \times 100\%. \tag{10.1}$$

Equations are provided in Table 10.2.

In fact, *HSI* is identical to the required skin wettedness value (i.e. $HSI = w_{req}$) although they were derived independently. The *HSI* as an index therefore is related to strain, essentially in terms of body sweating, for values between 0 and 100. At $HSI = 100$, sweating required is the maximum that can be achieved and thus represents the upper limit of the prescriptive zone. For $HSI > 100$, there is body heat storage and allowable exposure times are calculated based on a 1.8 °C rise in 'core' temperature (heat storage of 264 K J). For $HSI < 0$ there is mild cold strain, for example, when workers recover from heat strain (see Table 10.3).

An upper limit of 390 W m^{-2} is assigned to E_{max} (sweat rate of 1 litre per hour, taken to be the maximum sweat rate maintained over 8 h). Simple

Table 10.2 Equations used in the calculation of the heat stress index (*HSI*) and allowable exposure times (*AET*)

			Clothed	Unclothed
Radiation loss	$R = k_1(35 - t_r)\text{Wm}^{-2}$	for k_1	4.4	7.3
Convection loss	$C = k_2v^{0.6}(35 - t_a)\text{Wm}^{-2}$	for k_2	4.6	7.6
Maximum evaporative loss	$E_{max} = k_3v^{0.6}(56 - P_a)\text{Wm}^{-2}$ (upper limit of 390 Wm^{-2})	for k_3	7.0	11.7
Required sweat loss	$E_{req} = M - R - C$			
Heat stress index	$HSI = (E_{req}/E_{max}) \times 100$			
Allowable exposure time	$AET = 2440/(E_{req} - E_{max})\,\text{min}$			

Table 10.3 Interpretation of heat stress index (*HSI*) values

HSI	Effect of eight-hour exposure
−20	Mild cold strain (e.g. recovery from heat exposure)
0	No thermal strain
10–30	Mild to moderate heat strain – little effect on physical work but possible effect on skilled work
40–60	Severe heat strain, involving threat to health unless physically fit – acclimatization required
70–90	Very severe heat strain – personnel should be selected by medical examination, adequate water and salt intake assured
100	Maximum strain tolerated daily by fit acclimatized young men
Over 100	Exposure time limited by rise in deep body temperature

assumptions are made about the effects of clothing (long sleeved shirt and trousers), and the skin temperature is assumed to be constant at 35 °C.

The index of thermal stress (ITS)

Givoni (1963, 1976) provided the index of thermal stress (*ITS*) that was an improved version of the heat stress index (*HSI*). The heat balance equation became

$$E_{req} = H - (C + R) - R_s, \tag{10.2}$$

where R_s is the solar load and metabolic heat production H is used instead of metabolic rate to account for external work. An important improvement is the recognition that not all sweat evaporates (e.g. some drips) hence required sweat rate is related to required evaporation rate by

$$S_w = \frac{E_{req}}{\eta_{sc}}, \tag{10.3}$$

where η_{sc} is the efficiency of sweating.

Used indoors, sensible heat transfer is calculated from:

$$R + C = \alpha v^{0.3}(35 - T_g). \tag{10.4}$$

For outdoor conditions with solar load, T_g is replaced with t_a and allowance made for solar load (R_s) is given by

$$R_s = -E_s K_{pe} K_{cl}(1 - a(v^{0.2} - 0.88)). \tag{10.5}$$

The equations used fit experimental data and are not strictly rational.

Maximum evaporation heat loss is

$$E_{max} = K_p v^{0.3}(56 - P_a), \tag{10.6}$$

and efficiency of sweating is given by

$$\eta_{sc} = \exp\left[-0.6\left(\left(\frac{E_{req}}{E_{max}}\right) - 0.12\right)\right],$$ (10.7)

but if $\dfrac{E_{req}}{E_{max}} < 0.12,$ $\eta_{sc} = 1$

and if $\dfrac{E_{req}}{E_{max}} > 2.15,$ $\eta_{sc} = 0.29.$

McIntyre integrates the equation to give *ITS* (in $g\,h^{-1}$) as

$$ITS = \frac{(H - (R + C) - R_s)}{0.37\eta_{sc}},$$ (10.8)

where the factor 0.37 converts $W\,m^{-2}$ into $g\,h^{-1}$.

Required sweat rate (SW_req)

A further theoretical and practical development of the *HSI* and *ITS* was the required sweat rate (SW_{req}) index (Vogt *et al.*, 1981; Cena and Clark, 1981). This index calculated sweating required for heat balance from an improved heat balance equation but, most importantly, also provided a practical method of interpretation of calculations by comparing what is required with what is physiologically possible and acceptable in humans. Extensive discussions and practical evaluations (Mairiaux and Malchaire, 1988) of this index led to it being accepted as an International Standard (ISO 7933, 1989; see Mairiaux and Malchaire, 1990 and Chapter 14). Similar to the other rational indices SW_{req} is derived from the six basic parameters (air temperature t_a, radiant temperature t_r, relative humidity ϕ, air velocity v, clothing insulation I_{cl}, metabolic rate (M) and external work (W)). Effective radiation area values for posture (sitting $= 0.72$, standing $= 0.77$) are also required. From this the evaporation required is calculated from:

$$E_{req} = M - W - C_{res} - E_{res} - C - R.$$ (10.9)

Equations are provided for each component (see Chapter 1 and Table 10.4). Mean skin temperature is calculated from a multiple linear regression equation or a value of 36 °C is assumed.

From the required evaporation (E_{req}) and maximum evaporation (E_{max}) and sweating efficiency (r), the following are calculated.

$$\text{Required skin wettedness } w_{req} = \frac{E_{req}}{E_{max}}$$ (10.10)

$$\text{Required sweat rate } SW_{req} = \frac{E_{req}}{r}.$$ (10.11)

Table 10.4 Equations used in the calculation of the SW_{req} index and assessment method of ISO 7933 (1989)

$SW_{req} = E_{req}/r_{req}$
$E_{req} = M - W - C_{res} - E_{res} - C - R$
$C_{res} = 0.0014M(35 - t_a)$
$E_{res} = 0.0173M(5.624 - P_a)$
$C = h_c F_{cl}(t_{sk} - t_a)$
$R = h_r F_{cl}(t_{sk} - t_r)$
$w = E/E_{max}$
$r = 1 - \dfrac{w^2}{2}$
$h_c = 2.38|t_{sk} - t_a|^{0.25}$ for natural convection
$\quad\ = 3.5 + 5.2\ \text{var}$ for var $< 1\,\text{m s}^{-1}$
$\quad\ = 8.7\ \text{var}^{0.6}$ for var $> 1\,\text{m s}^{-1}$
$\text{var} = v_a + 0.0052\,(M - 58)$
$h_r = \sigma \epsilon_{sk} A_r / A_D \dfrac{[(t_{sk}+273)^4 - (t_r+273)^4]}{(t_{sk}-t_r)}$
$F_{cl} = 1/[(h_c + h_r)I_{cl} + 1/f_{cl}]$
$f_{cl} = 1 + 1.97 I_{cl}$
$E_{max} = (p_{sk,s} - P_a)/R_t$
$R_t = 1/(h_e F_{pcl})$
$h_e = 16.7 h_c$
$F_{pcl} = 1/[1 + 2.22 h_c(I_{cl} - (1 - 1/f_{cl})/(h_c + h_r))]$
$t_{sk} = 30 + 0.093 t_a + 0.045 t_r - 0.571 v_a + 0.254 P_a + 0.00128M - 3.57 I_{cl}$
or $t_{sk} = 36\,°C$ for an approximation or when values are beyond limits for which the t_{sk} equation was derived.

Interpretation of SW_{req}

Reference values in terms of what is acceptable, or persons can achieve, are used to provide a practical interpretation of calculated values (see Table 14.3).

First, a prediction of skin wettedness (w_p), evaporation rate (F_p), and sweat rate (SW_p), are made. Essentially, if what is calculated as required can be achieved then these are predicted values (e.g. $w_p = w_{req}$). If they cannot be achieved the maximum values can be taken (e.g. $SW_p = SW_{max}$). More detail is given in a decision flow chart (see Chapter 14).

If the required sweat rate can be achieved by persons and it will not cause unacceptable water loss, then there is no limit due to heat exposure over an eight-hour shift. If not, the duration limited exposures (DLE) are calculated from the following:

when $E_p = E_{req}$ and $SW_p \le D_{max}/8$,

then $DLE = 480$ min and SW_{req} can be used as a heat stress index.

If the above are not satisfied then $DLE_1 = 60Q_{max}/(E_{req} - E_p)$, and $DLE_2 = 60\,D_{max}/SW_p$, and the DLE is the lower of DLE_1 and DLE_2. Fuller details are given in ISO 7933 (1989) and in Chapter 14.

Predicted heat strain (PHS)

Malchaire *et al.* (1999) report a programme of research to improve the analytical method of the assessment of hot environments based upon the SW_{req} index (ISO 7933, 1989). This led to the predictive heat strain (PHS) method which is proposed in the revision of ISO 7933, 1989 (i.e. ISO CD 7933, 2001). Laboratories from Belgium, Italy, Germany, the Netherlands, Sweden, and the UK carried out European research (BIOMED) to design a practical strategy for heat stress assessment. Guidance was required on how to improve the working environment: to improve models of clothing; to include environments with high radiation, high humidity and high air velocity; and to improve criteria for interpreting the analysis, in particular, to take account of individual differences in response.

A number of laboratory and field evaluations of the SW_{req} index and its interpretation had identified limitations. Bethea and Parsons (1998b, 2000) found differences between predicted responses (sweat loss, 'core' temperature) and actual responses of subjects exposed to a range of hot conditions in a climatic chamber. They also applied the method in paper mills, steel mills and for forestry workers and found limitations in terms of usability and validity. McNeill and Parsons (1996) considered heat stress in nightclubs and Kampmann and Piekarski (2000) provided extensive evaluations of ISO 7933 (1989 – SW_{req}) in German coal mines. He identified discontinuities (prediction changes incrementally for a very small change in conditions) and contradictions in assessment. He also found that ISO 7933 (1989) would limit work in conditions where German miners had been working (over many years) without problem. This and other research led to a change to the standard ISO 7933 (1989) to provide cautionary notes regarding validity and scope and the publication of the (almost identical) European standard EN 12515 (1996) for which Germany claimed exemption (A-Deviation). McNeill and Parsons (1999) considered ISO 7933 (1989) for application in Industrially Developing Countries and found in both laboratory (simulated tea picking in a thermal chamber) and field studies (in Ghana) that the standard had limitations in validity and usability.

Malchaire *et al.* (1999) constructed a large database (1113 files) of the responses of people to hot conditions from both laboratory and field studies. The database was selectively divided into two parts, one half was used to develop a new model and the other half was used in its evaluation. Theoretical and practical considerations and empirical modelling provided modified equations and methods that led to an improved model for SW_{req} with significant changes for it to be a new method: Predicted Heat Strain (PHS). Malchaire *et al.* (1999) describe modifications brought to the required sweat rate index to provide the predicted heat strain assessment method. These include: modification to respiratory heat loss; introduction of mean body temperature; distribution of heat storage in the body; prediction of rectal temperature; exponential averaging for mean skin temperature and sweat rate; evaporative efficiency of sweating; w_{max} limits for non-acclimatized subjects; maximum sweat rate; increase of core temperature with activity; limits of internal temperature; maximum dehydration and water loss; influence of radiative protective clothing and the inclusion of the effects of ventilation on clothing insulation.

Determination of the predicted heat strain

The predicted heat strain method is similar to the method proposed in ISO 7933 (1989) but with modifications and additions. A list of symbols are provided in Table 10.5.

As before, inputs to the method are the six basic parameters. The required evaporation is calculated from:

$$E_{req} = M - W - C_{res} - E_{res} - C - R - S_{eq}, \tag{10.12}$$

where M is metabolic rate and is derived from ISO 8996. W is mechanical work and can be neglected.

$$C_{res} = 0.00152 \, M(28.56 - 0.885 \, t_a + 0.641 \, P_a) \tag{10.13}$$

$$E_{res} = 0.00127 \, M(59.34 + 0.53 \, t_a - 11.63 \, P_a) \tag{10.14}$$

$$C = h_{cdyn} \times f_{cl} \times (t_{cl} - t_a) \tag{10.15}$$

$$R = h_r \times f_{cl} \times (t_{cl} - t_r) \tag{10.16}$$

$$E = \frac{w(P_{sk,s} - P_a)}{R_{tdyn}} \tag{10.17}$$

$$dS_{eq} = C_{sp} \times (t_{cr,eq\,i} - t_{cr,eq\,i-1}) \times (1 - \alpha). \tag{10.18}$$

Heat storage (dS_{eq}) associated with an increase in core temperature (due to metabolic rate increase) is derived from

$$t_{cr} = 36.6 + (t_{cr,eq} - 36.6) \times \left(1 - \exp\left(\frac{-t}{10}\right)\right), \tag{10.19}$$

where core temperature rises to equilibrium temperature, heat storage = S_{eq}

$$t_{cr,eq} = 0.0036M + 36.6. \tag{10.20}$$

h_{cdyn} is the greatest value of:

$2.38|t_{sk} - t_a|^{0.25}$

$3.5 + 5.2 Var$

$8.7 Var^{0.6}$

$$h_r = \frac{5.67 \times 10^{-8} \varepsilon A_r}{A_D} \times \frac{(t_{cl} + 273)^4 - (t_r + 273)^4}{t_{cl} - t_r}. \tag{10.21}$$

Table 10.5 Additional symbols used in equations for the predicted heat strain method

Symbol	Term	Unit
α_i	skin-core weighting at time i	dimensionless
α_{i-1}	skin-core weighting at time $(i-1)$	dimensionless
τ	time constant	min
θ	angle between walking direction and wind direction	degrees
A_p	fraction of the body surface covered by the reflective clothing	dimensionless
c_e	water latent heat of vaporization	Joules per kilogram
$C_{orr,cl}$	correction for the dynamic clothing insulation	dimensionless
$C_{orr,Ia}$	correction for the dynamic boundary layer thermal insulation	dimensionless
$C_{orr,tot}$	correction for the dynamic clothing insulation as a function of the actual clothing	dimensionless
$C_{orr,E}$	correction for the dynamic permeability index	dimensionless
c_p	specific heat of dry air at constant pressure	Joules per kilogram of dry air Kelvin
c_{sp}	specific heat of the body	Watts per square metre per Kelvin
$D_{lim\ tre}$	allowable exposure time for heat storage	min
$D_{limloss50}$	allowable exposure time for water loss, mean subject	min
$D_{limloss95}$	allowable exposure time for water loss, 95% of the working population	min
dS_i	body heat storage during the last time increment	Watts per square metre
dS_{eq}	body heat storage rate for increase of core temperature associated with the metabolic rate	Watts per square metre
$F_{cl,R}$	reduction factor for radiation heat exchange due to wearing clothes	dimensionless
F_r	emissivity of the reflective clothing	dimensionless
h_{cdyn}	dynamic convective heat transfer coefficient	Watts per square metre Kelvin
$I_{a\ st}$	static boundary layer thermal insulation	square metres kelvin per Watt
$I_{cl\ st}$	static clothing insulation	square metres kelvin per Watt
$I_{tot\ st}$	total static clothing insulation	square metres kelvin per Watt
$I_{a\ dyn}$	dynamic boundary layer thermal insulation	square metres kelvin per Watt
$I_{cl\ dyn}$	dynamic clothing insulation	square metres kelvin per Watt
$I_{tot\ dyn}$	total dynamic clothing insulation	square metres kelvin per Watt
i_{mst}	static moisture permeability index	dimensionless

i_{mdyn}	dynamic moisture permeability index	dimensionless
$incr$	time increment from time $(i-1)$ to time i	min
R_{tdyn}	dynamic total evaporative resistance of clothing and boundary air layer	square metres kilopascals per Watt
S_{eq}	body heat storage for increase of core temperature associated with the metabolic rate	Watts per square metre
$SW_{p,i}$	predicted sweat rate at time i	Watts per square metre
$SW_{p,i-1}$	predicted sweat rate at time $(i-1)$	Watts per square metre
SW_{req}	required sweat rate	Watts per square metre
$t_{cr,eq\ i}$	core temperature as a function of the metabolic rate at time i	degrees celsius
$t_{cr,eq\ i-1}$	core temperature as a function of the metabolic rate at time $(i-1)$	degrees celsius
$t_{cr,i}$	core temperature at time i	degree celsius
$t_{cr,i-1}$	core temperature at time $(i-1)$	degree celsius
$t_{re,max}$	maximum acceptable rectal temperature	degrees celsius
$t_{re,i}$	rectal temperature at time i	degrees celsius
$t_{re,i-1}$	rectal temperature at time $(i-1)$	degrees celsius
$t_{sk,eq}$	steady state mean skin temperature	degrees celsius
$t_{sk,eq\ nu}$	steady state mean skin temperature for nude subjects	degrees celsius
$t_{sk,eq\ cl}$	steady state mean skin temperature for clothed subjects	degrees celsius
$t_{sk,i}$	mean skin temperature at time i	degrees celsius
$t_{sk,i-1}$	mean skin temperature at time $(i-1)$	degrees celsius
$Walksp$	walking speed	metres per second

$A_r/A_D = 0.67$ for crouching subject, 0.70 seated, 0.77 standing. In the case of reflective clothing h_r is corrected using:

$$h_{r_{corrected}} = h_r \times F_{cl,R},\qquad(10.22)$$

where

$$F_{cl,R} = (1 - A_p)0.97 + A_p \times F_r.\qquad(10.23)$$

Mean temperature of clothing (t_{cl}) is calculated by iteration in the usual way (see Chapter 1).

Steady state mean skin temperature is given by:
For nude subjects:

$$t_{sk,eqnu} = 7.19 + 0.064t_a + 0.061t_r - 0.348v_a + 0.198p_a + 0M + 0.616t_{re}.\qquad(10.24)$$

For clothed subjects:

$$t_{sk,eqcl} = 12.17 + 0.020t_a + 0.044t_r - 0.253v_a + 0.194p_a + 0.005346M + 0.51274t_{re}.\qquad(10.25)$$

For I_{cl} values between 0.2 and 0.6 Clo:

$$t_{sk,eq} = t_{sk,eq\,nu} + 2.5 \times \left(t_{sk,eq\,cl} - t_{sk,eq\,nu}\right) \times (I_{cl} - 0.2). \qquad (10.26)$$

The skin temperature at any time i is then given by

$$t_{sk,i} = 0.7165\, t_{sk,i-1} + 0.2835\, t_{sk,eq}. \qquad (10.27)$$

Dynamic insulation of clothing is determined by correcting the total (including air layer) clothing insulation using empirical equations:

$$I_{tot\,st} = I_{cl\,st} + \frac{I_{a\,st}}{f_{cl}}, \qquad (10.28)$$

where the increase in surface area due to clothing:

$$f_{cl} = 1 + 1.97 I_{cl\,st} \quad \left(I_{cl\,st} \text{ in } m^2\,K\,W^{-1}\right)$$

$$I_{tot\,dyn} = C_{orr,tot} \times I_{tot\,st} \qquad (10.29)$$

$$I_{a\,dyn} = C_{orr,I_a} \times I_{a\,st} \qquad (10.30)$$

$$I_{cl\,dyn} = I_{tot\,dyn} - \frac{I_{a\,dyn}}{f_{cl}}, \qquad (10.31)$$

where

$$C_{orr,tot} = C_{orr,cl} = e^{(0.043 - 0.398\,Var + 0.066\,Var^2 - 0.378\,Walksp + 0.094\,Walksp^2)}$$
$$(10.32)$$

for $I_{cl} \geq 0.6$ Clo.
For nude person or adjacent air layer:

$$C_{orr,tot} = C_{orr,I_a} = e^{(-0.472\,Var + 0.047\,Var^2 - 0.342\,Walksp + 0.117\,Walksp^2)} \quad (10.33)$$

for 0 Clo $\leq I_{cl} \leq 0.6$ Clo

$$C_{orr,tot} = (0.6 - I_{cl})C_{orr,I_a} + I_{cl} \times C_{orr,cl}. \qquad (10.34)$$

With *Var* limited to $3\,m\,s^{-1}$ and *Walksp* limited to $1.5\,m\,s^{-1}$. When walking speed is undefined or the person is stationary: $Walksp = 0.0052\,(M - 58)$ with $Walksp \leq 0.7\,m\,s^{-1}$.

The evaporative resistance of clothing is derived using the clothing permeability index i_m where I_{mst} is i_m for static conditions and i_{mdyn} is i_{mst} corrected for the influence of air and body movement.

$$i_{mdyn} = i_{mst} \times C_{orr,E}, \qquad (10.35)$$

where

$$C_{orr,E} = 2.6 \times C_{orr,tot}^2 - 6.5 \times C_{orr,tot} + 4.9 \tag{10.36}$$

If $i_{mdyn} > 0.9$ then $i_{mdyn} = 0.9$

Dynamic evaporative resistance:

$$R_{tdyn} = I_{totdyn}/i_{mdyn}/16.7. \tag{10.37}$$

Required sweat rate:

$$SW_{req} = \frac{E_{req}}{r_{req}}, \tag{10.38}$$

and required skin wettedness:

$$W_{req} = \frac{E_{eq}}{E_{max}} \tag{10.39}$$

$$E_{max} = \frac{(P_{sk,s} - P_a)}{R_{tdyn}}. \tag{10.40}$$

Interpretation of analysis for the predicted heat strain model

Two criteria of thermal stress (w_{max}, SW_{max}) and two criteria for thermal strain ($t_{re,max}$, D_{max}) are used to interpret the analysis. Suggested limit values are provided in Table 10.6.

In Table 10.6 it is assumed that maximum sweat rate in acclimatized subjects is, on average, 25 per cent greater than for non-acclimatized subjects. Dehydration limits (D_{max}) are based upon a maximum dehydration rate of 3 per cent (for industry, not the Army or sportsmen). Even when water is available, workers tend not to drink as much as they lose in sweat. For an exposure lasting 4–8 h, a rehydration rate of 60 per cent is observed on average (in 50 per cent of workers) regardless of the total amount of sweat produced, and is greater than 40 per cent in 95 per cent of cases.

Table 10.6 Suggested limit values used in the predicted heat strain method

	Unacclimatised	*Acclimatised*
Maximum wettedness W_{max}	0.85	1.0
Maximum sweat rate SW_{max} Wm^{-2}	$(M - 32) \times A_D$	$1.25 \times (M - 32) \times A_D$
Maximum dehydration and water loss $D_{max\,50}$ $D_{max\,95}$	7.5% × body mass 5% × body mass	7.5% × body mass 5% × body mass
Rectal temperature limit $T_{re_{max}}$	38 °C	38 °C

Hence based upon the rehydration rate D_{max50} and D_{max95} are set (i.e. 60 per cent replacement of 7.5 per cent loss $= 4.5$ per cent so 7.5 per cent $-$ 4.5 per cent $= 3$ per cent dehydration. 40 per cent replacement of 5 per cent loss $= 2$ per cent, so 5 per cent $- 2$ per cent $= 3$ per cent dehydration). Rectal temperature is derived from heat storage S where

$$S = E_{req} - E_p + S_{eq}. \qquad (10.41)$$

Heat storage leads to an increase in core temperature taking into account the increase in skin temperature. The fraction of the body mass at the mean core temperature is given by:

$$(1 - \alpha) = 0.7 + 0.09(t_{cr} - 36.8), \qquad (10.42)$$

where $(1 - \alpha)$ is limited to 0.7 for $t_{cr} < 36.8\,^\circ\mathrm{C}$ and 0.9 for $t_{cr} > 39.0\,^\circ\mathrm{C}$.

$$t_{co} = \frac{1}{1 - \frac{\alpha}{2}} \left[\frac{dSi}{C_p W_b} + t_{co0} - \frac{t_{co0} - t_{sk0}}{2} \alpha_0 - t_{sk} \frac{\alpha}{2} \right], \qquad (10.43)$$

Rectal temperature is then estimated as:

$$t_{re,i} = t_{re,i-1} + \frac{2t_{cr,i} - 1.962t_{re,i-1} - 1.31}{9}. \qquad (10.44)$$

Determination of predicted values

Figure 10.3 shows a flow chart for the calculation of predicted sweat rate (SW_p), predicted wettedness (w_p) and predicted evaporation rate (E_p). It is interesting to note that:

$$r_{req} = 1 - \frac{w_{req}^2}{2} \quad \text{for} \quad W_{req} \le 1 \qquad (10.45)$$

and

$$r_{req} = \frac{(2 - w_{req})^2}{2} \quad \text{for} \quad W_{req} \ge 1 \qquad (10.46)$$
$$\text{limited to } r_{req} \ge 0.05.$$

That is, the required skin wettedness is allowed to be theoretically greater than 1 for computation of the predicted sweat rate.

Determination of the allowable exposure duration (DLE)

The maximum allowable exposure duration, D_{lim}, is reached when either the rectal temperature or the accumulated water loss reaches the

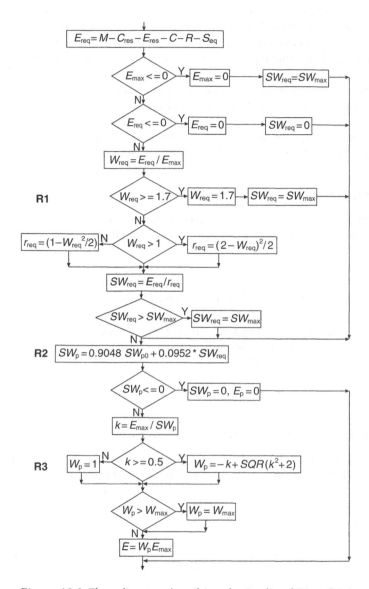

Figure 10.3 Flow diagram describing the Predicted Heat Strain model (from ISO CD 7933, 2001).

corresponding limits (Table 10.6). If E_{max} is negative (i.e. condensation) or estimated allowable exposure time is less than 30 min, then the method is not applicable.

To allow calculation using the PHS method, an electronic copy of a computer programme can be downloaded from the worldwide web.

Piette and Malchaire (1999) and Mehnert *et al.* (2000) present a validation of the PHS method using the other half of the database used in its derivation. They found good correlations between predicted and observed sweat rates and rectal temperatures. Kampmann *et al.* (1999) compare the PHS model with the SW_{req} method used in ISO 7933 (1989). They note that as the PHS method provides improvements that were required, a comparison has limited use. However comparison of predicted and observed responses where the models could be compared, showed that the PHS model was more accurate. Malchaire and Piette (1999) provide a similar analysis and draw similar conclusions when comparing the PHS model with the WBGT index as used in ISO 7243.

Other rational indices

The SW_{req} index and ISO 7933 (1989) and developments leading to the PHS method, provide the most sophisticated form of rational method based on the heat balance equation. The method of interpretation and practical guidance provided were major advances. More developments with this approach can be made, however a major change in approach is to use a thermal model. The *ET** and *SET* provide indices based upon the two-node model of human thermoregulation and Givoni and Goldman (1973) provide prediction models for assessment of heat stress. The *ET** and *SET* indices are discussed in Chapter 8 and the Givoni/Goldman model and the two-node model are discussed in Chapter 15.

Empirical indices

Effective temperature and corrected effective temperature

The effective temperature (*ET*) index (Houghton and Yagloglou, 1923) was originally established to provide a method for determining the relative effects of air temperature and humidity on comfort; see Chapter 8. Three subjects judged which of two climatic chambers was warmer by walking between the two. Using different combinations of air temperature and humidity (and later other parameters), lines of equal comfort were determined. Immediate impressions were made so that the transient response was recorded. This had the effect of over-emphasizing the effects of humidity at low temperatures and underestimating it at high temperatures (when compared with steady-state responses). The use of the black globe temperature to replace dry bulb temperature in the *ET* nomograms (see Chapter 8) provided the corrected effective temperature, *CET* (Bedford, 1946). Although originally a comfort index, research reported by Macpherson (1960) suggested that the *CET* predicted physiological effects of increasing mean radiant temperature. *ET* and *CET* are now rarely used as comfort indices but have been used as heat stress indices. The report by Bedford (1946) *Environmental Warmth and its Measurement*, proposed *CET* as an index of warmth, upper limits being a *CET* of 34 °C for 'reasonable efficiency' and 38.6 °C for tolerance. Further investigation however showed that *ET* had serious disadvantages for use as a heat stress index although it is still used by some industries (e.g. mining). Their limitations led to the development of the predicted four-hour sweat rate (*P4SR*) index.

Predicted four-hour sweat rate (P4SR)

The predicted four-hour sweat rate index (*P4SR*) was established in Climate Chambers at the National Hospital for Nervous Diseases, London by McArdle *et al.* (1947) and evaluated in Singapore in seven years of work summarized by Macpherson (1960). The P4SR is the amount of sweat secreted by fit, acclimatized young men exposed to the environment for four hours. Activity was the predicted activity pattern of ratings serving guns with ammunition during a naval engagement. The single number (index value) which summarizes the effects of the six basic parameters, is an amount of sweat from the specific population, but it should be used as an index value and not as an indication of an amount of sweat in an individual group of interest. The *P4SR* is an empirical index and steps taken to obtain the index value are summarized by McIntyre (1980) as follows:

1   If $t_g \neq t_a$, increase the wet bulb temperature by 0.4 $(t_g - t_a)$ °C.
2   If the metabolic rate $M > 63 \text{ W m}^{-2}$, increase the wet bulb temperature by the amount indicated in the chart (see Figure 10.4).
3   If the men are clothed, increase the wet bulb temperature by 1.5 I_{clo} (°C).

The modifications are additive. The basic four-hour sweat rate (*B4SR*) is determined from Figure 10.4. The *P4SR* is then:

$$P4SR = B4SR + 0.37I_{clo} + (0.012 + 0.001I_{clo})(M - 63). \tag{10.47}$$

It was acknowledged that outside of the prescriptive zone (e.g. *P4SR* > 5 l), sweat rate was not a good indicator of strain. The *P4SR* nomograms were adjusted to attempt to account for this. The *P4SR* appears to have been useful over conditions for which it was derived, however the effects of clothing are oversimplified and it is most useful as a heat storage index. McArdle *et al.* (1947) proposed a *P4SR* of 4.5 litres for a limit where no incapacitation of any fit, acclimatized young men occurred.

Heart rate prediction

Fuller and Brouha (1966) proposed a simple index based on the prediction of heart rate (*HR*) in beats per minute:

$$HR = 22.4 + 0.18M + 0.25(5t_a + 2P_a), \tag{10.48}$$

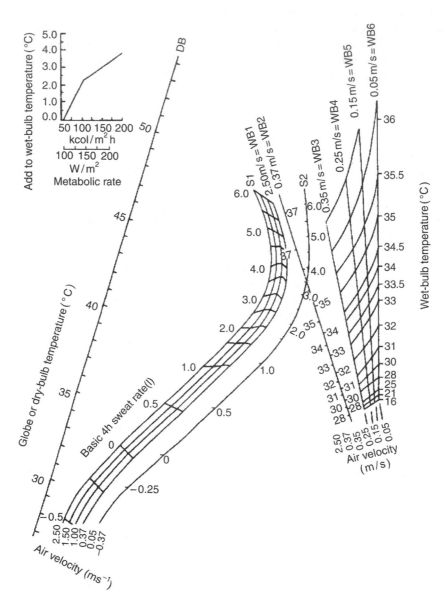

Figure 10.4 Nomogram for calculating P4SR.

where M is metabolic rate $W m^{-2}$, t_a is air temperature in °C and P_a is vapour pressure in mb.

The relationship is originally formulated with metabolic rate in BTU/h and partial vapour pressure in mm Hg. This provided a simple prediction of heart rate from $(T + p)$, hence the $T + p$ index.

Givoni and Goldman (1973) provide equations for predicting heart rate of persons (soldiers) in hot environments. They define an index for heart rate (IHR) from a modification of predicted equilibrium rectal temperature t_{ref}.

$$T_{ref} = 36.75 + 0.004(M - W_{ex}) + (0.025/I_{clo})(t_a - 36)$$
$$+ 0.8e^{0.0047(E_{req} - E_{max})} \tag{10.49}$$

and

$$IHR = 0.4M + (2.5/I_{clo})(t_a - 36) + 80e^{0.0047(E_{req} - E_{max})}, \tag{10.50}$$

where

M = metabolic rate (Watts)
W_{ex} = mechanical work (Watts)
I_{clo} = thermal insulation of clothing (Clo)
t_a = air temperature (°C)
E_{req} = total metabolic and environmental heat load (Watts)
E_{max} = evaporative cooling capacity for clothing and environment (Watts).

Equilibrium heart rate (HR_f in beats per minute) is then given by

$$HR_f = 65 + 0.35(IHR - 25) \qquad \text{for} \quad IHR \le 225 \tag{10.51}$$
$$HR_f = 135 + 42\left[1 - e^{-(IHR-225)}\right] \quad \text{for} \quad IHR > 225, \tag{10.52}$$

where

HR_f = equilibrium heart rate (bpm)
(The assumed resting heart rate in comfortable conditions is 65 bpm.)

that is, a linear relationship exists between rectal temperature and heart rate, for heart rates up to about 150 beats per minute, and then an exponential relationship exists as the heart rate approaches its maximum.

Givoni and Goldman (1973) also provide equations for changing heart rate with time and also corrections for degree of acclimatization of subjects.

A method of work and recovery heart rate is described by NIOSH (1986) – from Brouha (1960) and Fuller and Smith (1980, 1981). Body temperature and pulse rates are measured during recovery following a work cycle or at specified times during the working day. At the end of a work cycle the worker sits on a stool, oral temperature is taken and the following three pulse rates are recorded:

P_1 – Pulse rate counted from 30 s to 1 min

P_2 – Pulse rate counted from 1.5 to 2 min

P_3 – Pulse rate counted from 2.5 to 3 min.

Ultimate criterion in terms of heat strain is an oral temperature of 37.5 °C. If

$$P_3 \leq 90 \, \text{bpm} \quad \text{and} \quad P_3 - P_1 \simeq 10 \, \text{bpm},$$

this indicates work pattern is high but there will be little increase in body temperature. If

$$P_3 > 90 \, \text{bpm} \quad \text{and} \quad P_3 - P_1 < 10 \, \text{bpm},$$

stress (heat + work) is too high and action is needed to redesign work. Vogt *et al.* (1981) and ISO 9886 (1992) provide the following model of heart rate for assessing thermal environments;

$$\text{Total heart rate } HRt = HR_0 \text{rest thermal neutrality}$$
$$+ HR_M \text{ work}$$
$$+ HR_S \text{ static exertion}$$
$$+ HR_T \text{ thermal strain}$$
$$+ HR_N \text{ emotion (psychological)}$$
$$+ HR_e \text{ residual.}$$

The component of thermal strain HR_T is sometimes referred to as 'thermal beats'. A possible heat stress index can be calculated from

$$HR_T = HR_r - HR_0,$$

where HR_r is heart rate after recovery and HR_0 is the resting heart rate in a thermally neutral environment.

Direct heat stress indices

The wet bulb globe temperature (WBGT) index

The wet bulb globe temperature ($WBGT$) index is by far the most widely used heat stress index throughout the world. It was developed in a US Navy investigation into heat casualties during training (Yaglou and Minard, 1957) as an approximation to the more cumbersome corrected effective temperature (*CET*), modified to account for the solar absorptivity of green military clothing. It is given by

$$WBGT = 0.7t_{nwb} + 0.2t_g + 0.1t_a,$$

for conditions with solar radiation, and

$$WBGT = 0.7t_{nwb} + 0.3t_g,$$

for indoor conditions with no solar radiation, where

t_{nwb} = temperature of a naturally ventilated wet bulb thermometer

t_a = air temperature

t_g = temperature of a 150 mm diameter black globe thermometer.

WBGT limit values were used to indicate when military recruits could train. It was found that heat casualties and time lost due to cessation of training in the heat were both reduced by using the WBGT index instead of air temperature alone. The WBGT index was adopted by both NIOSH (1972) and ISO 7243 (1982, 1989) and is still proposed today (see Chapter 14 and ISO 7243, 1989) which provides a method easily used in a hot environment to provide a 'fast' diagnosis. The specification of the instrument is provided in Chapter 14, as are *WBGT* limit values for acclimatized or non-acclimatized persons (Table 14.1). For example, for a resting acclimatized person in 0.6 Clo the limit value is 33 °C *WBGT*. The limits provided in ISO 7243 (1989) and ACGIH (1992) are almost identical. However, the ACGIH includes TLV *WBGT* correction factors for clothing (see Table 10.7).

It is noted that for people wearing special clothing, TLV values should be established by an expert. The application of heat stress assessment methods (including *WBGT* and SW_{req}) to people wearing special protective clothing is considered by BS 7963 and presented in Chapter 14. Whether an index such as *WBGT* (with high weighting to natural wet bulb representing the exposed sweating body) should be used to assess heat stress in impermeable clothing is debatable.

The simplicity of the index and its use by influential bodies has led to its widespread acceptance. Like all direct indices it has limitations when used to simulate human response and *WBGT* should be used with caution in

Table 10.7 TLV *WBGT* correction factors (°C) for clothes

	Clo *value*	WBGT *correction* (°C)
Summer work uniform	0.6	0
Cotton overalls	1.0	−2
Winter work uniform	1.4	−4
Water barrier permeable	1.2	−6

practical applications. It is possible to buy portable instruments which determine the *WBGT* index (Olesen, 1985b) (see Chapter 5).

Physiological heat exposure limit (PHEL)

Dasler (1974, 1977) provides *WBGT* limit values based on a prediction of exceeding any two physiological limits (from experimental data) of impermissible strain. The limits are given by

$$PHEL + \left(17.25 \times 10^8 - 12.97M \times 10^6 + 18.61M^2 \times 10^3\right) \times WBGT^{-5.36}.$$

$$(10.52)$$

This index therefore uses the *WBGT* direct index in the environmentally driven zone (see Figure 10.2) where heat storage can occur.

Wet globe temperature (WGT) index

The temperature of a wet black globe of appropriate size can be used as an index of heat stress. The principle is that it is affected by both dry and evaporative heat transfer as is a sweating man and the temperature can then be used, with experience, as a heat stress index. Olesen (1985b) describes WGT as the temperature of a 2.5 inch diameter black globe covered with a damp black cloth. The temperature is read when equilibrium is reached after about 10–15 min of exposure. NIOSH (1986) describe the Botsball (Botsford, 1971) as the simplest and most easily read instrument. It is a 3 inch diameter copper sphere covered by a black cloth kept at 100 per cent wettedness from a self-feeding water reservoir. The sensing element of a thermometer is located at the centre of the sphere and the temperature is read on a colour-coded dial.

A simple equation relating *WGT* to *WBGT* (in °C) is

$$WBGT = WGT + 2,$$

$$(10.53)$$

for conditions of moderate radiant heat and humidity (NIOSH, 1986) but of course this relationship cannot hold over a wide range of conditions.

The Oxford index (WD)

Lind *et al.* (1957) proposed a simple direct index used for storage limited heat exposure and based on a weighted summation of aspirated wet bulb temperature (t_{wb}) and dry bulb temperature (t_{db}).

$$WD = 0.85t_{wb} + 0.15t_{db}.$$

$$(10.54)$$

Tolerance times for mine rescue teams were based on this index. It is widely applicable but is not appropriate where there is significant thermal radiation.

ACCLIMATIZATION

Persons not recently exposed to hot environments may initially find them very stressful but after a few days there will be a significant increase in tolerance. This is because of behavioural changes such as reduced level of activity and changes in clothing and diet, but also because there are physiological changes. Acclimatization refers to the substantial physiological changes that take place after prolonged exposure to heat. In simple terms, the major change is due to the 'training' of the sweat glands to produce more sweat. Internal body temperature and heart rate can therefore be controlled within acceptable limits as heat stress is combated by the increased evaporative heat loss due to increased sweating. Other physiological changes include an increased blood volume, fall in internal body temperature and a fall in NaCl content of sweat and urine. Lind and Bass (1963) demonstrated this effect (Figure 10.5).

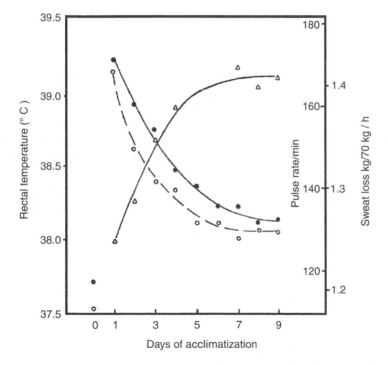

Figure 10.5 Typical average rectal temperatures (●), pulse rates (○) and sweat losses (△) of a group of men during the development of acclimatization to heat. On day 0, the men worked for 100 min at an energy expenditure of 300 kcal h^{-1} in a cool climate; the exposure was repeated on days 1 to 9, but in a hot climate with dry and wet bulb temperatures of 48.9 °C and 26.7 °C (120 °F and 80 °F).

Source: Adapted from Lind and Bass (1963).

Persons do not have to go to a hot climate to become acclimatized, although some behavioural changes will have to be learnt there. A simple description of the process is that it is due to the training effect on the sweat glands. The more frequent the stimulation, the greater the volume of sweat produced (Fox *et al.*, 1964). It was also shown that exercise was not required. Controlled hyperthermia of persons resting in a hot environment, e.g. in a special ventilated suit allowing sweating and evaporation, showed that the degree of acclimatization was directly proportional to the extent of the body temperature rise and its duration each day.

Investigation into individual and group responses shows wide variation in acclimatization. Inhabitants of New Guinea have much lower sweat rates than the Nigerians in West Africa. Women appear to have lower sweat rates than men, although physically active (sports) women can have sweat rates equivalent to those of acclimatized men. There are clearly differences between races and it is probable that genetic factors have some effect in variations in sweat rate (Edholm and Weiner, 1981). In general, people become acclimatized to the conditions to which they are exposed, so those working in hot climates may not be 'fully acclimatized' to hotter conditions.

Artificial acclimatization (or acclimation) is induced by making people sweat for at least one hour per day and usually achieved through raising body temperature during exercise in the heat. An effective method is to ensure the subject achieves profuse sweating (and evaporation) as soon as possible and retains this level throughout the session while consuming fluid at regular intervals. Exercise itself will have some effect and top athletes who train regularly will be partly acclimatized. It is debatable if acclimatization in a climatic chamber provides great physiological advantage to top athletes who are about to go to a hot climate to compete, as is common practice, although there is some evidence. There may also, however, be significant psychological and behavioural benefits. Acclimatization takes place over a number of days with greatest benefits in the first few days. Fit people acclimatize more rapidly (and acclimatization can help fitness), and cessation of alcohol consumption and taking of vitamin C are thought to help. Clearly, drinking water is important and will replace fluid loss.

Acclimatization programmes are essential to a successful transfer of people from temperate to hot climates if they are to perform efficiently. Turk (1974) described a method used by the British Army to acclimatize large groups of soldiers who are about to be sent to a hot climate. Internal body temperature is raised rapidly by exercise in a hot environment (47 °C DB (dry bulb)/32 °C WB (wet bulb)). It is important to ensure that the dew point of the environment is below that of skin temperature, so that sweat can evaporate. When sublingual temperature is raised to around 38.8 °C, it is maintained at that level by the soldier controlling the rate of exercise (step tests), hence maintaining profuse sweating. Substantial acclimatization can then be obtained by repeating the test for one hour per day for four days (although greater than one hour and longer than four days would give greater effect but may not be as convenient). The effectiveness of these methods has been conclusively shown in practical cases in terms of performance and avoidance of heat illness when soldiers arrive in the hot climate. Such studies also demonstrate the importance of motivation and good leadership in maintaining human performance.

A drastic reduction in heat casualties and death was found in the South African Gold Mines after the introduction of an acclimatization programme (Wyndham and Strydom, 1969). Screening out of those who never achieve tolerance to heat was also used. Step climbing for four hours per day for eight days in large chambers acclimatizes up to 200 men at a time. A high level of acclimatization is achieved for about one week after exposure to heat ceases and declines to complete loss after about one month. Fitness greatly assists in maintaining acclimatization.

Clark and Edholm (1985) describe the acclimatization programmes used in the experiments to develop the *P4SR* heat stress index. Eighteen subjects spent four hours on each of five days in the week in the hot room. Values of sweat produced in the four hours ranged from $0.8–1.55 \, \mathrm{l \, m^{-2}}$ on the first day to $1.5–3.4 \, \mathrm{l \, m^{-2}}$ on the final day. Sweat rates approximately doubled in each subject, with those with the lowest sweat rate remaining lowest throughout the experiment and the highest remaining the highest.

Working practices for hot environments

NIOSH (1986) provide a comprehensive description of working practices for hot environments, including preventative medical practices. McCaig (1992a) provides a proposal for medical supervision of individuals exposed to hot or cold environments. It should always be remembered that it is a basic human right that, when possible, persons can withdraw from any extreme (or any other) environment without need of explanation (Declaration of Helsinki, WMA, 1985). Where exposure does take place, defined working practices will greatly improve safety.

It is a reasonable principle in environmental ergonomics and in industrial hygiene that, where possible, the environmental stressor should be reduced at source. NIOSH (1986) divide control methods into five types. These are presented in Table 10.8.

There has been a great deal of military research into so-called N.B.C. (nuclear, biological and chemical) protective clothing. In hot environments it is not possible to remove the clothing and working practices are very important. Acclimatization is an effective method for increasing heat tolerance. If total protective clothing is worn, however, increased sweat loss may not increase evaporative loss and increased water loss will prove a disadvantage. Methods of cooling soldiers quickly so that they are able to perform again include sponging the outer surface with water and blowing dry air over the clothing. Organizational solutions such as cool protected

Table 10.8 Working practices for hot environments

	Example
Engineering controls	
1. Reduce heat source	Move away from workers or reduce temperature. Not always practicable
2. Convective heat control	Modify air temperature and air movement. Spot coolers may be useful
3. Radiant heat control	Reduce surface temperatures or place reflective shield between radiant source and workers. Change emissivity of surface. Use doors which open only when access required
4. Evaporative heat control	Increase air movement; decrease water vapour pressure. Use fans or air conditioning. Wet clothing and blow air across person
Work and hygiene practices and administrative controls	
1. Limiting exposure time and/or temperature	Perform jobs at cooler times of day and year. Provide cool areas for rest and recovery. Extra personnel; worker freedom to interrupt work; increase water intake
2. Reduce metabolic heat load	Mechanization. Redesign job. Reduce work time. Increase work force
3. Enhance tolerance time	Heat acclimatization programme. Keep workers physically fit. Ensure water loss is replaced and maintain electrolyte balance if necessary
4. Health and safety training	Supervisors trained in recognizing signs of heat illness and in first aid. Basic instruction to all personnel on personal precautions, use of protective equipment and effects of non-occupational factors (e.g. alcohol). Use of a buddy system. Contingency plans for treatment should be in place
5. Screening for heat intolerance	History of previous heat illness. Physically unfit
Heat alert programme	
1. In spring establish heat alert committee (industrial physician or nurse, industrial hygienist, safety engineer, operation engineer, high-ranking manager).	Arrange training course. Memos to supervisors to make checks of drinking fountains, etc. Check facilities, practices, readiness, etc.
2. Declare heat alert in predicted hot weather spell.	Postpone non-urgent tasks. Increase workers; increase rest. Remind workers to drink. Improve working practices

Auxiliary body cooling and protective clothing	Use if it is not possible to modify worker, work or environment and heat stress is still beyond limits. Individuals should be fully heat acclimatized and well trained in use and practice of wearing the protective clothing. Examples are water cooled garments, air cooled garments, ice packet vests and wetted overgarments
Performance degradation	It must be remembered that wearing protective clothing that is providing protection from toxic agents will increase heat stress. All clothing will interfere with activities and may reduce performance (e.g. reducing the ability to receive sensory information hence impairing hearing and vision for example)

Source: Adapted from NIOSH (1986).

rest areas may be necessary. The transfer of military clothing technology to industrial situations is a new innovation (Crockford, 1991), but much is known and appropriate working practices can greatly reduce risk.

House (1994, 1996) demonstrated a technique for reducing heat strain in Navy personnel while wearing protective clothing. The hands are placed in 12 °C water and because the skin, including that of the hands, is vasodilated, sufficient heat can be lost to have a significant contribution to whole-body cooling. Active cooling systems are becoming widely available, as are physiological measurement systems (see Chapters 5 and 7). Ice vests, liquid cooling and air cooling garments are all commonly available and must be assessed in terms of their practical application and overall contribution to heat gain and loss (thermal audit). Heart rate and internal body temperature measuring equipment is now widely available and can be used in personal monitoring of workers in hot environments. It is important to recognize that such systems must be implemented into working practices with appropriate expertise in terms of use of the equipment and interpretation of measurements.

The increased interest in heat stress assessment in industry and the requirements for organizations to carry out risk assessments have led to an interest in useable methods. Thermal indices and assessment methods could be regarded as complex and academic and not directly applicable to any specific work situation. Malchaire *et al.* (1999) have prepared a methodology for practical application. It is based upon a three level approach; 'Observation', 'Analysis', 'Expertise' (see also ISO CD 15265 and Chapter 14). The 'Observation' stage is carried out by workers and managers from the company to collect information, make 'straightforward' improvements and judge whether further analysis is required. The 'Analysis' stage is conducted by a specialist to quantify the risk, determine the optimum work organization and determine whether 'Expertise' is required, where a full work assessment and design will be carried out. Bethea and Parsons (2000) conducted a programme of laboratory and field research to develop a practical heat stress assessment methodology for use in UK industry. They validated existing and proposed standards (ISO 7933, 1989; and the Predicted Heat Strain method) and the risk assessment strategy proposed in ISO CD 15265. Based upon focus groups of those involved in risk assessment in industry and at professional conferences concerned with occupational hygiene, a proposal was made for a useable approach to risk assessments in hot environments. Field experiments were then carried out comparing risk assessment strategies as performed by workers responsible for heat stress assessment in hot environments in paper mills and a steel foundry. Based upon usability criteria (functionality, ease of use, etc.) an improved methodology was determined. Whichever methodology is eventually adopted, it is clear that it must be compatible with the culture and requirements of the organization and that training will be required as well as careful implementation of the methodology into working practices in hot environments.

11 Cold stress

INTRODUCTION

In air environments, cold stress generally produces severe discomfort before any effect on health occurs. There is therefore a strong behavioural reaction to cold and many methods used for its avoidance: clothing, activity, shelter, etc. Human thermal environments that can be described as cold would lead to a tendency for heat loss (or negative heat storage) from the body. There is sometimes some semantic confusion. An environment with an air temperature of 5 °C may be described as 'cold'. However, an active, heavily clothed person in that environment may be hot and sweating into clothing in an attempt to lose heat. When the person rests then the previously warm to hot human thermal environment of 5 °C air temperature, becomes cold and heat loss and discomfort are exacerbated by damp clothing. In human thermal environment terms, the person has gone from a hot to a cold environment, whereas the air temperature has not changed.

As well as severe discomfort, cold stress can lead to a fall in body temperature. For example, deep body temperature can fall below 35 °C (hypothermia) relatively rapidly if a person is immersed in cold water or (less rapidly) in air especially if air velocity is high and the person is wet. Skin temperature can also fall, leading to non-freezing and freezing cold injury. It is particularly important to protect the hands, feet and exposed skin on the face and head, including the nose, eyes, ears and cheeks due to large surface area to mass ratio, and the lips due to moisture. The lowering of body temperature, particularly of the heart, can lead to death, especially in those not able to withstand severe stress.

Much investigation into cold stress has been into military and expedition type activities and in working outdoors. There is increasing interest however in working indoors, particularly in freezer rooms (and in some 'kitchens' or food preparation areas where food is kept below 4 °C until it is heated then eaten very soon after). Effects of cold on health, cold stress indices, acclimatization, protective clothing and working practices for cold environments are all topics of interest and are discussed below.

PHYSIOLOGICAL RESPONSES TO COLD

When the body becomes cold, vasoconstriction reduces blood flow to the skin and hence heat loss. Where there is a tendency for body temperature to fall, non-shivering thermogenesis (muscle tensing, feeling of stiffness and enhanced metabolism) will increase heat production. As the body temperature falls (skin, skin with core, or both) and thermoregulates, shivering begins; there may also be some 'reaction' type shivering (at higher body temperatures) due to psychological response and rate of fall in skin temperature due to sudden exposure. Initially, there is asynchronous firing of muscle fibres to produce heat but no work. After further cooling the muscle discharges synchronize to produce the 10–12 Hz oscillation associated with shivering and hence producing heat. Heat production can be increased by up to six times resting levels over short periods and around double for longer durations. During exercise shivering is inhibited. There are large individual variations in response. It has been suggested that the maximum oxygen uptake during shivering is about 50 per cent of VO_2 max.

Shivering is an effective method of increasing heat production. In the limbs some heat is lost to the skin but some will be transferred to maintain body temperature. Slonim (1952) showed that the neck muscles in animals (including man) are the first to shiver. This will help maintain brain temperature. Shivering is inhibited in some illnesses and by some drugs, e.g. insulin. It is also affected by levels of O_2/CO_2 and is inhibited by any form of anaesthesia. During surgical operations for example, falls in deep body temperature can be marked if the operating theatre is not kept warm (Clark and Edholm, 1985).

The effectiveness of human thermoregulation in the cold is such that an environment which produces any fall in deep body temperature below about 36 °C can be regarded as severe and a fall below 35 °C (hypothermia) potentially dangerous. In 'controlled' hypothermia (e.g. for surgical operations), powerful techniques (e.g. ice cooling or the use of lytic drugs) in anaesthetized patients must be used. These are specialized conditions however, and under normal conditions environments must be extreme. There are also great individual differences in response. Violent shivering occurs in an attempt to maintain body temperature. If temperature begins to fall muscles become stiff and blood viscosity increases so movements become clumsy. There may be a clouding of consciousness (e.g. confusion and sometimes apathy), a loss of sensory information (e.g. blurring of vision) and unconsciousness. There is a large individual variation but almost all persons will be unconscious at an internal body temperature of 30–31 °C, and at these levels and below there will be major risk of death due to ventricular fibrillation (asynchronous behaviour of cardiac muscle – probably due to the direct effects of low temperature on the pacemaker). The effects of reduced internal body (core) temperature are summarized in Table 11.1.

Table 11.1 Progressive clinical presentations of hypothermia

Core temperature		Clinical signs
(°C)	(°F)	
37.6	99.6	'Normal' rectal temperature
37.0	98.6	'Normal' oral temperature
36.0	96.6	Metabolic rate increases in an attempt to compensate for heat loss
35.0	95.0	Maximum shivering
34.0	93.2	Victim conscious and responsive, with normal blood pressure
33.0	91.4	Severe hypothermia below this temperature
32.0	89.6	Consciousness clouded; blood pressure becomes difficult to measure
31.0	87.8	Pupils dilated but react to light; shivering ceases
30.0	86.0	Progressive loss of consciousness; muscle rigidity increases
29.0	84.2	Pulse and blood pressure difficult to measure; respiratory rate decreases
28.0	82.4	Ventricular fibrillation possible with myocardial irritability
27.0	80.6	Voluntary motion ceases; pupils nonreactive to light; deep tendon and superficial reflexes absent
26.0	78.8	Victim seldom conscious
25.0	77.0	Ventricular fibrillation may occur spontaneously
24.0	75.2	Pulmonary oedema
22.0	71.6	Maximum risk of ventricular
21.0	69.8	fibrillation
20.0	68.0	Cardiac standstill
18.0	64.4	Lowest accidental hypothermia victim to recover
17.0	62.6	Isoelectric electroencephalogram
9.0	48.2	Lowest artificially cooled hypothermia patient to recover

Low body temperatures can lead to death; however, survival will depend upon the individual. The brain will survive for longer with reduced blood flow at low temperatures as metabolic rate is reduced. Open-heart surgery can be successfully completed at heart temperatures of 18–20 °C. At 15 °C the heart will cease to function but even at body temperatures as low as 10 °C for short periods rewarming may be successful. Individual cases and highly controlled conditions however should not confuse the overall picture which is that it takes extreme conditions to reduce body temperature below 36 °C and any temperature below 35 °C can be dangerous.

In cold conditions where vasoconstriction reduces blood flow and hence heat loss, there is an apparently anomalous reaction where at around 12 °C (or below, depending upon rate of cooling, muscle and skin temperatures) vasodilation occurs. This is called cold induced vasodilatation (CIVD) and was described by Lewis (1930). The response is shown in Figure 11.1.

The so-called 'hunting phenomenon' is due to vasoconstriction followed by vasodilation as vasoconstriction cannot be effected at low temperatures. As temperature rises with blood flow, so vasoconstriction is applied again

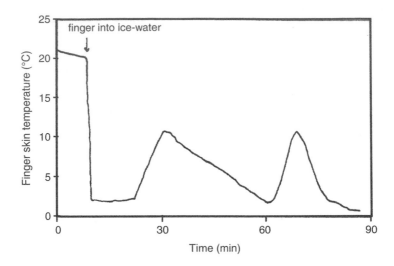

Figure 11.1 The effects of cold-induced vasodilation (CIVD) on finger skin temperature, measured with a thermocouple under adhesive tape.
Source: Lewis (1930).

and skin temperature falls. Hence there is a cyclic change in skin temperature. This can easily be demonstrated by immersing the hand in cold water, or is commonly experienced when 'playing' in snow for example.

There is some debate over exact physiological mechanisms or 'biological' advantages which CIVD offers. Hands become warm and can perform as increased blood flow supplies the hands and limbs. However, there is a net increased heat loss – leading rapidly to death in cold water for example (Keatinge, 1969) at a time when the body is attempting to preserve heat for survival.

Cabanac (1995) describes 'selective brain cooling' (SBC) in animals and (controversially) humans. This is a mechanism of thermoregulation where evaporative heat loss from the nose and mouth (due to breathing/panting) is, 'diverted from the upper airways towards the endocranial activity via special venous arrangements to cool the brain directly'. He hypothesizes that humans also exhibit selective brain warming (SBW). He suggests that the brain selectively preserves its metabolic heat similar to the way para-doxical sleep is hypothesized to aid stability of brain temperature. One reason for this may be to maintain the stability of retinal temperature. The claimed reversal of the emissary venous blood flow in the angularis occuli veins (direction depending upon whether the person is hot or cold: blood flows rapidly from face to brain during hyperthermia and sluggishly from brain to face during hypothermia) ... 'might also contribute to the warming of the human eye in a cold environment ... '. Whether this occurs and how effective it is in humans remains a matter for debate.

Blood pressure is a function of cardiac output and total peripheral resistance to blood flow through the blood vessels. The body has a mechanism for controlling blood pressure that responds to changes (e.g. caused by changes in blood distribution), however rapid changes in posture, exercise or thermal conditions can show dramatic changes in blood pressure. This can have severe consequences for the body, including death.

Tochihara *et al.* (1998) note that cold exposure and rapid exposure to large temperature changes can cause serious health hazards. They hypothesize that the reason for the large number of sudden deaths in Japan during and after bathing at home might lie in the combination of cold pre- and post-bathroom temperatures with high bath water temperatures. In a study of 12 male subjects, they found significant increase in systolic blood pressure when changing room air temperatures were below 15 °C and rapid decreases in blood pressure after entering bath water at 40 °C. They conclude that a safety range for bathroom temperature is 22–30 °C and that further research is required into the responses of the elderly.

White *et al.* (1994) note that the majority of hyperthermia-related deaths occur in hot tubs, spas and jacuzzis and that alcohol ingestion, a vasodilator, promotes orthostatic intolerance and is associated with most of these deaths.

Toner and McArdle (1988) note that cutaneous blood vessels are under neural control that is influenced by skin and core temperatures as well as baroreflexes. Exposure to cold causes vasoconstriction and re-distribution at the core of the body. If the exposure to cold is rapid (especially to the face, due to cold wind or water) cardiac output, stroke volume and blood pressure can be elevated. The practice of taking a cold plunge in 'ice' water may have psychological benefits, however it can be dangerous and is not generally recommended for health.

PSYCHOLOGICAL RESPONSE TO COLD

Psychological responses to cold can be large and include behavioural reaction, due to increased discomfort, and direct effects on psychological performance in terms of arousal, reduced memory capacity, perception, etc. There may also be changes in mood and personality with consequent effects on the social dynamics of a group (Rivolier *et al.*, 1988) or apathy generated by low body 'core' temperatures (Collins, 1983) and interactions with other environmental components (Griefahn, 1988).

A fuller discussion of psychological responses given in Chapter 3, demonstrates how it is important when describing psychological responses to identify the psychological model used. Rivolier *et al.* (1988) describe the psychological responses of scientists experimenting on themselves while living in tents in the Antarctic, and they conclude that it is important to consider the whole man. Although their bodies did not often become cold within the micro-climates of their clothing, there were still major psychological effects due to living in relative isolation in the cold environment.

In industrial contexts, offices or schools, cold can produce discomfort that can affect behaviour in terms of absenteeism from workplaces, desks or even from work overall. Distraction and discomfort not only affects absenteeism but can lead to a loss in attention and a breakdown of discipline.

The effects of thermal environments on comfort are discussed in Chapter 8 and on human activity, performance and productivity in Chapter 12.

Cold injury and illness

When the body becomes cold – particularly specific local parts such as the hands, feet and face – then cold injury can occur. Because of the nature of cold effects, duration of exposure is important. There has been much concern about such injuries especially during wars, including the First and Second World Wars. In the Falklands War, where a British Task Force was sent to relieve the Falkland Islands from Argentinean occupation, there was serious concern that the war would be lost because of cold injury to the feet of British soldiers. In industrial work there is also a possibility of cold injury and illness, for example, when working in freezer rooms on hand contact with frozen food. Traditionally, cold injury is divided into two types: *frostbite* (and frostnip) occurs where cooling lowers temperature such that tissue fluid freezes; and non-freezing cold injury or *immersion foot* (also called trench foot) which occurs where reduced blood flow after chilling and low temperatures (1–15 °C) causes damage to nerves – hence sometimes called peripheral vasoneuropathy (Oakley and Lloyd, 1990). Less severe injuries are cracked skin, chilblains (pernio) – as a result of chilling usually of fingers, toes, and even ears (Edholm and Weiner, 1981). There is little evidence of cold injury of the respiratory tract due to cold. There are reports of children's tongues becoming 'stuck' to metal surfaces as the result of a rather risky game in cold climates.

McCaig (1992b) describes frostnip as a white spot on tissues exposed to cold surfaces or wind. This may lead to frostbite. Non-freezing cold injury is described as a numbness during exposure and painful swelling on rewarming.

Contact between skin and a cold surface can cause sticking and hence freezing of the tissue and damage on renewal. At what surface temperatures this occurs and in which material types are yet to be determined. This information is important for the design of freezer rooms and other working environments.

Hamlet (1998) suggests that both environmental and host factors combine to produce cold injuries. Environmental factors include temperature, precipitation, and wind. Host factors include smoking, previous cold injuries, race, rank, malnutrition, fatigue, environmental protection, leadership, other injuries, and drugs. He classifies cold/wet injuries as chilblain (pernio) and trench foot (immersion foot). Cold/dry injuries are frostnip and frostbite. Hamlet (1988) considers the effects of hypothermia in terms of cardiovascular, respiratory, CNS and blood finger responses. Autopsy findings of hypothermia victims include pancreatic lesions, gastric erosion, microscopic degeneration of the myocardium, cyanotic red discoloration of skin, pulmonary change, pulmonary oedema, interstitial haemorrhage and focal emphysema. Hamlet (1998) considers Raynaud's disease (abnormal maintenance of vasoconstriction of the fingers or toes associated with emotional stress, vibration, or cold) for his list of non-freezing cold injuries but concludes that, although cold urticaria, cold-induced paraethesia, Raynaud's, and cold-induced asthma have a relation to cold exposure, they are not usually considered to be cold injuries.

Collins (1983) investigated statistics of seasonal mortality and considers that in the last twenty years there has been a general decrease in winter mortality, particularly in the older age group, due to the introduction of central heating – and a reduction in pollution. An increase of diseases in winter still occurs; however with such statistics it is difficult to identify an exact cause. Keatinge and Donaldson (1998) note that mortality during winter forms one of the largest groups of preventable deaths and the size of the group varies greatly between countries. They report on the 'Eurowinter Project' in which 1000 homes were surveyed across eight regions in Europe from the Arctic to the Mediterranean. Analysis showed that people in warmer countries (e.g. Greece) had higher 'excess winter deaths' than those in colder countries (e.g. Finland). People in cold regions were much more likely to heat their bedrooms and living rooms, wear a hat, gloves, and more windproof and waterproof clothes outdoors, keep moving when outside and, in the case of women, to wear trousers rather than a skirt. Mortality from ischaemic heart disease rises rapidly to a peak within two days after the peak in a cold spell and mortality due to respiratory disease, twelve days after. The results suggest that both indoor heating and outdoor protection against cold are important in preventing winter mortality. Windproof bus shelters for the elderly and the provision of a limited core area of a house that can be heated to full comfort levels are practical proposals.

The UK has experimented with a number of financial schemes to support the elderly during the winter. These have included a fixed sum to all elderly and a varying sum based upon the severity of outside conditions. Other schemes include a fixed energy 'bill', irrespective of consumption level. Co-ordinated policy may combine the requirement to insulate homes (sustainable buildings) with effective use of resources and financial incentives. The evidence from countries with 'cold' climates suggests that there is sufficient understanding of how to avoid excess winter mortality and that organization, education, training and culture changes would contribute to a successful strategy in areas for which extremes of cold are unusual.

Wide ranging epidemiological studies among cold-room workers and their handling of frozen food, for example, have yet to be conducted, and few details can be provided about the nature and extent of possible injuries or of consequent working practices. Griefahn *et al.* (1997) have conducted extensive field trials of workers in the cold in German industry. They found that there were many thousands of workers who work in the cold ranging from cold store temperatures ($<-20\,°C$) to chilled environments where food is prepared ($<4\,°C$). Of particular importance was the complaint of exposure to draught. Griefahn (1997) considers cold when combined with other physical stressors. She notes that cold is essential for prevention of premature spoiling of food, chemical products and drugs, and there is a continuous stress on workers often combined with shift work, noise, vibration and poor light. Cold is a seasonal stressor in construction work, forestry, agriculture, horticulture and navigation often combined with noise, whole-body and hand-arm vibration, poor light and electromagnetic fields. Vibration white-finger is exacerbated by cold (probably due to vasoconstriction) as are muscular-skeletal and cumulative trauma disorders. Griefahn (1997) also considers lumbago, which is prevalent in carriers exposed to cold, hearing problems and cardiovascular diseases.

Gravelling and Flemming (1996) surveyed work in the cold in UK industry. They found that most work was associated with the food industry, that there were few reported contact injuries, and that little was established in terms of working practices for cold environments. This work, and work in Germany, has led to the standards DIN 33 405–5 (1994) and BS 7915 (1998) and the proposed International Standard ISO CD 15743 (2002) concerned with working practices for cold environments.

COLD STRESS INDICES

Wind chill index (*WCI*)

The wind chill index (*WCI*) combines the effects of air temperature and air velocity into a single index or wind chill value. It was derived empirically by Siple and Passel (1945) who conducted experiments into the heat loss of uninsulated cans of water under Antarctic conditions. Measurements were made at Little America, Antarctica, by members of the United States Antarctic Service in 1941.

They constructed an atmospheric measuring device consisting of a sealed (pyrolin) cylinder (5.875 inches long, 2.259 inches diameter) containing 250 g of water and a thermohm measuring device, a 'naked' thermohm for measuring air temperature and a cup anemometer for measuring air velocity. Measurements were made in a series of experiments at various places in the Antarctic under conditions of freezing temperatures and darkness. The cooling rate of the water (melted snow) was determined by the length of time taken for the water to freeze and give off its latent heat of fusion. This was established by identifying when the water reached $0\,°C$ and began to freeze and when the temperature (of the water) began to drop again. The amount of heat given up therefore being the latent heat of fusion of water ($797.1\,\mathrm{kg\,cal\,m^{-2}}$). The rate of cooling is therefore this value divided by the time to freeze the water. Still air measurements were made in an enclosed pit, but all other measurements were made under existing winter conditions in Antarctica where outgoing radiation exceeded incoming radiation and absolute humidity was very low. Both factors were not considered to have significant effect. The rate of cooling provided by the environments was compared with that predicted by the Kata cooling time (Hill, 1919) and the equation of Winslow *et al.* (1936) based on the square root of the air velocity. Using curve fitting, an improved equation was derived (see below).

To determine the effects of cooling on man, numerous simultaneous observations of the time required for the freezing of normal human flesh exposed in the path of cold wind were made by the Medical Officer, Frazier. 'The subject, bareheaded but otherwise warmly clothed, was faced into the wind and the length of his exposure (before a sharp twinge of pain announced that the moment of freezing began) was timed. Sudden blanching of skin was usually apparent.'

Around twenty separate subjects took part in the experiments and almost all exhibited freezing of the nose with additional freezing of the eyelids, cheeks, wrist, side of temple and chin. A scale relating cooling (in cal/m²/hr) to effects was constructed. For example, at around 1400 cal/m²/hr, 'Freezing of human flesh begins, depending upon degree of activity, amount of solar radiation, character of skin and circulation. Travel and life in temporary shelter becomes disagreeable.' An equation was derived to estimate the rate of cooling of exposed skin:

$$WCI = \left(10\sqrt{v} + 10.45 - v\right)(33 - t_\mathrm{a})\ \mathrm{kg\,cal/m^2/h}. \tag{11.1}$$

This is now more commonly converted to SI units in the form:

$$WCI = 1.16\left(10\sqrt{v} + 10.45 - v\right)(33 - t_\mathrm{a})\,\mathrm{W\,m^{-2}}. \tag{11.2}$$

Although for the above it is important for v to be in $\mathrm{m\,s^{-1}}$ and t_a in °C, *WCI* is generally regarded as an index value and not as an indication of heat loss from the body. The index units need not therefore be converted to SI and can be ignored. Recent formulations of the index have considered the heat transfer properties and provide consistency with other heat transfer evaluation using SI units and $\mathrm{Wm^{-2}}$.

A related index often quoted is the chilling temperature, t_ch. This is the temperature of calm conditions ($v = 1.8\,\mathrm{m\,s^{-1}}$) which would provide equivalent effect as the actual environment (i.e. same *WCI* value). Re-arranging the *WCI* equation gives:

$$t_\mathrm{ch} = 33 - \frac{WCI}{22}\ °C, \tag{11.3}$$

Table 11.2 Wind chill index (*WCI*), chilling temperature (t_{ch}) and corresponding sensation of people dressed in arctic clothing

WCI	$t_{ch}(°C)$	Effect
50		hot
200		pleasant
400		cool
800		cold
1000	−12	very cold
1200	−21	bitterly cold
1400	−30	exposed flesh freezes
1600	−40	exposed flesh freezes within 1 h
1800	−49	
2000	−58	exposed flesh freezes within 1 min
2200	−67	
2400	−76	exposed flesh freezes within 30 s
2500		intolerable

or from the SI version with *WCI* in Wm^{-2}

$$t_{ch} = 33 - \frac{WCI}{25.5} °C. \tag{11.4}$$

For Antarctic-like conditions the indices can be interpreted on the scale presented in Table 11.2.

The wind chill index is the most widely used cold stress index. This is despite its theoretical limitations. Molnar (1960) argues that the *WCI* is limited in that it applies only for unprotected surfaces, does not consider respiratory heat loss, cannot be used at wind speeds greater than $20\,m\,s^{-1}$, is strictly empirical and is not derived from thermodynamic laws, overestimates cooling power for naked surfaces and underestimates cooling power for clothed surfaces. Burton and Edholm (1955) conclude that it is theoretically impossible to express the effects of wind on heat loss without referring to the amount of clothing. They attribute the success and applicability of the *WCI* not to its heat loss predictions but because the tolerance of a person exposed to cold is largely determined by exposure of bare skin such as on the face and hands. The *WCI* is more applicable for these areas not covered by clothing. For a totally clothed body (including hands and face), the *WCI* greatly exaggerates the importance of wind and would unnecessarily restrict exposure. As with all indices however, theoretical predictions provide only one guide to the likely effects of exposure. Experience with the use of an index in any context provides the guide to working practices.

Equivalent still air temperature (ESAT)

Burton and Edholm (1955) take a more rational approach and consider the effect of wind as a correction to the air temperature to give equivalent

Table 11.3 Relationship between W (insulation decrement) and v (air velocity)

Air velocity		W	I_a
Ft min^{-1}	v, m s^{-1}		
20	0.10	0.0	1.0
35	0.18	0.2	0.8
50	0.25	0.3	0.7
75	0.38	0.4	0.6
120	0.61	0.5	0.5
210	1.07	0.6	0.4
425	2.16	0.7	0.3
1050	5.34	0.8	0.2
4500	22.80	0.9	0.1

thermal demand in still air. This corrected temperature is called the equivalent still air temperature. A decrement (W) for the insulation of the air layer (I_a) for the effects of wind gives:

$$I_a = I_{sa} - W. \tag{11.5}$$

The standard value of I_{sa} for the insulation of still air is chosen as 1.0 Clo. Insulation decrement W then depends upon air velocity as shown in Table 11.3. From a heat balance equation the heat production H can be represented as

$$H = \frac{0.11(t_s - t_o)}{I_{cl} + I_a}, \tag{11.6}$$

in terms of wind decrement

$$H = \frac{0.11(t_s - t_o)}{(I_{cl} + I_{sa} - W)}, \tag{11.7}$$

or re-arranging

$$H = 0.11 \frac{(t_s - (t_o - HW/0.11))}{(I_{cl} + I_{sa})}. \tag{11.8}$$

So the thermal wind decrement, TWD (in $°C$), is given by

$$TWD = \frac{HW}{0.11}. \tag{11.9}$$

The thermal wind decrements for a total heat loss of 1 Met (58.15 W m^{-2}) are given in Table 11.4.

Table 11.4 Thermal wind decrements for a total heat loss of 1 Met

Air movement		Thermal wind decrement (TWD)
Ft min^{-1}	m s^{-1}	°C
20	0.10	0
35	0.18	1.8
50	0.25	2.7
75	0.38	3.6
120	0.61	4.5
210	1.07	5.5
425	2.16	6.4
1050	5.34	7.3
4500	22.80	8.2

For values of 2, 3, 4, 5 Met etc. simply multiply by the thermal decrement, e.g. for 2 Met at 0.38 m s^{-1} (75 ft min^{-1}); the thermal wind decrement is $(2 \times 3.6 =)$ 7.2 °C. The thermal wind decrement is therefore the product (in appropriate units) of the total heat loss and the insulation wind decrement. It is emphasized that the thermal decrement is multiplied by the heat loss in Mets. For example, a person in comfort in still air at 20 °C would be uncomfortable if the wind greatly increased to provide a heat loss of 1 Met and hence the equivalent temperature would be about $(20 - 8 =)$ 12 °C. In an environment of −18 °C air temperature and still air, if air movement is greatly increased, a heat loss may be as great as 5 Met. The temperature decrement would be around $(5 \times 8.2 =)$ 41 °C giving an equivalent temperature of −59 °C. Burton and Edholm (1955) state that the equivalent still air temperature is more theoretically correct than the *WCI*. Having calculated the equivalent still air temperature, the effects of clothing insulation have to be taken into account. This is not theoretically possible with the *WCI*.

The equivalent still air temperature is not widely used in the above form and, presumably because of simplicity, the *WCI* and t_{ch} have been widely adopted. Burton and Edholm's index however did demonstrate that it is the six basic parameters that should contribute to a thermal index. The use of the heat balance equation and of the interaction of metabolic heat production with other parameters was later developed by Holmér (1984) and ISO TR 11079 (1993) to form the *IREQ* cold stress index.

The equivalent shade temperature (EST) and the still shade temperature (SST)

Burton and Edholm (1955) extended the equivalent still air temperature to include a solar radiation element and hence proposed equivalent shade temperature (*EST*) and the still shade temperature (*SST*). The *SST* is obtained by subtracting the thermal wind decrement and adding the radiation increment to the air temperature (see Youle *et al.*, 1990).

The radiation increment, *TRI* (in °C), is determined by rational analysis and is given by

$$TRI = 0.42(1 - 0.9x)aI_a \, °C,$$ (11.10)

where

TRI = thermal radiation increment

   x = average cloudiness (in tenths, e.g. 3/10 would mean that *x* = 0.3 and an estimated 3/10 of the sky is covered by cloud). Radiation effect will also depend upon the region of the earth under consideration. For example cloud is thinner in arctic regions allowing greater penetration for the same cloud cover.

   a = absorbing power of clothing (per cent) – e.g. 88 per cent for black, 57 per cent for khaki and 20 per cent for white.

   I_a = insulation of air layer, taking account of air velocity (see Figure 7.2).

A chart is presented in Figure 11.2, for estimating the thermal radiation increment to be added to the air temperature to give equivalent shade temperature.

(e.g. *x* = 0.5, *v* = 0.61, *a* = 57, I_a = 0.5, *TRI* = 6.6°C)

Figure 11.2 shows that radiation may increase effective temperature for military clothing by as much as 20°C in calm air and unclouded sky. Blum (see Burton and Edholm, 1955) made calculations from meteorological data and showed increments of 4–8 °C were common. These are only estimates however as there are many factors involved.

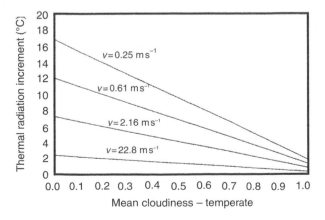

Figure 11.2 Thermal radiation increment for temperate climates (khaki clothing).
Source: Calculated from equations given by Burton and Edholm (1955).

Burton and Edholm (1955) make the interesting point that using the greenhouse effect and translucent clothing an inactive man in 3.0 Clo and still air could sit comfortably at $-18\,°C$ air temperature in an unclouded sky.

The still shade temperature (*SST*) is obtained by subtracting the thermal wind decrement (*TWD*) from air temperature to obtain the equivalent still air temperature (*ESAT*) and adding the thermal radiation increment.

In summary:

$$ESAT = t_a - TWD \qquad\qquad (11.11)$$

$$EST = t_a + TRI \qquad\qquad (11.12)$$

$$SST = t_a + (TRI - TWD). \qquad\qquad (11.13)$$

Burton and Edholm (1955) give the example that if wind speed is 5 mph, for a resting man, TWD = TRI for full sunshine. If the wind speed is greater than 5 mph then it will be colder to stand outside in the sun than inside a shelter at the same temperature. For a man active at 2 Mets, the critical wind speed is less than about 3 mph and so on.

The required clothing insulation (IREQ) index

Holmér (1984) developed the idea of a thermal index based on the rational calculation (i.e. from a calculation of the heat balance equation for a clothed person) of clothing insulation required for heat balance and for comfort. For any combination of environmental conditions and human activity, where there is a tendency for negative storage in the body (i.e. cold), there is a theoretical clothing insulation value that will meet these conditions. This is similar in concept to the use of required sweat rate (SW_{req}) as a heat stress index (Chapter 10).

Required values to meet certain criteria are commonly used as rational indices and required clothing insulation had been proposed previously (see Figure 11.3). The idea is a practical one, as it was acknowledged that although measurements or estimates of environmental conditions were useful, they were often used to determine appropriate clothing as the main application of any analysis. Burton and Edholm (1955) provided the following equation:

$$I_t = \frac{0.082(91.4 - t_a)}{M}, \qquad\qquad (11.14)$$

where

t_a = air temperature (°F)

M = metabolic rate (Mets)

I_t = total clothing insulation, $I_{cl} + I_a$ (Clo).

How many Clo units are needed?   What 1 Clo of clothing insulation is good for.

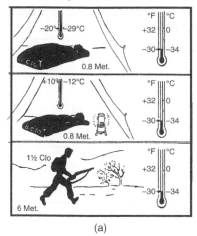 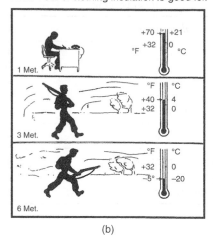

(a)   (b)

Figure 11.3 Pictograms to illustrate the importance of the metabolic rate in determining the clothing required in different environments (Burton and Edholm, 1955). (a) Clothing insulation required; (b) What 1 Clo is 'good for'.

Holmér (1984) provided the modern version of this index and ISO TR 11079 (1993) provides a method for calculating IREQ in terms of the up-to-date heat balance equation. Two indices are proposed: clothing insulation required for heat balance ($IREQ_{min}$); and clothing insulation required to provide comfort ($IREQ_{neutral}$).

$IREQ_{min}$ defines a minimal thermal insulation required to maintain body thermal equilibrium at a subnormal level of mean body temperature. The minimal IREQ represents the highest level of physiological strain to which humans should be exposed when performing occupational work.

$IREQ_{neutral}$ is defined as the thermal insulation required to provide conditions of thermal neutrality, i.e. thermal equilibrium maintained at a normal level of mean body temperature. This level represents zero or low physiological strain.

Physiological criteria for $IREQ_{min}$ and $IREQ_{neutral}$ are provided in Table 11.5.

IREQ is the resultant clothing insulation required during actual environmental conditions and activity of the body to maintain physiological requirements. It can be applied as:

1   a cold stress index, integrating t_a, t_r, rh, v and M;
2   a method of identifying the relative effects of specific parameters and evaluating measures of improvement; or
3   a method of specifying clothing insulation requirements and selection of clothing to be used.

Table 11.5 Suggested physiological criteria for determination of *IREQ*, *DLE* and local cooling

General cooling	Minimal IREQ	Neutral IREQ
IREQ	'high strain'	'low strain'
t_{sk} (°C)	30	$35.7 - 0.0285\,M$
w (ND)	0.06	$w = 0.001\,M$
DLE		
Q_{lim} (Wh m^{-2})	−40	−40

Local cooling	'High strain'	'Low strain'
hand temperature (°C)	15	24
WCI Wm^{-2}	1600	–
Respiratory tract and eye (°C)	$t_a < -40$	–

IREQ is calculated from the heat balance equation

$$M - W = E_{res} + C_{res} + E + K + R + C + S,$$

K is assumed to be negligible and heat exchange through clothing is given as

$$R + C = M - W - E_{res} - C_{res} - E = \frac{t_{sk} - t_{cl}}{I_{clr}},$$

$$IREQ = \frac{t_{sk} - t_{cl}}{M - W - E_{res} - C_{res} - E}, \tag{11.15}$$

and

$$M - W - E_{res} - C_{res} - E = R + C, \tag{11.16}$$

but equation (11.15) contains two unknowns *IREQ* and t_{cl} therefore:

$$t_{cl} = t_{sk} - IREQ(M - W - E_{res} - C_{res} - E)$$

is used in (11.16), where formulae for R and C contain t_{cl}. The value of *IREQ* that satisfies (11.16) is then calculated by iteration. ISO TR 11079 (1993) provides a computer program listing for ease of use.

If the *IREQ* value calculated cannot be achieved or is not available then there will be negative heat storage (net heat loss). For these conditions allowable exposure times, called duration limited exposures (*DLEs*), are calculated from:

$$DLE = \frac{Q_{lim}}{S}, \tag{11.17}$$

where $Q_{\lim}$ is the heat storage limit (in Wh m^{-2}) (see Table 11.5) and S is calculated from equation (11.19) in which t_{cl} is calculated using available clothing insulation (I_{clr}) and

$$t_{cl} = t_{sk} - I_{clr}(M - W - H_{res} - E - S) \tag{11.18}$$

and substitutes into

$$S = M - W - H_{res} - E - R - C \tag{11.19}$$

The recovery time *RT* is the time required to restore body heat balance to thermo-neutral conditions. *IREQ, DLE* and *RT* values are easily calculated using a simple computer program. Figures 11.4 and 11.5 provide graphs of *IREQ* values over a range of conditions. By far, the most effective method of calculation however is to use a computer.

The *IREQ* index is still under development and has yet to be adopted for practical application. O'Leary (1994) and O'Leary and Parsons (1994) investigated the application of *IREQ* to the selection of clothing and found *IREQ*$_{min}$ as a useful starting position as a basis for more refined aspects of design that depend upon the specific task and working conditions. Aptel (1988) compared *IREQ* values with clothing insulation actually worn by workers in freezer rooms. He found *IREQ* to be inadequate; however, the use of resultant and not intrinsic clothing insulation for *IREQ* was not considered. Also, as for any rational index, the estimates of clothing insulation and metabolic heat production are subject to inaccuracy, yet are influential in the results of heat balance calculations.

The usefulness of *IREQ* in practical applications is yet to be determined. It is commonly thought that tolerance to cold is dominated by local skin

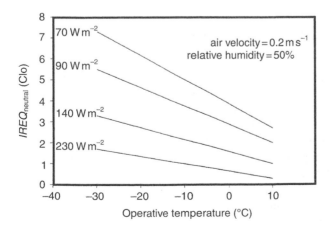

Figure 11.4 IREQ neutral as function of ambient operative temperature at four levels of metabolic heat production.

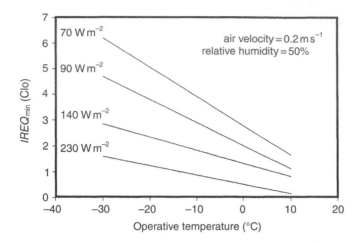

*Figure 11.5 IREQ*min as a function of ambient operative temperature at four
levels of metabolic heat production.

temperatures – hands, feet, face – and problems occur due to sweating in
clothing when working and particularly subsequent problems when resting.
ISO TR 11079 (1993) provides some guidance for practical application
(see Figure 11.6). The publication of this standard in the first instance as a
technical report allowed experience with the method to determine whether
it was more widely adopted in the future. This has now been confirmed and
the *IREQ* index has become a standard method for the assessment of cold
environments. It has been adopted in BS 7915 (1998) and DIN 33–403–5
(1994) and forms the basis of the proposed International Standard ISO
15743.

A new wind chill index and a universal climate index

The Wind Chill Index (*WCI*) of Siple and Passel (1945) has been universally
accepted and used throughout the world, in particular in terms of determining when exposed flesh freezes and effects of cold on extremities. It has both
theoretical and practical limitations. Siple (1939) had proposed a Wind
Chill Index where the equivalent temperature was calculated by multiplying
the (sub-freezing) air temperature (°C) by the air velocity (m s^{-1}). However
this is meaningful only below 0 °C and has not been shown to be a valid
indicator of human response. Boutelier (1979) made the sensible suggestion
that to include the effects of radiation, globe temperature should replace air
temperature in the *WCI* equation. However, the effects of air velocity on
globe temperature may confound the effect.
   There have been a number of attempts to improve the *WCI* formula
and some have been adopted (e.g. Environment Canada, 2001) have

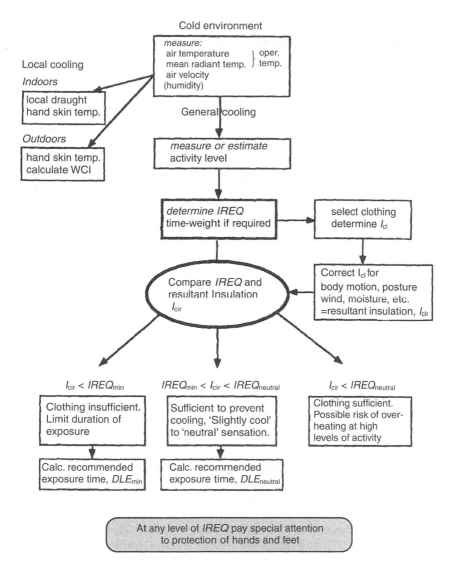

Figure 11.6 Procedure for the evaluation of cold climates.

adopted a new Wind Chill Index (Osczevski and Bluestein, 2001) based upon a model of how fast a human face loses heat. They provide the following form (SI units) of the equations directly from Siple and Passel (1945):

$$C = 0.323\left(18.97\sqrt{v} - v + 37.62\right)(33 - t_a),$$ (11.20)

where

> C = wind chill factor (cooling rate) $W\,m^{-2}$
> v = wind velocity $km\,h^{-1}$
> t_a = air temperature   °C.

They provide the following form of Siple and Passel's equation for equivalent temperature $(Te, °C)$. They note that the given reference wind speed $(Vr, km\,h^{-1})$ is typically an average person's walking speed.

$$Te = 33 - \left[\frac{18.97\sqrt{v} - v + 37.62}{18.97\sqrt{v_r} - v_r + 37.62}\right](33 - t_a). \tag{11.21}$$

There were three criticisms of the Siple and Passel equations: the reference wind speed; the wind speeds were speeds at face height (yet the standard weather observation at 10 m is usually used); and the skin temperature is assumed to be 33 °C, even though skin freezing is an outcome and skin freezes around −1 °C. A new equation was proposed as follows:

$$W = 13.12 + 0.6215\,t_a - 11.37\,V_{10m}^{0.16} + 0.3965\,t_a \times V_{10m}^{0.16}, \tag{11.22}$$

where

> W = Wind Chill Index   °C
> t_a = air temperature   °C
> V_{10m} = wind speed at 10 m $km\,h^{-1}$
>   (standard anemometer height).

Solar radiation is not included at this stage. It is emphasized that although the new wind chill index is expressed in °C, it is an index of human sensation and not a measure of temperature. Environment Canada (2001) notes that the new wind chill index uses wind speed calculated at the average height of the human face (1.5 m) by correcting the 10 m value with a multiplication of 2/3, it is based upon heat transfer from the face, it uses a calm wind threshold of observed walking speed of 4.8 km h^{-1}, and it uses a consistent standard for skin resistance to heat loss.

The new index was validated at the Defence and Civil Institute of Environmental Medicine (DCIEM) in Canada involving physiological measures (including rectal and face skin temperatures) for six male and six female subjects walking for 90 min in a range of wind (4.8, 8, 18, 29 km h^{-1}), temperatures (−10, 0, +10 °C), and 'drizzle' conditions. An equation to approximate minutes to frostbite is as follows:

$$t_f = \{(-24.5[(0.667V_{10m}) + 4.8]) + 2111\}(-4.8 - t_a)^{-1.668}, \tag{11.23}$$

where

t_f = time to frostbite (minutes) for the 5 per cent most susceptible
   segment of the population

V_{10m} = wind speed (km h^{-1}) at the standard anemometer height of 10 m
   (as reported in weather observations)

t_a = air temperature (°C).

The equation is valid for winds of more than 25 km h^{-1} and times less than 15 min.

A consistent approach to a new wind chill index would be to formulate it in terms of convective heat transfer from the human body. Siple and Passel (1945) made early attempts at this (based upon $\sqrt{v}$) but used a modified equation to better fit with experimental data. From Chapter 1 the equation for convective heat transfer is as follows:

$$C = h_c(t_{cl} - t_a). \tag{11.24}$$

If we assume exposed skin, then

$$\begin{aligned} C &= h_c(t_{sk} - t_a) \\ &= 8.3\,v^{0.53}(33 - t_a)\,\mathrm{W\,m^{-2}}, \end{aligned}$$

for a comfortable skin temperature. A further simplification (possibly – not necessary) would be

$$C = 8\sqrt{v}\,(33 - t_a)\,\mathrm{W\,m^{-2}}, \tag{11.25}$$

where C is a *WCI* for the whole-body.

If one assumes still air of 0.25 m s^{-1} (a more appropriate value – especially indoors – than 1.8 m s^{-1} used in the original index) then the equivalent chilling temperature is

$$t_{ch} = 33 - \frac{C}{4}\,°\mathrm{C}. \tag{11.26}$$

WCI in terms of C is, correctly, a cooling rate of the environment taking account of air temperature and air velocity. Interpretation of the effects on people must therefore be made in terms of the other four basic parameters, including clothing and radiation.

A continuation of the use of convective heat transfer as a *WCI* is to consider heat loss from extremities. Danielsson (1998) notes for common climatic conditions for a cylinder, in cross airflow

$$h_c = 4.47d^{-0.38}v^{0.62}, \tag{11.27}$$

where d is the diameter of the cylinder (m). Wilson and Goldman (1970) have used this approach to predict the temperature when a finger will freeze and this has been further developed by Tikuisis (1994). If an assumption is made of a generalized local part of the body then

$$h_c = 20\sqrt{v}, \tag{11.28}$$

may be reasonable, giving:

$$WCI = 20\sqrt{v}(33 - t_a)\,\mathrm{W\,m^{-2}}, \tag{11.29}$$

and for $v = 0.25\,\mathrm{m\,s^{-1}}$,

$$t_{ch} = 33 - \frac{WCI}{10}\,{}^\circ\mathrm{C}, \tag{11.30}$$

demonstrating the great influence of air velocity on cooling when compared with true still air.

For whole-body heat loss the units $\mathrm{W\,m^{-2}}$ refer to the surface area of the whole-body, however, care must be taken in interpretation when considering body parts. If the WCI is to be used as a simple, practical index then the simplification provided above will have value. For a more detailed analysis there is little point in embellishing the simple index with complications. In that case, all six parameters should be considered and a thermal model should be used to predict human response. The answer to which WCI should be chosen is 'one of them'. No index of this type will be entirely valid. It should therefore be simple. The existing WCI has the great advantage that it is standard and used.

Siple and Passel (1945) suggested that the WCI could be plotted across the continent of Antarctica and that it would be useful to map out other continents in this way. The severity of environments could then be detailed in a global map. The WCI has been influential in weather forecasting and reporting throughout the world and discussion of an improved WCI for weather forecasting has led to proposals for a Universal Climate Index. Höppe (2001) has proposed such an index based upon a model of human thermoregulation. A universal climate index may also relate conditions to the temperature of a standard environment that would give equivalent effect on people. However, should the standard environment be different for people in hot climates and cold climates? Whether a universal climate index can be found that will be universally valid remains to be seen.

Acclimatization to cold

There is little doubt that people learn to behave in cold climates such that they can survive and keep warm. In terms of human thermal environments therefore the body is generally exposed to a 'warm' micro-climate even when environmental temperatures are very low. The question of whether there is physiological acclimatization is therefore difficult to show and evidence is inconclusive.

Rivolier *et al.* (1988) categorize possible adaptation (of people who live permanently in cold climates) of the body into three types along a continuum.

1   Hypothermic – 'allowing' body temperature to fall thus reducing heat loss.
2   Insulative – enhanced insulation hence preventing cooling.
3   Metabolic – Increased heat production (e.g. non-shivering thermogenesis).

There is some evidence of these types of adaptation; however results are inconclusive. Aborigines live in hot day-time climates and sleep in cold night-time climates. At night they are said to suppress shivering and allow body temperature to fall. In these conditions people from temperate climates fail to sleep. There are numerous examples of such adaptations but also other possible explanations. For example, Aborigines may have adapted their behaviour to receive radiant heat from nearby fires during the night. This has also been suggested as an explanation for the ability of the (now extinct) Indians of the Tierra del Fuego to maintain heat balance while lightly clothed and in cold climates. There are numerous studies of adaptation of those who live and work in cold environments and of comparisons and possible physiological acclimatization in those exposed after being in temperate regions. The results have been generally inconclusive. An interesting adaptation is reported concerning Tibetan monks in the Himalayas. By imagining heat production within the body (in this case to melt imagined ice in the brain), it is said that internal heat production can be greatly increased. This effect however, was not repeated in the laboratory. The example provides some evidence that cultural (and religious) practices could be adopted to have further practical utility in the avoidance of cold stress for example or in controlled loss of body weight.

Rivolier *et al.* (1988) review the problem and describe an experiment where 12 men were studied before, during and after return from a six-month Antarctic expedition. They were interested in whether it was possible to induce acclimation (i.e. in a laboratory) by immersion in cold baths, whether this had any advantage while living in the cold, whether a short stay in the Antarctic induces acclimatization, and whether acclimation (laboratory induced) and acclimatization are similar. After the extensive 'experiment' there was no clear evidence of acclimation or acclimatization to cold. It can be concluded that if there is acclimatization of the whole body to the cold, evidence for it is not outstanding and, for practical purposes, it can be assumed that any advantages will be gained from behavioural changes and learning to live in the cold. It may be expected that any acclimatization would lead to lower preferred temperatures. This is not necessarily the case. There is some evidence that people adapted to living in extremes, heat or cold prefer lower and high temperatures, respectively. It may be that acclimatization involves an early reaction, anticipation and compensation mechanism. Mercer (1989) contains detailed discussion.

There is evidence of local acclimatization to cold of the fingers and hands. It is often cited that people whose hands are regularly exposed to cold exhibit less vasoconstriction and more prolonged cold induced vasodilatation (CIVD). This has the advantage of maintaining manual performance. Edholm (1978) describes the experiments of Mackworth (1953) which show that people who regularly work outdoors in the cold show less drop in finger sensitivity when exposed to cold. Other studies cite fishermen, Eskimos, and pearl divers who are all exposed to cold water but maintain hand temperature. Whether this response is local acclimatization or whether hands have become damaged and restrict the ability to vasoconstrict is yet to be established.

Working practices for cold environments

Workers may become uncomfortable in the cold and loss in manual dexterity and distraction effects, for example, can contribute to accidents. Correct working practices should prevent hypothermia, cold injury, illness and death from cold. The American Conference of Government and Industrial Hygienists (ACGIH, 1996) provide threshold limit values (TLVs) for exposure to cold based on the wind chill index (*WCI*). These are limits within which nearly all workers can be repeatedly exposed without adverse health effects. ACGIH (1996) suggests that protection must be provided to workers in terms of:

1   adequate insulating clothing – to maintain core temperature above 36 °C; and
2   special precautions for older workers or workers with circulatory problems – providing extra clothing insulation or reduction in the exposed period, and taking the advice of a medical officer.

Guidance for evaluation and control

General

1   For exposed skin, cutaneous exposure should not be permitted at equivalent chill temperatures of −32 °C or less.
2   At air temperatures below 2 °C, workers who become wet should be given a change of clothing and treated for hypothermia.
3   Recommended limits for periods of work are given in Table 14.18.

Manual dexterity

Special protection for the hands is required to maintain manual dexterity for the prevention of accidents, e.g. warm air jets, radiant heaters, warm plates, and thermal insulation on tool handles.

If air temperature is below 16 °C, 4 °C or −7 °C for sedentary, light or moderate work, respectively, and fine manual dexterity is not required, then gloves should be used.

Contact injury

1 When cold surfaces are <−7 °C and within reach, a warning should be given to prevent inadvertent contact by bare skin.
2 If air temperature is <−17.5 °C, wear mittens and design machine controls, etc. accordingly.

Protective clothing and protective practices

1 Wind: protect from air velocity by using windshields or wind-break garments.
2 Wet clothing: in light work use water impermeable outer garment; in heavy work, outwear should be changed as it becomes wetted. Design for easy ventilation. If sweating has occurred change into dry clothing before entering cold area. Change socks as required.
3 Extremities: if handwear, footwear, and facemasks cannot prevent excessive cold, then supply in auxiliary heated versions.
4 Limited clothing insulation: if available clothing is inadequate then redesign work to reduce exposure or wait until conditions are less cold.
5 Working with liquids: special care taken if evaporative liquids (e.g. alcohol, gasoline) may spill on the hands in cold temperatures.
6 Work-warming regimen: use heated warming shelters (tents, cabins, rest rooms, etc.) and loosen clothing in shelter or change to warm clothing including shoes. Provide warm sweet drinks. Limit intake of coffee due to diuretic and circulation effects.
7 Very cold environments <−12 °C equivalent chill temperature: worker kept under constant observation. Use buddy system. Avoid sweating. Avoid periods of rest or low work level. Instruct workers on current practices and first aid.

Workplace recommendations

1 Air velocity: keep below $1 \, \text{m s}^{-1}$ in refrigerated rooms. Provide wind protection clothing. Use safety goggles out of doors especially in sleet and snow.
2 Monitoring: monitor air temperature, air velocity, and equivalent chill temperature.

3   Screen workers: exclude workers suffering from diseases or taking
     medication which interferes with normal body temperature regulation
     or reduces tolerance to cold. For very cold conditions medical certifi-
     cation is required.

The above provide guidelines for developing working practices for cold
exposure and are based on a summary of those presented in ACGIH (1992).
Specific working practices will depend upon the practical application and
context and should be developed within any industry. It should also be
remembered that all problems of people working in the cold are not fully
quantified or understood.

A full analysis of task requirements and the above working practices can
be used in job design or redesign for working in cold conditions. Any
comprehensive analysis including organizational aspects and available technology may lead to a great reduction or loss altogether of the necessity for
people to be exposed at all.

Parsons (1998b) and Rintamaki and Parsons (1998) present a systematic
review and some critical comment on issues relevant to the development of
working practices in cold environments. Working practices could be regarded
as a system of procedures that ensure the objectives of the work are achieved.
The objectives would include the health and safety, well being and productivity
of an individual and organization. Working practices for cold environments
would therefore be oriented towards ensuring that the effects of cold are sufficiently alleviated to ensure objectives are achieved. These are considered below.

Do the workers have to be exposed to the cold?

As cold can have significant effects on workers, a serious and primary
consideration should be to whether it is necessary that the workers are
exposed to cold. That is, can the objectives be met in some other way. An
example would be the use of robotics in warehouse storage and retrieval.
Another example is the design of the work to locate a product or process in
the cold but maintain a higher temperature for the workers. In the food
packing industry for example, there is a move from large cooled rooms
where workers work packing food on trays, to conveyor belt and cooled
tray systems where workers work in 16 °C air temperatures but food is
maintained below 4 °C. This, however, causes local cooling of the hands
and asymmetric environments that have not been investigated in terms of
human comfort and health. The general point is that if it is possible to
remove the worker from the cold exposure then this should be given high
priority. If it is necessary for the worker to be exposed to cold then
consideration should be given as to how exposure time can be kept to
a minimum. Can some of the jobs be performed outside of the cold environment? Increasing the size of the workforce to allow shorter shifts, extended
breaks and job rotation may also reduce exposure time. Work should be

designed to avoid periods of inactivity such as waiting and resting in the cold. Work breaks may be useful but it is not clear that breaks are welcome as rewarming takes a considerable time and workers do not usually relish returning to the cold.

Selection of workers for cold work

Part of a system of working practices will include procedures for selection of workers. That is, as well as the normal procedures for selection there will be additional requirements related to cold work. Evidence for selecting particular personnel and populations for work in the cold is incomplete. General guidance suggests that younger workers (e.g. 25–45 years) are more tolerant than old workers and that those with medical complaints are more affected by cold. It is sometimes suggested that cold store operators are more healthy than the general population but evidence is inconclusive. Cold workers are a self selected population and it is misleading to imply that work in the cold promotes health. An additional consideration is that people should be able to respond appropriately to remove themselves from a cold environment if necessary. Restricted mobility and mental impairment will inhibit their ability to do this. It is also an advantage if workers selected are 'team players' as it is beneficial to use a buddy system (workers watch out for each other) and team working to enhance worker's health and safety, morale and job satisfaction.

Screening

Workers selected for work in the cold should be screened by qualified personnel before exposure. Knowledge of how medical disorders are affected by cold is incomplete. Some specific disorders are consistently used in screening as indicators that will increase risk. The British Refrigerated Food Industry Confederation (RFIC) lists the following:

heart or circulation problems
diabetes
thyroid problems
blood disorders
kidney or urine disorders
any kind of arthritis or bone disease
any infection including ear, nose and throat
lung function problems or asthma
chronic gastro-enteritis or acute diarrhoea or vomiting (must be notified the same day)
neurological (nerve) malfunction
psychological problems
eyesight or hearing difficulty
prescribed medication

Table 11.6 Work in very cold environments (−10 to −40 °C). Pre-employment fitness requirements

---

1   Age – preferably 18–35. Either sex but NOT pregnant females.
2   Physique – preferably mesomorphic with adequate adipose tissue. Good physical fitness beneficial. Beware very tall-may not fit fork lift trucks.
3   No history of chronic respiratory disease, sinusitis or allergies. Asthma is a contraindication. Normal pulmonary functions by P.E.F.R. or spirometry.
4   No history of cardiovascular disease. Normotensive – beware hypertensives on treatment. Myocardial infarction is a contraindication. No circulatory disorders or vascular insufficiency, Raynaud's Disease, etc.
5   Anaemia is a contraindication and beware haemoglobinaemias, e.g. sickle cell trait/disease in certain ethnic groups.
6   No chronic gastro-intestinal disease – class as food handlers.
7   No genito-urinary disease, including infections.
8   No neurological disorders and no history of mental disease. Emotionally stable and mentally alert in view of product-handling with pallet trucks and fork lift trucks.
9   No history of rheumatoid disease or osteoarthritis.
10 Presence of endocrine disease, a contraindication. Changed requirement for insulin in diabetes. Euthyroid.
11 Exclude all chronic infections, including eye and ear infections.
12 Good personal hygiene is important, including dental hygiene.
13 Preferably no alcohol at least 12 h prior to cold store work. Moderation in tobacco intake.
14 Eyesight and hearing must be adequate (define standard).
15 Spectacle wearers will have great difficulty in seeing on leaving cold environment because of fogging of glasses. Nothing really known about contact lens wearers – caution advised.
16 Take careful record of drug treatment – body thermoregulation can be altered by many drugs, e.g. barbiturates, phenothiazines, benzodiazapines, B-blockers, etc. People on these may not be able to work in the cold.
17 Where breathing apparatus is used, bearded men may be unable to wear masks correctly.

Although sensible, the above list is a general list and provides little detail on interpretation. A more detailed list which is used in an organization with cold warehouses down to −28 °C air temperature, is given in Table 11.6.

System of reporting

A system of reporting provides a method for monitoring health of workers. If they feel symptoms of dizziness, abnormal cold, pain in hands and feet, heavy legs or other abnormal responses then the workers should report it. They should be trained in the systems for reporting and recording their symptoms. A system of monitoring and responding to reports will also be required.

Advice

General advice can be provided to workers and a system will be required to ensure that the advice is followed. It is usually recommended that alcohol is not consumed eight to twelve hours before a shift. Coffee intake should be restricted and workers are often recommended to eat protein. Advice on behaviour and procedures is essential and particularly of the clothing worn.

CASE STUDY OF COLD WORK IN A HOSPITAL 'PLATING AREA'

Background

A Regional Health Manager, responsible for working conditions in hospitals, wished to establish that work design in a hospital 'plating' area was satisfactory. He approached the climatic ergonomist and, in a preliminary meeting, the ergonomist described his expertise and work and the manager described his requirements. After the meeting and completed actions, for the manager to produce a description of requirements and the ergonomist a proposal and costs, a project was agreed (Parsons, 1998a).

The food preparation, cooking and serving methods in the new hospital had moved away from the traditional 'hot' kitchen method to a chilled system which maintained food at around 2 °C until it was ready to be eaten when it was heated. This was for reasons of hygiene. Food arrived in refrigerated lorries, already cooked and chilled from the factory. It was unloaded through a sealed entrance into a chilled area in the hospital. 'Kitchen' staff manned workstations around a conveyor belt and served an item of chilled food from large containers (e.g. potatoes, rice, peas). Trays containing menu, fruit juice and plates are placed on the conveyor at one end and food served onto plates as the trays pass along to the other end of the conveyor where they are removed and stacked in trolleys. The closed trolleys are also heated ovens such that by appropriate timing and temperature a trolley is taken to a ward and the patient is given the requested meal at an optimum temperature and condition, the food having been above 2 °C for a minimum period of time. The workers have therefore to work in the plating area at 2 °C air temperature for over one hour and there was a lack of guidance on correct working practices.

The project

The scope of the project was agreed in writing as follows:

1 To carry out an objective assessment of the chilled environment.
2 To advise the Health Authority on thermal comfort, welfare, clothing and other areas thought relevant with respect to staff working in this environment.
3 To carry out a subjective assessment of staff.

Method

The ergonomist visited the hospital for one day and conducted the survey. The manager was present for a short period and introduced the catering managers for the hospital and the region. Preliminary (structured) interviews revealed that there was a clear interest in the assessment being undertaken. Over all shifts, there were a total of 3 men and 38 women who work in the area. Activity in the plating room involves cleaning the area, delivering and removing trolleys and food, serving food and preparing trays. Cleaning the area involved the use of a specialist machine but when the ergonomist was there he noticed that this was done by hand. When not in the plating area the staff prepare special meals and wash up. They put on extra clothes and go into the plating area for about one hour. When finished they take off the extra clothes and have a hot drink before continuing work. The selection of clothing had involved staff. It consisted of normal underwear, long john trousers, long john vest, T-shirt, blue trousers, white smock top, quilted jacket, quilted leggings, thin gloves, thin impermeable overgloves, neckerchief (optional), hair net, light trilby hat, shoes and own socks. When not in the plating area, the overjacket was removed. The thermal underwear was worn at all times. An estimated Clo value from ISO 9920 (1995) of 2.48 Clo in the plating area with jacket and trousers, 2.3 Clo with jacket and outside the plating area without quilted clothes, 2.0 Clo.

Objective assessment

All eight workplaces in the plating area were measured in terms of air temperature, radiant temperature, humidity and air velocity. Equipment was calibrated before and after the measurements. It included eight 150 mm diameter black globe thermometers, eight thermistor sensors linked to a data logger, a whirling hygrometer, a hot wire anemometer, and a (child's) bubble kit. Measurements were made at ankle, chest and head height at each workplace. The room was empty for the period of about 110 min when the assessment was made. The engineer was asked to create the conditions experienced when at work. The empty room was acceptable but not ideal and the general lack of direct contact with the workers was an indication that it was a sensitive issue. A blueprint plan of the rooms was provided by the manager. All work was light arm work with an estimated metabolic heat production of $70 \, \text{Wm}^{-2}$.

The results showed that the air temperature had been successfully maintained in a cyclical pattern between 1 and 3 °C and was mixed by fans providing a turbulent and gusting air velocity of between 0.2 and $1.2 \, \text{m s}^{-1}$ over the workplaces. The globe temperatures and wet bulb temperatures were similar to the air temperatures and therefore humidity was close to 100 per cent and there were no radiant effects. A basic clothing insulation of 2.3 Clo and metabolic heat production of $70 \, \text{Wm}^{-2}$ were assumed. The

IREQ and wind chill indices were calculated according to ISO TR 11079 (1993). For the worst case of 1.6 °C air temperature and 1.2 m s^{-1} air velocity, the workers could work for 2.22 h. If comfort was to be preserved then this exposure should be reduced to 1.33 h and if the fans were turned off, to 1.7 h. The wind chill index of 783 Wm^{-2} indicated that the environment was cold but that there would be no damage to extremities. It is also within ACGIH (1996) guidelines.

Subjective assessment

A standard single sheet questionnaire was administered under instruction, by the hospital manager in a structured interview. The questions included thermal sensation, preference and satisfaction scales and a 'catchall' question for comments. It would be preferable for the ergonomist to have administered the questionnaire while staff were working but this was not practicable or allowed. Eighteen responses were obtained. While in the plating area eight subjects were cold and ten wished to be warmer with seven no change and one cooler. Generally at work outside the chilled area, 14 staff felt warm to hot and six indicated that they would like to be cooler. Comments were made on local cooling of the hands, nose and ears, that it was too warm outside of the chilled area, putting on and taking off clothing was inconvenient, and that fans caused discomfort and were noisy.

Summary of recommendations

A full report was provided including all data, analysis, conclusion tables of IREQ values for a range of conditions, and guidelines for cold work. There were seven recommendations.

1 *IREQ*$_{min}$ should be used as a starting point for designing and assessing future workplaces.
2 Engineering control should allow the fans to be turned off during the work period.
3 Wet skin and wet clothing should be avoided when cleaning.
4 The use of easily donned and doffed additional layers of clothing (over normal clothing and no thermal underwear) to *IREQ*$_{min}$ levels should be used, with individual selection of available garments for extremities.
5 A relatively light, tight-fitting hat to cover hairnet and ears is recommended. Appearance will be important.
6 Staff should be encouraged to contribute to the design of their own system. Teamwork is important when working in the cold.
7 If conversion from traditional to chilled work continues, a systematic approach to implementation should be developed. Methods exist and should be used. A system could include prototyping and evaluation in a climatic chamber away from the work.

Feedback and clarification were provided in a telephonic discussion of the report. The report was accepted, the ergonomist paid, and no further contact made.

Clothing dynamics and work in the cold

Work in the cold often involves a range of activities and movement from warm to cold conditions and vice versa. When moving from warm to cold conditions, or during sweating, condensation may occur within clothing and on the surface of clothing when the temperature reaches the dew point. This will change the thermal properties of clothing as well as cause chilling sensations in the workers. Clothing ventilation will be important to allow sweat to evaporate, but must not allow excessive heat loss. Heated and active clothing systems, including heated gloves to maintain hand skin

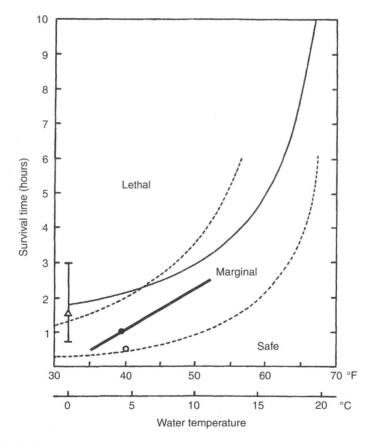

Figure 11.7 Estimated survival time in cold water. ------- Three zones (safe, marginal, lethal); ———— US Navy Survival based upon hypothermia; ▬▬▬ Estimate of survival based upon Dachau data; ○ Useful activity for clothed person; ● 50% survival.

Sources: Modified from Hayward *et al.* (1975); Toner and McArdle (1988).

temperature, may become more widely used in industrial applications as technology advances and costs reduce (see Chapter 7).

Cold water survival

Water has a thermal conductivity of around 25 times that of air and hence unclothed people immersed in cold water lose heat rapidly which can lead to hypothermia and death. Toner and McArdle (1988) review studies of hypothermia in water and survival predictions. The heat balance equation for the human body will apply involving water velocity across the subject (currents and movement destroy the boundary layer), water temperature, clothing, and metabolic heat production. Personal factors will also include body fat and physical condition of the person and other physical factors will be the depth of immersion and duration of exposure. Toner and McArdle (1988) present a summary curve which has been modified from Hayward (1975). This is presented in Figure 11.7. Safe zone – no likelihood of death; Marginal zone – some fatalities (50 per cent of persons experience unconsciousness resulting in drowning); Lethal zone – no survivors.

Use of heat balance calculations and thermal models will increase accuracy of prediction where greater numbers of individual factors can be considered (e.g. Wissler, 1982 – see Chapter 15).

12 Interference with activities, performance and productivity

INTRODUCTION

There is no doubt that hot, moderate or cold environments can interfere with human activities, affect task performance, and influence productivity. The most obvious of these effects are those of distraction and on the ability to carry out tasks involving manual dexterity in the cold. What is not clear, despite numerous studies, are details of the mechanisms involved and hence a comprehensive underlying model which relates cause, human thermal environment (i.e. in terms of the six basic parameters) to effects on activity, performance, or productivity.

When considering specific tasks or jobs it may not seem necessary to understand underlying mechanisms. What is important is task or job performance and productivity. No two situations will be identical however and a general performance model is a desirable long-term goal.

Despite limited integration of knowledge much is known about how hot and cold environments may affect human performance, particularly physiological mechanisms. In extreme conditions of hypothermia and hyperthermia, clouding of consciousness, confusion, illness and collapse will have obvious implications. Within these extremes there are also effects that can often be related to thermoregulatory responses. As the body cools, vasoconstriction reduces skin blood flow causing cooling of the skin and, for example, a loss in sensitivity and stiffness of the fingers. Synovial fluid viscosity increases and joints become stiff, nerve conduction rate reduces and there is a loss in strength as muscles cool. In addition, discomfort and shivering provide distractions and cause behavioural changes due to over-arousal for example. When the body is hot, vasodilation enhances ease of body movement; sweating may affect grip however and there may be distraction effects and psychological strain as body temperature rises especially in sedentary subjects.

Whether the above reactions will have effects on activity, performance or productivity will depend upon the specific job or tasks of interest and what is actually meant by those terms. This is very important and any investigation or 'performance model' must address this point. Possible descriptions are as follows.

Activity is what people do. It may involve psychological or physiological factors, and that these factors are successfully carried out means that the activity is done. That is, the activity is carried out, in terms of the person; however, the term 'activity' is independent of achieving a goal. Thermal stress may interfere with an activity therefore, by affecting the processes involved in carrying out that activity.

Performance is the extent to which activities have been carried out to achieve a goal. The term 'performance' when used in isolation has no meaning. It must always refer to a task and relate to a goal i.e. 'performance at something'. That a thermal environment interferes with an activity, may or may not affect performance depending upon how important that activity is in the overall task.

Productivity is the extent to which activities have provided performance in terms of system goals. This usually relates to the goals of an organization. Productivity in a manufacturing industry may be measured in terms of how much of a product is produced but it may also refer to quality. In a school, it may be measured in terms of number of examination successes and in a restaurant or a welfare payments office, how many customers or clients are served, how satisfied they are and whether their 'wants' have been sufficiently satisfied for them to return. Clearly there are many factors involved. The role of the thermal environment, although it may be very important, will not necessarily be obvious. Other factors related to the organizational culture and structure and other environmental components – noise, lighting, etc. – will also contribute to overall productivity.

The general descriptions provided above demonstrate that thermal environments may affect physiological and psychological processes, which may, in turn, affect activities that may affect performance at tasks that may interact with other factors to affect overall productivity. The overriding point is that to answer a question about whether a thermal environment will affect activities, performance or productivity, a framework must be set up for deciding upon what is meant by the terms in the context of interest. Fitts and Posner (1967) make a similar point when considering human performance *per se*; 'Before we can understand the complexities of human performance there must be a unified framework for studying it.' Griffin (1990) considers activity interference caused by vibration and noted that because two situations will rarely be identical, fundamental research should consider reasons for interference not the extent of interference. It could be argued that this presents a rather academic approach to the subject. Where time is short and an answer required, performance at specific tasks and situations may be successfully investigated in laboratory and field studies. It should be remembered however that results of such studies may not be generalized and evidence seems to suggest that if an identical study were carried out into the same conditions it is not certain that the same conclusions would be drawn.

Studies of factory output and accidents are presented below as well as a summary of effects of hot, moderate, and cold environments on cognitive and manual performance. The lack of a framework for considering performance, as described above, is an obvious problem in interpreting results and this is discussed further at the end of the chapter.

EARLY STUDIES: FACTORY OUTPUT AND ACCIDENTS

McIntyre (1980) summarizes the results of early field studies into 'temperature and performance'. The often cited results of Vernon (1919a) are presented in Figure 12.1, and general results relating thermal conditions to overall factory output are provided in Table 12.1, while Figures 12.2 and 12.3 show how accident rates increased as temperatures move away from 'optimum' levels.

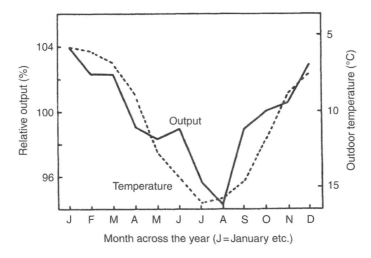

Figure 12.1 Seasonal variation of output of tinplate workers.
Source: Vernon (1919a).

Table 12.1 Summary results of studies that investigated the relationship between thermal conditions and factory output

Study	Industry	Summary of results
Farmer *et al.* (1923)	Glass industry	Output lower in summer
Vernon (1919, 1920)	Steel industry	Output lower in summer
Vernon (1919a)	Tinplate workers	Output lower in summer
Wyatt *et al.* (1926)	Weaving linen	Output fell if $t_a > 24\,°C$ at 80% *rh*
Weston (1922)	Weaving linen	Output fell if $t_{wb} > 23\,°C$
Vernon *et al.* (1927)	Coal miners	Work rate fell as temperature rose from 17 °C to 32 °C at 80% *rh*
Vernon (1919b)	Munitions	Accident rate increased as temperature moved away from 20 °C

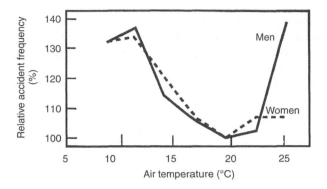

Figure 12.2 Accident frequency related to workplace temperature (munitions
     workers).
Source: Adapted from Chrenko (1974).

There are two important conclusions that can be made about these studies:

1   there appears to be strong evidence that output and accident rates can
    be affected by thermal conditions; and
2   results cannot be conclusive nor underlying mechanisms identified
    because of the many factors involved in field studies.

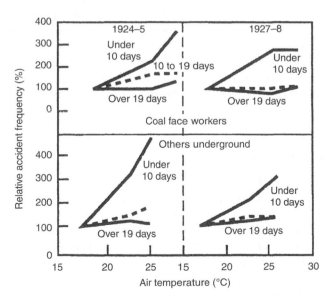

Figure 12.3 Accident frequency related to workplace temperature depends
     upon temperature and experience (mining).
Source: Adapted from Chrenko (1974).

Effects of heat on cognitive performance

In hot environments the body will sweat and 'core' temperature may rise. The resulting distress or discomfort may lead to behavioural changes and effects on cognitive performance, for example mental performance, information processing, memory and so on. Effects can be deleterious or enhancing and generally the many studies in this area have been inconclusive. A summary of investigations and their findings is provided in Table 12.2. Many tasks and task components require both physical and cognitive function. For example reaction time type tasks may require perception, information processing, and physical action (manual) as the person detects the stimulus, decides on a response (e.g. whether and how to respond), and responds by pressing a button for example. Wing (1965) reviewed a number of studies (see Table 12.2) and estimated *ET* (later *WBGT*) temperature values, in terms of exposure time, below which a fall in mental performance would not be expected for sedentary jobs (Figure 12.4).

There appears to be some general agreement that performance 'falls off' above about 33 °C WBGT (30 °C ET) – McIntyre (1980). The evidence however is far from conclusive.

Ramsey and Kiron (1988) reviewed over 150 studies and concluded that the evidence concerning the effects of heat on performance has been extensive and contradictory. They note that mental or very simple tasks show little decrement and are frequently enhanced during brief exposures (Figure 12.5). They recommend a simple decision rule as categorizing the task of interest into mental or simple tasks and perceptual motor tasks. Results are reported in terms of estimated *WBGT* values and it is noted that details are often not available of the six basic parameters that make up the thermal conditions.

There is therefore a lack of conclusive evidence of the effects of hot environments on cognitive performance. There is evidence of the effects of mild heat stress and this is discussed later under moderate environments (see also Table 12.2).

Effects of heat on manual performance

The effects of heat on physical work have been considered under heat stress assessment (Chapter 10) and heat stress indices are used to provide limits, taking account of metabolic heat production due to activity. Work in the heat (or not) will cause fatigue which should also be considered, when determining work/rest regimens. Thermoregulatory mechanisms will cause sweating which may affect activities requiring grip and hence may influence performance at some tasks. A rise in internal temperature and the increased workload of attending to the stress may also affect manual performance and can also lead to unsafe behaviour (Ramsey *et al.*, 1983) (see Figure 12.6).

There are also possible explanations as to why performance should improve. Increased arousal level due to stress and increased blood flow to

Table 12.2 Investigations into the effects of heat on cognitive performance

Reference	Study	Findings
Wilkinson (1974)	Laboratory study – vigilance task where subjects had to detect a longer tone	Response time longer at 37.3 °C body temp. – improvement at higher temps due to arousal
Bursill (1958)	Tracking task: 41 °C, 35 °C WB	Decrement of perceptual awareness in the heat
Mackworth (1952)	Singapore: Morse code operations	Unskilled workers more affected than skilled
Wing (1965)	Reviewed studies of effects on a number of mental tasks	Limit values to preserve mental performance in terms of ET and exposure times
Poulton (1976)	Speed and vigilance performance	Often improves in the heat
Alnutt and Allan (1973)	Reasoning ability; core temp: 38.5, skin temp. kept low and comfortable	Reasoning ability not affected but 15% quicker
Macpherson (1949)	Field study of 'tropical fatigue' (psycho-neurotic illness); men from hot temperatures to hot climates	Men more irritable, less inclined to exert themselves, less amiable, more reported sick; psychological cause!
Pepler and Fabrijio (1958)	Survey of teachers	75% of teachers thought performance and behaviour of pupils improved with air conditioning
Mayo (1955)	Two groups of US navy technical trainees: one group 24 °C; other 33.6 °C	No difference in test scores after 2 or 4 weeks, but half group thought they performed worse due to heat
Shoer and Shaffron (1973)	Schools and classrooms: 22.5 °C and 26 °C; students' learning tasks, 19 tests	26 °C class performance worse at complex tasks
Pepler (1971)	School tests	Individual student performance related to temperature experienced
Pepler and Warner (1968)	Students in a climatic chamber	25.6 °C optimum learning temperature, but not conclusive
Ryd and Wyon (1970) Hamberley and Wyon (1967)	Language laboratory in Swedish schools observation classroom	Comprehension and oral performance worse at 27 °C; mild heat is de-arousing
Wyon (1969, 1970) Johansson (1975)	Tsai–Partington test Cue-utilization task	English schoolboys improved performance in mild heat Improved performance when relaxed in mild heat

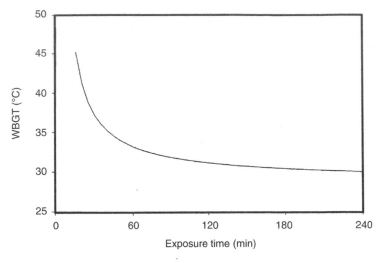

Figure 12.4 The maximum temperature for undiminished mental performance as a function of exposure time.

Sources: Wing (1965) and McIntyre (1980).

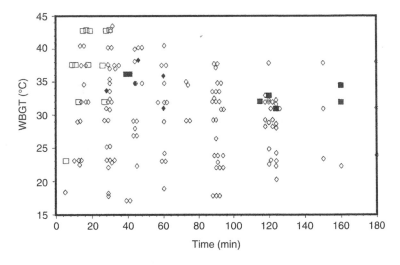

Figure 12.5 Summary review of studies of the effects of thermal environments on mental or simple task performance (■ significant decrement; ◆ partial decrement; ◇ no change; □ enhancement).

Source: Ramsey and Kwon (1988).

muscles may enhance attention, 'warming up' and movement. It is not expected that there will be significant detrimental effects on movement so there have been few studies. Weiner and Hutchinson (1945) found signifi-

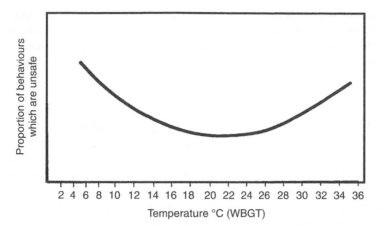

Figure 12.6 General relationship between the proportion of unsafe work
behaviours and thermal conditions.

Source: Ramsey *et al.* (1983).

cant decrement in performance and Meese *et al.* (1984) found improvement
then a fall in manual dexterity. Studies of effects on psychomotor performance
have been summarized by Ramsey (1986). He concludes that decrements in
performance occur around the upper limit of the prescriptive zone (30–33 °C

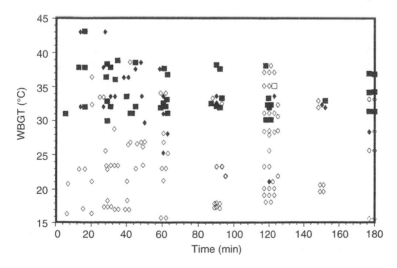

Figure 12.7 Summary review of studies of the effects of thermal environments
on perceptual motor task performance (■ significant decrement;
◆ partial decrement; ◇ no change; □ enhancement).

Source: Ramsey and Kwon (1988).

WBGT) consistent with the onset of physiological strain (Figure 12.7). These results, although useful, demonstrate the limitations of using a simple index such as WBGT where clothing and activity must be considered separately, instead of the six basic parameters where conditions can be more comprehensively defined.

Effects of moderate environments on cognitive performance

Although there is little evidence of a fall in cognitive performance in hot environments, moderately warm environments have shown effects particularly on tasks requiring vigilance. This has almost always been attributed to a reduction in level of arousal (mild vasodilation, relaxation, etc.) in these conditions (Wilkinson, 1974; Poulton, 1976) although distraction effects due to discomfort may also contribute. Where 'moderate' environments have become sufficiently warm or cool to cause discomfort then arousal may increase and performance may improve. These mechanisms are obviously task and context specific and will be discussed later. It is probably unwise to create conditions for discomfort in the hope that productivity will increase.

A series of studies was conducted in South Africa where a climatic chamber was taken to a factory and groups of workers (to a total of around 1000) worked over a full-day's shift on a variety of seventeen tasks (Meese *et al.*, 1984). A range of cold, moderate and hot conditions were studied (see Table 12.3).

Most tasks were manual or psychomotor in nature and particular attention was paid to ethnic and gender differences (black males, white males, black females, white females). Memory recognition tended to be the highest at 26 °C for white men and 32 °C for white women. This effect was attributed to low arousal aiding short-term memory (Wyon *et al.*, 1979). Over all subjects however there was no obvious pattern. The work of Meese *et al.* (1984) is particularly useful because of experimental control, good descriptions of methodology and tasks and the detailed reporting of environmental conditions allowing the six basic parameters to be quantified.

Effects of moderate environments on manual performance

Where a moderate thermal environment is defined in terms of the six basic parameters then metabolic heat production and clothing will be implicitly considered and it seems unlikely that there will be significant effects on

Table 12.3 Thermal conditions used in the investigations of Meese *et al.* (1984)

	Cold series (May to August)				Hot series (September to April)			
Air temperature (°C)	6	12	18	24	20	26	32	38
Relative humidity (%)	65	45	45	35	40	30	20	15

manual performance, i.e. it will be near to optimum – to some extent moderate conditions will be used as a standard for comparison so the question of effects has little meaning in this context. Minor deviations from optimum can however have significant effects.

Meese *et al.* (1984) found that manual dexterity performance improved from 20 °C to 32 °C air temperatures and there was improvement at a simulated welding task as air temperatures increased from 20 °C to 26 °C. There were however differences between groups of workers. In the cold, manual dexterity and finger strength decreased over a range of tasks as temperature reduced from 24 °C to 18 °C. This is a more pronounced reduction than often reported in the literature, e.g. Fox (1967).

Effects of cold environments on cognitive performance

Although the human thermal environment is defined by six basic parameters it must be recognized that a person working in 0 °C air temperature wearing 4.0 Clo of clothing insulation may be operating under similar thermal conditions as a sedentary person in light clothing at 22 °C but the situations are different. Despite this there have been few studies that have shown an effect of cold on cognitive performance. Horvath and Freedman (1947) investigated the visual discrimination reaction time of soldiers who lived for fourteen days in a climatic room at −20 °F (−29 °C) and found no difference between performance in those cold conditions and in moderate conditions. A similar result was found by Enander (1987) on a simple reaction time task conducted in air temperature of 5 °C. Teichner (1954) also found reaction time to be unchanged by cold conditions, but the presence of a cold wind did lengthen reaction time and this was attributed to a distraction effect. Payne (1959) noted that in cold conditions, subjects showed hostility and a lack of attention. Such distraction effects can therefore reduce performance at mental tasks. An additional influence will also be caused by increased workload caused by donning and doffing clothing, etc.

Effects of cold environments on manual performance

There have been numerous studies into the effects of cold on manual performance and there are clear repeatable findings that performance decreases. In the cold, vasoconstriction and lowering of tissue temperatures causes numbness, a decrease in manual dexterity and strength. In an extensive study by Horvath and Freedman (1947) where subjects were exposed to cold for fourteen days, dexterity of fingers and hand strength were markedly diminished even after short exposures. Fox (1967) in an extensive review, concluded that there is a clear relationship between hand skin temperature and manual performance. The critical hand skin temperatures '... below which there is a precipitous decline' are 8 °C for tactile sensitivity and 12–16 °C for manual dexterity.

More recent studies have shown significant effects in more moderate cold (Enander, 1987; Meese *et al.*, 1982, 1984) and so-called distraction and arousal models have been used to explain results. The American Conference of Government and Industrial Hygienists (ACGIH, 1992) suggest special protection for the hands, to maintain manual dexterity and prevent accidents, if fine work is performed for more than 10–20 min and air temperatures are less than 16 °C. Rapid cooling of the hands can be caused by windchill when driving a motor cycle for example (high relative air velocity and low air temperature) causing difficulty in operating controls.

In the cold series of experiments conducted by Meese *et al.* (1984) (see Table 12.3), finger strength and speed and manual dexterity (pegboard, screwplate, block threading and knot tying and an assembly line task) all decreased as the temperature fell from 24 °C through 18 °C, 12 °C to 6 °C.

A number of studies have investigated the effects of cold on manual dexterity and compared them with the effects of gloves (Meredith, 1978; Parsons and Egerton, 1985; Wagstaff, 1983). Parsons and Egerton (1985) also investigated different glove designs ranging in number of digit compartments from the standard five finger glove to the mitten, and attempted to determine the relative importance of cold and glove design to the overall effects on manual performance.

Todd (1988) investigated the relationship between hand skin temperature (MBTST, °C) and manual performance in an undergraduate project at Loughborough University, UK. The following relationship was proposed based on a rotating blocks test.

$$\text{Performance decrement} = 125 - \frac{887}{\text{MBTST}}\,\%, \tag{12.1}$$

where

$$\text{Performance decrement} = \frac{\text{performance at specified MBTST}}{\text{performance at a MBTST of } 30\,°\text{C}}$$

This equation can be used with the 25-node model of human thermo-regulation (Stolwijk and Hardy, 1977), to predict possible effects on manual performance from environmental conditions.

Human performance in transient conditions

Studies of the effects of hot, moderate and cold environments on manual and cognitive performance have mainly been reported for steady state conditions, although in field studies people will have experienced some changing environments. In practice there is a dynamic interaction between a person and his thermal environment involving changing environmental conditions, change in activity and clothing and changes in behaviour, posture and position where people may move from one place to another. Despite this, there appears to have been no controlled experiments into the effects of changing environments on cognitive or manual performance. Consider for example, the effects on manual performance of moving from a hot to a cold environment. Hands may be initially warm, but lose heat rapidly, especially if the body was sweating. The decrease in hand skin temperature would then begin to affect manual performance. Condensation of moisture in the cold conditions may lead to chilling effects and, when combined with physiological and behavioural thermoregulation, the overall challenge may significantly affect comfort and provide distraction. A changing environment could, therefore, have specific effects on performance due to change in physiological condition and significant effects of distraction as the person attends to the new conditions and responds to the challenges of a new situation.

Other changing environments (e.g. open window, switch on fan, cold to hot, wet to dry) can be subjected to a similar analysis as presented above. Although there have been no studies of the effects of change, it is likely that significant effects occur and it would be a fruitful area for research.

Distraction revisited

The above discussions have identified that distraction may affect human performance in hot, moderate and cold environments. A simple practical approach is to consider that if a person is distracted, then productivity will fall in relation to the 'time off the task'. As this is a quantifiable and direct measure, it is worth exploring further. While the issue can be complex, to a practical approximation, if a worker is paid £80 for an 8 h day, and he or she spends 30 min paying attention to (complaining, changing posture, opening and closing windows, thinking about, adjusting clothing, altering controls and monitoring results, etc.) the thermal environment, then the reduction in productivity may cost around £5. At £5 per worker, per day, etc., it can be seen that a robust estimate of 'time off task' provides an index of productivity with practical value. It should also be noted that this will complement assessments of work in heat or cold where exceeding environmental limits for unacceptable physiological strain will lead to a cessation of work or work-rest regimes and 'time off task' can then be determined and related directly to loss in productivity. Clearly 'time off task' due to distraction and 'time off task' due to unacceptable strain are additive.

Teichner (1967) suggested that distraction could account for a drop in cognitive performance in the context of cold stress acting as a type of secondary task and hence increasing workload. An obvious extension of this is in terms of temporal effect. Does the heat or cold stress interfere directly with the cognitive task (e.g. perception, central processing, output) and have a continuous effect, or does the attention to the strain 'take over' as in lapses of concentration or, as in the practical case above, does the person cease performing the cognitive task and switch totally to concentrate on the stressor or a combination of these. A changing environment will cause greater distraction as people readily attend to change.

Brooke and Ellis (1992) review the effects of cold on performance and discuss what they call 'distraction versus arousal'. They state that distraction reduces cognitive performance by inducing lapses in attention rather than by producing a general decrement in performance. The 'arousal hypothesis' predicts some continuous effects which may be improvements or impairments. They report laboratory experiments (involving serial choice reaction time and verbal reasoning tasks) which support the arousal hypothesis in that continuous effects were found with some drop and some improvement in performance. The argument is rather subtle and it is not clear that explanations in terms of arousal and distraction are mutually exclusive. It could be argued that there is always a level of discomfort which will distract a person from performing a task. In laboratory experiments, where controlled conditions apply and the subject knows that he or she will leave, and not return day to day as for people at work, then the subject may feel that there is little they can do, or think that they ought not to do anything, about the conditions. Distraction may then occur in short lapses but the context will be different from a person at work; when total attention may be given to the conditions and doing something about it.

If the practical consequences of distraction (i.e. time off task) are used as a performance measure then it is useful to know why distraction occurs, how much distraction can be caused by how much thermal stress (six basic parameters), and what other factors are important.

A definition of thermal distraction could be 'a tendency for a person to attend to a thermal state (hot, uncomfortable, cold) when performing a task.' In practical terms, attending to a thermal state will involve 'time off task' and hence is a measure of task performance. By definition then distraction occurs when a person attends to a thermal stimulus. This would usually involve a change in physiological state, for example a change or rate of change of skin temperature. A person would attend to it because of the thermal discomfort (or an estimate of the predicted consequences of not responding) it produces. The degree of distraction will depend upon the strength of the stimuli (e.g. discomfort) and the motivation and commitment to the primary task. If workers are motivated they will be distracted less than if they are not, for the same strength of the stimulus. It will also depend upon a perception of consequences of not attending to the distracting stimuli and of attending to the stimuli. For example, if there is a perception of no adaptive opportunity, although there is an unacceptable thermal environment, there is little point in stopping work as there is little one can do about it.

The link between adaptive behaviour and distraction is important for designing acceptable thermal environments. If people must continually adapt to maintain thermal comfort, then adaptive opportunity (including personal control systems and changes in posture, clothing, etc.) will provide 'time off task' that will affect productivity. Lack of adaptive opportunity however may lead to discomfort and loss of concentration (or even refusal to carry on) as the discomfort is attended to, even if there is little one can do about it.

The relationship between the thermal environment and distraction can, to a first approximation, be predicted from existing thermal indices and models, however this would have to be confirmed in both laboratory and field studies. A degree of distraction could be linked to the *PMV* index (as for the *PPD*) and then interpreted in terms of likely 'time off task', depending upon an estimate of motivation and commitment of workers to the task. A practical link is then made between human thermal environments (six basic parameters) and productivity.

PERFORMANCE MODELS AND INDICES

A standard environmental approach

The lack of a comprehensive understanding of how human thermal environments affect human performance has delayed the development of a thermal performance index. If one considered a traditional approach to index development then the performance decrement caused by thermal conditions could be related to the air temperature that would give equivalent performance in a standard environment. For example: the Performance Temperature (*PT*) is the temperature of a standard environment ($t_a = t_r$, $v = 0.15 \, \text{m s}^{-1}$, 50% rh) in which a person with the same level of motivation, clothing and tasks to carry out would give the same level of performance at those tasks as in the actual environment under consideration. The data required to determine such equivalence would have to be determined, or it could be derived from a physiological model of human thermoregulation (e.g. same mean or hand skin temperature, etc.). The PT index could then be related to productivity.

Individual control and other issues

Ramsey and Kiron (1988) suggested 'perceived control' as a possible reason for a fall in performance due to thermal stress. If operators believe that they have no control over conditions then they become disenchanted or give up. Designing workplaces so that an individual can control conditions (e.g. with fans, heaters, etc.) will avoid the problem and may even enhance performance. Wyon (1996, 1999) suggests that, 'bringing the user back into the loop is far more important for health, comfort and productivity than optimizing uniform conditions to accord with group average requirements, yet indoor environmental research has always concentrated on the group and ignored individual choice.' He presents the '3 I' principle (Insight, Information, Influence) – Insight into how the building works and Information about the consequences of their control; then they will be able to Influence their conditions.

Designers do not need to provide conditions for the comfort of the average of a group, but only a range of conditions (e.g. 6 K for 99 per cent comfortable) within which individuals can control. Wyon (1996, 2000) simplifies and summarizes the thermal effects on performance as:

Thinking   Individual performance is assumed to be at 100 per cent at temperatures up to individual neutrality, to decrease linearly with temperature over the next 6 K and to remain at 70 per cent at higher temperatures.

Typing   Individual performance at 100 per cent up to individual neutrality, to decrease linearly to 70 per cent over the next 4 K and to remain at 70 per cent at higher temperatures.

Skill   Individual performance at 100 per cent down to 6 K above individual neutrality, and to decrease linearly with temperature to 80 per cent at temperatures 12 K or more below individual neutrality.

Speed   Speed of individual finger movements is 100 per cent at temperatures down to 6 K above individual neutrality, and to decrease linearly down to 50 per cent at temperatures 12 K or more below individual neutrality.

Wyon then compares performance of subjects with no individual control and those with individual control (IC) of ±3 K. He finds improvements in performance at all tasks in line with the general guidance provided above. He also cites the West Bend Mutual Insurance Company study which demonstrated an increase in productivity when individual control was operative in comparison with when it was installed but inoperative.

Reductionist approach

A reductionist approach to considering the effects of cold on manual performance would be to determine factors that make up manual performance, identify tasks which require performance at those factors, and from a knowledge of the effects of cold on individual factors, attempt to predict the effects of cold on the more complex or realistic task of interest. The approach is reductionist in the sense that in practice one starts with a complex, realistic job or task, conducts a task analysis which divides the job into components (factors) then, from a knowledge of performance at tasks representative of the components, a prediction is made about the effects of cold on overall 'job' performance. Fleishman and Ellison (1962) conducted a factor analysis into manual tasks (see Table 12.4 and Figure 12.8).

Table 12.4 Components that make up all manual tasks

Manual component	Definition
Manual dexterity	The ability to make skilful well co-ordinated arm and hand manipulations; it does not emphasize finger movements
Finger dexterity	The ability to make skilful manipulations with the fingers
Wrist-finger speed	As name suggests; e.g. paper and pen task
Speed of arm movement	As name suggests
Aiming or positioning	Ability to perform quickly and precisely a series of movements requiring eye-hand co-ordination
Hand steadiness	Ability to make precise steady arm and hand movements of the kind that minimize speed and strength
Grip strength	The maximum force that can be exerted with a fist grip and thumb-on-side-of-forefinger usually measured with a grip dynamometer
Tactile sensitivity	Cutaneous sensitivity such as the ability to distinguish between two pinpoints on the skin in close proximity

Source: Fleishman and Ellison (1962).

Meredith (1978) used this approach and successfully predicted performance at a 'real' military task from a knowledge of performance at component tasks. This provides validation of the method of using simple laboratory tasks to make predictions about practical application. The reductionist approach however usually ignores interactions between effects, and that tasks are performed using a very complex system will never be entirely represented in this way. It does however provide an underlying philosophy to studying effects on simple tasks in a laboratory with a view to predicting effects on real tasks. This point is often missed by those interested in application and surprisingly also by many of those who conduct the laboratory experiments.

The reductionist approach has potential but it can be argued that it is impossible to divide tasks in this artificial way and that accurate predictions should not be expected.

Underlying models – explanations and factors

Ramsey (1988) proposes four underlying explanations for the results of the effects of thermal environments on task performance: arousal, physiological, distraction and perceived control models.

Arousal

Performance at a task will depend upon a person's arousal level compared with that required for optimum performance. A task that is boring (e.g.

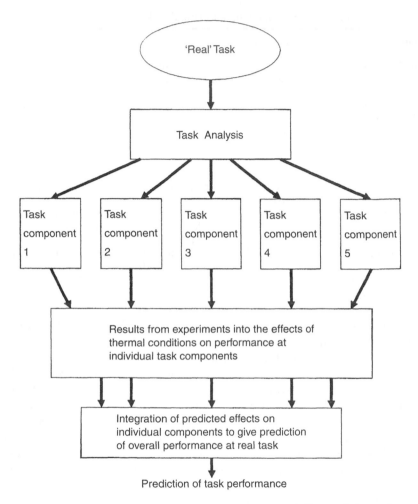

Figure 12.8 Reductionist method for predicting the effects of thermal conditions on 'real' task performance.

vigilance tasks) will be de-arousing and a person will perform better if arousal can be raised to an optimum level by the stimulation caused by a stressor such as heat or cold. A warm environment however will reduce arousal level and hence performance at a vigilance task. If the task is demanding and arousing then thermal stress may over-arouse a person and performance will fall compared with that in a moderate thermal environment where arousal may be at an optimum for that task (see Figure 12.9).

This explains the results obtained over many studies where performance has been found to improve, not change or decrease for similar thermal stimuli as well as explaining reasons for differences in results between individuals.

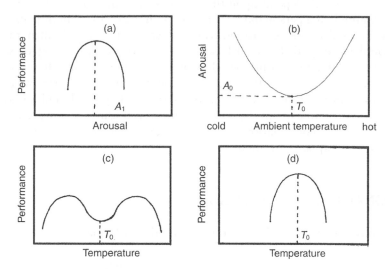

Figure 12.9 The arousal model for explaining effects on task performance. (a) Performance is a function of arousal level. Optimum arousal level, A_1, depends upon task complexity. (b) Temperature, T_0, for minimum arousal, A_0, is higher than that for thermal neutrality. (c) Theoretical performance versus temperature curve of a simple task where $A_0 < A_1$. (d) Theoretical performance versus temperature curve, where $A_0 > A_1$.

Physiological (body temperature)

This explanation suggests that the physiological functions involved are directly affected by the thermal conditions and hence so are overall activity and performance, for example sweating, elevated 'core' and skin temperatures in hot environments, mild vasodilation in warm environments, and low muscle and skin temperatures in the cold. Manual performance in the cold is particularly related to physiological responses.

Distraction

Extreme thermal stress requires attention that distracts the person from the task at hand and in effect increases workload and task difficulty with possible effects on performance. For example, if a task is very demanding or the stressor very distracting (e.g. as in the cold), then there will be a form of work overload and performance will fall.

Perceived control

Performance falls in a thermal environment because the operators 'give up', as they perceive that they cannot control the thermal stress. It is also possible that this will depend upon the person's attitudes to achieving the task. For example, if a person did not identify with the goals of an organization and an inadequate reward/motivational structure existed, then the effects could be amplified due to lack of motivation to counteract these effects. Greater affects on groups with low morale or inter-group differences (e.g. found by Meese *et al.*, 1984) could possibly be explained in this way.

The above 'explanations' represent the most widely used underlying themes or 'models' to explain the wide variation in findings. They are not independent and a combination of these explanations may be involved. None, however, explains all findings.

A RATIONAL PERFORMANCE MODEL

It is a desirable goal to establish a rational model of human performance such that it can be used to investigate the effects of thermal environments. This would involve identifying relevant factors and determining their effect on the model and hence predicting effects on performance. This is a similar approach to that of a model of human thermoregulation. For example, the relevant factors are the six basic parameters (t_a, t_r, rh, v, Clo, Act) which act on the model (passive body acted on by a human thermoregulatory system) to provide outputs in terms of temperature, heat transfer rates, sweat loss, etc. A thermal index would integrate the factors into a single index number, representing effects: heat stress, comfort, cold stress, etc. A performance index such as the performance temperature (PT – see above) could do the same. A possible performance model is proposed in Figure 12.10.

Discussion

Productivity

A summary of the evidence presented above leads to the conclusion that thermal environments can affect human functions and hence activities, which in turn may affect performance and productivity. There can be no general conclusion regarding effects on productivity, as individual and contextual factors will always dominate.

Productivity is related to the goals of the organization and any measurement should be related to those goals. They are usually stated in general terms as part of the 'corporate mission' of a company and interpreted by individual departments and groups within the company in terms of their own work. An individual may perform well, but in this context, he may or may not be productive. If the thermal environment causes a decrement in individual performance, and performance is related to productivity, then productivity will fall.

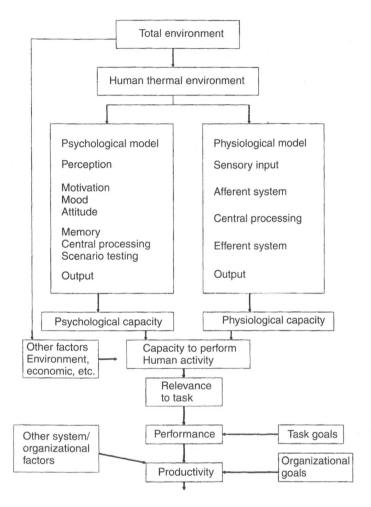

Figure 12.10 Model for considering the effects of the thermal environment on human activity performance and productivity.

Determining influences of climate on individual performance may be a rather subtle way of investigating productivity. More important effects may be behavioural – in terms of sickness or accidents, absenteeism from the workplace or from work altogether, and how easy it is to distract the person from the main task.

Motivation

Scientists experienced in applied human sciences generally agree that level of motivation and morale can dominate effects on performance even in very extreme environmental conditions. At a recent workshop a presenter noted that the three most important factors influencing human performance are 'motivation, motivation and motivation' and the three ways to achieve high motivation and morale were 'leadership, leadership and leadership'. The arousal model indicates that in some circumstances one can be over-motivated; however, the point should be well taken.

Cooke *et al.* (1961) investigated productivity in South African mines and showed that a 'good bossboy' would produce much greater productivity from workers than a poor one or none at all, even in high environmental temperatures. A military investigation into acclimatization compared the performance of acclimatized and unacclimatized soldiers in a hot climate. While acclimatized soldiers maintained lower body temperatures than unacclimatized soldiers, performance was dominated by the quality of leadership provided.

That motivation can dominate effects, is important. However, physiological and psychological responses to thermal conditions also affect performance. Motivation levels will vary and can interact with effects of thermal environments. The interactions should be studied and quantified. However, the role of thermal conditions, which is the topic of discussion here, should not be ignored or dismissed.

Level of skill

The effect of an environmental stressor on performance will depend upon the level of skill of the person at the task. Mackworth (1950, 1952) demonstrated, with Morse code operators, that heat stress generally had less effect on performance, the higher the level of skill of the operator, even up to quite extreme conditions (see Figure 12.11). This is explained by a simple human performance model based on workload.

The thermal stress overloaded the less-skilled operator, as he required a greater proportion of his maximum work capacity to perform the task. A skilled worker was less affected as he required a lower proportion of his maximum capacity to achieve the same level of performance and could cope with the extra workload caused by the stress. Interaction between skill level and thermal environments may be important and training in the heat or cold may be of benefit.

Summary models

The model presented in Figure 12.10 is a conceptual model of a person that can be used to provide a framework for describing effects of thermal conditions on human performance. Using both laboratory and field studies, effects can be quantified leading to a working predictive model that has validity and great practical utility. Both laboratory and field experiments

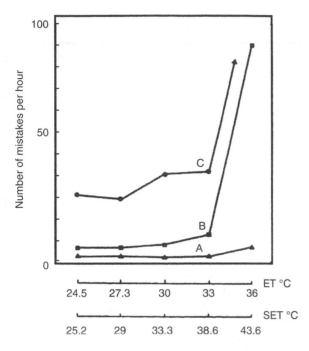

Figure 12.11 Mistakes made by Morse code operators. All eleven subjects
were trained operators. Group A contained the best three, B the
next five, and C the worst three.

Source: Mackworth (1952).

and studies must contribute to such a model that is an effective 'way
forward' in this subject.

A great deal of research has been conducted, unfortunately not systematically or within such a framework. However summary results can be
produced. Wyon (1986) reviews the literature and provides a tentative
summary of effects. Figure 12.12 provides such a summary and presents
a predicted drop or increase in performance with air temperature for light
and sedentary work and light and normal clothing. While such summary
charts are useful they can be misleading. The effects are presented in a
deterministic way and simplifies the problem, ignoring context for example
and not showing the wide range of individual responses. If caution is taken
in interpreting such summary models they can be of great benefit in focuss-
ing research into the problem and providing general guidance.

Environmental design for productivity

It is sometimes argued that environmental conditions should vary from
those producing thermal comfort so that performance may improve. For

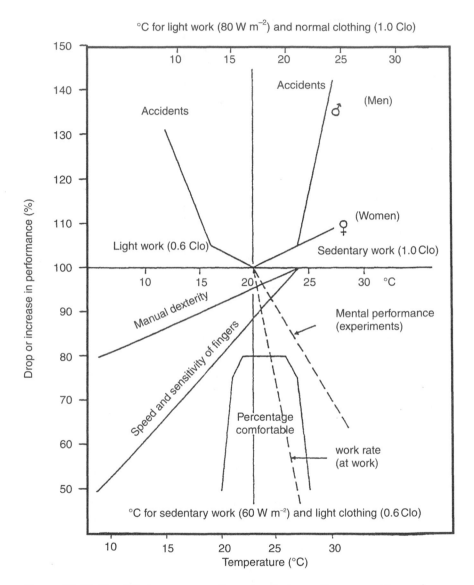

Figure 12.12 Simplified summary of results: indoor climate, accidents, human efficiency and comfort.

Source: Wyon (1986).

example, a cool environment may be more arousing and may improve performance at a boring task. There is some rationale behind this. For example, for a driver who is tired and tending to fall asleep in a car, a comfortable-to-warm environment should be avoided. Discomfort in general however, causes dissatisfaction and complaint and can affect performance. There is no substantial evidence that one should design for discomfort to improve performance.

A more recently accepted principle is that, to maintain motivation and provide satisfaction and to reduce complaints and absenteeism, workers should be involved in the design and control of their own environment. In surveys, or design and assessment of buildings therefore workers' opinions and suggestions are encouraged. An extension of this principle is to provide individual control of environments at workstations (e.g. Pacink, 1990). This requires a new overall philosophy of designing for individual comfort and the heating and ventilation of buildings. There are many human-factors issues concerning individual control systems and claims of increased worker productivity by their introduction are yet to be verified.

13 Human skin contact with hot, moderate and cold surfaces

INTRODUCTION

When human skin comes into contact with a hot surface the skin temperature rises. This causes a reaction that can vary from local vasodilation and sweating to pain sensations and physical damage of the skin (burn). If the skin comes into contact with surfaces of moderate temperature there will be no skin damage due to heat loss or gain, but there may be vasodilation and vasoconstriction leading to sensations ranging from hot through neutral to cold and associated discomfort or feelings of pleasantness. When it comes into contact with a cold surface, there will be vasoconstriction, pain, maybe sticking if the skin is wet and skin damage due to frostbite. This chapter reviews methods and data concerned with the relationship between skin contact with hot, moderate and cold surfaces and human response to the contact.

A method of determining skin reaction to contact with hot, moderate and cold surfaces would be to expose the skin of human subjects to those surfaces and observe what happens. While this is acceptable for moderate surfaces, for ethical reasons this is unacceptable for extremes of heat and cold and hence less direct methods must be used. These include empirical methods, e.g. attempting to simulate conditions using animal skin, mathematical methods, e.g. of heat transfer, and physical models, e.g. developing artificial devices that react like human skin.

Knowledge of human thermal response to contact with solid surfaces has practical application in the design and assessment of products and workplaces. In the design of children's toys, devices in the home such as cookers, toasters and kettles and machinery used in industry, hot surfaces can cause burns and cold surfaces in cold stores, for example, can cause frostbite. Handrails should be designed to provide comfortable and acceptable sensations when gripped; adverse reaction when attempting to operate a control could contribute to an accident, and so on. Detailed analysis of the problem is presented below in terms of contact with hot surfaces. Many of the principles described also apply for contact with moderate and cold surfaces for which data are presented towards the end of the chapter.

SKIN CONTACT WITH A HOT SURFACE – ANALYSIS OF THE EVENT

Analysis of the event of human skin coming into contact with a hot surface can be considered in terms of three components: the human skin, the hot surface and the nature of the contact. Each of these is described below.

Human skin

Properties of human skin will vary across different areas of the body and will also change with time. In addition to intra-subject differences there will be inter-subject differences. Despite these differences however, human skin, both between and 'within' humans, has a common structure and most skin is similar in function. For a detailed description of the structure and function of skin the reader is referred to Montagna and Parakkal (1974) and Wood and Bladon (1985).

The specific structure of any particular area of skin will depend upon the function of that skin. For example, a vital area for thermoregulation will have a rich blood supply and many sweat glands. Human skin is made up of layers: an outer horny layer of dead cells, the epidermis and the dermis (see Figure 13.1). Under the dermis is a layer of fat: panniculus adiposus. The cutaneous epithelium, at the base of the epidermis, continually generates epidermal cells that move to the surface of the skin, die and eventually are removed from the skin surface.

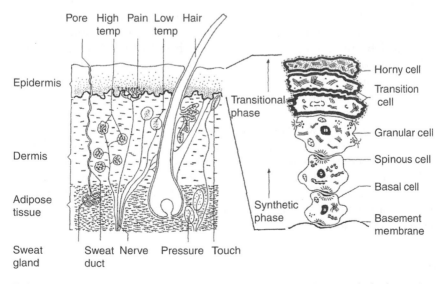

Figure 13.1 The structure of human skin. A superficial, partial thickness burn damages the epidermis, but not the basement membrane or below.

The function of the skin can be considered to act mainly as a protective and containing barrier while allowing necessary interaction between the body and the environment. Two important interactions are the regulation of heat exchange and the sensory perception of the environment. To aid in performing these functions, human skin contains systems for supporting surface hair and mechanisms for controlling moisture on the skin surface (sweat). These systems are based in the dermis of the skin but penetrate the epidermis to the surface of the skin. The dermis is also supplied with blood vessels and mechanisms for controlling the lymphatic vessels and nerve receptors which are sensitive to temperature, superficial touch and pressure, and for controlling the flow of blood through them. The skin therefore functions as a dynamic system, changing its condition depending upon the requirements of the body. There are many other structures and functions of the skin; however the simple model provided above is sufficient for the present discussion of the effects of human skin contact with hot surfaces. The principles also apply for contact with moderate and cold surfaces for which more specific detail is provided later in this Chapter.

Skin reaction to heating

Thermoregulatory response

The skin plays a fundamental role in the homeothermic function of maintaining internal body temperature at around 37 °C. If the whole-body becomes 'too hot', blood flows through the dermis (vasodilation) to release heat through the epidermis to the environment. If greater heat loss is required then the surface of the skin is moistened with sweat so that the latent heat of vaporization may be lost through evaporation. Whole-body thermoregulation is controlled by the thermoregulatory system based in the hypothalamus. Local heating of skin however will also cause vasodilation and sweating, given a sufficient response time. Reaction of skin to contact with hot surfaces may depend upon the initial condition of the skin. From the above discussion, this may vary from cold conditions where there are low skin temperatures, low blood supply, and 'dry' skin to hot conditions involving relatively high skin temperatures, rich blood supply and possibly wet skin due to sweating.

The above describes the reaction of the skin to allow 'normal' thermoregulation. It provides a framework for identifying the possible initial skin condition under which contact with a hot object may occur. Some physical properties of human skin are provided in Table 13.1. Important factors that will affect intra- and inter- human variation in skin condition are provided in Table 13.2.

Pain

The sensation of pain can be considered as a warning that undesirable effects are occurring, or may occur, to the body. This simple model of pain

Table 13.1 Thermal properties of human skin

Dimension	Units	Values	
Approximate values of physical dimensions			
Mass	kg	4	
Surface area	m^2	1.8	
Volume	l	3.6	
Water content	%	70–75	
Specific gravity	ND	1.1	
Thickness	mm	0.5–5	
Approximate values for thermal properties			
Density (p)	kg m^{-3}	860	
Specific heat (c)	J kg^{-1} K^{-1}	5021	
		Vasoconstricted	Vasodilated
Thermal conductivity (k)	W m^{-1} K^{-1}	0.2–0.3	0.4–0.9
Thermal diffusivity ($a = k/pc$)	(m^2 s^{-1}) $\times 10^{-8}$	4.63–6.95	9.26–20.84
Thermal penetration coefficient [$b = (kpc)^{1/2}$]	J m^{-2} s$^{-1/2}$ K^{-1}	929–1138	1314–1971

Sources: Data adapted from *Bioastronautics Data Book*, Webb (1964), Ray (1984), McIntyre (1980), Giancoli (1980), Monteith and Unsworth (1990), Stolwijk and Hardy (1965).

Note
Because of the dynamic and 'living' nature of skin it is only possible to provide approximate 'static' values for its properties.

Table 13.2 Factors which influence variation in human skin

Factor	Explanation
Intra-subject factors	
Area of the body	Regional difference in epithelium structure and thickness; water content; pigmentation
State of vasodilation/vasoconstriction	Instantaneous state of local capillary blood flow
Wet or Dry (e.g. state of thermoregulatory sweating)	Presence of hair, surface oil and contaminants
Inter-subject factors	
Age	Children, adults
Occupation	Use of skin – manual/office worker
Sex	Males/females
Ethnic differences	

would prove useful in establishing skin damage caused by contact with hot surfaces. Surfaces that cause pain would, under this model, indicate temperatures of surfaces which would cause skin damage. There are, however, a number of important additional factors that affect sensations of pain. If the nerves which detect or transmit pain signals are destroyed then severe damage may occur with little or no pain. Also, reporting of pain and pain severity will depend upon the disposition of the 'whole human'. For a similar sensation, some individuals may amplify the effect and report severe pain and others (or the same individual at different times) may augment the effect and report little pain.

There are a number of theories of pain and its reporting (see Keele and Neil, 1971) for a summary of these. Here, it is sufficient to identify that skin contact with hot surfaces may cause pain, and note that it would be naïve to determine 'pain thresholds' and relate them directly to conditions which cause skin damage.

Skin burns

The constitution of human cells is such that at temperatures above around 43 °C, damage can begin to occur if exposure to that temperature is sufficiently long. It is generally true therefore, that if skin temperature in contact with a solid surface is below about 43 °C, discomfort and pain sensations will be avoided and no skin damage will occur. Note that this applies to *local* skin temperatures; if the whole-body were at 42 °C then there would be a serious breakdown in thermoregulation, since 'safe upper limit' levels for internal body temperature are less than around 38.5 °C.

There are a number of methods of classifying skin burns and all are based on the extent of damage to the different layers of the skin described above. These methods have been reviewed and a summary is provided below in Table 13.3.

Solid surfaces

Skin reaction to contact with a hot solid surface will depend upon the rate at which heat transfers from the surface to the skin. This will depend both on the nature of the skin and the nature of the surface. Metals, for example, will 'give up' heat more easily than wood, for similar conditions. Factors relating to the

Table 13.3 Different classification systems for burns

(A)	(B) (Scotland)	(C) (USA)	(D)
Partial thickness skin destruction	1st degree	1st degree	Superficial partial thickness skin destruction
		2nd degree	Deep partial thickness skin destruction
Whole thickness skin destruction	2nd degree	3rd degree	Whole thickness skin destruction

Table 13.4 Thermal properties of materials

Material	Thermal conductivity (50 °C) (k) $W\,m^{-1}\,K^{-1}$	Density (p) $Kg\,m^{-3}$	Specific heat (c) $J\,Kg^{-1}\,K^{-1}$	Thermal diffusivity $(a = kp^{-1}\,c^{-1})$ $m^2\,s^{-1} \times (10^{-8})$	Thermal penetration coefficient $(b = (kpc)^{1/2})$ $J\,m^{-2}\,s^{-1/2}\,K^{-1}$
Aluminium	204	2700	900	8395	22265
Copper	382	8900	390	1005	36413
Gold	293	19300	126	12049	26693
Iron and Steel	45	7800	450	1282	12568
Lead	34	11300	126	2388	6958
Silver	416	10500	230	7226	31696
Glass	0.76	2600	840	35	1288
Concrete	0.8–1.4	2300	879	40–60	1271–1682
Granite	1.7–4.0	2700	816	77–181	1935–2969
Pine wood	0.11–0.15	432–641	2803	6–12	365–519
Oak	0.17–0.21	481–609	2385	12–18	442–552

Sources: Adapted from Giancoli (1980) and Pitts and Sissom (1977).

Notes
For solids: density and specific heat are only weakly dependent on temperature. Change in the thermal conductivity depends approximately linearly on temperature over the small range of interest. For 50 °C values a mean was taken of 0 °C and 100 °C values. (Applies to metals above.)

solid surface that may, on contact, affect heat transfer to the skin, are: number of layers, surface roughness, wet or dry, surface temperature, thermal conductivity, specific heat, density, material thickness and surface 'cleanness'. Thermal properties of materials, which will affect heat transfer, are provided in Table 13.4.

Analysis of the contact between skin and a hot solid surface

Table 13.2 and the list above present relevant factors and properties of human skin and hot surfaces. To determine the reaction of skin to contact with a hot surface, it is necessary to establish the nature of the contact. Heat will flow from the hot surface to the 'cooler' skin (this is the first law of thermodynamics). The rate of heat flow will depend upon the nature of the two surfaces and amount of heat flow will depend upon the contact time and how 'perfect' the contact is, which will be related to the pressure. Contact surface areas will also affect the overall skin reactions.

The above discussion analyzes the event of skin contact with a hot surface into three components (human skin, solid surface and nature of contact). The 'whole' event however is the interaction of all of these and the relative importance of the factors will depend upon the context. There are a number of ways of studying this interaction to determine likely skin reaction. A review of studies is provided below in terms of empirical methods, mathematical models and physical models.

Review of investigations into human skin contact with hot surfaces

Empirical methods

An obvious approach to determine skin reaction to contact with hot solid surfaces is to conduct empirical studies where contact is made between hot surfaces and human skin and the reaction is observed and quantified. A relatively 'complete' database could be established if exposures were made over ranges of relevant factors. A problem with this 'reductionist' approach is that it is impractical to investigate very large numbers of combinations of conditions and that there are ethical issues regarding the study of pain and skin damage on human subjects. Despite these difficulties, empirical studies have been conducted and, although not comprehensive, provide an indication of the likely reaction of human skin to contact with hot surfaces.

Use of animal skin

Leach, Peters and Rossiter (1943)

Leach *et al.* (1943) considered the origin of toxins in burns and looked at microscopic and macroscopic damage caused by exposure of skin to a heated brass cylinder (iron) of 25.4 mm diameter. The cylinder was lagged with asbestos and flowing water of controlled temperature, allowed the base of the cylinder (thin metal, tinned surface) to heat skin up to 80 °C, when placed against it. A review of the literature suggested that immersion of rabbits' ears in water at greater than 50 °C, produced severe damage (Cohnheim, 1873) and a number of studies had shown that cells will die at temperatures of 42–47 °C if heating time is long enough.

Leach *et al.* (1943) used anaesthetized guinea pigs (500–700 g), and some rats (100–200 g), with shaved backs. Burns were made in the skin with an iron for various combinations of temperatures (45–80 °C) and exposure times (10 s to 6–10 min), and 2–4 observations were made for each condition. It was found that rat skin and guinea pig skin gave similar reactions. Two conditions were considered in detail: one minute exposure time over a temperature range of 45–80 °C; and 55 °C over a range of exposure times. Skin reaction was considered in terms of erythema (redness), flare (speed of redness), blanching, blueing, heat fixation, incipient blister formation, oedema, and edge wheal. In addition, eight stages of microscopic change were identified ranging from stage A – swelling of the epithelial nuclei etc. (reversible) – to stage H – heat coagulation of the epithelium with gross distortion of epithelial cells and their nuclei and marked damage to collagen fibres of the dermis (irreversible).

It was concluded that for guinea pig (and rat) skin, (iron) temperatures of 47 °C, for up to 6 min exposure, produce no visible change. Temperatures of

50–55 °C applied for one minute produce cellular damage and temperatures of 70–80 °C for 10–20 s will produce severe scabbing. It is noted that the temperatures are for 'iron' temperatures and reactions for guinea pig (and rat) skin. It is expected that more severe reactions would occur in human skin at these temperatures.

Moritz and Henriques (1947)

The most extensive and influential studies concerned with the effects of hyperthermia in human skin were by Moritz and Henriques (1947). Data from these studies form the basis of a number of standards and limits proposed for surface temperatures of equipment. A discussion of previous studies by Hudack and McMaster (1932) and by Leach *et al.* (1943) noted that results were not directly related to human skin. In addition, there were problems with the heating 'iron' used by Leach *et al.*, in that the water temperature in the device was higher than skin temperatures (due to stagnant water near the metal surface in contact with the skin and the temperature gradient through the metal). In addition, the metal iron would cause vascular occlusion due to pressure caused by its weight on the skin.

For these reasons, an apparatus was developed which allowed skin to be exposed to a flowing stream of liquid at a range of temperatures, and pressures over the range of 700–860 mm Hg (see Figure 13.2). A thermocouple allowed temperature to be measured directly at the surface of the skin. Both oil and water were used to investigate effects on skin. However,

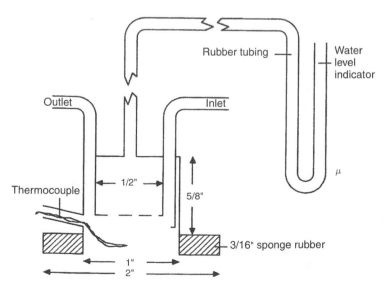

Figure 13.2 Apparatus used by Moritz and Henriques (1947) to expose skin to a flowing stream of hot liquid.

no difference was found in skin reaction, so water was used for the majority of studies. An extensive discussion and experimental comparison of human and pig skin showed that there were major similarities in structure.

Closely clipped anaesthetized young pigs (8–10 kg) were exposed to a 25.4 mm (1 inch) circle of hot flowing water on the lateral body surface; it had been found that different areas of the pigs' body responded differently. One hundred and seventy nine exposures from 1 s to 7 h to varying temperatures from 44 °C to 100 °C were made. Skin reactions to exposures were categorized as first degree (insufficient intensity to cause complete destruction of the epidermis), to second and third degree (full thickness destruction of the epidermis) 'according to the depth to which irreversible injury was estimated to have occurred'.

With the apparatus shown in Figure 13.2 (13 with oil, 20 with water) 33 exposures were made on eight human volunteers to determine the extent to which the results from pig skin were applicable to human skin. Skin exposed was on the anterior thoracic region and the ventral aspect of the forearm. Exposures were over a temperature range of 44–60 °C, for durations from 5 s to 6 h. Skin reactions were interpreted as sub-threshold (first degree – involving hyperaemia without loss of epidermis) to threshold and supra-threshold (second and third degree – involving complete epidermal necrosis). Data for exposure of human skin are presented in Table 13.5. For both experiments on pigs and humans, a threshold exposure represented the shortest time, at a given temperature, for complete destruction of the epidermis. A comparison of results indicated that '...there is little or no quantitative difference in the susceptibility of human and porcine epidermis to thermal injury' (Moritz and Henriques, 1947). Two 'threshold curves' of temperature and exposure time were produced (Figure 13.3): an upper curve showing the likely division between sub- and supra-threshold (as defined above); and a lower curve below which no appreciable injury occurred. Exposures lying between the curves resulted in epidermal damage but not trans-epidermal necrosis.

An interesting point is that the lowest surface temperature that was responsible for cutaneous burning was 44 °C for an exposure of 6 h. In addition, for each degree rise in surface temperature from 44 to 51 °C the time required to produce irreversible cellular injury was approximately halved. For exposures above 51 °C it was apparent that different mechanisms apply where exposure time is not sufficient to reach steady-state conditions, i.e. there was a division into transient (>51 °C) and steady-state (<51 °C) conditions.

Exposures described above were at normal atmospheric pressure. It can be expected that if (relatively cool) blood flows at the site of hyperthermia, this would protect the skin against burning. Moritz and Henriques (1947) investigated the effects of pressure on the skin by exposing pig skin both at atmospheric pressure and at 80 mm Hg above atmospheric pressure. It was found that there was no evidence of increase in severity of burn with

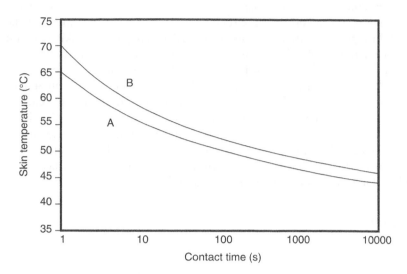

Figure 13.3 Skin temperatures above which a burn would be expected (curve B) and below which no appreciable injury occurred (curve A).

Source: Moritz and Henriques (1947).

pressure, at burn threshold conditions, 49 °C for 7 min exposure and 51 °C for 2 min. It was concluded that pressure sufficient to prevent superficial dermal blood flow did not increase sensitivity to epidermal thermal injury.

A final series of experiments involved three young pigs exposed to 49 °C water at atmospheric pressure. All exposures for less than 7 min provided no necrosis of the epidermis; however, all exposures of 9 min led to complete necrosis.

Using these data (18 exposures) as a control, the cumulative effects of exposures were investigated. For example, three 3 min exposures repeated with a 3 min interval between them provided complete and irreversible epidermal necrosis. Overall the results showed that some epidermal injury takes place over the first three minutes of exposure and that at least 24 min was required for appreciable recovery and at least four hours for complete recovery. As expected, the longer the initial exposure, the greater the recovery time. For example, one 5 min exposure produced mild vascular reaction whereas two 5-min exposures, separated by a 4-hour recovery time, produced epidermal necrosis.

Sevitt (1949)

Sevitt (1949) used a heated 'iron' cylindrical device, similar to that used by Leach *et al.* (1943), to burn the backs of (clipped, shaved or depilated) anaesthetized guinea pigs. Exposures (25.4 mm diameter circles) were made over the temperature range 43–80 °C for exposure times of 10 s to 20 min. Edge and centre skin temperatures were measured in some of the burns and dyes were used to investigate the effects on circulation. The 'visual appearance' of the exposed skin (plus dye) after exposure was used to categorize the skin reaction ('burn'), cessation of blood flow (stasis) being an important factor.

A very high correlation was found, for skin reaction in exposures of 10, 20, 30 and 60 seconds, between when stasis sets in within 4 h of burning, analgesia of the skin occurs, and whole skin loss develops. Curves are plotted of exposure time against the temperature of the burning iron. Extrapolation implies that for very long exposure times a temperature of 50–55 °C will still cause whole skin loss: the 'dermal threshold'. Two mechanisms for skin necrosis are proposed: one caused by direct heat; and the other due to the ischaemic change caused by stasis.

Human skin

There are ethical problems in conducting controlled laboratory experiments that involve pain and especially damage to human skin. Limited data, which are available, have usually involved exposures of the experimenters themselves (or colleagues).

Studies of exposures which occur outside the laboratory have not been reported in sufficient detail to determine cause and effect. The extent of skin damage (burn) can be well defined, in a hospital for example. However, detail about relevant factors and nature of contact with the hot surface are not available.

Moritz and Henriques (1947) exposed the skin of eight human volunteers to hot flowing water or oil. These experiments were mainly to confirm that the reaction of porcine skin was similar to that of humans. The experiments are as described above and the results are presented in Table 13.5. Although limited, these data are probably the most extensive available.

Stoll, Chianta and Piergollini (1979)

Stoll *et al.* (1979) conducted experiments with two male and two female human subjects (with 'normal' pain thresholds) to establish the highest temperatures materials may attain without causing pain or burn on contact. This was to allow selection of 'thermally safe' construction materials for use in aircraft cockpits.

Exposure times to 'pain thresholds' were determined for over 2000 observations, for finger contact with six heated (45–195 °C) materials (aluminium, steel, hercuvit, glass, Teflon and Masonite). Initial skin temperature of the finger was 32.5 ± 0.5 °C. The interface temperature between finger and material was recorded by a fine wire thermocouple.

Although some consistency was noticed, the tests demonstrated the difficulty in determining pain thresholds. Large differences in exposure times to pain, both within and between subjects, were attributed to differences in

Table 13.5 Reaction of human skin (ventral forearm and anterior thoracic) to hot flowing water (*oil)

Exposure	Temperature (°C)	Exposure time h:mm:ss	Sub-threshold 1st deg.	Threshold and Supra-threshold 2nd & 3rd deg.	Subject
1	44	5:00:00	×		BF
2*		5:00:00	×		BF
3		6:00:00		×	BF
4*		6:00:00		×	BF
5*	45	2:00:00	×		KL
6*		3:00:00		×	KL
7		3:00:00		×	HA
8*	47	0:18:00		×	RK
9*		0:20:00	×		KL
10*		0:20:00	×		AM
11*		0:20:00	×		PG
12		0:25:00		×	RK
13*		0:40:00		×	AM
14		0:40:00		×	PG
15		0:45:00		×	RK
16	48	0:15:00	×		PG
17		0:15:00		×	AR
18		0:18:00		×	AM
19*	49	0:08:00	×		AM
20		0:08:00	×		AM
21		0:09:30		×	AM
22*		0:10:00		×	AM
23		0:11:00		×	AM
24		0:15:00		×	AM
25	51	0:02:00	×		AM
26		0:04:00		×	AM
27		0:06:00		×	AM
28	53	0:00:30	×		AM
29		0:01:30		×	AM
30	55	0:00:20	×		PG
31		0:00:30		×	AR
32*	60	0:00:03	×		FH
33*		0:00:05		×	FH

Source: Moritz and Henriques (1947).

epidermal thickness. Differences between reactions to materials correlated well with the thermal inertias (product of thermal conductivity (k), density (p) and specific heat (c) of the materials).

Exposure times to 'blister' were predicted by multiplying the 'averaged best fit' curve of exposure times to pain, by 2.5. Using curves of the reciprocal of the calculated exposure time to blister, an estimate was provided for the temperature of each material that could cause a minimal blister on contact for 0.3 s (0.2 s reaction time and 0.1 s for good contact).

To verify predicted blister temperatures, materials were heated to the predicted level and the back of the finger (of one subject) exposed to the material for an estimated 0.3 s. Four materials were used; aluminium, hercuvit, glass, and Masonite. The results confirmed the predicted temperatures very well, even though extrapolation of average data had been used in their calculation. Curves of material temperature, for both pain and predicted blister thresholds, and 0.3 and 1.0 second exposures, are plotted against the reciprocal of the square root of the material thermal inertia and were found to be linear. It was concluded that accuracy should not be expected to be better than ±10 per cent.

If materials must be used but temperatures cannot be kept below those indicated to produce a blister, then it is suggested that insulating coating can be used and the 'safe' temperature calculated on the basis of a two-layer heat transfer model. Further calculations are made to predict the effect of epidermal thickness on pain thresholds. It is noted however that, because subjective test techniques are used and the data were collected over a number of years, there is significant variation in the data. Further curve fitting to the data provides equations by which one can calculate threshold temperatures (e.g. for pain) from material type ($1/kpc$), exposure time and epidermal thickness.

An example calculation is provided. Possible errors in the calculations are discussed but the method is thought to be reliable for exposure times from 1 to 5 s. A practical point is that from the calculations it is concluded that relatively insulating materials will deform or disintegrate before they reach temperatures required to produce a blister after 0.3 s exposure.

Other studies of human skin

A number of authors have considered skin reaction to contact with hot surfaces. These include Bull (1963), who considered burns and although no new data are produced, he summarizes data based on the work of Moritz and Henriques (1947) and Sevitt (1949).

Bull (1963) proposes curves of surface temperature and exposure time for partial thickness and full thickness burns. McIntyre (1980) provides a practical review of the subject and Lawrence and Bull (1976) extend the curves of Bull (1963) to include thresholds for discomfort (see Figure 13.4). Eight males and eight females held a hot copper pipe (handle) through which thermostatically controlled water was circulated. It was found that discomfort thresholds occurred when the skin/handle interface reached 43 °C. A review also showed that average bath temperatures of a group of 20 subjects (temperatures tolerated without discomfort) was 40.5 °C (36–42.5 °C). Average shower temperatures of seven subjects were found to be 40 °C (38.5–41.0 °C). Fishman and Jenne (1981) found that females prefer shower temperatures 1 °C higher than males, however the mean maximum tolerable temperature was 44 °C for both sexes. No subject was prepared to withstand shower temperatures in excess of 50 °C. The question of thermal

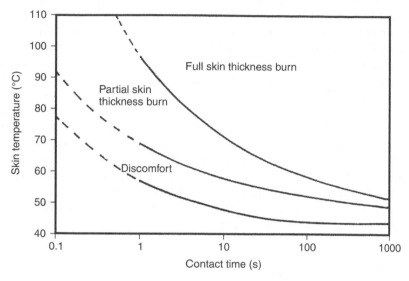

Figure 13.4 Skin temperatures that cause discomfort and burns.
Source: Lawrence and Bull (1976).

discomfort and solid surfaces is discussed later in this Chapter. It is useful to have an indication of which temperatures will not produce skin damage in order to establish damage thresholds.

One of the few 'applied ergonomics' studies in this area is presented by Ray (1984). He considers heat transfer theory, not in the context of burns, but in terms of what would be acceptable safe surface temperatures for products. This involved subjective measures of comfort on 48 female subjects, who touched heated discs of three materials as well as exerted a light grip on a heated saucepan. A main part of the work was to investigate the use of a mathematical model involving the use of contact temperature, i.e. the temperature at the interface between the skin and the hot surface. This model and others are discussed below.

MATHEMATICAL MODELS OF HEAT TRANSFER

The difficulty in obtaining comprehensive empirical data relating to skin damage and hot surfaces gives emphasis to the use of mathematical models. In practical applications, the contextual factors in any contact will mean that no mathematical model will provide a perfect representation. However, some models will be useful. Empirical models which 'fit' equations involving identified relevant factors to empirical data (such as those produced by Stoll *et al.*, 1979) may be of value for the conditions for which they were derived. However, the lack of reliable empirical data and of an underlying 'rational' model makes them of limited use.

Rational models involving heat transfer between a solid surface and human skin, involving (dynamic) models of the properties and reactions of the solid surface and the human skin, could provide solutions; for example, to questions such as at what temperature will the whole epidermis be destroyed? Or when will dermal tissue reach 44 °C? Rational models fall into two types: those which provide simple representations of the system with prediction 'sufficiently accurate' for many applications; and those which attempt to provide a detailed underlying dynamic model of the system. Naturally, models fall on a continuum between the two extremes and it should also be noted that the more detailed models may be more, but will be no less accurate than the more 'crude' simple models. The practical point is, under which circumstances more detailed models are more accurate in any situation, so that an informed choice can be made when addressing specific problems.

Simple models

The most commonly presented simple model is based on two semi-infinite slabs of material brought into perfect thermal contact. This is represented in BS PD 6504 (1983) and shown in Figure 13.5. When the two slabs are brought into contact, heat flows from the hotter to the cooler slab until equilibrium temperature is reached. However, at the interface between the slabs this temperature (contact temperature) is achieved instantaneously.

The rate of flow of heat, and hence when equilibrium will be reached, will depend upon the properties of the two materials. McIntyre (1980) provides an equation for calculating contact temperature:

$$T_{con} = \frac{b_1 t_1 + b_2 t_2}{b_1 + b_2},\tag{13.1}$$

$$b_i = (kpc)^{1/2}.$$

where $b_i (i = 1, 2)$ is the thermal penetration coefficient (in $J\,s^{-1/2}\,m^{-2}\,K^{-1}$), i.e. square root of thermal inertia.

If one assumes that contact temperature is the temperature between human skin and a solid surface, then using empirical data (e.g. Moritz and Henriques, 1947) skin reaction can be predicted for any contact, material type and skin condition. For example, suppose skin came into contact with aluminium at 80 °C for 4 s, using data in Tables 13.1 and 13.4, contact temperature is calculated as follows:

$$t_c = \frac{b_s \times t_s + b_h \times t_h}{b_s + b_h},\tag{13.2}$$

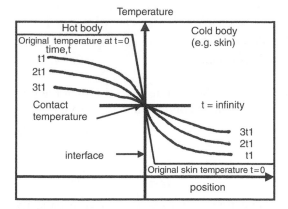

Figure 13.5 Temperature changes when two semi-infinite slabs of material, at different temperatures, come into perfect contact.

where b_s and b_h are the thermal penetration coefficients for skin and the hot material, and t_s and t_h are the temperatures of the skin and hot material respectively.

$$t_c = \frac{(1000 \times 33) + (22\,265 \times 80)}{23\,265} = 78.0\,°C.$$

Comparing this with the curves in Figure 13.3 for a skin temperature of 78 °C, predicts that skin will be subjected to a partial thickness burn. Ray (1984) cites Van der Held (1939) and provides the following derivation of the above equation from Fourier's law:

$$\frac{dt}{dr} = \frac{k}{pc}\frac{d^2t}{dx^2},$$

where

$\dfrac{dt}{dr}$ = rate of change of temperature over time

$\dfrac{d^2t}{dx^2}$ = first derivative of temperature gradiant

c = specific heat capacity

p = density

k = thermal conductivity

t_c = contact temperature.

For the following equations suffix h or s represent properties for hot material and skin respectively.

Consider heat transfer between an outer layer of a hot object, thickness x, and a layer of skin, thickness y.

$$\text{mass per unit area of material} = xp_h.$$
$$\text{mass per unit area of skin} = yp_s.$$

Assume heat loss by hot object equals heat gained by skin, then the rate of heat conducted (q) equals the rate of heat conducted to skin from hot object ($k \times$ temperature gradiant), i.e.

$$q = \frac{k_h}{x}(t_h - t_c) = \frac{k_s}{y}(t_c - t_s),$$

and net heat loss or gain in unit time is mass $\times$ specific heat capacity $\times$ temperature change, so

$$q = xp_h \times c_h \times (t_h - t_c) = yp_s \times c_s \times (t_c - t_s)$$
$$q^2 = \frac{k_h x p_h c_h}{x}(t_h - t_c)^2 = \frac{k_s y p_s c_s}{y}(t_c - t_s)^2$$
$$q = (k_h p_h c_h)^{1/2}(t_h - t_c) = (k_s p_s c_s)^{1/2}(t_c - t_s),$$

but the thermal penetration coefficient $b = (kpc)^{1/2}$

$$b_h(t_h - t_c) = b_s(t_c - t_s)$$
$$b_h t_h - b_h t_c - b_s t_c + b_s t_s = 0$$
$$(b_h t_h + b_s t_s) = (b_h + b_s)t_c$$
$$t_c = \frac{b_h t_h + b_s t_s}{b_h + b_s}, \tag{13.3}$$

i.e. as for the equation from McIntyre (1980).

Both Ray (1984) and McIntyre (1980) provide values for the thermal penetration coefficient of different materials including skin. Comparison of the values show that they differ, and if b values are calculated for materials, from Tables 13.1 and 13.4, then they will also be slightly different.

One consideration about models is whether the representation is sufficiently detailed for practical application. For example, it is probably not a reasonable assumption that there is perfect contact between the skin and the hot surface. Representing the skin (and the solid surface) as a single semi-infinite slab is also an approximation. In addition only contact temperature is calculated; any predictions of effects on human skin still rely on the limited empirical data available. An important use of the model is in the prediction of the relative effect of surface temperatures of different material types. Hot metals will produce greater contact temperatures and effects on the skin than wood at the same surface temperature for example. Using data for surface temperatures which provide burn thresholds for metals, one can calculate burn threshold temperatures for other materials. However, using this model, burn thresholds for materials with relatively high thermal insulation give excessively high temperature values.

A problem with use of the model is selecting the thermal properties of the materials (as discussed above). Clearly the selection of the values is of great importance. However, suppose we consider effects of contact with hot wood on the skin; The thermal conductivity of hard wood can be around 40 times that of soft wood and four times as dense. Similar ranges in variation can be found within the properties of metals. Properties will also depend upon temperature. In addition, only approximate estimates can be made for human skin; little is known about the effects of skin condition on thermal properties and the skin will change its properties as it becomes damaged.

Another point regarding all models, therefore, is that however sophisticated their representation system, the accuracy of predictions made will be greatly influenced by the accuracy of data supplied to the model.

Comprehensive models

Comprehensive models of skin in contact with hot surfaces consider a detailed representation of the skin/surface system and its dynamic behaviour. A thorough review of such models is provided by Diller (1985). He lists thermal boundary conditions during a burn, constitutive and physiological properties of skin and criteria to define the thresholds for levels of injury, as the requirements for an effective model. Diller (1985) describes the first successful analytical model by Henriques and Moritz (1947) based upon the one-dimensional heat conduction equation (similar to equation 13.1) used by Ray (1984). When the temperature behaviour of the system has been solved, a damage function is used based on the biochemical behaviour of cells to increased temperature. This model neglects the effects of local blood perfusion and assumes constant skin layer thickness and thermophysical properties. Diller (1985) goes on to describe models of increasing complexity which, with the increase of power and availability of digital computers, can be used in practical applications. These include non-linear finite element models which can simulate the behaviour of human skin (epidermis, dermis, fat) during heating and cooling phases and can be used to study the likely behaviour of burns.

Radiation burns have been described by Elton (1998) who developed a model of radiation transfer through clothing (ICARUS) including high levels of radiation from fire and nuclear blast. They linked the heat transfer model and a model of human thermoregulation and used the data of Moritz and Henriques (1947) to establish prediction of burns. The model was evaluated using manikins wearing military uniforms.

Physical models

Mathematical models can be used in practical application by measuring (predicting or estimating) values of the parameters required to run the model (usually on a digital computer). For example, if a prediction of skin reaction to contact with a hot oven door is required, then even the simple semi-infinite slab model (above) requires values for surface temperature, thermal properties of material and skin, and likely exposure times, to predict a skin reaction in terms of burns. An alternative is to construct a physical model of human skin and observe the reaction on the physical model. This method is widely used in many areas of work, for example in determining thermal properties of clothing (whole-body copper men, artificial foot, etc.) or properties of acoustic equipment (head manikins) etc. The principle is that the manikin responds, in the area of interest, in a way similar to that of the human body. It can then be used instead of the human body to determine likely responses.

The thermesthesiometer

Marzetta (1974) produced an instrument (the thermesthesiometer) which 'behaved' like the human finger to 'make burn hazard measurements in consumer products'. The instrument is based on a simple heat transfer model such that the contact temperature achieved between the thermesthesiometer and the hot surface is an indication of the tissue temperature if human skin were in contact with the hot surface. Predicted injury can then be determined from experimental data, e.g. Moritz and Henriques (1947) and Stoll *et al.* (1979). A schematic diagram of the thermesthesiometer probe is presented in Figure 13.6.

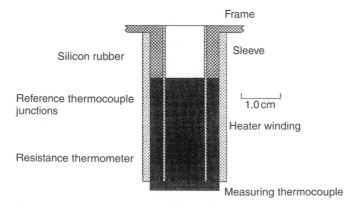

Figure 13.6 Schematic drawing of thermesthesiometer probe.
Source: Marzetta (1974).

The 'core' of the probe is maintained at a temperature of 33 °C. The operation of the instrument is to manually bring the probe into contact with the hot surface (firm, square and positive contact, 4–8 Newton force – although degree of pressure does not noticeably affect the readings). The stated accuracy is ±3 °C. However, after experience for a one-second contact '. . . the operator can expect a repeatability of measured temperature to within two degrees for more than 90 per cent of trials.'

An interesting point is that the introduction of a water or oil film to the probe face makes a significant difference to the measurements. However, experiments with 5 mm thickness Teflon tape on hot brass showed that contact temperature increased with contact time. An example is quoted of steel covered with a plastic coat. For a surface temperature of 102 °C, contact temperature is 60 °C for a one-second contact and 73 °C for a four-second contact. The work of Marzetta (1974) was conducted at the National Bureau of Standards, USA. The studies of the performance of the thermesthesiometer, its calibration and its practical use are extensive. If used correctly it appears to provide a reasonably accurate estimate of contact temperature.

Siekmann (1989, 1990)

In an introductory paper, Siekmann (1989) presents the 'semi-infinite slab' model reported by Wu (1972, 1977) and notes that a calculation is difficult for cases where the thermal properties of surfaces and thickness of any additional coatings, are not known. Use of the thermesthesiometer is recommended and also, the comparison of contact temperatures with the data of Moritz and Henriques (1947). A discussion of 'safe' surface temperatures is also provided.

Siekmann (1990) used the thermesthesiometer to measure the contact temperatures of heated discs of a range of materials. The discs were heated on their undersurface, using a metal cylinder containing flowing water (oil for temperatures over 100 °C). The thermesthesiometer was clamped and brought onto the surface of the disc making good contact at 10 N force. Surface temperatures of a wide range of bare and coated materials are provided which will produce contact temperatures of 43–65 °C, the burn threshold for a contact time of one second. Steel coated with various thicknesses of varnish showed that, for one-second contacts, varnish has an insulating effect which increases linearly with thickness. As with the results of Marzetta (1974), the insulative coating 'worked better' for shorter contact times. For contact times over one minute, the varnish had negligible insulating effect. This work has formed the basis of a proposed European standard for assessing the effects of contact with touchable surfaces (EN 563, 1994).

Other physical models

The thermesthesiometer provides an estimate of contact temperature for a finger in contact with a hot surface. Similar devices have been used to simulate bare foot contact with hot floors, mainly in the investigation of thermal discomfort (Olesen, 1977; Yoshida *et al.*, 1989). Yoshida *et al.* (1989) also describe a 'warmth tester' device. Both artificial foot and warmth tester measure heat flow between the device and the solid surface.

A possible method of determining contact temperatures (and maybe effects) between living human skin and hot solid surfaces would be to use (recently) dead animal or human tissue (plus skin). This however, does not appear to have been considered.

Artificial skin has been used to assess the effects of heat (e.g. molten metal splash) through working clothes. The method is to place the skin over a whole-body manikin and expose the clothed manikin to the hazard – molten metal, fire, etc. When the test is over the skin (plastic film) is inspected and the degree of damage can be related to skin temperature and possibly to likely effects on human skin – e.g. Maggio (1956) and Mawby and Street (1985). This method may be useful for assessing contact between a clothed body and hot surfaces; however, the representation of the skin is probably inadequate for a reasonable assessment of the effects of hot solid surfaces on bare human skin.

Data bases, standards, and limits

Data concerned with the reaction of skin to contact with hot surfaces are of great interest in practical application for the specification of safe surface temperatures (Parsons, 1993; Ray, 1984; Siekmann, 1989, 1990). These data, along with many other factors, have been used to provide limiting values for surface temperatures of industrial and domestic appliances, toys, etc. The establishment of limiting values for a particular 'machine' in a particular context involves many factors other than data on likely skin reaction. Some standards and limits which have been proposed and produced are summarized below.

British Standards

BS 4086 (1983) provides 'recommendations for maximum surface temperatures of heated domestic equipment'. It is an amended version of BS 4086 (1966) and was confirmed only after some debate, due to lack of agreement on technical grounds. A 'companion' document, BS PD 6504 (1983) was published to provide 'medical information on human reaction to skin contact with hot surfaces'. It is intended that BS 4086 (1983) assists in the production of standards for heated domestic equipment and that BS PD 6504 (1983) provides guidance where heat reaction to skin contact has to be taken into account.

BS 4086 (1983) provides maximum surface temperatures for three types of material (metals, porcelain and vitreous materials and plastics, wood or rubber). Three functional types of surface are considered which can be

Table 13.6 Recommended maximum surface temperatures of heated domestic equipment

	Handles, parts held during use °C	Knobs, handles etc.: touched for short periods °C	Other surfaces accidental contact or very short periods °C
Metals	55	60	105
Porcelain, vitreous materials	65	70	120
Plastics, wood or rubber	75	85	125

Source: BS 4086 (1983).

related to exposure time. These are: handles of appliances which are integral with the appliance and intended for holding (relatively long contact time); handles and knobs which are touched for short periods but not gripped (short contact time); and other surfaces which could be touched accidentally but are not 'normal' working surfaces of the appliance (momentary contact). Table 13.6 summarizes the recommended maximum surface temperatures for each category of conditions. It is important to note that for momentary accidental contact conditions, the standard acknowledges that it may not be possible to prevent some surfaces rising above a 'touchable temperature' and practical design methods (positioning guards, etc.) may be the only method of dealing with this hazard. The implication therefore is that maximum limit temperatures provided are not solely based on thresholds for adverse skin reaction. The medical information presented in BS PD 6504 (1983) is shown in Figures 13.7, 13.8 and 13.9.

A review of the properties of human skin is provided. The simple semi-infinite slab model involving contact temperature is also described. Discomfort is described as the threshold beyond which subjects thought that pain or injury would occur. It is also considered that minimum practical contact times in many situations will depend upon reaction time: 0.25–0.5 s, and that results using this simple model are conservative and higher temperatures may be tolerated in some cases (due to imperfect contact for example). It is noted that analytical solutions to realistic situations are unlikely to be accurate. However, they could be usefully used to indicate the effects of possible solutions to build up experience for any particular application.

European Standards

During 1989 and 1990 a European ergonomics standards committee was formed to produce a standard concerned with hot surface temperatures. This was in anticipation of the single European Market to be established in

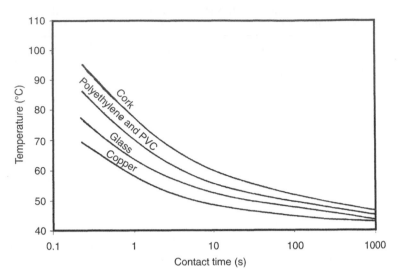

Figure 13.7 Discomfort thresholds for skin contact with solid surfaces of different materials.

Source: BS PD 6504 (1983).

1993 and involved active participation from France, Germany, Sweden and the UK. The committee produced a European Standard (EN 563) which provided 'Ergonomics data to establish temperature limit values for hot

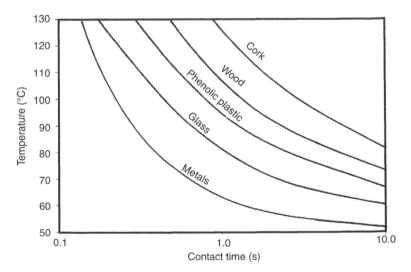

Figure 13.8 Pain thresholds for fingertip contact with solid surfaces of different materials.

Source: BS PD 6504 (1983).

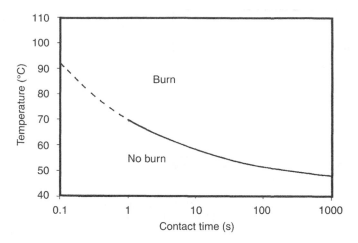

Figure 13.9 Threshold for the onset of skin burning after contact with a brass block.

Source: BS PD 6504 (1983).

surfaces' CEN (1990). The standard is a 'B' type standard which provides an ergonomics database; it does not provide limit values for specific appliances. So-called 'C' type standards can use data in 'B' type standards to provide guidance in establishing limits for specific machinery (appliances).

Based on the work of Moritz and Henriques (1947) and on the thermesthesiometer work of Siekmann (1990), data are provided to allow the assessment of the risk of burning.

It is recognized that data concerning skin contact with hot surfaces and burns, are not complete in terms of all important factors (see Table 13.2). For this reason, three 'areas' are presented on curves of material surface temperature versus exposure time. These show an area where a burn is possible and within which lies the 'burn threshold' for a superficial partial thickness burn. The area above this is where a burn would be expected, and the area below this is where no burn would be expected. Curves showing the three areas are provided for metals, plastics, ceramics, glass, stone and wood. Figure 13.10 shows the overall philosophy. 'Correction' curves are also provided for coated metals. Curves showing the important exposure times up to 10 s are presented in Figures 13.11, 13.12, 13.13, and 13.14.

The data presented in EN 563 (1994) are intended for the determination of temperature limit values, for machines for example. It is interesting that EN 563 (1994) provides data for a minimum contact time of 1.0 s. It is debatable whether realistic contact times of less than 1.0 s occur, and also how the ranges of temperatures provided in the proposed standard should be used to establish temperature limit values.

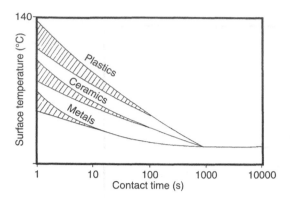

Figure 13.10 Method of representing burn threshold data for different materials according to EN 563 (1994). The single line curve continues down to 45 °C for up to eight hours of exposure.

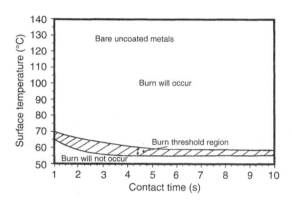

Figure 13.11 Burn thresholds for bare, uncoated metals. Source: EN 563 (1994).

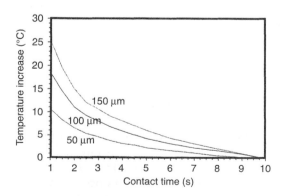

Figure 13.12 Increase in burn threshold for contact with metals coated with 'lac'. Source: EN 563 (1994).

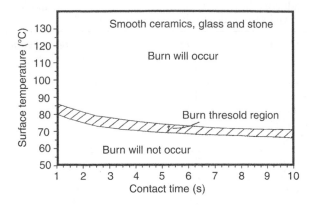

Figure 13.13 Burn thresholds for smooth ceramics, glass and stone.
Source: EN 563 (1994).

Other guidelines and standards

There has been little work on International Standards (e.g. ISO) in the area of human skin reaction to contact with solid surfaces. The ergonomics standard committee, within ISO, has however recently decided to produce standards in this area. Hot, moderate and cold surfaces will be considered as well as contact by naked and covered skin. For hot surfaces it is likely that there will be close 'harmonization' between ISO and CEN standards.

Yoshida *et al.* (1989) review the subject to provide a basis for establishing International Standards. The simple one-dimensional model of heat transfer is used to derive the equation for contact temperature (see (13.1) above).

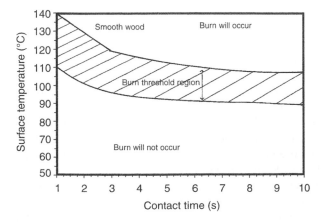

Figure 13.14 Burn thresholds for smooth wood.
Source: EN 563 (1994).

Table 13.7 Pain from conductive heating. Light touch pressure. Elbow and knee sometimes received second degree burns without pain

Body area	Clothing worn	Metal surface temperature (°C)	Average tolerance time (s)
Hand	Bare skin	49	10–15
Kneecap	Bare skin	47	34
Fingertip	Leather gloves	66	12.6
	Leather gloves	71	7.3
Hand-palm	Leather gloves	66	25.2
	Leather gloves	79	9.7
	Leather gloves	85	8.0
Forearm	SAC alert suit	66	20.6
	SAC alert suit	79	8.0
Upper arm	Flight coverall	66	7.5
	SAC alert suit	66	31.3
	Alert suit plus string underwear	149	7.2
	K-2B suit	66	18.1

Source: *Bioastronautics Data Book* (Webb, 1964).

A brief review of the work of Stoll *et al.* (1979) and of British Standards is also provided, as are devices for simulating heat flow into the body, e.g. using an artificial foot.

The *Bioastronautics Data Book* (Webb, 1964) provides data concerning pain (and some burns) from conductive heating. Some of these data are summarized in Table 13.7. Metal surface temperatures and average tolerance times are provided for contact between clothed and unclothed skin and solid surfaces.

DISCUSSION

Human data revisited

It can be seen from the above reviews that there is a limited amount of recorded data where human skin has contacted hot surfaces to cause a burn, and much reliance has been placed on the work of Moritz and Henriques (1947), so it is useful to consider these data in some detail. For each burn the details were summarized in Table 13.5 and are now also presented in Figure 13.15. The data of Stoll *et al.* (1979) are shown in Table 13.8.

Moritz and Henriques (1947) categorized first degree burns as those which ranged from transient hyperaemia to those with severe and prolonged erythema followed by the formation of miliary vesicles which did not coalesce. Burns where there was complete necrosis of the epidermis over the entire target area were categorized as second or third degree burns

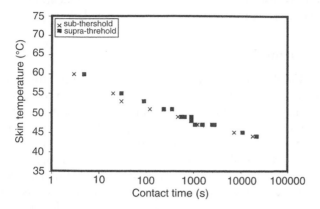

Figure 13.15 Burn data from individual subjects from the experiments of Moritz and Henriques (1947). ■ – supra-threshold, 2nd and 3rd degree burn; × – sub-threshold, 1st degree burn.

depending upon their depth. No consistent relationship was found between level of discomfort and severity of burn. It is interesting to note that one subject had a significantly lower threshold to thermal injury than others, showing complete epidermal necrosis for exposure to a skin temperature of 47 °C for 18, 25 and 45 min. These data are inconsistent with the curves shown in Figure 13.3.

The data from Stoll *et al.* (1979) provide important information regarding 'momentary contact', often considered useful for providing surface temperature limits where accidental contact may occur. For these very short contact periods however, it is difficult to conduct controlled experiments and the nature of contact is very important. The selection of 0.3 s as a

Table 13.8 Exposure of human skin (back of finger)

Material	Temperature (°C)	Exposure time (s)	Skin reaction
Aluminium	132.0	0.2	Transient erythema, no visible effect
		0.45	Full blister
Masonite	194.4	0.29	Barely perceptible darkening of skin
	194.6	0.35	Blister slightly larger than minimal
Hercuvit	149.3	0.2	Erythema
	149.6	0.29	Full blister
Glass	160.0	0.46	Full blister
	160.0	0.29	Full blister
	157.0	0.26	Erythema
	158.0	0.3	Threshold blister

Source: Stoll *et al.* (1979).

minimum contact time is also debatable in any practical application and will have a great influence on any temperature limit selected.

Classification of skin damage

Classification systems for skin burns have naturally been oriented towards medical considerations, and this has led to three broad categories (see Table 12.3). For the assessment of the effects of human skin contact with hot surfaces it would be useful to consider the possibility of a more 'sensitive' classification system, particularly with respect to less severe skin reactions. If this were possible then it would allow a more comprehensive evaluation of any 'model' for predicting severity of skin damage. A more sensitive categorization system would depend upon a consistent and detailed description of skin reaction. A summary of the description and classification used in empirical studies is described in the next section.

Diller (1985) provides a detailed description of skin reaction to burning and this is summarized in Table 13.9.

Prediction of skin reaction

The discussion above considered both simple and more comprehensive models for predicting skin reaction to contact with hot surfaces. The simple one-dimensional heat transfer model involved calculation of contact temperature (equation 13.1). However, it was noted that predicted contact temperatures were found to be higher than 'measured' and that the overestimation increased with decreasing thermal conductivity of material.

It seems possible that a reason for the inaccuracy of prediction is the unrealistic assumptions made by the model. Although this simple model does not make accurate predictions, it can form the basis of a practical method for calculating contact temperatures and hence determining likely skin reaction. Two important considerations are not taken into account by the simple model: the nature of the contact and the skin condition.

Contact between skin and a hot surface

The simple model described above assumes two semi-infinite slabs in perfect contact. In practice, contact will not always be perfect. The degree of contact will depend upon the pressure of contact and the effect will depend upon contact time. In practice an estimate of contact will be possible in terms of such descriptors as 'minimum', 'low', 'medium', 'high' and 'maximum' contact.

Skin condition

Skin condition will be largely dictated by whole-body response to thermal conditions. Under cold conditions at a temperature of lower than 30 °C,

Table 13.9 Description and classification of skin damage during burns

Classification		Macroscopic effects (appearance)	Microscopic effects
1st degree	I	General erythema, redness of tissues; possible irritation of nerve endings	Vasodilation of sub-papillary wall in infected area
2nd degree Partial thickness	IIa	Oedema with blistering; waterproof cover over wound; regeneration without scarring	Increased permeability of sub-papillary plexus leading to fluid loss; most basal cells not injured; cells swollen and distorted
Deep	IIb	Blistering not widespread; eschar of plasma and necrotic cells over wound; scarring	Necrosis of much of basal cell layer; widespread stasis; cells swollen and distorted
3rd degree Full thickness	III	Inflammation of periphery of burn; overlying eschar highly permeable to water and bacteria; resurfacing from margins of wound or by skin graft	All epidermal and dermal structure destroyed; large volume of extra-cellular fluid below wound
4th degree	IV	Similar to 3rd degree burn but greater complications in healing	Incineration of tissue; injury to bone

Source: Diller (1985).

Note
In addition to local effects, there are also whole-body responses including shock from loss and strain on many organs.

there will be vasoconstriction and skin may be dry. Under 'normal' moderate conditions at a temperature of around 33 °C, the skin will show some vasodilation and be generally dry.

In hot conditions at a temperature of around 36 °C, blood vessels will be dilated and skin will be wet.

A simple practical model

Estimated correction values for the skin condition and the nature of contact between skin and a hot object can be used with equation (13.1) to calculate an equivalent contact temperature (T_{ceq} – Parsons, 1992b). If the body is hot then skin temperature can be estimated to be 36 °C, neutral 33 °C, and cold 30 °C. If skin is hot or neutral then it can be assumed to be vasodilated. If the body is cold then skin is vasoconstricted. Appropriate thermal penetration coefficient values (b) can be selected from Table 13.1. Contact temperature is then calculated as before from

$$t_c = \frac{b_h t_h + b_s t_s}{b_h + b_s}.$$

However, this will apply for two semi-infinite slabs in perfect contact. Practical correction can be made for this by considering the nature of the contact and the possible presence of materials (sweat, grease, etc.) on the skin. A method of correction can be similar to that proposed, for correcting from bare metals to coated metals, in EN 563 (1994). The following equation can be used as an approximation to T_{ceq}

$$T_{ceq} = t_c - (t_c - t_{cc})e^{-(b_c/b_s)t}, \tag{13.4}$$

where

t_c = contact temperature

t_{cc} = contact temperature if the solid surface were made entirely of the coating on the skin and corrected for type of contact

b_c = thermal penetration coefficient of the coating on the skin and corrected for type of contact

b_s = thermal penetration coefficient for the skin

t = contact time in seconds.

It can be seen that (13.4) will satisfy the boundary conditions that:

1 when t tends to zero, contact temperature will tend to that of the coating; and
2 when t tends to infinity, contact temperature will tend to that of the material.

Rate of reduction on correction will depend upon the thermal properties of the coating. The lower the thermal penetration coefficient, the greater the effect of the coating.

Equation (13.4) will give a practical correction to calculated contact temperature, taking account of nature of contact and skin condition. A practical example is provided below. The correction applies only for skin covered with thin coatings of material such as grease or sweat. For relatively thick coatings of the material the model could be extended to include the thickness of the coating.

Practical example

Suppose the skin of a machine operator comes into contact for about one second, with bare steel at 70 °C, when operating a machine in a hot environment. The skin will be vasodilated and sweating, and the contact will be light.

Procedure for calculating equivalent contact temperature (T_{ceq})

1. *Calculate contact temperature* $(in\ ^{\circ}C)$

$$t_c = \frac{b_h t_h + b_s t_s}{b_h + b_s}$$
$$= \frac{12\,568 \times 70 + 1500 \times 36}{12\,568 + 1500}$$
$$= 66.4$$

2. *Estimate thermal penetration coefficient for coating* (b_{cl}) Assume skin sweating, b for water at $60\,^{\circ}C$ is 1636 (in $J\,m^{-2}\,s^{-1/2}\,K$)
3. *Calculate* b_c (*in* $J\,m^{-2}\,s^{-1/2}\,K$) *using a weighting factor for level of contact from*:

Level of contact	Weighting factor (WF)
Minimum	0.2
Low	0.4
Medium	0.6
High	0.8
Maximum	1.0

Note
Level of contact will also involve roughness of surface, since a rough surface, for example, may allow air between the skin and the material surface, producing imperfect contact. This will interact with pressure of contact.

$$b_c = WF \times b_{cl}$$
$$= 0.4 \times 1636$$
$$= 654.4.$$

Although the above is an approximation, it provides a practical method of taking account of the possible significant effects of skin condition and level of contact. For example, for light (low) contact there may be a layer of air between some of the skin surface and the hot material surface. Note that for clean, dry skin the b_c value should be as if the material were coated with material and the appropriate b value used. A weighting factor can then be used as above. If perfect contact were achieved; then from (13.4) $t_{cc} = t_c$ and $T_{ceq} = t_c$.

In the example therefore,

$$t_{cc} = \frac{654.4 \times 70 + 1500 \times 36}{654.4 + 1500}$$
$$= 46.3.$$

From (13.4)

$$T_{ceq} = t_c - (t_c - t_{cc}) \exp - \left(\frac{b_c}{b_s}\right) t$$

$$= 66.4 - (66.4 - 46.3) \exp - \left(\frac{654.4}{1500}\right) \times 1$$

$$= 66.4 - 20.1 \times 0.646$$

$$= 53.4.$$

4. *Interpretation*   The T_{ceq} value of 53.4 °C can be compared with skin temperature curves for skin damage (Figure 13.3) and it can be seen that at 53.4 °C for 1 s of exposure, the contact is not expected to produce a burn.

Clearly the above model is not sufficient for a three-layer model of heat transfer; however it does provide a simple method for use in application. A definition of equivalent contact temperature may be 'the temperature between two semi-infinite slabs of material in perfect contact, one slab of hot material and the other of human skin, that would produce equivalent effect on human skin as the actual contact between human skin and a hot surface of interest.' For example, for conditions with poor thermal contact, the equivalent contact temperature would be lower than the calculated contact temperature, had the contact been perfect.

SURFACES OF MODERATE TEMPERATURE

When skin is in contact with surfaces of a moderate temperature (between approximately 5 °C and 40 °C or a wider range depending upon material type) no damage or pain will occur. There will however be thermal sensations ranging from hot to cold and related subjective responses concerning comfort, pleasantness, acceptability (e.g. of a handrail), and so on. Methods of determining human responses to contact with moderate surfaces have included empirical methods and the use of physical and mathematical models.

Empirical research has been conducted in a series of experiments where people in a thermally neutral state touched (light grip) horizontal handrails through a hole in a thermal chamber and gave their immediate subjective impressions (sensation, comfort) and their impression after 20 s of contact. Halabi and Parsons (1995) present results of eight male and eight female subjects for temperatures 15 °C, 20 °C and 25 °C. This range was extended from 5 °C to 35 °C in further experiments conducted by Herrman *et al.* (unpublished) who used 16 males and 16 females at 5, 10, 30 and 35 °C in an otherwise identical experiment to that reported by Halabi and Parsons (1995). The combined results are presented in Figures 13.16 and 13.17.

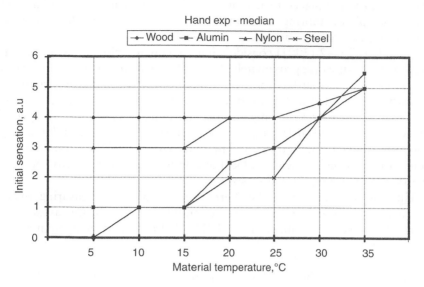

Figure 13.16 Initial sensation on contact between the hand and a handrail
(0 – very cold, 1 – cold, 2 – cool, 3 – slightly cool, 4 – neutral,
5 – slightly warm, 6 – warm).

In a further series of experiments, Herrman and Parsons (unpublished)
carried out experiments into contact between the bare foot and floor
materials. Subjects dressed in clothing for thermal neutrality took off right

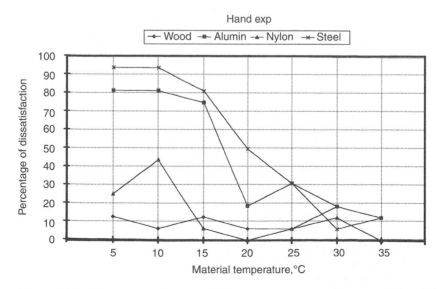

Figure 13.17 Percentage of dissatisfied subjects on contact between the hand
and a handrail.

shoe and sock and placed their foot through a hole in an insulated box and onto a sample of flooring (0.45 cm × 0.45 cm × 3.5 cm) made of ceramic tile, wood, concrete or carpet.

ISO TR 13732–2 (2001) considers human contact with surfaces at moderate temperatures (approximately 10–40 °C). It is noted that, in a warm environment, a cool surface may feel comfortable and in a cool environment a warm surface may feel comfortable. The thermal state of the person as well as the body part, skin temperature, environmental temperature, type of object contacted and surface material will all be important. The relationship between material temperatures and thermal sensation of the hand on initial contact is provided, as well as, local discomfort caused by warm and cool floors when wearing normal shoes.

There is no mention of the thermesthesiometer for measuring finger contact temperature, even though it should be useful for assessment of moderate surface temperatures. DIN 52614 is cited as a method for measuring energy loss $(kJ\,m^{-2})$ from an artificial foot in contact with the floor. That is a water-filled cylinder (15 cm diameter) with a rubber membrane at the base. The heat loss is determined for a contact time of 1 and 10 min. Results for typical floor construction and recommended floor temperature ranges are provided. Mathematical models of heat exchange are also presented in ISO 13732–2 (2001). The simple model of contact temperature (see above) and related thermal diffusivity are shown to be good predictors of human response; taking account of material properties. Some discussion and guidance on sitting on electrically heated floors is also provided. The principle of human reaction to hot surfaces provided earlier in this chapter apply to contact with moderate (and cold) environments and there seems little limitation in using the equivalent contact temperature index (T_{ceq}) for predicting responses to contact with moderate or cold surfaces.

COLD SURFACES

A proposal was made in 1989 to prepare an ergonomics database and a corresponding European standard concerned with skin damage caused by contact with cold surfaces. After a worldwide search for data, it was concluded that insufficient (almost no) information was available. A dedicated call for research led to a programme of European research (ColdSurf – Sweden, Belgium, The Netherlands, Germany, Finland and UK) which conducted surveys of industry and literature, empirical research, modelling and the development of a measuring instrument (Holmér and Geng, 2000). Building on the work of Chen (1997), human skin reaction (cooling curves) to finger touching of metal cubes and gripping of vertical bars (wood, nylon, stone, steel and aluminium materials) was investigated. Whole-body experiments were used where subjects were exposed in thermal chambers, as well as experiments where subjects placed their hands in cabinets. The experiments were carefully planned and, although individual studies were conducted in different laboratories, identical equipment and methodologies were used. Over all laboratories a database of 1657 tests over 24 exposure conditions was made from finger touching and 581 tests for 21 exposure conditions for gripping. The effects of pressure of contact were also investigated. The results were specifically required for the production of a European standard. A wide individual variation was found and a series of complex 'rules' for interpolation and extrapolation were used. Although, therefore, the results will not apply to individuals, they are presented in a useful generalized form and are presented in a proposed joint ISO and European standard – EN ISO 13732-3, 'Ergonomics of the thermal environment – Touching of cold surfaces Ergonomics data and guidance for application'. A practical generalization based upon previous literature was to choose skin temperatures of 15 °C, 7 °C and 0 °C as temperatures below which pain, numbness and freezing would occur. The task was then to identify cooling times to reach those landmarks. This was achieved using non-linear (exponential) empirical modelling involving the thermal penetration coefficient ($F_c = \sqrt{p\,k\,c}$) to account for the effects of material type. The idealized curves are presented in Figures 13.18, 13.19 and 13.20 for different contact durations and different materials in terms of touching and gripping thresholds for pain, numbness and frostnip. The thermal properties of materials were measured during the programme of research and values are included in a table in the proposed standard. Examples of application are presented in an annex to the standard.

Contact between the hands and cold surfaces has been the subject of a number of Doctoral theses. Chen (1997) (see also Chen *et al.*, 1994, 1996)

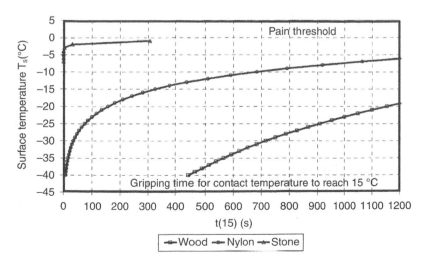

Figure 13.18 Predicted time to reach pain on gripping a bar.

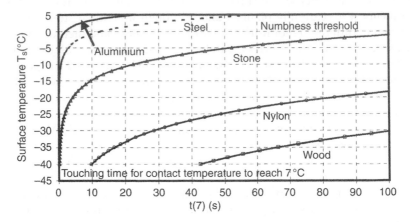

Figure 13.19 Predicted time to numbness on finger contact with a smooth
surface.

constructed a series of laboratory based experiments involving human subjects in hand and finger contact on cold surfaces for the bare and gloved
hand. Subjects placed their hand through a hole in a small climate chamber
and temperature cooling curves and subjective responses were recorded.
Factors investigated include contact force, surface temperature, material
property, surface mass, the thermal condition of the whole-body and the
morphology of the fingers. Cooling curves were established based upon a
Newtonian model with two time constants. For metals, contact at 7 °C

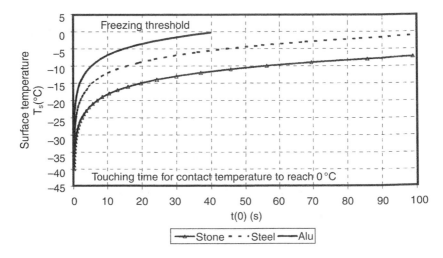

Figure 13.20 Predicted time to skin freezing on finger contact with a smooth
surface.

surface temperature can impair manual dexterity, 4 °C is a limit for tactile sensitivity and −4 °C for cold injury. Geng (2001) (see also Holmér and Geng, 2000) continued this work as part of a coordinated European project. She conducted experiments on hand/finger cooling responses of human subjects with and without gloves. The simple contact temperature model (see above) was used to establish cooling curves and criteria based upon contact temperature of 0 °C (freezing), 7 °C (numbness), and 15 °C (pain). An artificial finger for the assessment of cold surfaces was developed and the effects of gloves, including heated gloves, were determined in terms of manual dexterity and tactile sensitivity. The data, along with that of other European laboratories, contributed to the draft international standard ISO CD 13732–3 (2001). Jay and Havenith (2000) found that there were no differences between male and female responses for cooling of the finger but there were effects of size, structure and shape of the hand. Rissaner and Rintamäki (2000) found that hand dimensions were important during rapid cooling but not during slow cooling. Jay (2002) found that there were no significant differences between responses of dominant and non-dominant hand. Piette and Malchaire (2000) determined tolerance times for gripping cold materials with bare hands and found that they vary inversely as a function of the contact temperature and linearly as a function of the temperature of the material. Powell and Havenith (2000) found that severe decreases in manual dexterity can occur after contact with cold materials.

14 International Standards

INTRODUCTION

Optimum indoor air temperatures have often been the subject of debate, and suggested limit or guidance values for buildings have been proposed over many years by a number of professional institutions and in legislation. There is an increasing international interest in providing guidance to ensure protection and good practice in areas where people are exposed to hot, moderate and cold environments and there has been recognition that air temperature is only one component of a human thermal environment.

The requirements of regional (e.g. European) and international (global) markets and the recognition that systems, services and products should be designed for human use have raised the profile of ergonomics and led to a proliferation of ISO Standards, including those in the area of the ergonomics of the thermal environment. Both published standards and those in development are described below. For full detail, the reader is referred to the original standards. For standards in development, the current position is described which may be subject to change due to national and international comment and voting.

ISO STANDARDS

The collection of ISO (International Organization for Standardization) Standards and documents, concerned with the ergonomics of the thermal environment, can be used in a complementary way to provide an assessment methodology. The subject is divided into three principal areas (hot, moderate and cold environments) and remaining standards are divided into human reaction to contact with solid surfaces, supporting standards and standards concerned with specific populations and areas of application (Figure 14.1).

For the assessment of hot environments a simple method based on the *WBGT* (wet bulb globe temperature) index is provided in ISO 7243. If the

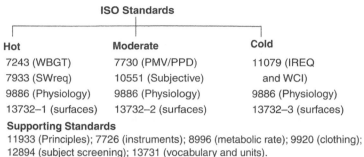

ISO Standards

Hot	Moderate	Cold
7243 (WBGT)	7730 (PMV/PPD)	11079 (IREQ
7933 (SWreq)	10551 (Subjective)	and WCI)
9886 (Physiology)	9886 (Physiology)	9886 (Physiology)
13732–1 (surfaces)	13732–2 (surfaces)	13732–3 (surfaces)

Supporting Standards
11933 (Principles); 7726 (instruments); 8996 (metabolic rate); 9920 (clothing); 12894 (subject screening); 13731 (vocabulary and units).
Application
Vehicles: 14505–1 Principles,14505–2 Teq, 14505–3 human subjects; 14415 (Disabled, aged . . . .); 15265 (risk assessment); 15743 (working practices in cold); 15742 (Combined envs).

Figure 14.1 ISO Standards for assessing thermal environments.

WBGT reference value is exceeded, a more detailed analysis can be made (ISO 7933) involving calculation, from the heat balance equation, of sweating required in a hot environment. If the responses of individuals or of specific groups are required (for example in extremely hot environments) then physiological strain should be measured (ISO 9886).

ISO 7730 provides an analytical method for assessing moderate environments and is based on the predicted mean vote and predicted percentage of dissatisfied (*PMV/PPD*) index, and on criteria for local thermal discomfort. If the responses of individuals or specific groups are required, then subjective measures should be used (ISO 10551).

ISO TR 11079 provides an analytical method for assessing cold environments involving calculation of the clothing insulation required (IREQ) from a heat balance equation. This can be used as a thermal index or as a guide to select clothing.

ISO work on contact with solid surfaces is divided into hot, moderate, and cold surfaces and standards are in final stages of development (ISO 13732 Parts 1, 2, and 3). Supporting standards include an introductory standard (ISO 11399) and standards for estimating the thermal properties of clothing (ISO 9920) and metabolic heat production (ISO 8996). Other standards consider instruments and measurement methods (ISO 7726) and standards concerned with vocabulary, symbols and units (ISO 13731), medical screening of persons to be exposed to heat or cold (ISO 12894) and a standard that considers the responses of disabled persons (ISO TS 14415). Standards under development include ISO 14505 Parts 1, 2 and 3 for the assessment of vehicle environments; ISO 15742, concerned with the combined stress of environmental components (including thermal); a standard (ISO 15743) concerned with working practices in cold environments; and a standard providing an overall philosophy of application including risk assessment (ISO 15265).

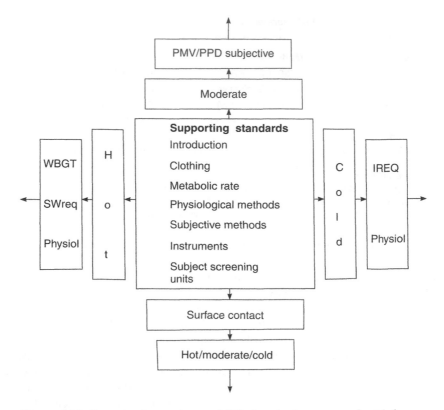

Figure 14.2 Organization and use of ISO Standards concerned with human thermal environments.

The ISO working system showing how the collection of standards can be used in practice, is presented in Figure 14.2 (see also Parsons, 2001a).

ISO 7243: Hot environments – estimation of the heat stress on working man, based on the WBGT index

This standard provides a simple convenient method, and uses the WBGT heat stress index to assess hot environments. Inside buildings and outside buildings without solar load

$$WBGT = 0.7t_{nw} + 0.3t_g. \qquad (14.1)$$

While outside buildings with solar load

$$WBGT = 0.7t_{nw} + 0.2t_g + 0.1t_a, \qquad (14.2)$$

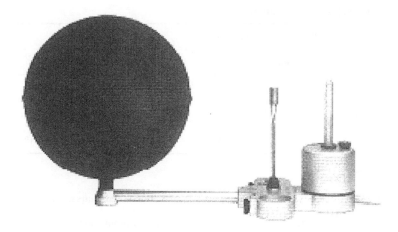

Figure 14.3 WBGT transducers.

where

t_{nw} is the natural wet bulb temperature
t_g is the temperature of a 150 mm diameter black globe
t_a is the air temperature.

Equipment used must be within specification. For example, if the globe size is incorrect or the air temperature is not shielded from radiation, this may have significant consequences for the outcome of the assessment. The following summarizes the specification for the sensors (see Figure 14.3).

The *natural wet bulb sensor* is cylindrical in shape (6 ± 1 mm diameter and 30 ± 5 mm long), with a measuring range of 5–40 °C and accuracy of ± 0.5 °C. The support of the sensor is 6 mm in diameter and a clean white wick of high water absorbent material (e.g. cotton) covers (as a sleeve fitted with precision) the whole of the sensor and 20 mm of the support.

The *globe temperature* is the temperature at the centre of a thin, matt black globe (mean emission coefficient of 0.95) with a measuring range of 20–120 °C with an accuracy of ± 0.5 °C to 50 °C and ± 1 °C to 120 °C. It is important that the globe is of 0.15 m in diameter.

The *air temperature sensor* should be shielded from the effects of radiation by a device that does not restrict air circulation. It should measure over the range of 10–60 °C with an accuracy of ± 1 °C.

The *WBGT* value used in the standard is a weighted average, over time and space, and is measured over a period of maximum heat stress. The weighting for spatial variation is given by:

$$WBGT = \frac{WBGT \text{ head} + 2 \times WBGT \text{ abdomen} + WBGT \text{ ankles}}{4}.$$

Table 14.1 WBGT reference values

Metabolic rate (W m^{-2})	WBGT reference value			
	Acclimatized (°C)		Not acclimatized (°C)	
Resting $M < 65$	33		32	
$65 < M < 130$	30		29	
$130 < M < 200$	28		26	
$200 < M < 260$	25	(26)*	22	(23)*
$M > 260$	23	(25)*	18	(20)*

Source: ISO 7243 (1989).

Notes
The values given have been established allowing for a maximum rectal temperature of 38 °C for the persons concerned.
* Figures in brackets refer to sensible air movement; figures without brackets refer to no sensible air movement.

For time variations (e.g. in metabolic rate, *WBGT*, globe temperature) a time-weighted average is taken over a period of work/resting of one hour. This is calculated from the beginning of a period of work.

The *WBGT* value of the hot environment is compared with a *WBGT* reference value, allowing for a maximum rectal temperature of 38 °C (Table 14.1).

ISO 7933: hot environments – analytical determination and interpretation of thermal stress using calculation of required sweat rate

This standard specifies a rational method for assessing hot environments by calculating and interpreting required sweat rate (SW_{req}). The SW_{req} index is a development of the heat stress index (HSI – Belding and Hatch, 1955) and of the index of thermal strain (ITS – Givoni, 1976). It is derived from the work of Vogt *et al.* (1981) in the CNRS laboratories in Strasbourg, France. During the development of the standard, a number of investigations were carried out into its validity and practical use (e.g. Wadsworth and Parsons, 1986; Parsons, 1987). In particular, an extensive programme of work was undertaken by the European Iron and Steel community, involving researchers from many European countries (CEC, 1988). The results of these studies, involving both laboratory and industrial investigations, led to significant modifications to the proposed standard and it was eventually published in 1989.

Measurement of the hot environment in terms of air temperature, mean radiant temperature, humidity and air velocity, and estimates of factors relating to clothing, metabolic rate and posture, are used to calculate the heat exchange between a standard person and the environment. This allows the calculation of the required sweat rate (for the maintenance of the thermal equilibrium of the body) from the following equation:

$$E_{req} = M - W - C_{res} - E_{res} - C - R \qquad (14.3)$$

and

$$SW_{req} = \frac{E_{req}}{r_{req}}, \qquad (14.4)$$

where

M = metabolic power
W = mechanical power
C_{res} = respiratory heat exchange by convection
E_{res} = respiratory heat exchange by evaporation
K = heat exchange on the skin by conduction
C = heat exchange on the skin by convection
R = heat exchange on the skin by radiation
E_{req} = required evaporation for thermal equilibrium
SW_{req} = required sweat rate for thermal equilibrium
r_{req} = evaporation efficiency at required sweat rate.

Metabolic and mechanical power are estimated, although W is often taken as zero if detailed information about the task is not known. They can be determined using methods provided in ISO 8996. K is regarded as having negligible effect and the following equations are used to calculate the remaining terms. Table 14.2 gives a description of terms used.

$$C_{res} = 0.0014M(35 - t_a) \qquad (14.5)$$
$$E_{res} = 0.0173M(5.624 - P_a) \qquad (14.6)$$
$$C = h_c F_{cl}(t_{sk} - t_a) \qquad (14.7)$$
$$R = h_r F_{cl}(t_{sk} - t_r), \qquad (14.8)$$

where

$$w = \frac{E}{E_{max}}$$
$$r = 1 - \frac{w^2}{2}$$
$$h_c = 2.38|t_{sk} - t_a|^{0.25} \quad \text{for natural convection}$$
$$h_c = 3.5 + 5.2 \, var \quad \text{for } var < 1 \, \mathrm{m\,s^{-1}}$$
$$h_c = 8.7 \, var^{0.6} \quad \text{for } var > 1 \, \mathrm{m\,s^{-1}}$$

Table 14.2 Description of terms used in ISO 7933 (1989)

Symbol	Term	Units
M	metabolic power	$W\,m^{-2}$
W	mechanical power	$W\,m^{-2}$
C_{res}	respiratory heat loss by convection	$W\,m^{-2}$
E_{res}	respiratory heat loss by evaporation	$W\,m^{-2}$
K	heat exchange on the skin by conduction	$W\,m^{-2}$
C	heat exchange on the skin by convection	$W\,m^{-2}$
R	heat exchange on the skin by radiation	$W\,m^{-2}$
E	heat flow by evaporation at skin surface	$W\,m^{-2}$
E_{req}	required evaporation for thermal equilibrium	$W\,m^{-2}$
SW_{req}	required sweat rate for thermal equilibrium	$W\,m^{-2}$
w	skin wettedness	ND
w_{req}	skin wettedness required	ND
r_{req}	evaporative efficiency at required sweat rate	ND
t_a	air temperature	°C
P_a	partial vapour pressure	kPa
h_c	convective heat transfer coefficient	$W\,m^{-2}\,K^{-1}$
F_{cl}	reduction factor for sensible heat exchange due to the wearing of clothes	ND
t_{sk}	mean skin temperature	°C
h_r	radiative heat transfer coefficient	$W\,m^{-2}\,K^{-1}$
t_r	mean radiant temperature	°C
$p_{sk,s}$	saturated vapour pressure at skin temperature	kPa
R_t	total evaporative resistance of limiting layer of air and clothing	$m^2\,kPa\,W^{-1}$
E_{max}	maximum evaporative rate which can be achieved with the skin completely wet	$W\,m^{-2}$
v_{ar}	relative air velocity	$m\,s^{-1}$
v_a	air velocity for a stationary subject	$m\,s^{-1}$
σ	Stefan–Boltzmann constant, 5.67×10^{-8}	$W\,m^{-2}\,K^{-4}$
E_{sk}	skin emissivity (0.97)	ND
A_r/A_{du}	fraction of skin surface involved in heat exchange by radiation	ND
f_{cl}	ratio of the subject's clothed to unclothed surface area	ND
F_{pcl}	reduction factor for latent heat exchange	ND
h_e	evaporative heat transfer coefficient	$W\,m^{-2}\,kPa^{-1}$
I_{cl}	basic dry thermal insulation of clothing	Clo or $m^2\,°C\,W^{-1}$

$$\text{var} = v_a + 0.0052(M - 58)$$

$$h_r = \sigma \varepsilon_{sk} A_r/A_{Du} \frac{\left[(t_{sk} + 273)^4 - (t_r + 273)^4\right]}{(t_{sk} - t_r)} \tag{14.9}$$

$$F_{cl} = \frac{1}{(h_c + h_r)I_{cl} + \frac{1}{f_{cl}}}$$

(14.10)

$$f_{cl} = 1 + 1.97 I_{cl}$$

$$E_{max} = \frac{P_{sk,s} - P_a}{R_t}$$

$$R_t = \frac{1}{h_e F_{pcl}}$$

$$h_e = 16.7 h_c$$

$$F_{pcl} = \frac{1}{\left(1 + 2.22 h_c \left(I_{cl} - \frac{(1 - 1/f_{cl})}{(h_c + h_r)}\right)\right)}$$

(14.11)

$$t_{sk} = 30.0 + 0.093 t_a + 0.045 t_r - 0.571 v_a + 0.254 p_a + 0.00128 M - 3.57 I_{cl}.$$

This regression equation can be used for the following ranges for each individual parameter

t_a	22.9–50.6	(°C)
t_r	24.1–49.5	(°C)
P_a	0.8–4.8	(kPa)
v_a	0.2–0.9	(m s^{-1})
M	46.4–272	(W m^{-2})
I_{cl}	0.1–0.6	(Clo)
t_{sk}	32.7–38.4	(°C).

See Mairiaux *et al.* (1987). An approximation for t_{sk} can be made using 36 °C, although 'crude', this may be a more sensible value to use in many applications.

Predicted values for evaporation from the subject (E_p), sweat rate (SW_p) and skin wettedness (w_p) are determined for the standard subject by a method shown in Figure 14.4. Predictions are made taking into account required values (w_{req}, E_{req} and SW_{req}) and limit values (w_{max} and SW_{max}).

The required sweat rate is compared with the maximum limit values for skin wettedness (w_{max}) and sweat rate (SW_{max}) which can be achieved by persons. These are presented for acclimatized and non-acclimatized persons at work and rest (see Table 14.3).

In the case of no thermal equilibrium there will be heat storage and hence the body core temperature will rise. In terms of heat storage, limiting values are presented for warning and danger. They are also presented in terms of the maximum allowable water loss compatible with the maintenance of the hydromineral equilibrium of the body.

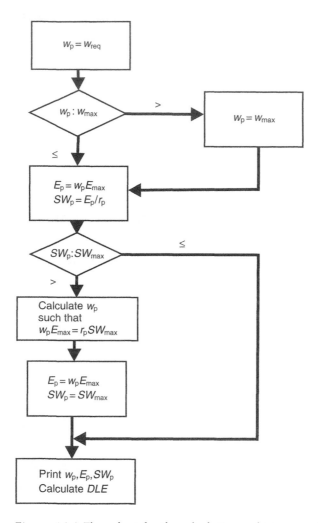

Figure 14.4 Flow chart for the calculation and interpretation of values used in
ISO 7933 (1989).

The predicted sweat rate can be determined from the required sweat rate
and the limit values. If the required sweat rate can be achieved by persons
and it will not cause unacceptable water loss, then there is no time limit due to
heat exposure, over an eight-hour shift. If this is not the case then allowable exposure times – duration limited exposures (*DLEs*) – are calculated
from the following equations.
When

$$E_{\mathrm{p}} = E_{\mathrm{req}} \text{ and } SW_{\mathrm{p}} < \frac{D_{\max}}{8} \qquad (14.12)$$

Table 14.3 Reference values for criteria of thermal stress and strain used in ISO 7933 (1989) for the analytical assessment of hot environments

Criteria	Non-acclimatized		Acclimatized	
	Warning	*Danger*	*Warning*	*Danger*
Maximum skin wettedness				
w_{max}:	0.85	0.85	1	1
Maximum sweat rate				
rest ($M < 65\,W\,m^{-2}$)				
SW_{max}: $\quad W\,m^{-2}$	100	150	200	300
$g\,h^{-1}$	260	390	520	780
work ($M > 65\,W\,m^{-2}$)				
SW_{max}: $\quad W\,m^{-2}$	200	250	300	400
$g\,h^{-1}$	520	650	780	1040
Maximum heat storage				
Q_{max}: $\quad W\,h\,m^{-2}$	50	60	50	60
Maximum water loss				
D_{max}: $\quad W\,h\,m^{-2}$	1000	1250	1500	2000
g	2600	3250	3900	5200

then $DLE = 480\,min$ and SW_p can be used as a heat stress index. If the above conditions are not satisfied then:

$$DLE_1 = \frac{60Q_{max}}{(E_{req} - E_p)} \qquad (14.13)$$

$$DLE_2 = \frac{60D_{max}}{SW_p}. \qquad (14.14)$$

DLE is the lower value of DLE_1 and DLE_2. If DLE is determined by DLE_1 (i.e. heat storage) then the worker must rest until there is no longer a risk of heat stress. If DLE is determined by DLE_2 (i.e. dehydration), then no further exposure is allowed during the day.

If workers carry out a number of types of work during the day and under different thermal conditions, ISO 7933 provides a method for assessing sequences of 'tasks' (including work and rest) based on a time weighting of E_{req} and E_{max} values. An example of the use of ISO 7933 in practical application is provided later.

If E_{max} is negative (i.e. condensation will occur) or if exposure time is short (i.e. < 30 min), then the method used in ISO 7933 is inappropriate. Physiological measurements on individuals should be taken according to ISO 9886.

A computer program is provided to allow ease of calculation and efficient use of the standard. This rational method of assessing hot environments allows identification of the relative importance of different components of the thermal environment and hence can be used in environmental design.

Revision of ISO 7933

A number of laboratory and field studies had identified limitations in ISO 7933 (Kampmann *et al.*, 1999; Wadsworth and Parsons, 1986; Haslam and Parsons, 1989b). To improve the method, a series of studies was conducted by a European Research Programme (BIOMED) which led to the Predicted Heat Strain method (see Chapter 10). Limitations which were identified and improvements made included: the prediction of skin temperature; heat transfer through clothing, including ventilation; the increase in core temperature linked to activity; the prediction of sweat rate in very humid conditions; limiting criteria (alarm and danger levels); and maximum water loss allowed (Malchaire, 1999). The research led to a proposal for a new ISO 7933, 'Ergonomics of the thermal environment – Analytical determination and interpretation of heat stress using calculation of Predicted Heat Strain.' The scope, principles and general methodology are similar to the original standard. However the detailed method (see Chapter 10) and the consequent outcome of assessment can be significantly different. A computer programme listing is provided in an annex to the standard.

ISO 7730: moderate thermal environments – determination of the *PMV* and *PPD* indices and specification of the conditions for thermal comfort

This standard considers whole-body thermal sensation and local thermal discomfort caused by draughts. It is based on the predicted mean vote (*PMV*) and the predicted percentage of dissatisfied (*PPD*) indices (Fanger, 1970), and more recent work concerning draughts (Olesen, 1985a; Fanger *et al.*, 1989). These methods have been widely 'marketed' throughout the world and integrating instruments have been developed which allow the direct measurement of the indices (Olesen, 1982) (see Figure 14.5).

The *PMV* is the predicted mean vote of a large group of persons, on the following thermal sensation scale, if they had been exposed to the thermal conditions under assessment.

+3   hot
+2   warm
+1   slightly warm
 0   neutral
−1   slightly cool
−2   cool
−3   cold

The *PMV* is calculated from the air temperature, mean radiant temperature, humidity and air velocity of the environment and estimates of metabolic rate and clothing insulation. It is derived from a heat balance equation

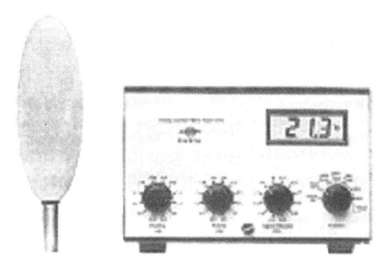

Figure 14.5 Thermal comfort metre type 1212.

for the human body combined with empirically determined equations which define sweat rates and mean skin temperatures which are within comfort limits. The equation is provided in Chapter 8. The *PPD* is derived from the following equation:

$$PPD = 100 - 95 \exp\left[-\left(0.03353 \times PMV^4 - 0.2179 \times PMV^2\right)\right].$$

It is based on data from the exposure of 1300 subjects to various thermal environments and is shown in Figure 14.6.

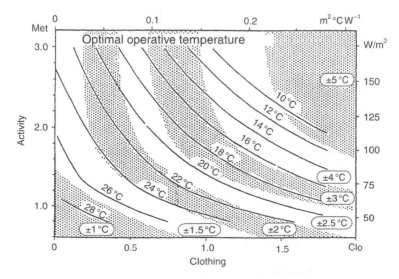

Figure 14.6 Optimum temperature for minimal dissatisfaction caused by whole-body discomfort.

The draught rating (DR) is expressed as the percentage of people to be bothered by draughts, where

$$DR = (34 - t_a)(v - 0.05)^{0.62}(0.37vT_u + 3.14),$$ (14.15)

where

t_a = local air temperature (°C)
v = is local mean air velocity (°C)
T_u = is local turbulence intensity (per cent) defined as the ratio of the standard deviation of the local air velocity to the local mean air velocity.

The DR model is based on experiments on 150 human subjects for the following range of conditions.

$t_a : 20\text{--}26\,°C$

$v : 0.05\text{--}0.4\,m\,s^{-1}$

$T_u : 0\text{--}70\%$

It applies to people performing mainly sedentary activity with whole-body sensations close to neutral. Risk of draught is lower at higher activities and if people are warmer than neutral.

Thermal comfort is defined in the standard as 'that condition of mind which expresses satisfaction with the thermal environment'. Dissatisfaction may be caused by whole-body or local discomfort. An annex (included, but

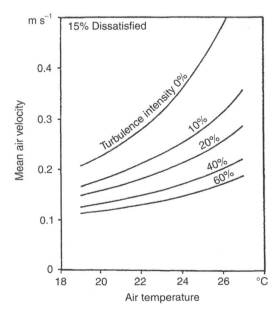

Figure 14.7 Dissatisfaction (15 per cent) caused by draughts.

labelled as not part of the standard) provides guidance in terms of levels of dissatisfaction. This includes dissatisfaction caused by whole-body discomfort (Figure 14.6) and by draughts and other local effects (see Figure 14.7 and Table 14.4). Tables of metabolic rate and clothing insulation values are included. For more detailed estimates one can use ISO 8996 and ISO 9920.

A computer program is provided to allow ease of calculation and efficient use of the standard.

Revision of ISO 7730

The proposed revision of ISO 7730 extends the information provided and orientates the standard towards application by including design criteria (Olesen and Parsons, 2002). There is no proposed change to the PMV/PPD index nor draught rating. The proposed revised standard (ISO 7730: Ergonomics of

Table 14.4 Guideline recommendations (i.e. in an Annex and not an integral part of the Standard) for the avoidance of local thermal discomfort

Parameter		Recommendations for light sedentary activity	
		Winter conditions (heating period)	*Summer conditions (cooling period)*
Operative temperature	(°C)	20 ± 2	24 ± 1.5
Vertical air temperature difference between head and ankles (1.1 m and 0.1 m above the floor)	(°C)	<3	<3
Mean air velocity	(m s^{-1})	Less than specified from data given in the Standard (Figure 14.7)	
Relative humidity	(%)	30–70	30–70
Surface temperature of the floor (Floor heating systems may be designed for 29 °C)	(°C)	19–26	
Radiant temperature asymmetry	(°C)		
From windows or other cold vertical surfaces (in relation to a small vertical plane 0.6 m above the floor)		<10	
From a warm heated ceiling (in relation to a small horizontal plane 0.6 m above the floor)		<5	

Source: ISO 7730 (1994).

the thermal environment – Analytical determination and interpretation of thermal comfort using calculation of the PMV and PPD indices and local thermal comfort) has identical scope to the previous standard and includes modification and additions in terms of increased air speed (to offset an increase in t_a), adaptation (although discussed, no model is proposed – see Chapter 9), humidity and the long term evaluation of general thermal comfort conditions.

A weighting factor (wf) is defined as

$$wf = \frac{PPD_{\text{actual } PMV}}{PPD_{PMV \text{ limit}}}. \tag{14.16}$$

The weighting time is summated for a characteristic working period over one year.

Warm period : $\sum wf \times$ time hours where $PMV > PMV_{\text{limit}}$

Cold period   : $\sum wf \times$ time hours where $PMV < PMV_{\text{limit}}$

The weighting time may be used for the evaluation of long term comfort conditions. An acceptable weighting time of around 100–150 h may be specified (Olesen, 2001). A significant proposed change to the standard is the introduction of categories of space. High quality space (Category A) would be maintained between a PMV value of ± 0.2; Category B, ± 0.5; and Category C, ± 0.7.

ISO 7726: thermal environment – instruments and methods for measuring physical quantities

This standard provides definitions of the basic parameters (air temperature, mean radiant temperature, humidity, air velocity) and derived parameters (natural wet bulb temperature, globe temperature). It also provides methods of measurement and specifications of measuring appliances (see Tables 5.7 and 14.5).

No specific instrument is standardized, only specifications. The standard can therefore serve as a guide to manufacturers of instruments as well as specifying measuring requirements, in a contract between investigator and a client for example. An updated version of the standard includes plane radiant temperature, surface temperature and directional radiation.

ISO 8996: ergonomics – determination of metabolic heat production

This standard provides methods and data for estimating the metabolic heat production of humans. It provides fundamental support to other ISO Standards in the series, for assessing hot, moderate, and cold environments (e.g. Parsons, 2001b). The standard can also be used for the assessment of working practices, the metabolic cost of specific jobs or sports activities, the total metabolic cost of activity, and for other applications.

Table 14.5 Characteristics of measuring instruments (see also Table 5.7)

Temperature	Measuring range	Accuracy	Response time (90%)	Comment
Natural wet bulb (t_{nw})	Not recommended for comfort, 5–40 °C for stress	Not recommended for comfort; ±0.5 °C for stress	Not recommended for comfort; for stress, value to be specified as characteristic of the measuring appliance	Characteristics of the sensor prescribed
Globe (t_g)	Use not recommended as a comfort index; 20–120 °C for stress	Not for comfort; for stress: 20–50 °C ±0.5 °C; >50–120 °C ±1 °C	The shortest possible; value to be specified as characteristic of the measuring appliance	Characteristics of the sensor prescribed; the globe temperature may also be used in the cold, moderate and hot temperature zone for estimating mean radiant temperature
Wet globe (t_{wg})	Use not recommended as a comfort index; 0–80 °C for stress	Not for comfort; for stress: ±0.5 °C	The shortest possible; value to be specified as characteristic of the measuring appliance	The measuring accuracy for the sphere temperature for determining t_r is not necessarily the same as that for measuring the globe temperature as a derived value; characteristics of the sensor prescribed

The methods are derived from a number of studies concerned with determining metabolic rate, and some are well established. The data are mainly from the work of Spitzer and Hettinger (1976) in the laboratories of the University of Wupertaal, Germany.

Six methods of estimation are presented (Table 14.6) in three types. The first type is by use of tables, where estimates are provided based on a description of activity (see Tables 6.2–6.5). These range from general description (e.g. low, high, etc.) to specific descriptions of occupations (e.g. bricklayer) and methods of summing components of tasks (e.g. basal metabolic rate plus posture component plus movement component, etc.). Examples of the components methods involving the use of tables are provided in Tables 14.7, 14.8 and 14.9.

Table 14.6 Six methods for estimating metabolic heat production

Level	Method		Accuracy	Inspection of the workplace
I	A	Classification according to kind of activity	Rough information where the risk of error is very great	Not necessary
	B	Classification according to occupation		Information on technical equipment, work organization
II	A	Use of tables of group assessment	High error risk; accuracy ±15%	Time study necessary
	B	Use of estimation tables for specific activities		
	C	Use of heart rate under defined conditions		Not necessary
III		Measurement	Risk of errors within the limits of the accuracy of the measurement and of the time study; accuracy ±5%	Time study necessary

Source: ISO 8996 (1990).

Table 14.7 Metabolic rate by group assessment

Metabolic rate	= Basal	+ Posture	+ Work	+ Motion
Example: Raking leaves on a lawn				
	= Basal	+ n/a	+ Light two arm work	+ Walking
	= 44	+ 0	+ 65	+ 60
	= 169 W m^{-2}			

Source: ISO 8996 (1990).

Table 14.8 Metabolic rate by group assessment

Metabolic rate	= Basal	+ Posture	+ Work	+ Motion
Example: Planing planks by hand				
	= Basal	+ standing stooped	+ Light trunk work	+ n/a
	= 44	+ 30	+ 125	+ 0
	= 199 W m^{-2}			

Source: ISO 8996 (1990).

Table 14.9 Metabolic rate by group assessment

Metabolic rate	= Basal	+ Metabolic rate/m s^{-1}	× speed m s^{-1}
Example: Walking upstairs			
	= Basal	+ W m^{-2}/m s^{-1}	× speed upstairs
	= 44	+ 1725	× 0.22
	= 424 W m^{-2}		

Source: ISO 8996 (1990).

Table 14.10 Estimation of metabolic heat production using heart rate

For the range 120 to (HR_{max} − 20):
    $HR = HR_0 + RM(M − BM)$bpm
where
    HR = heart rate
    M = metabolic rate
    BM = basal metabolic rate
    RM = increase in heart rate per unit of metabolic rate. This can be determined
          experimentally for individual subjects or groups performing relevant tasks
    HR_0 = heart rate at rest (in prone position) under thermo-neutral conditions

Source: ISO 8996 (1990).

The second type of method is by the use of heart rate. The total heart rate is regarded as a sum of several components and in general is linearly related to the metabolic heat production for heart rates above 120 beats per minute. This method is shown in Table 14.10.

The third type of method is to calculate the metabolic heat production from measures of oxygen consumption and carbon dioxide production during activity and recovery. An example of this type of method is shown in Table 14.11.

The methods and data provided in the standard are comprehensive so the implementation of the standard into a computer system is beneficial (e.g. Parker and Parsons, 1990). Although one of the most extensive databases

Table 14.11 Estimation of metabolic heat production from the collection and analysis of expired gases

---

Measure and record:
(a) Subject's sex, weight, height and age
(b) Method of measurement
(c) Duration of measurement
    partial method: main period
    integral method: main and subsequent period
(d) atmospheric pressure
(e) volume of expired air
(f) temperature of expired air
(g) fraction of oxygen in expired air
(h) fraction of carbon dioxide in expired air

$$EE = (0.23RQ + 0.77) \times 5.88 : \quad RQ = \frac{VCO_2}{VO_2}$$

$$M = EE \times VO_2 \times \frac{1}{A_D},$$

where
    EE = Energy equivalent in $W\,h\,1O_2^{-1}$
    RQ = Respiratory Quotient (ND)
    VO_2 = Oxygen consumption in litres of O_2 per hour
  VCO_2 = Carbon dioxide production in litres of CO_2 per hour
    A_D = Dubois body surface area in m^2.

Notes
VO_2 and VCO_2, above, are values at standard conditions of temperature and pressure, dry (STPD): t = 0 °C, P = 101.3 kPa, dry gas. Additional calculation of these standard values is required from those values measured (see Chapter 6).

available on this topic, the inherent errors in use of the methods and derivation of the data should be taken into account (see Parsons and Hamley, 1989; Parsons, 2001b). The standard provides guidance on the level of accuracy one could expect with each method (see Table 14.6).

All metabolic rate values are provided in units of $W\,m^{-2}$; i.e. watts per square metre of the body surface area, and are based on the standard man (see Table 14.12). They should be corrected for non-standard individuals or populations; for example, this will be particularly relevant when the activity

Table 14.12 Standard persons used when estimating values of metabolic rate presented in ISO 8996 (1990)

		Male	*Female*
Height	(m)	1.7	1.6
Weight	(kg)	70	60
Surface area	(m^2)	1.8	1.6
Age	(yrs)	35	35
Basal metabolic rate	(W m^{-2})	44	41

Table 14.13 Example calculation of average metabolic rate over a period where a number of tasks and activities are carried out

	Duration (s)	*Metabolic rate* ($W m^{-2}$)
Walk in factory $4 km h^{-1}$	35	165
Carry sack of 30 kg	50	250
Standing	25	70

Source: ISO 8996 (1990).

Note
Time-weighted average $= 200 W m^{-2}$.

involves tasks such as walking upstairs (overcoming gravity) where human body weight will be important. For conditions where the physical level of work varies, a time-weighted average procedure is recommended (an example is shown in Table 14.13).

Revision of ISO 8996

The revision of ISO 8996 is in its early stages of development. Current proposals use the format proposed in ISO 15265 (see below). The types of method for estimating metabolic rate are replaced by four levels. Level I (Screening) includes classification according to occupation and activity; Level II (Observation) uses tables of group assessment and specific activities; Level III (Analysis) uses heart rate under defined conditions; and Level IV (Expertise) uses the measurement of oxygen consumption, the double-labelled water method and direct calorimetry. The description of the methods for determining overall metabolic rate in a combination of activities using activity recording and time-weighted average is enhanced by a diary method. Consideration was given to a method for estimating average metabolic rate from concentration of carbon dioxide in a room, however little practical information is available and a method is not included at this stage.

ISO 9886: evaluation of thermal strain by physiological measurements

In extreme environments, or for other reasons such as research, it may be necessary to measure the physiological strain on humans exposed to thermal environments. This standard describes methods for measuring and interpreting body core temperature, skin temperatures, heart rate and body mass loss.

Annex I of the standard presents a comparison of the different methods concerning their field of application, their technical complexity, their discomfort, and the risks that measurement might involve. Measurement methods are described in Annex II and limit values are proposed in Annex III

Table 14.14 Physiological measures considered by ISO 9886 (1992)

Physiological response	Measure considered
Body core temperature	Oesophageal temperature
	Rectal temperature
	Gastro-intestinal temperature
	Oral (mouth) temperature
	Tympanic temperature
	Auditory canal temperature
	Urine temperature
Skin temperature	Local skin temperature
	Mean skin temperature
	ISO four-point method
	ISO eight-point method
	ISO 14-point method
Heart rate	The partial method is used to identify the component due to thermal stress
Body mass loss	Due to respiration and sweating
	Take account of body inputs (food and drink) and body outputs (urine and stools)

of the standard. Much of the information in the standard is similar to that presented in Chapter 5 (see also Table 14.14).

The principle of the standard is to present information to allow the informed selection and correct application and interpretation of physiological measures.

ISO 9920: estimation of the thermal characteristics of a clothing ensemble

This International Standard presents methods for estimating the thermal characteristics (resistance to dry heat loss and evaporative heat loss) of a clothing ensemble based on values for known garments, ensembles and textiles. It does not take into account the influence of rain and snow on the thermal characteristics and special protective clothing (water-cooled suits, ventilated suits, heated clothing) is not considered.

The main part of the standard is a large database of clothing insulation values that have been measured on heated manikins. The data is mainly from the work of McCullough *et al.* (1985) and Olesen *et al.* (1982). Values are provided for dry thermal insulation and resistance to water diffusion. Dry insulation is given in terms of basic thermal insulation (I_{cl}). Resistance of clothing to water vapour diffusion is provided in terms of the (non-dimensional) permeability index, i_m. The i_m value ranges from around 0.5 for a nude person to around 0.2 for impermeable like clothing. A typical value would be around 0.4 (see Chapter 7).

The tables of thermal insulation values of clothing are comprehensive; an example is provided in Table 14.15. Values for total ensembles are supplied as well as for dry insulation values for individual garments (I_{clu}) which make up ensembles. If the thermal insulation value of a total ensemble is not provided in the tables, then a summation procedure is provided for estimating the insulation provided by the ensemble from the I_{clu} values.

The summation procedure to obtain I_{cl} from the insulation values of individual garments (I_{clu}) is:

$$I_{cl} = \Sigma I_{clu}.$$

The I_{clu} values are effective thermal insulation values for garments. That is they do not account for the increase in surface area for heat exchange over the body due to clothing. If basic thermal insulation values for garments (I_{cli}) are known then

$$I_{cl} = 0.82\Sigma I_{cli}.$$

For example in ISO 7730, thermal insulation values are provided for garments in terms of I_{cli}. As well as information about garment style (e.g. long sleeves, short sleeves, etc.), fabric type and thickness are also supplied.

The thermal insulation of an individual garment (Clo) may also be estimated from the area of the body covered using

$$I_{clu} = 0.61 \times 10^{-2} A_{cov}.$$

When the thickness of the fabric (H_{fab}) is also known then

$$I_{clu} = 0.43 \times 10^{-2} A_{cov} + 1.4 H_{fab} \times A_{cov},$$

Table 14.15 Example of clothing insulation values for a clothing ensemble

Garment	No.	Type	Weight (g)	f_{cl}	I_{clu} (Clo)	*$m^2 \, °C \, W^{-1}$
Underpants	23	Briefs	80		0.04	0.006
Undershirt	31	T-shirt	180		0.10	0.016
Coverall	120	Work	890		0.51	0.079
Overtrousers	191	Heat protective felt	1300		0.33	0.051
Over Jacket	193	Heat protective felt	1620		0.42	0.065
Socks	254	Ankle length	61		0.02	0.003
Shoes	255	Suede, rubber soles	499		0.02	0.003
Total ensemble	489	Heat protective clothing	4630	1.50	1.55	0.240
					(I_{cl})	

Source: ISO 9920 (1995).

Notes
* Value calculated from Clo values and *not* value for material.
  Values for ensemble are measured. Not by Summation method.

where

A_{cov} = body surface area covered (%)
H_{fab} = thickness of fabric (m) measured according to ASTM D1777 using a 7.5 cm diameter presser foot and 69.1 N m^{-2} pressure.

The estimate of f_{cl} (ratio of clothed to nude surface area) is

$$f_{cl} = 1 + 0.31 I_{cl}.$$

It is noted that the pumping effect may reduce the thermal insulation by between 5 per cent and 50 per cent. A typical reduction in thermal insulation of 20 per cent is recommended as an estimate of the effects of wind penetration. This emphasizes that the I_{cl} values provided in the standard are very much a starting point for determining the insulation provided by clothing in practical applications.

The evaporative resistance of clothing (R_T) is the sum of the resistance of the external air layer (R_a) and the clothing layer (R_{cl}) and can be estimated from I_{cl} for 'normal permeable clothing' by

$$R_T = 0.06 \left[\frac{1}{h_c} + 2.22(I_{cl} - I_a)\left(1 - \frac{1}{f_{cl}}\right) \right] \text{ m}^2 \text{ kPa W}^{-1}, \tag{14.17}$$

where R_T and I_{cl} are as defined above and

$$I_a = \frac{1}{h_r + h_c}$$

or

$$R_T = \frac{I_T}{i_m L}$$
$$= \frac{0.06}{i_m}\left(\frac{I_a}{f_{cl}} + I_{cl}\right),$$

where I_{cl} is in m^2 °C W^{-1}. For most normal clothing i_m has a value of about 0.38. So,

$$R_T = 0.16\left(\frac{I_a}{f_{cl}} + I_{cl}\right) \tag{14.18}$$

and an approximation for R_{cl} gives

$$R_{cl} = 0.18 \times I_{cl} \text{ m}^2 \text{ kPa W}^{-1}. \tag{14.19}$$

The data provided in this standard are the most comprehensive available and have developed in parallel with the development of the standard. An interesting point however is that the database became so large that it is difficult to use (e.g. see McCullough *et al.*, 1985). Parker and Parsons (1990) describe a computer-based system that allows efficient use of the standards.

Revision of ISO 9920

Revision of ISO 9920 is at an early stage of development. Proposals have included: increasing the database of clothing (to include cold weather clothing, to increase the range of clothing worn in moderate environments, etc.); introducing equations for clothing ventilation effects; rationalising description of clothing (total versus intrinsic for vapour resistance); increasing data concerning vapour permeation properties; increase sections on thermal manikins; and splitting the standard into two parts – one for clothing data and the other presenting methods of determining clothing properties.

Technical Report, ISO TR 11079: evaluation of cold environments – determination of required clothing insulation, *IREQ*

This Technical Report is to propose methods for the assessment of cold environments, encourage experimental work to validate and elaborate methods, and to identify research needs. After the Technical Report had been available for three years a decision was to be taken as to whether it should be developed as an ISO Standard. That decision has been taken and a full International Standard is under development, based upon the methods in this Technical Report.

The report is concerned with local effects on the body (e.g. cold hands) and more general whole-body effects. The whole-body effects are assessed using a rational, heat balance equation, method and the calculation of required clothing insulation (*IREQ*) based on Holmér (1984).

The clothing insulation required for thermal equilibrium (*IREQ*$_{min}$), and that required for thermal comfort (*IREQ*$_{neutral}$), are calculated by satisfying the following equations:

$$IREQ = \frac{t_{sk} - t_{cl}}{M - W - H_{res} - E} \qquad (14.20)$$

and

$$M - W - H_{res} - E - R - C = 0, \qquad (14.21)$$

where

t_{sk} = mean skin temperature
t_{cl} = clothing surface temperature
M = metabolic energy production
W = rate of mechanical work
H_{res} = respiratory heat loss
E = evaporative heat loss
R = radiative heat loss
C = convective heat loss.

For persons who wear clothing insulation that is less than $IREQ_{min}$, there is a risk of progressive body cooling. If clothing is worn with an insulation of greater than $IREQ_{neutral}$, then there will be an increasing feeling of warmth. The interval between $IREQ_{min}$ and $IREQ_{neutral}$ is the clothing regulatory zone where persons will adjust clothing insulation. The $IREQ$ values are most conveniently calculated using a computer program. However guidance is provided in graphical form (see Figures 11.4 and 11.5).

The calculated $IREQ$ value can be used as a required clothing insulation value, for example, to allow the selection of clothing for work in a cold environment. It can also be used as a cold stress index. The higher the value of $IREQ$, at any given activity level, the greater is the cooling power of the environment.

If the $IREQ$ is used to select appropriate clothing, then it is emphasized that insulation provided by clothing is a dynamic property that varies with such factors as body posture, activity, moisture content and wind. If $IREQ$ cannot be met, then a procedure is presented for calculating maximal exposure times and required recovery times with available insulation.

Local cooling of the hands, head and feet, is also considered by the Technical Report. It is noted that knowledge is incomplete in this area. For indoor environments, draughts and lower limits for hand skin temperatures are discussed. For outdoor environments, the Wind Chill Index (WCI) and the chilling temperature (t_{ch}) are used as indices (see Chapter 11). A procedure for the evaluation of cold environments, based on methods proposed in the Technical Report, is provided (see Figure 11.6).

The usefulness of $IREQ$, both as a cold stress index and as a specification for required clothing, has yet to be established. For example, it is not clear how resultant thermal insulation should be determined. In addition, the criteria upon which $IREQ$ is based are debatable. Maybe a mean skin temperature of 30 °C is not unusual nor a sign of cold stress, as implied by $IREQ$. A drop in internal body temperature however is unusual for clothed persons in air, and criteria based upon this may be of limited practical value. Behavioural responses of persons in cold environments are very dominant. A more pragmatic approach may be to concentrate on the effects of cold upon extremities of the body (e.g. hands and feet) and on exposed skin. These observations are generally supported by the work of Aptel (1988), who investigated workers in freezer rooms. O'Leary and Parsons (1994) evaluated the $IREQ$ index and found $IREQ_{min}$ a useful starting point for clothing selection and design and as a basis for developing working practices for cold environments.

Revision of ISO TR 11079

ISO TR 11079 was established as a Technical Report and not a full International Standard because the $IREQ$ (clothing insulation required) index had not been validated in practice. It has now been widely used and accepted as a useful index for the assessment of cold environments and it has been agreed that a full international standard will be produced. The revised standard (ISO 11079: 'Ergonomics of the thermal environment – Determination and interpretation of cold stress when using required clothing insulation (*IREQ*) and local effects') is at an early stage of development and is currently similar to the original Technical Report. There is a possibility of integrating this standard with a standard for working practices in cold environments (ISO 15743 – see below). It is likely that further information will be provided on selection of appropriate clothing from *IREQ* values and the inclusion of the effects of clothing ventilation.

ISO 10551: assessing the influence of the thermal environment using subjective judgement scales

This standard presents the principles and methodology behind the construction and use of subjective scales, and provides examples of scales that can be used to assess thermal environments (see Table 14.16).

A practical example and a discussion of methods of data analysis are provided. The principle of the standard is to provide background information to allow ergonomists to construct and use subjective scales as part of the assessment of thermal environments.

ISO 11399: ergonomics of the thermal environment: principles and application of International Standards

This standard provides background information to allow the correct, effective, and practical use of International Standards concerned with the ergonomics of the thermal environment.

This includes a description of each International Standard and how they can be used together, a description of the underlying principles used in each International Standard, and a description of the underlying principles

Table 14.16 Subjective scales considered in ISO 10551 (1995)

Judgement	Example	Related to
Perceptual	How do you feel now? (e.g. hot)	Personal
Affective	How do you find it? (e.g. comfortable)	Thermal
Thermal preference	How would you prefer to be? (c.g. warmer)	State
Personal acceptance	Is the environment acceptable/unacceptable?	Environment
Personal tolerance	Is the environment tolerable?	

concerning the ergonomics of the thermal environment. The standard will aid in the selection and application of the standards.

Revision of ISO 11399

The revision of ISO 11399 is at an early stage of development. The current proposal is that the standard will take the same format as the present standard but will be updated to include a significant number of new standards. There is an opportunity to include a more extensive philosophy and description of use of standards but this has yet to be decided.

Example of a consent form for a volunteer subject in an ergonomic investigation involving exposure to heat or cold

IN CONFIDENCE

Name ....   Age............years   Sex: Male/Female

Normal medical adviser. Name ....

Address ....

....

1.   I am willing to participate as an experimental subject in the study of

....

to be conducted by ....

at..

2.   I have received an explanation of the nature and purpose of this study and of any risks to my health which are foreseen.

3.   I agree to provide accurate information about my health and to be medically examined if this is considered necessary. I agree that my normal medical adviser can provide information about my medical history to the authorized adviser to the study (independent medical officer). I understand that all information about my health will be treated in confidence.

4.   I agree to cooperate fully with the investigators and not knowingly to do anything which might invalidate the results.

5.   During the course of the investigation to which I am now giving my consent, I will not participate as a subject in any other study, without first informing the investigators and obtaining their approval, which may be withheld.

6.   I understand that I am free to withdraw my consent to participate in the study at any time without the need to give an explanation for my decision.

Signed..   Date..

Statement by investigator

In connection with the study described above, I have explained to............ the nature and purpose of the study and the foreseeable risks from participation in the study. I have explained that the decision to volunteer does not affect the right to compensation in the event of illness or injury.

Signed..   Date..

Figure 14.8 Subject consent form (ISO 12894, 2001).

ISO 12894: ergonomics of the thermal environment: medical supervision of individuals exposed to extreme hot or cold environments

This standard provides a method of screening of persons who are, or may be, exposed to hot or cold thermal environments. The method allows a decision to be made as to whether or not it is acceptable for a person to be exposed (screening) and also surveillance methods during exposure. Both laboratory and field investigations are included and guidance is provided on the degree and availability of medical expertize, ethical considerations, measurement methods and some guidance on physiological limit values.

The standard provides advice to those concerned with the safety of human exposures to extreme hot ($WBGT > 25\,°C$) or cold ($t_a < 0\,°C$) environments. The range provided is an approximation as effects will depend upon all six basic parameters including activity level and clothing. The standard is intended to assist those responsible for implementing the appropriate level of medical supervision in laboratory and occupational exposures, including scientific investigations and demonstrations for teaching purposes. A series of useful annexes to the standard provide: general principles (including an example of a subject consent form, Figure 14.8); medical effects (and treatment) of exposure to heat and cold; medical supervision, including medical examinations and examples of medical fitness and assessment questionnaires (see Figures 14.9 and 14.10); practical medical supervision of volunteers not normally exposed; and minimum levels of medical supervision for occupational exposures to extreme heat and cold.

ISO 13731: ergonomics of the thermal environment – vocabulary and symbols

ISO 13731 provides vocabulary, definitions, symbols and units that are used in standards concerned with the ergonomics of the thermal environment. The standard therefore provides a reference that can be of general use in the application of standards as well as to those who write standards in order to provide consistency. There are 261 terms provided with symbols and units.

STANDARDS UNDER DEVELOPMENT

ISO 13732: ergonomics of the thermal environment: Method for the assessment of human responses to contact with hot, moderate or cold surfaces

The purpose of this three-part International Standard is to present a method for predicting the thermal sensation and skin damage caused by contact between naked, and covered, skin and solid surfaces. The standard concerning hot surfaces (ISO 13732, Part 1) will be greatly influenced by European

Medical fitness assessment questionnaire prior to hot exposure

IN CONFIDENCE

This questionnaire should be completed prior to exposure to hot conditions. It is recommended that it is administered by someone with appropriate knowledge, for example, a nurse or trained laboratory scientist.

Please circle the appropriate responses

Name.. Sex: Male/Female   Date....../....../......

Age................years.

Present occupation.

1. Have you ever experienced episodes of fits or faints, or loss of consciousness (apart from concussion)? .... Yes/No

2. Do you suffer from diabetes mellitus or any general medical condition, for example, affecting the bowel or kidneys? .... Yes/No

3. Do you suffer from any disease of the heart or blood vessels, including high blood pressure? .... Yes/No

4. Do you suffer from any chest disease, e.g. asthma? .... Yes/No

5. Have you been treated for any serious mental ill health, or do you suffer from anxiety or depression? .... Yes/No

6. Do you suffer from any disease of the skin? If yes, please specify .... Yes/No

7. Have you had any treatment which reduces your ability to sweat, e.g. sympathectomy? .... Yes/No
   If yes, please specify ....

8. Are you currently taking any medication? .... Yes/No
   If yes, please specify ....

9. Have you ever had any illness due to the heat e.g. faints or collapse? .... Yes/No
   If yes, please state what happened and describe the circumstances
   ....
   ....

   Were you treated in hospital for this? .... Yes/No
   Has this, or anything similar, ever happened again? .... Yes/No

10. If female, is it possible that you are now pregnant?   Yes/No

NOTE If the reply to any question from 2–10 is yes, refer to the medical officer for advice. Note that prescribed or self-administered medication may impair normal physiological responses to the heat.

11. How often do you undertake exercise which leaves you out of breath?
    never/occasional/regular/daily

12. How often do you drink alcohol?   never/occasional/regular/daily

13. Please give any other relevant comments here;
    ....
    ....
    ....

BRIEF EXAMINATION DETAILS

1. Height ....cm
2. Weight ....kg
3. Assessment of height and weight:
   e.g. Body Mass Index (weight in kg/height in m^2) ....
   % greater or less than recommended weight for height ....
4. Resting heart rate ....bpm
   sitting/lying (record posture)
5. Resting blood pressure ....../...... mm Hg
   sitting/lying (record posture)

Figure 14.9 Medical fitness assessment form for exposure to hot conditions (ISO 12894, 2001).

Medical fitness assessment questionnaire prior to cold exposure

IN CONFIDENCE

This questionnaire should be completed prior to exposure to cold conditions. It is recommended that it is administered by someone with appropriate knowledge, for example, a nurse or trained laboratory scientist.

Please circle the appropriate response.

Name .... Date......./......./.......

Age .... years   Sex: Male/Female

Present occupation ....

1. Have you ever experienced episodes of fits or faints, or loss of consciousness (apart from concussion)?   Yes/No

2. Do you suffer from thyroid or other general medical disease, for example, diabetes mellitus?   Yes/No

3. Do you suffer from any disease of the heart or blood vessels, including high blood pressure?   Yes/No

4. Do you suffer from Raynaud's phenomenon, or other peripheral vascular disease?   Yes/No

5. Do you suffer from any chest disease, e.g. asthma or chronic bronchitis?   Yes/No

6. Have you been treated for any serious mental ill health, or do you suffer from anxiety or depression?   Yes/No

7. Do you suffer from any disease of the skin? If yes, please specify   Yes/No

8. Do you suffer from any rheumatism or diseases of the joints?   Yes/No

9. Do you currently take any medication?   Yes/No

   If yes, please specify ....

10. Have you ever experienced any general or local allergic reaction to cold?   Yes/No

    If yes, please specify ....

11. Have you ever suffered from any freezing or non-freezing cold injury?   Yes/No

    If yes, please specify ....

12. Have you ever suffered an episode of low body temperature requiring medical treatment?   Yes/No

    If yes, please specify ....

13. If female, is it possible that you are now pregnant?   Yes/No

NOTE   If the reply to any question from 2-13 is yes, refer to the medical officer for advice. Note that prescribed or self-administered medication may impair physiological responses to the cold.

14. Do you smoke cigarettes or other tobacco?   Yes/No

    If yes, please specify ....

15. How often do you drink alcohol?

    never/occasional/regular/daily

16. How often do you undertake exercise which leaves you out of breath?

    never/occasional/regular/daily

17. Please give any other relevant comments here.

....

....

BRIEF EXAMINATION DETAILS

1. Height .... cm

2. Weight .... kg

3. Assessment of height and weight:
   e.g. Body Mass Index (weight in kg/height in m^2) ....
   % greater or less than recommended weight for height ....

4. Resting heart rate .... bpm
   sitting/lying (record posture)

5. Resting blood pressure ..../........ mm Hg
   sitting/lying (record posture)

Figure 14.10 Medical fitness assessment form for exposure to cold conditions (ISO 12894, 2001).

Standards (EN 563, EN 13202) and the standard concerning cold surfaces (ISO 13732, Part 3) is being developed simultaneously with a European standard.

ISO 13732 Part 1: Hot surfaces are in final stages of development. It provides an integration of European Standards EN 563, which provides ergonomics data in terms of burn thresholds; EN 363: 94/PrA1 which is an amendment to EN 563 to include short contact times of less than one second; and EN 13202 which provides guidance for establishing surface temperature limit values which will be used for products such as machines, toys, cookers, etc. (see Chapter 13). The standard is applicable to the healthy skin of adults and is not restricted to a specific area of application such as machines. Examples are provided of how to apply the burn threshold data, provided in the standard, to set temperature limit values for products, take protective measures against burning, and to assess the risk of burning.

ISO TS 13732 Part 2: Human contact with surfaces at moderate temperature provides principles and methods for predicting the sensation and discomfort where parts of the body are in contact with moderate surface temperatures (approximately 10–40 °C). Empirical data, mathematical models of heat transfer and a physical model of a foot (DIN 52614) are used to present data for contact by the hand (e.g. on handrails) and feet (on floors). Data are also provided on responses to contact between a seated person and an electrically heated floor (see Chapter 13).

ISO 13732 Part 3: Cold surfaces is in final stages of development. It is based upon the results of a European research programme (ColdSurf – see Chapter 13) and describes methods for the assessment of the risk of pain, numbness and cold injury (frostnip), when a cold surface is touched by a bare finger or hand. It can be used to set temperature limits for other standards (e.g. for products such as shelving or tools used in cold stores) and for risk assessment. Threshold data in terms of surface temperatures are provided for pain (skin temperature below 15 °C), numbness (skin temperature below 7 °C), and freezing (skin temperature below 0 °C) in terms of idealized curves for different materials (wood, nylon, stone, steel, aluminium) and a range of contact durations for finger touching and hand gripping. The standard also provides guidance on risk interpretation and protective measures. Production of the idealised curves involved the use of the simple model of contact temperature (see Chapter 13) for which the thermal properties of materials are provided. Examples of cold risk assessment are provided in an annex to the standard.

ISO 14505: ergonomics of the thermal environment – thermal environment in vehicles, parts 1 to 3

ISO 14505 is a standard under development in three parts. Part 1 describes the principles of assessment; Part 2 describes the determination of equivalent temperature, which is a thermal index proposed for assessment of vehicle environments, and its interpretation; and Part 3 describes the evaluation of vehicle thermal environments using human subjects. Much of the content of the standards is derived from the results of European programmes of research in the assessment of vehicle environments (EQUIV and AUTOGLAZE – see Chapter 9).

ISO 14505 Part 1: Principles and methods for assessment of thermal stress provides guidance in the assessment of the thermal conditions in a vehicle during steady state conditions. It considers hot, moderate and cold environments in terms of principles of assessment and the appropriateness of methods used in existing International Standards (ISO 7730, ISO 7243, ISO 7933, ISO TR 11079 and others). Estimates of metabolic rate for driving are $70 \, \text{W m}^{-2}$ for good roads, $80 \, \text{W m}^{-2}$ for rugged roads and $90 \, \text{W m}^{-2}$ for off-road driving. It is noted that clothing is compressed by the vehicle which will reduce insulation, however the seat itself will provide additional insulation. Solar radiation is identified as an important source of heat load, particularly when the sun is low and the vehicle has a large area of glazing. Whole-body and local equivalent temperatures were considered as indicators of thermal sensation. Examples of vehicle evaluation are provided.

ISO 14405 Part 2: Determination of equivalent temperature, provides guidance on the assessment of the mean thermal comfort vote of occupants of vehicles using equivalent temperature (t_{eq}) defined as:

> The uniform temperature of the imaginary enclosure with air velocity equal to zero in which a person will exchange the same dry heat by radiation and convection as in the actual environment.

It is assumed that in the imaginary and actual (vehicle) environment, the person has the same posture, activity and clothing. Whole-body equivalent temperature describes levels of thermal neutrality and local equivalent temperature determines to what extent a body part falls within a range of acceptable levels of heat loss. The equivalent temperature is determined by measuring the heat transfer across the body between a heated body and the environment. A thermal manikin is usually used to measure heat transfers but heated discs can also be used. The t_{eq} has been found to be 'instrument sensitive' and some interpretation of results is required, particularly when there is heat gain from the environment to the heated manikin. Examples of existing thermal manikins and other measuring instruments and their specifications and calibration are provided in annexes to the standard.

Interpretation of equivalent temperature has yet to be written. The standard will provide an interpretation of results that will allow an assessment to be made of the effects of the vehicle environment on the vehicle occupants.

ISO 14505 Part 3: Evaluation of thermal comfort using human subjects is in an early stage of development. A brief description of objective, subjective and behavioural methods is provided in an introduction. However the standard specifically provides guidance for the assessment of vehicle comfort using subjective judgements. It is not restricted to any particular vehicle and it is applicable to both laboratory and field assessments. The principles of individual sensation scales are presented with guidance on the design of assessment methods and analysis and interpretation of results. A standardized method for the assessment of vehicle thermal environments is presented which can be used by vehicle manufacturers and others who wish to assess vehicle environments. A standardized questionnaire (subjective form) for collecting subjective data is also provided.

ISO TS 14515: ergonomics of the thermal environment: the application of International Standards for people with special requirements

This Technical Specification provides information concerning the needs of people with special requirements so that International Standards concerned with the assessment of the thermal environment can be appropriately applied for their benefit. It includes a description of responses and adaptation to thermal environments of people with special requirements (physical disabilities, aged) and a consideration of the application of standards for thermal comfort, heat stress and cold stress (see Chapter 9). A summary table of thermal disabilities is provided in an annex to the standard.

ISO 15265: ergonomics of the thermal environment: risk assessment strategy for the prevention of stress or discomfort in thermal working conditions

The standard was originally intended to provide practical and useable methodology for the assessment of hot environments. In its present form, it has been extended to also cover moderate and cold environments and provide an overall philosophy or strategy for assessment. It is therefore an opportunity to provide structure to the application of other standards. A three-stage process of assessment is proposed. Stage 1: Observation, is to be conducted by people who have knowledge of the working conditions but are not experts in ergonomics. They collect information, identify simple methods that can reduce risk and decide whether a more thorough analysis is required. A series of scales are provided related to the six basic parameters (a scale for each) and a workers opinion scale. A score sheet then allows a quantification of risk. Stage 2: Analysis, quantifies the thermal risk as a function of the mean and maximum climatic parameters, determines the optimum work organization and determines whether an expert is needed. Stage 3: Expert, provides a more detailed assessment using specific measurements, characterizes the exposure of the worker and identifies special prevention/control measures. Examples of prevention measures are provided in an annex.

ISO 15742: guidance on the determination of the combined effects of environmental components on human health, comfort and performance

ISO 15742 has made little progress since it was first accepted as a new project to produce an International Standard. While there is recognition that people occupy 'total' environments and are simultaneously and continuously stimulated by a range of environmental components, existing assessment methods consider environmental components separately and independently. This proposed report (probably a Technical Report) considers the interaction of thermal environments (hot, moderate, cold), air quality, noise and light on the health, comfort and performance of people. Human responses to air quality, for example, will be influenced by the thermal environment (e.g. air temperature, humidity). Discussions of the scope of the standard suggest the inclusion of other environmental components such as vibration. Although this standard is under development it may be some time before it reaches publication.

ISO 15743: working practices in cold environments

ISO 15743 is in an early stage of development. It is an application standard that builds upon the more fundamental standards ISO 11079 (Clothing insulation required and wind chill index) and ISO 12894 (Medical screening). Recent versions of the standard have adapted the risk assessment strategy presented in ISO 15265. The aim of the standard is to provide guidance on the assessment and management practices for cold workplaces. It provides methods for risk assessment, medical screening, methods of cold risk management practices and guidance on the role and application of existing International Standards. Examples are provided of assessment for both indoor and outdoor work.

STANDARDS IN THE USA

American Society of Heating, Refrigerating and Air-conditioning Engineers (ASHRAE)

ASHRAE produce two major documents concerned with human response to the thermal environment: a standard entitled *Thermal environmental conditions for human occupancy* which is ANSI/ASHRAE Standard 55; and *Thermal Comfort* (previously called *Physiological principles, comfort and health*) which is Chapter 8 of the ASHRAE handbook *Fundamentals*. These documents were reviewed and published every five years. However, recently the standard has been considered under continuous review.

ANSI/ASHRAE Standard 55 (1989b, 1992) specifies thermal environmental conditions for the comfort of healthy people. Acceptable ranges are provided for light mainly sedentary activity for both winter and summer conditions (e.g. clothing) in terms of operative temperature and humidity; see Figure 14.11. Corrections are provided for the effects of clothing, air movement, and activity. Environmental measurement methods are also presented which are similar to those presented in ISO 7726.

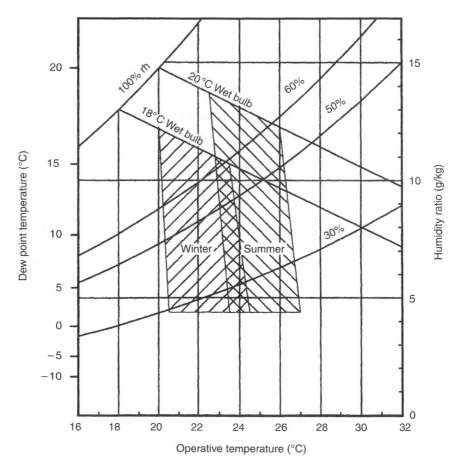

Figure 14.11 Acceptable conditions for persons conducting light activity. Source: ASHRAE (1997).

Chapter 8 of the ASHRAE handbook *Fundamentals* provides a comprehensive review of the fundamental background information upon which the Standard is based. Recent co-operation between members of ISO and ASHRAE has led to similar standards (ASHRAE 55 and ISO 7730 (revised)).

The American Conference of Governmental Industrial Hygienists (ACGIH)

The ACGIH publish annually a booklet entitled *Threshold Limit Values and Biological Exposure Indices* for practical use by trained industrial hygienists. Threshold limit values for physical agents in the work environment include those for heat stress and cold stress. 'These threshold limit values (*TLVs*) refer to levels of physical agents and represent conditions under which it is believed that nearly all workers may be repeatedly exposed day after day without adverse effect.' The *TLVs* are supplemented by much practical information and advice regarding working practices in hot and cold environments.

Heat stress

TLVs for heat stress are provided in terms of *WBGT* values (Table 14.17). It is no coincidence that *WBGT* is defined and measured in an identical way to that defined in ISO 7243 (see above) and that the *WBGT TLV* values are very similar to the reference values provided in ISO 7243 (see Table 14.1).

The values are based upon... 'the assumption that nearly all acclimatized, fully clothed workers with adequate water and salt intake should be able to function effectively under the given working conditions without exceeding a deep body temperature of 38 °C.' The values are for physically fit, acclimatized workers wearing light summer clothing (0.6 Clo). ACGIH (1996) provides WBGT correction factors for cotton coveralls (1.0 Clo) – take 2 °C off the WBGT limit value; winter work uniform (1.4 Clo) – take off 4 °C and for water barrier, permeable suits take off 6 °C. For special clothing or other significant deviations, an expert is required to provide assessment. Guidance is also provided in terms of measurement of the

Table 14.17 WBGT threshold limit value (°C)

Work-rest regimen	Work load		
	Light	Moderate	Heavy
Continuous work	30.0	26.7	25.0
75% work + 25% rest; each hour	30.6	28.0	25.9
50% work + 50% rest; each hour	31.4	29.4	27.9
25% work + 75% rest; each hour	32.2	31.1	30.0

Source: ACGIH (1996).

Table 14.18 Threshold limit values for cold stress

Air temperature – sunny sky			No noticeable wind		5 mph wind	
°C (approx.)		°F	Max. work period	No. of breaks	Max. work period	No. of breaks
1.	−26 to −28	−15 to −19	(Norm. breaks)	1	(Norm. breaks)	1
2.	−29 to −31	−20 to −24	(Norm. breaks)	1	75 min	2
3.	−32 to −34	−25 to −29	75 min	2	55 min	3
4.	−35 to −37	−30 to −34	55 min	3	40 min	4
5.	−38 to −39	−35 to −39	40 min	4	30 min	5
6.	−40 to −42	−40 to −44	30 min	5	Non-emergency work should cease	
7.	−43 & below	−45 & below	Non-emergency work should cease			
			10 mph wind		15 mph wind	
1.	−26 to −28	−15 to −19	75 min	2	55 min	3
2.	−29 to −31	−20 to −24	55 min	3	40 min	4
3.	−32 to −34	−25 to −29	40 min	4	30 min	5
4.	−35 to −37	−30 to −34	30 min	5	Non-emergency work should cease	
5.	−38 to −39	−35 to −39	Non-emergency			
6.	−40 to −42	−40 to −44	work should			
7.	−43 & below	−45 & below	cease			
			20 mph wind			
1.	−26 to −28	−15 to −19	40 min	4		
2.	−29 to −31	−20 to −24	30 min	5		
3.	−32 to −34	−25 to −29	Non-emergency work should			
4.	−35 to −37	−30 to −34	cease			
5.	−38 to −39	−35 to −39				
6.	−40 to −42	−40 to −44				
7.	−43 & below	−45 & below				

Sources: ACGIH (1996). Adapted from Occupational Health & Safety Division, Saskatchewan Department of Labour.

Notes
1 Schedule applies to moderate to heavy work activity with warm-up breaks of ten minutes in a warm location. For light to moderate work (limited physical movement): apply the schedule one step lower. For example, at −35 °C with no noticeable wind (Step 4), a worker at a job with little physical movement should have a maximum work period of 40 min with 4 breaks in a four-hour period (Step 5).
2 The following is suggested as a guide for estimating wind velocity if accurate information is not available: 5 mph: light flag moves; 10 mph: light flag fully extended; 15 mph: raises newspaper sheet; 20 mph: blowing and drifting snow.
3 If only the wind chill cooling rate is available, a rough rule of thumb for applying it rather than the temperature and wind velocity factors given above would be: (1) special warm-up breaks should be initiated at a wind chill of about $1750\,\mathrm{W\,m^{-2}}$; (2) all non-emergency work should have ceased at or before a wind chill of $2250\,\mathrm{W\,m^{-2}}$. In general the warm-up schedule provided above slightly under-compensates for the wind at the warmer temperatures, assuming acclimatization and clothing appropriate for winter work. On the other hand, the chart slightly over-compensates for the actual temperatures in the colder ranges, since windy conditions rarely prevail at extremely low temperatures.

environment, assessment of workload, work-rest regimen, water and salt supplementation, clothing and acclimatization and fitness.

Cold stress

The *TLV*s for cold stress are presented in terms of air temperatures for conditions of varying wind velocities; see Table 14.18. They are intended to protect workers from the severest effects of cold stress and injury. That is to prevent core body temperature falling below 36 °C and cold injury to body extremities (hands, feet, and head).

Much information is provided regarding evaluation and control of cold environments and the wind chill index (*WCI*) and the equivalent chill temperature (*ECT*) are used to provide limits (see Table 14.19, ISO TR 11079, and Chapter 11). For example, for continuous exposure, skin should not be exposed to an *ECT* of less than −32 °C. It is also noted that tissue freezes at −1.0 °C, regardless of wind speed. When cold surfaces are below −7 °C, warning should be provided to each worker to avoid contact with bare skin.

The importance of clothing is emphasized and limits provided are for workers wearing clothing appropriate for the level of cold and physical activity. Guidance is also provided on work-warming regimen. For refrigeration rooms air velocity should be minimized and should not exceed $1 \mathrm{~m~s}^{-1}$ at the job site.

For both heat stress and cold stress, the ACGIH provides a practical assessment and evaluation methodology which is used widely in the USA and throughout the world.

National Institute for Occupational Safety and Health (NIOSH)

NIOSH (1986) considered criteria for a recommended standard for occupational exposure to hot environments. NIOSH provide criteria documents to recommend standards for 'promulgation by an appropriate regulating body, usually the Occupational Safety and Health Administration (OSHA) and the Mine Safety and Health Association (MSHA)'. NIOSH (1986) reviews standards used in the USA and internationally. A review of heat stress standards is also provided by Henschel (1980).

The work of OSHA, ACGIH, ISO 7243 and AIHA (American Industrial Hygiene Association) all provide similar standards and recommendations based on the *WBGT* index. All of the branches of the armed services provide limits that are based on *WBGT* values and are compatible with the other standards. The American College of Sports Medicine (ACSM) also provides limits in terms of *WBGT* values but with specific recommendations for the holding of sports events. All of the standards provide guidance and working practices for their own particular areas of application: military, sport, industry, etc. However the use of the *WBGT* index is widespread.

Table 14.19 Equivalent chill temperature values

Estimated wind speed (in mph)	Actual temperature reading (°F)											
	50	40	30	20	10	0	−10	−20	−30	−40	−50	−60
	Equivalent chill temperature (°F)											
calm	50	40	30	20	10	0	−10	−20	−30	−40	−50	−60
5	48	37	27	16	6	−5	−15	−26	−36	−47	−57	−68
10	40	28	16	4	−9	−24	−33	−46	−58	−70	−83	−95
15	36	22	9	−5	−18	−32	−45	−58	−72	−85	−99	−112
20	32	18	4	−10	−25	−39	−53	−67	−82	−96	−110	−121
25	30	16	0	−15	−29	−44	−59	−74	−88	−104	−118	−133
30	28	13	−2	−18	−33	−48	−63	−79	−94	−109	−125	−140
35	27	11	−4	−20	−35	−51	−67	−82	−98	−113	−129	−145
40	26	10	−6	−21	−37	−53	−69	−85	−100	−116	−132	−148

(Wind speeds greater than 40 mph have little additional effect)

Little danger
In <h with dry skin. Maximum danger of false sense of security

Increasing danger
Danger from freezing of exposed flesh within one minute

Great danger
Flesh may freeze within 30 s

Trenchfoot and immersion foot may occur at any point on this chart

Source: ACHIH (1998).

Note
* Developed by U.S. Army Research Institute of Environmental Medicine, Natick, MA. For degrees Celsius $C = \frac{5}{9}(F - 32)$.

NIOSH (1986) also reviews national heat stress standards from Finland, Sweden, Rumania, USSR, Belgium, Australia and Japan. Most of these countries use the *WBGT* index in their recommended assessment procedures for hot environments.

STANDARDS IN THE UK

The Factories Act 1961, Section 3 and the Offices, Shops and Railway Premises Act 1963 Section 6, require that a reasonable temperature should be maintained in the workplace (HMSO, 1963). If the work generally involves low levels of physical activity then the 'temperature' should be at least 16 °C after one hour and a thermometer must be prominently displayed. There is no upper limit for thermal comfort although it is often discussed, particularly during summer heat waves. The Health and Safety at Work Act (1974) does not stipulate working temperatures, however there are a number of associated regulations from a variety of industries. The Control of Substances Hazardous to Health (COSHH) Regulations 1988 do not directly involve the thermal environment although they may interact with other requirements.

The Chartered Institute of Building Services Engineers provides environmental criteria for designing indoor environments to provide thermal comfort (CIBSE, 1986). The recommended values are given in terms of resultant temperature (t_{res}), which is the temperature recorded by a thermometer at the centre of a blackened globe 100 mm in diameter. An equation is provided for calculation of t_{res} from basic parameters:

$$t_{res} = \frac{t_r + t_a\sqrt{10v}}{1 + \sqrt{10v}},$$
(14.22)

where

t_a = inside air temperature
t_r = mean radiant temperature
t_{res} = dry resultant temperature
v = air velocity.

For 'still' indoor air

$$t_{res} = \frac{t_r + t_a}{2}.$$
(14.23)

Design conditions and recommended design values are presented in Tables 14.20 and 14.21. CIBSE (1986) also provides guidance on design of the thermal environment and criteria for local thermal discomfort.

Table 14.20 Design condition

Country	Season	Occupancy/ category	Resultant temperature (°C)	Relative humidity (%)
UK	Summer	Continuous	20–22	50
		Transient	23	50
UK	Winter	Continuous	19–20	50
		Transient	16–18	50
Tropics	Summer	Continuous (optimum)	23	50
		Continuous (maximum)	25	60
			25	45
		Transient (humid climate)	25	70
			26	50
		Transient (arid climate)	27	45
			28	40
Tropics	Winter	Short winter (as in humid climate)	Generally no heating required	
		Long winter (as in arid climate)	22	45

As a revision to CIBSE guidance for thermal comfort, Oseland *et al.* (1998) prepared a report, 'Building design and management for thermal comfort'. Three complementary approaches are taken; the heat balance model, the New Adaptive Algorithm, and Empirical Customary Temperatures. Of particular interest is the practical application of the adaptive approach. The New Adaptive Algorithm is given neutral temperature (t_n) as follows:
Free running buildings

$$t_n = 0.44\,t_{out} + 15 \quad 8\,°C < t_{out} < 36\,°C. \tag{14.24}$$

Heated/cooled buildings

$$t_n = 0.25(t_{out} - 18) - 0.0004(t_{out} - 18)^3 + 23 \quad 4\,°C < t_{out} < 32\,°C, \tag{14.25}$$

where t_{out} is the mean of outdoor daily maximum and minimum temperatures. Corrections to clothing, activity level and air velocity are provided in tables (together with consequences for t_n) to take account of adaptive behaviour. Consideration is also provided for personal control of local environments. This ambitious proposal is the first practical proposal for an adaptive approach to design and requires experience with the method and validation.

BS 7915 (1998): ergonomics of the thermal environment: guide to design and evaluation of working practices for cold, indoor environments

BS 7915 was developed after a survey of the requirements of British industry (Flemming and Graveling, 1997). It provides guidance on ways in which

Table 14.21 Design condition

Type of building	t_{res}(°C)	Type of building	t_{res}(°C)
Art galleries and museums	20	Hotels:	
		Bedrooms (standard)	22
Assembly halls, lecture halls	18	Bedrooms (luxury)	24
		Public rooms	21
Banking halls:		Staircases and corridors	18
Large (height > 4 m)	20	Entrance halls and foyers	18
Small (height < 4 m)	20		
		Laboratories	20
Bars	18		
		Law Courts	20
Canteens and dining rooms	20		
		Libraries:	
Churches and chapels:		Reading rooms (height > 4 m)	20
Up to 7000 m²	18	(height < 4 m)	20
>7000 m²	18	Stack rooms	18
Vestries	20	Store rooms	15
Dining and banqueting halls	21	Offices:	
		General	20
Exhibition halls:		Private	20
Large (height > 4 m)	18	Stores	15
Small (height < 4 m)	18		
		Police stations:	
Factories:		Cells	18
Sedentary work	19		
Light work	16	Restaurants and tea shops	18
Heavy work	13		
		Schools and colleges:	
Fire stations: ambulance stations:		Classrooms	18
Appliance rooms	15	Lecture rooms	18
Watch rooms	20	Studios	18
Recreation rooms	18		
		Shops and showrooms:	
Flats, residences, and hostels:		Small	18
Living rooms	21	Large	18
Bedrooms	18	Department store	18
Bed-sitting rooms	21	Fitting rooms	21
Bathrooms	22	Store rooms	15
Lavatories and cloakrooms	18		
Service rooms	16	Sports pavilions:	
Staircases and corridors	16	Dressing rooms	21
Entrance halls and foyers	16		
Public rooms	21	Swimming baths:	
		Changing rooms	22
Gymnasia	16	Bath hall	26
Hospitals:		Warehouses:	
Corridors	16	Working and packing spaces	16
Offices	20	Storage space	13
Operating theatre suite	18–21		
Stores	15		
Wards	18		
Waiting rooms	18		

cold stress or discomfort in cold, indoor environments can be evaluated and cold strain reduced. The clothing insulation required (*IREQ*) and the wind chill indices are used in practical examples of how to assess the risk from cold environments, minimizing effects, work organization and selection of clothing. Case studies with commentary are provided for parking pallets, chicken processing, food processing on a conveyor belt, and cold store work including forklift truck driving. The standard complements German Standard DIN 33 403-5 (1994) which also applies the *IREQ* index and will contribute to a proposed European and ISO Standard (ISO 15743).

BS 7963 (2000): ergonomics of the thermal environment: guide to the assessment of heat strain in workers wearing personal protective equipment

BS 7963 was developed after a survey of requirements of British industry (Hanson and Graveling, 1999). It identified the need for guidance for people wearing protective clothing in 'hot' environments which is beyond the scope of existing International and European Standards. The standard gives guidance on assessing the effects of personal protective equipment (PPE) on heat stress and its consequences for possible heat-related health problems (heat strain). It is intended for use by those responsible for the health and safety of workers wearing PPE in situations which could lead to heat strain. The effects of PPE on the heat balance equation, metabolic rate (see Chapter 10) and clothing insulation are presented as well as examples of how to extend the application of ISO 7243 (WBGT – BS EN 27243) and ISO 7933 (SW_{req} – BS EN 12515) for conditions where workers wear PPE. Examples are provided for hot conditions involving water barrier, water vapour permeable clothing worn while cleaning debris from a vessel; light inspection work of a hot process; and workers wearing chemical protective clothing.

Proposed British Standard: specification of physiological measuring instruments

For measurements of the physical environment, ISO 7726 provides a specification of environmental measuring instruments to ensure they are taken with appropriate accuracy, reliability, sensitivity and so on. ISO 9886 presents physiological measures for monitoring thermal strain. Although instruments for measuring physiological strain (e.g. heart rate, internal body temperature) are now inexpensive and widely available, there was no guidance on specifications, use and interpretation of measurements of those instruments. This proposed British Standard specifies the minimum characteristics of systems for the measurement of physiological variables used to quantify thermal strain in humans. It is intended for use by persons responsible for the health and safety of workers, including managers and instrument manufacturers and persons conducting laboratory and field studies. Principles of measurement of physiological variables, monitoring of physiological variables, system integration and case study examples are provided. A provisional table of instrument specifications is provided, however it is emphasized that this is a proposed standard and will be subject to revisions.

An important development in the UK, with respect to standards, is that as a member of the European community, there will be no national standards in the 'single market'. This is discussed below.

EUROPEAN STANDARDS (CEN)

The European committee for standardization (Comité Européen de Normalisation – CEN) provides a major development in the area of standards. To facilitate the operation of the single European market (Treaty of Rome, CEC, 1986) for implementation in 1992, a number of European Standards have been developed to satisfy 'Directives of the European Community' (e.g. CEN, 1989). A CEN Technical Committee was set up to consider standards in the field of ergonomics (CEN/TC 122). A general agreement was that if appropriate International Standards were available then they should be proposed directly as CEN Standards. For this reason, ISO Standards concerned with the ergonomics of the thermal environment (see above) were proposed as CEN Standards. In addition, specific standards were developed concerned with human skin contact with hot and cold touchable surfaces (see Chapter 13). The importance is that once a CEN Standard is under development or consideration, then no national standard will be produced on that topic. Eventually there will be few national standards, only CEN Standards. The countries affected by this are the member states of the European Community (UK, France, Germany, Luxembourg, Belgium, Spain, Italy, Holland, Denmark, Greece, Portugal and Eire). In addition Sweden, Norway, Austria, Finland, Iceland and Switzerland are also actively involved with CEN Standards. In future many more countries will join.

The production of European Standards provides the opportunity to combine the best of the standards that have been produced in the past with new ideas and developments, so that the practitioner can effectively design and evaluate thermal environments with regard to human occupancy. Particular activity with regard to the 'Ergonomics of the thermal environment' includes the creation of European Standards working group CEN/TC 122 WG11 which shadows its 'sister' ISO working group (ISO/TC 159 SC5 WG1) and includes the same European (and international) experts ensuring that all work is coordinated and led by ISO activity. An example where an International Standard was not acceptable was with heat stress standard ISO 7933. Practices in the German coal mines suggested that ISO 7933 was not valid and, after consideration at the European level, EN 12515 was published (identical to ISO 7933 with some modification mainly in terms of cautionary notes).

CEN/TC 122 WG3 is concerned specifically with skin reaction to hot and cold surfaces (EN 563, EN 13202, ISO 13732 Parts 1 and 3) and CEN/TC 122 JWG 9 is a joint working group of a range of CEN committees concerned with the integration of clothing systems. The problem of total clothing systems, made up of an ensemble of individual garments and protective equipment (helmets, goggles, etc.) all specified by separate individual standards committees, is considered an Ergonomics issue. A six-part standard has been proposed (Pr EN 13921, Parts 1 to 6) including general requirements, anthropometric factors, biomechanical characteristics, thermal characteristics, chemical composition and sensory factors (Parts 1 to 6 respectively). Pr EN 13921 Part 4 – thermal characteristics – presents tests that should be performed for establishing the range of thermal conditions in which the personal protective equipment (PPE) is considered safe to use. A system is proposed from material testing to prototyping and field studies. The proposed standards are much needed, however they have proven difficult to produce (lack of data and agreement of interested parties) and have not been accepted without further revision.

EXAMPLE OF THE APPLICATION OF INTERNATIONAL (ISO) STANDARDS FOR THE ASSESSMENT OF A HOT ENVIRONMENT

Example

The following hypothetical example demonstrates how ISO Standards can be used in the assessment of hot environments. Workers in a steel mill perform work in four phases. They don clothing and perform light work in a hot radiant environment for one hour. They rest for 30 min and then perform the same light work shielded from the radiant heat, for one hour, then perform work involving moderate activity in a hot radiant environment for 30 min.

ISO assessment

ISO 7243 provides a simple method for monitoring the environment using the $WBGT$ index. If the calculated $WBGT$ levels are less than the $WBGT$ reference values given in the standard, then no further action is required. If the levels exceed the reference values then the strain on the workers must be reduced. This can be achieved by engineering controls and/or working practices. A complementary or alternative action is to conduct an analytical assessment as described in ISO 7933.

The $WBGT$ values for the work are presented in Table 14.22. The environmental and personal factors relating to the four phases of the work are presented in Table 14.23. It can be seen that for part of the work the

Table 14.22 WBGT values (°C) for four hour phases

Work phase (min)	WGBT*	WBGT reference
0–60	25	30
60–90	23	33
90–150	23	30
150–180	30	28

Note

* $\text{WGBT} = \dfrac{\text{WBGTankl} + \text{WBGTabd} + \text{WGBThd}}{4}$.

Table 14.23 Basic data for the analytical assessment

Work phase (min)	t_a (°C)	t_r (°C)	P_a (kpa)	v (m s^{-1})	Clo (Clo)	Act (W m^{-2})
0–60	30	50	3	0.15	0.6	100
60–90	30	30	3	0.05	0.6	58
90–150	30	30	3	0.20	0.6	100
150–180	30	60	3	0.30	1.0	150

Source: ISO 7933 (1989).

Table 14.24 Analytical assessment

Work phase (min)	Predicted values			Duration limited exposure (min)	Reason for limit
	t_{sk} (°C)	w (ND)	s_w (g h^{-1})		
0–60	35.5	0.93	553	423	water loss
60–90	34.6	0.30	83	480	no limit
90–150	34.6	0.57	213	480	no limit
150–180	35.7	1.00	566	45	body temperature
Overall	–	0.82	382	480	no limit

Source: ISO 7933 (1989).

WBGT levels exceed those of the reference values. It is concluded that a more detailed analysis is required.

The analytical assessment method presented in ISO 7933 was performed using the data presented in Table 14.23 and the computer program provided in the standard. The results for acclimatized workers in terms of an alarm level are presented in Table 14.24.

An overall assessment therefore predicts that unacclimatized workers suitable for the work could carry out an eight-hour shift without undergoing unacceptable (thermal) physiological strain. If greater accuracy is required, or individual workers are to be assessed, then ISO 8996 and ISO 9920 will provide more detailed information on metabolic heat production and clothing insulation. ISO 9886 provides methods for measuring physio-logical strain on workers and can be used to design and assess environments for specific workforces. For example, internal body temperature, mean skin temperature, heart rate and sweat loss may be of interest in this example.

EXAMPLE OF THE APPLICATION OF INTERNATIONAL (ISO) STANDARDS FOR THE ASSESSMENT OF MODERATE ENVIRONMENTS

Example

Occupants of an office complain of thermal discomfort. A survey is requested to investigate the problem and suggest improvements.

Method

Workplaces within the office are identified and assessed according to ISO 7730. An example of the results is provided below for one of the workplaces.

Input data

Metabolic rate $= 70\,\mathrm{W\,m^{-2}}$
Clothing insulation $= 0.155\,\mathrm{m^2\,^\circ C\,W^{-1}}$
Air temperature $= 20\,^\circ\mathrm{C}$
Mean radiant temperature $= 20\,^\circ\mathrm{C}$
Relative air velocity $= 0.2\,\mathrm{m\,s^{-1}}$
Partial vapour pressure $= 1014\,\mathrm{Pa}$

Calculated output

Predicted mean vote $\quad\quad\quad\quad\quad PMV = -0.6$
Predicted percentage of dissatisfied $\quad PPD = 12$ per cent

Interpreted on the following scale:

	PMV
hot	+3
warm	+2
slightly warm	+1
neutral	0
slightly cool	−1
cool	−2
cold	−3

A *PMV* value of −0.6 (*PPD* = 12 per cent) could be considered as acceptable although a rerun of the computer program shows that if the air velocity was reduced to $0.15\,\mathrm{m\,s^{-1}}$ and air and mean radiant temperature increased to 22.5 °C then the optimum of *PMV* = 0 can be achieved. That is:

Input data

Metabolic rate = $70\,\mathrm{W\,m^{-2}}$
Clothing insulation = $0.155\,\mathrm{m^2\,{}^\circ C\,W^{-1}}$
Air temperature = 22.5 °C
Mean radiant temperature = 22.5 °C
Relative air velocity = $0.15\,\mathrm{m\,s^{-1}}$
Partial vapour pressure = 1014 Pa

Calculated output

Predicted mean vote   *PMV* = 0
Predicted percentage of dissatisfied   *PPD* = 5 per cent

The workplace also has an air movement at ankle height of $0.25\,\mathrm{m\,s^{-1}}$ with a turbulence intensity (standard deviation of the fluctuating air velocity divided by its mean) of 40 per cent. The draught rating (*DR*) value is calculated at around 35. That is, a prediction that 35 per cent of people would be disturbed by draughts if exposed to those conditions. A suggestion for the workplace could therefore be to reduce air velocity.

For survey work, guidance on selection and use of measuring instruments is provided in ISO 7726. It is also important to take subjective measures. A short questionnaire could be designed based upon the principles outlined in ISO 10551.

EXAMPLE OF THE APPLICATION OF INTERNATIONAL (ISO) STANDARDS FOR THE ASSESSMENT OF COLD ENVIRONMENTS

Example

A large freezer room is kept at −23 °C and a worker is required to operate a forklift truck with a maximum speed of $1\,\mathrm{m\,s^{-1}}$ through the room. What clothing is required for the driver and for how long can he work safely?

Method

The ISO Technical Report (ISO TR 11079) considers the assessment of persons exposed to cold, based on the *IREQ* thermal index. Measurements can be made of the environment according to ISO 7726. It is also probable that the environmental conditions (temperatures) are closely monitored for the process involved (e.g. the storage of food). The values for the environmental conditions and worker's activity can be used in the calculation of *IREQ*. This is performed using a computer program. The calculation is of the *IREQ* min value, i.e. the minimum clothing required consistent with heat balance. The *WCI* is also calculated and provides guidance on the effects of the conditions of the extremities (hands, feet, etc.) and exposed flesh. These are:

Input data

Dry bulb temperature $= -23\,°C$
Wet bulb temperature $= -23\,°C$
Relative air velocity $= 1.0\,\mathrm{m\,s^{-1}}$
Mean radiant temperature $= -23\,°C$
Metabolic rate $= 70\,\mathrm{W\,m^{-2}}$

Calculated output

Min clothing insulation required $IREQ_{min} = 5.63\,Clo$
Wind Chill Index $WCI = 1089$

Wind Chill Index to be interpreted on the scale:

Pleasant	200
Cool	400
Cold	800
Very cold	1000
Bitterly cold	1200
Exposed flesh freezes	1400
Intolerable	2500

The $IREQ_{min}$ value is 5.63 Clo. The $IREQ_{neutral}$ value (to provide comfort) can be calculated from a similar procedure and for this example would be 6.27 Clo. These values are not achievable in practice even for a sitting subject. It can be concluded therefore that the work must be redesigned. A possible solution would be to build a cab on the vehicle. There may be other engineering and organizational solutions. A calculation of duration limited exposure (*DLE*) times would help. ISO 9920 describes clothing ensembles with the related Clo (I_{cl}) values. For cold protective clothing 2.48 Clo provides a relatively high insulation value achieved by wearing underpants, undershirt, shirt, insulated jacket, insulated trousers, overtrousers, overjacket, socks, shoes, gloves and a hat.

Calculation of allowable exposure times (*DLEs*) for the above clothing ensemble gives a value of 35 min. It is important to note that clothing insulation will also be affected by a number of additional factors. ISO 9886 provides methods for physiological measurements that would probably be recommended in this case to aid in overall job design.

15 Thermal models and computer aided design

INTRODUCTION

Stephen Hawking in *A Brief History of Time* (Hawking, 1988) began his book with the following account.

> A well known scientist (some say it was Bertrand Russell) once gave a public lecture on astronomy. He described how the earth orbits around the sun and how the sun, in turn, orbits around a vast collection of stars called our galaxy. At the end of the lecture, a little old lady at the back of the room got up and said, 'What you have told us is rubbish. The world is really a flat plate supported on the back of a giant tortoise.' The scientist gave a superior smile before replying, 'What is the tortoise standing on?' 'You're very clever, young man, very clever,' said the old lady. 'But it's turtles all the way down!'.

Hawking goes on to consider the nature of the universe. The model provided by the scientist is Newtonian; that by the old lady is a simplified version of that of ancient Hinduism. The point is that modelling (use of pictures, representations, mathematical equations, etc.) is an integral part of scientific (indeed human) activity. In addition, it is implicit that a model cannot be a perfect representation. Only the thing that we are trying to model can be that. There will always be differences between a model and what it represents. The question of whether a model 'works' or not becomes a question of whether the imperfections are significant in terms of the application to which the model is put.

The above has relevance to thermal modelling. There is often scepticism about attempting to model human systems. Physiological response to thermal conditions is complex and there are wide variations between individuals. In addition knowledge is incomplete. It is particularly true, therefore that models of humans in thermal environments will be imperfect. They can still however be useful for both research and in practical application.

The development of computers has provided an opportunity to develop and study thermal models and use their predictions as an aid to solving practical problems. The natural progression of this work is to computer aided environmental design. This area provides a new field for investigation and a challenge for the future.

THERMAL MODELS

Simple thermal models have been extensively used in the assessment of humans in thermal environments. For example simple thermal indices such as the effective temperature index (ET – Houghton and Yagloglou, 1923) or the wet bulb globe temperature index (WBGT – Yaglou and Minard, 1957) are thermal models as are rational indices such as the heat stress index (HSI – Belding and Hatch, 1955). They provide a method of integrating relevant factors to give a single number which is related to how humans respond to the combined effects of the factors. In this way the human response is modelled. The method of obtaining the index value varies from the use of graphs, charts and nomograms to the more recent use of digital computers. The method of obtaining the index is immaterial; however the development of the digital computer has provided a facility that allows more detailed models to be used in practical application.

Thermal models are often considered to be those which provide a rational representation of the human body involving both heat transfer between the body and the environment, the anthropometry and thermal properties of the body and a dynamic representation of the human thermoregulatory system. Wissler (1988) uses this definition of a thermal model to suggest that all models can be characterized by their representation of the temperature field within the body, thermoregulatory responses and garments and boundary conditions. However, a broad approach is presented below where a model is considered to be any method for predicting human response to the thermal environment.

Physical models

Physical models are like measuring instruments that respond to those factors of the environment to which humans respond. The response is usually in terms of temperature, although it may be in terms of mass or vapour loss or heat transfer, for example. The temperature or temperature profile of the model gives an indication of the temperatures of humans if they were exposed to that environment. Because physical models respond to important factors related to human response and simple physical models often provide a single temperature value that can be related to human response, simple physical models are often used to provide thermal index values, e.g. WGT, WBGT, etc. More elaborate thermal models closely represent the shape and response of the human body. The most sophisticated of these is the 'family' of thermal manikins.

Simple physical models

One of the first physical thermal models was produced by Dufton (1936), who used a heated cylinder, the eupatheoscope (Figure 15.1), as a sensor for feedback in a room heating system instead of a simple thermostat which would respond only to temperature. A more modern equivalent of this is the heated ellipse (Olesen, 1982 – see Figure 14.5) – which can be used in the calculation of the *PMV/PPD* thermal comfort index (Fanger, 1970).

A number of physical models have been based on globe thermometers. The wet globe temperature (*WGT*) has been used for assessing heat stress in the US steel industry (Botsford, 1971), and in a special form is called the botsball (see Figure 15.2). A similar device has been used for military

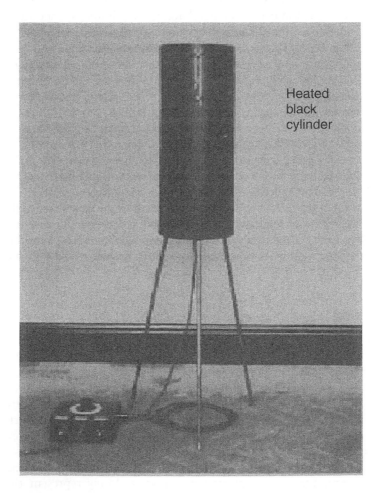

Heated
black
cylinder

Figure 15.1 The eupatheoscope.
Source: Bldg Res Bd (HMSO, 1938).

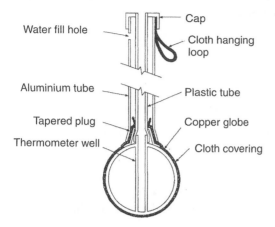

Water fill hole

Cap

Cloth hanging loop

Aluminium tube

Plastic tube

Tapered plug

Copper globe

Thermometer well

Cloth covering

Figure 15.2 Sectional sketch showing construction of a wet globe thermometer (not to scale).

applications (Goldman, 1988). Adding wet cloth (muslin) to the globe provides a representation of a sweating body. If the cloth has the colour of clothing worn (e.g. a uniform) then a more accurate representation of radiation heat exchange may be made.

A dry black globe indoors may represent a 'comfortable' person. This has been used to determine the dry resultant temperature (CIBSE, 1986).

Whether ellipses, cylinders or spheres (globes) are used as physical models, they are often derived from or justified using rational models. A classic calculation has been to determine the diameter of globe which would best represent the human body (Humphreys, 1977; Gagge *et al.*, 1967; BOHS, 1990). An example of this calculation is presented below.

Humphreys (1977) suggests that to assess the warmth of a room for human comfort a globe could be used which responds to radiation and convection in proportions similar to those for the human body. The temperature of an object can be considered as a weighted average of air temperature (t_a) and mean radiant temperature (t_r), i.e. for a globe (temperature t_g)

$$t_g = \frac{h_c}{h_c + h_r} t_a + \frac{h_r}{h_c + h_r} t_r. \tag{15.1}$$

The performance of the globe can thus be described by the factor $h_r/(h_c + h_r)$. Humphreys (1974) reviews the literature and derives equations for heat transfer for the human body and for globes.

For a globe, an approximation for h_r is $4\varepsilon\sigma T^3$. For emissivity $\varepsilon = 0.97$, Stefans constant $\sigma = 5.67 \times 10^{-8}\,\mathrm{W\,m^{-2}\,k^{-4}}$ and, a mean of the surface temperature of the globe and the mean radiant temperature of the surroundings of, $T = 293$ K, h_r, can be approximated by a value of 5.3 W m^{-2} °C^{-1}. From Hey (1968) for a globe of diameter D metres:

$$\frac{h_c D}{k} = 0.32 \left(\frac{DG}{\mu}\right)^{0.6}, \qquad (15.2)$$

where

$\qquad k$ = thermal conductivity of the film of air sorrounding the sphere

$\qquad G$ = mass velocity = density (kg m^{-3}) × velocity (m s^{-1})

$\qquad \mu$ = viscosity of air (kg m^{-1} s^{-1}).

h_c can therefore be calculated and hence $h_r/(h_c + h_r)$.

Values of $h_r/(h_c + h_r)$ are plotted against the diameter of the globe D for a range of air velocities.

For natural convection or for very low air velocities ($v = 0.1$–0.15 m s^{-1}), Humphreys (1974) concludes that a realistic estimate of $h_r/(h_c + h_r)$ for the human body is between 0.42 and 0.5. From the above equation, it can be calculated that a globe of diameter 40 mm would provide a similar value. A thermometer placed at the centre of a blackened table tennis ball (38 mm diameter) would therefore provide a convenient physical model which would also have the advantage of a relatively quick response time over that of globes of larger diameter, e.g. the traditional 150 mm diameter globe.

The use of a globe of 38 mm diameter will therefore provide a simple physical model to be used in practical application. It is important to consider the assumptions made in the derivation of the dimensions for the physical models. Humphreys argues that there is evidence that $h_r/(h_c + h_r)$ tends towards a value of 0.1 as air velocity increases to 1.0 m s^{-1} for the human body in certain types of clothing. Hence there may be general applicability of the 40 mm globe. However, other workers make different assumptions and hence arrive at different 'optimum' diameters for globes. In many practical applications, the rational arguments become rather 'thin' and must be viewed with caution.

Thermal manikins

McCullough and Jones (1984) provide a review and description of thermal manikins and identify a total of 12 laboratories which use them, throughout the world. In a similar review Holmér (1999) estimated over 80. They are mainly used for the evaluation and determination of the thermal properties of clothing. They range in sophistication from static, single-posture dry representations to multi-compartment, moving sweating versions (Umbach, 1988; Meinander, 1992 – see Figure 15.3). Manikins enable the use of

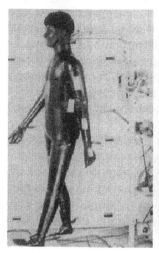

Figure 15.3 A movable copper manikin for the measurement of the thermophysiological properties of ready-made garment ensembles. The photograph on the right illustrates the measurement of the influence of wind speed on a garment ensemble's thermal insulation.

standard methods of assessment and they are also useful for applications where it is inconvenient to use human subjects, e.g. in investigating cold water immersion (Smallhorn, 1988; Allen, 1988), and for investigating responses of babies (Holmér, 1984) (see Figure 15.4).

Wyon *et al.* (1985) have used a dry thermal manikin (VOLTMAN) as a physical model in motor cars. The temperature distribution across the

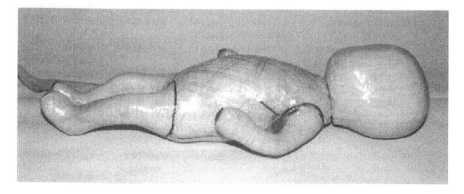

Figure 15.4 Thermal manikin used to investigate the heat transfer of babies.
Source: Holmér (1988).

manikin is maintained at similar to that of a comfortable person. The manikin in a sitting posture is positioned in a car seat and the heat loss from various parts of the manikin is determined and related to thermal discomfort. VOLTMAN has been used over realistic conditions for the design of heating and air conditioning systems in passenger motor cars. International activity in this area would suggest that with the advent of new technologies more sophisticated manikins will be developed and used. Of particular interest are the developments in computer control and the use of robotics.

Throughout the 1990s and into the twenty-first century there has been a proliferation of thermal manikins worldwide. As their role in environmental assessment, heat transfer measurement and determination of the thermal properties of clothing has been established, so have Human Thermal Environments Laboratories recognized their unique contribution. There are specialist conferences concerning thermal manikin testing (e.g. Nilsson and Holmér, 2000) and a reasonable estimate would be that there are more than 100 specialist thermal manikins in the world. Holmér (1999) reviews the development of thermal manikins and notes their requirement in a number of standards. He lists 12 milestones in the development of human-shaped thermal manikins (Table 15.1) and eight significant performance features of thermal manikins (Table 15.2). Attempts at breathing, moving, sweating thermal manikins have had some success. They are, however, not humans

Table 15.1 Milestones in the development of human-shaped thermal manikins

Milestone	Description	Material	Control	Features	Country
1	one-segment	copper	analogue		USA (1945)
2	multi-segment	aluminium	analogue		UK (1964)
3	radiation manikin	aluminium	analogue		France (1972)
4	multi-segment	plastics	analogue	moveable	Denmark (1973)
5	multi-segment	plastics	analogue	moveable	Germany (1978)
6	multi-segment	plastics	digital	moveable	Sweden (1980)
7	multi-segment	plastics	digital	moveable	Sweden (1984)
8	fire manikin	aluminium	digital		USA
9	immersion manikin	aluminium	digital	moveable	Canada (1988)
10	sweating manikin	aluminium	digital		Japan (1988)
	sweating manikin	plastic	digital	moveable	Finland (1988)
	sweating manikin	aluminium	digital	moveable	USA (1996)
11	female manikin	plastics single wire	digital, comfort regulation mode	moveable	Denmark (1989)
12	breathing thermal manikin	plastics single wire	digital, comfort regulation mode	moveable, breathing simulation	Denmark (1996)

Source: Holmér (2000).

Table 15.2 Performance features of thermal manikins

Relevant simulation of human body heat exchange: whole-body and local
Measurement of 3 dimensional heat exchange
Integration of dry heat losses in a realistic manner
Objective method for measurement of clothing thermal insulation
Quick, accurate and repeatable
Cost-effective instrument for comparative measurements and product development
Provide values for prediction models:
clothing insulation, evaporative resistance, heat loss

Source: Holmér (2000).

and as for all models, how successful they are depends upon how well they meet with the requirements of the application.

Empirical models – the Givoni/Goldman model

Empirical models can be developed by exposing human subjects to a range of thermal environments and 'fitting' mathematical models to the human response data obtained. Simple empirical models include the effective temperature thermal index (Houghton and Yagloglou, 1923). Probably the most widely used empirical thermal model is that by Givoni and Goldman (1972, 1973) for predicting responses to hot environments (computer program listings are provided in Berlin *et al.*, 1975). They provide equations for predicting heart rate and rectal temperatures based on the responses of US soldiers. The model assumes that for any combination of metabolic rate, environment and clothing there must be an internal body temperature and corresponding skin temperature at which the human body will reach equilibrium. The equilibrium rectal temperature is calculated as

$$t_{ref} = F_1(M_{net}) + F_2(H_{R+C}) + F_3(E_{req} - E_{max}).\qquad(15.3)$$

where

t_{ref} = final equilibrium rectal temperature (°C)

M_{net} = metabolic heat load (W)

H_{R+C} = sensible environmental heat load (W)

E_{req} = required evaporative cooling = $M_{net} + H_{R+C}$ (W)

E_{max} = maximum evaporative capacity of the environment (W).

F_1, F_2 and F_3 are experimentally derived functions that are applied to each component of the equation. The functions are provided below for work, rest and recovery from work. An example of the predicted response

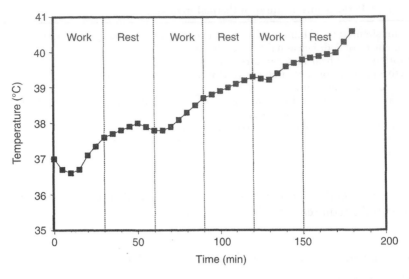

Figure 15.5 Example of the predicted rectal temperature of the model of Givoni and Goldman (1972). Hot conditions were for $t_a = t_r = 39.6\,°C$, $v = 0.1\,\mathrm{m\,s^{-1}}$, $rh = 35$ per cent, $I_{cl} = 0.7\,\mathrm{Clo}$, $f_{cl} = 1.2$, $i_m = 0.38$, $M = 190\,\mathrm{W\,m^{-2}}$, $W = 0\,\mathrm{W\,m^{-2}}$ for work periods and $M = 60\,\mathrm{W\,m^{-2}}$, $W = 0\,\mathrm{W\,m^{-2}}$ for rest periods.

to a hot environment is shown in Figure 15.5. A computer program of the model is presented in Haslam and Parsons (1989a).

Calculation of M_{net}

The metabolic heat load of subjects walking at $v\,\mathrm{m\,s^{-1}}$ up a gradient G (per cent) for a total mass of m_t (kg) is

$$M_{net} = M - (0.098\,m_t \times v \times G), \tag{15.4}$$

where total metabolic rate (Watts)

$$M = m_t\left[\left(2.7 + 3.2(v - 0.7)^{1.65}\right) + G(0.23 + 0.29(v - 0.7))\right]. \tag{15.5}$$

For heavy loads and unusual locations adjustments are needed. However, the equation is useful for rates up to 1400 Watts and down to walking speeds of $0.7\,\mathrm{m\,s^{-1}}$. For slower walking speeds the values should be measured. For subjects at rest a constant value of 105 Watts can be assumed.

Environmental heat load (R + C)

For an average man, $1.8\,\text{m}^2$ surface area,

$$R + C = \frac{11.6}{\text{Clo}}(t_a - 36). \tag{15.6}$$

Required evaporative cooling

$$E_{req} = M_{net} + (R + C). \tag{15.7}$$

Evaporative capacity

$$E_{max} = 25.5 \frac{i_m}{\text{Clo}}(44 - \phi P_a), \tag{15.8}$$

$i_m/$Clo is obtained from a static copper manikin.

44 mm Hg is the vapour pressure at $36\,°\text{C}$ skin temperature when condensation is occurring. This should be added to total heat load.

Predictive formula

For the 'fine tuning' of the formula for calculation of equilibrium rectal temperature t_{ref}, an experiment was conducted involving the exposure of eight male subjects to a range of thermal conditions. It was found that the predicted evaporative capacity was underestimated, as thermal properties of clothing had been determined on a static thermal manikin. To account for this an effective air speed (v_{eff}) term was determined where

$$v_{eff} = v_{air} + 0.004(M - 105)\,\text{m s}^{-1}. \tag{15.9}$$

The Clo and $i_m/$Clo values were then calculated from the v_{eff} value (see Table 15.3). t_{ref} is then given by

$$t_{ref} = 36.75 + 0.004 M_{net} + \frac{0.025}{\text{Clo}}(t_a - 36) + 0.8\,e^{0.0047(E_{req} - E_{max})}. \tag{15.10}$$

Table 15.3 Thermophysical properties of clothing, including its surrounding air layer, as a function of effective air velocity (v_{eff})

Clothing	Clo (total)	$i_m/$Clo
Shorts	$0.57\,v_{eff}^{-0.30}$	$1.20\,v_{eff}^{+0.30}$
Shorts and short sleeved shirt	$0.74\,v_{eff}^{-0.28}$	$0.94\,v_{eff}^{+0.28}$
STD fatigues	$0.99\,v_{eff}^{-0.25}$	$0.75\,v_{eff}^{+0.25}$
STD+OG	$1.50\,v_{eff}^{-0.20}$	$0.51\,v_{eff}^{+0.20}$
Breckenridge (1977)*	$1.50\,v_{eff}^{-0.15}$	$0.51\,v_{eff}^{+0.15}$

Note
* Later work.

Time pattern at rest under heat stress conditions

$$t_{re}(t) = t_{re}(0) + \Delta t_{re}(0.1) \exp 0.4^{(t-0.5)}, \tag{15.11}$$

where

$t_{re}(t)$ = rectal temperature at any time t

$t_{re}(0)$ = initial rectal temperature

$\Delta t_{re}(t) = t_{ref} - t_{re}(0)$

t = time (h) with $t - 0.5$ allowing 30 min for the initial lag in resting rectal temperature change when the elevation reaches 0.1 of the total change.

Time pattern of elevation during work

Working

$$t_{re}(t) = t_{re}(0) + \Delta t_{re}\left[1 - e^{-(2-0.5\sqrt{\Delta t_{re}})(t-58/M)}\right], \tag{15.12}$$

where

$t_{re}(t)$ = rectal temperature (°C) at any time t (h) after beginning work.

Note that during the computed time delay period at initiation of work after rest, the rectal temperature continues to follow the resting pattern and should be computed using resting $t_{re}(t)$ for this interval.

Recovery time pattern of rectal temperature after work

Recovery

$$t_{re}(t) = t_{rew} - (t_{rew} - t_{rer})\left[1 - e^{-a(t-t_{\Delta rec})}\right], \tag{15.13}$$

where

t_{rew} = rectal temperature at the beginning of decrease °C (Note: not necessarily equal to t_{re} at the end of work.)

a = time constant of recovery = $1.5(1 - e^{-1.5CP_{eff}})$

$t_{\Delta rec}$ = time lag of recovery = $0.25e^{-0.5CP_{eff}}$

CP_{eff} = effective cooling power = $0.27(i_m/\text{Clo})(44 - \phi P_a)$
$+ (0.174/\text{Clo})(36 - t_a) - 1.57.$

Validation of the predictive formulae

Givoni and Goldman (1972) use a number of experimental studies to validate their predictive formulae. It is interesting that they suggest for air temperatures of less than 30 °C, if an air temperature of 30 °C is used in the equations, then the formulae can be used for air temperatures down to 15 °C. It is recommended that the predictive formulae should not be used for air temperatures above 49 °C. It is also emphasized that the predictive equations are for predicting the average response of a group of young, healthy, reasonably fit acclimatized men.

Modifications for clothing

The Givoni/Goldman model requires clothing insulation values in terms of i_m and Clo; where i_m is measured in moving air $(1\,\mathrm{m\,s^{-1}})$ and Clo represents total insulation (clothing and air layer). Most databases of the thermal properties of clothing however present intrinsic clothing insulation values (i.e. independent of the air layer and hence the effects of the environmental conditions on the air layer). To give the model more general application, Haslam and Parsons (1989a) provided a method for converting intrinsic insulation values to those required by the Givoni/Goldman model.

To convert input clothing parameters as measured at $v = 0.1\,\mathrm{m\,s^{-1}}$ to those that would have been obtained at $v = 1.0\,\mathrm{m\,s^{-1}}$, a calculation is made on the following assumptions: $t_a = t_r = 24(°\mathrm{C})$, $h_c = 4.57\,\mathrm{W\,m^{-2}\,°C^{-1}}$ (calculated by subtracting h_r from h assuming $I_a = 0.71\,\mathrm{Clo} = 0.11\,\mathrm{m^2\,°C\,W^{-1}}$). The Lewis relation constant is $16.5\,\mathrm{K\,kPa^{-1}}$.

Dry clothing insulation

For still air, $v = 0.1\,\mathrm{m\,s^{-1}}$

$$h = h_c + h_r\,\mathrm{W\,m^{-2}\,°C}$$

$$h = \frac{1}{I_a} = \frac{1}{0.11} = 9.1\,\mathrm{W\,m^{-2}\,°C}$$

$$h_c + h_r = 9.1\,\mathrm{W\,m^{-2}\,°C}$$

$$h_c = 4.57\,\mathrm{W\,m^{-2}\,°C} \quad \text{and} \quad h_r = 4.53\,\mathrm{W\,m^{-2}\,°C^{-1}}.$$

For moving air $v = 1.0\,\mathrm{m\,s^{-1}}$, $h_c = 8.6v^{0.53} = 8.6\,\mathrm{W\,m^{-2}\,°C^{-1}}$. As h_r does not change: $h = h_r + h_c = 8.6 + 4.53 = 13.13\,\mathrm{W\,m^{-2}\,°C^{-1}}$,

$$I_a = \frac{1}{h} = 0.076\,\mathrm{m^2\,°C\,W^{-1}} = \frac{0.076}{0.155} = 0.49\,\mathrm{Clo}.$$

Total (I_t) and intrinsic (I_{cl}) clothing insulation are then calculated using the appropriate I_a value from

$$I_t = I_{cl} + \frac{I_a}{f_{cl}}.$$

Evaporative resistance

From Chapter 7:

$$I_{et} = I_{ecl} + \frac{I_{ea}}{f_{cl}}.$$

As air velocity changes, parameters can be related using:

$$i_m = \frac{I_t}{LR \cdot I_{ecl}} + \frac{I_{ea}}{f_{cl}},$$

and

$$I_{ea} = \frac{1}{LR \cdot h_c} \text{ where } h_c = 8.6 \, v^{0.53} \text{as above.}$$

Clothing values can be further modified to take account of body movement (see Table 15.3).

Values used in the Givoni/Goldman model can therefore be calculated from intrinsic clothing insulation values supplied by databases of clothing insulation (e.g. ISO 9920, 1995; McCullough *et al.*, 1985, 1989).

An example of the predictions of the program as modified by Haslam and Parsons (1989) are supplied in Figure 15.5.

Database models

One use of a thermal model is that it may be applied quickly and for conditions where there may be insufficient experimental data. Empirical models are however derived from experimental data and are usually 'curve fits' or summaries of data of human responses, e.g. predicting average responses. They do not necessarily provide 'accurate' predictions for conditions beyond which they were derived and often do not consider variation in response. They are also usually derived from one or a series of experiments from one laboratory. The body of experimental data cannot be exhaustive. A further consideration concerns methodology. Although an empirical model may be satisfactory, a rational model representing underlying mechanisms is to be preferred. An empirical model based entirely on regression and correlation will be of limited value and should be discouraged. Use of curve fitting within a rational context (e.g. to identify values of parameters of the heat transfer process within heat transfer equations) may contribute to understanding whereas correlation without mechanism will not. There are many combinations of factors which contribute to human responses to thermal environments. However, extensive laboratory and field studies have been conducted throughout the world and there are much published data (Haslam and Parsons, 1989).

It would be useful if these data could be made available to 'predict' or provide guidance on human response in an application of interest. Until recently this task would have been difficult and access to such a database would have been cumbersome. However, with the development of database technology, using data from actual human responses to provide the 'best' prediction of human response is feasible. Parsons and Bishop (1991) created a database of human responses to thermal environments, and a method of 'matching' the conditions for which responses are required with those in the database. An example of its use is provided in Figure 15.6.

A major part of constructing a database model is to determine which data should be presented and in what form. It is essentially a computer-based modelling procedure and consequently the design of the software interface is important. In addition, the method of matching (involving quantifying differences between data in the database and what is required), giving priority to matching factors, etc. is important. For example, is it 'better' to closely match air temperature or clothing worn? The internet and

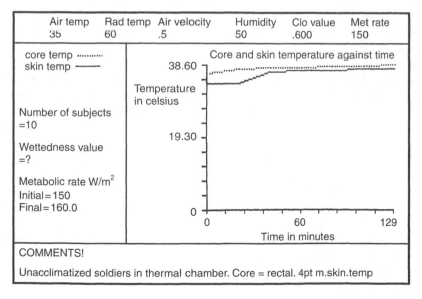

Figure 15.6 Example of the predictions provided by the database model of Parsons and Bishop (1991). Representation on the computer screen has greater resolution and use of colour.

worldwide web have provided opportunities for the construction of large databases of data collected internationally. Careful selection of data and quality control can provide empirical models. Why use a 'curve fit' as an approximation when the raw data can be made available in useable form?

Rational thermal models

Rational models provide a mathematical description of human responses to thermal environments. They could be considered as an extension of the heat balance equation used to determine rational thermal indices. However, the term thermal model usually refers to a dynamic mathematical simulation of the human body and its response to thermal environments, involving both a passive and controlling system for the body as well as mechanisms of heat exchange. These are described below using the model of Stolwijk and Hardy (1977) as an example.

Stolwijk and Hardy 25-node model

It is useful to describe the work of Stolwijk and Hardy (1977) as this provides an example of the components used in a thermal model and this model has provided the basis and inspiration for much of the work on thermal modelling.

The passive (controlled) system of the model is represented by five appropriately sized cylinders, representing trunk, arms, hands, legs and feet, and a sphere, representing the head. Each of these has four concentric layers or compartments representing core, muscle, fat and skin layers, and 'an additional central blood compartment, representing the large arteries and veins, exchanges heat with all other compartments via the convective heat transfer occurring with the blood flow to each compartment' (Stolwijk and Hardy, 1977). The model assumes that the body is symmetrical to reduce the number of calculations, i.e. one cylinder represents each pair of hands, arms, feet and legs; the values are doubled later (Figure 15.7).

The six segments (five cylinders and one sphere), four compartments per segment and the central blood compartment make a total of 25 nodes. To further define the passive system, the dimensions and thermal properties of each of the nodes are provided from reviews of previous studies. The model presented is based on a standard, 1.72 m, 74.4 kg man with a volume of $74.4 \times 10^{-3}\, \text{m}^3$. The quantities which define the passive system as used by Stolwijk and Hardy (1977) are presented in Table 15.4. It is emphasized that those values should be replaced for predictions for humans of different sizes for example.

The controlling system consists of a temperature sensing system, an integrating system and an effector system. It is a simple representation of the human thermoregulatory system based on set points (see Figure 15.8).

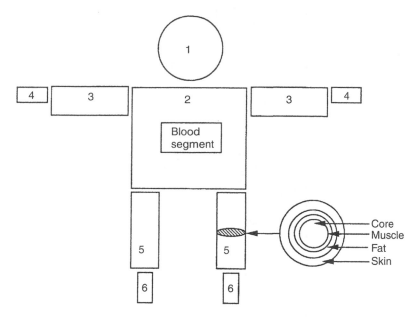

Figure 15.7 Representation of the passive (controlled) system of the 25-node model of Stolwijk and Hardy (1977). (6 body segments × 1 blood compartment = 25 nodes.)

The controlling system is defined in terms of controlling coefficients, for example to determine the strength of vasodilation or sweating (see Table 15.5). The quantities that define the passive and controlling system for this model are presented in Table 15.6. An example of the model's predictions is provided in Figure 15.9. In its original form the model represents only the nude body.

Pierce two-node model

A simplified version of the Stolwijk and Hardy model was also developed at the J. B. Pierce laboratory, USA. Nishi and Gagge (1977) present a two-node model for use in practical application, which includes equations for the thermal resistance provided by clothing over the body. The passive system is two concentric cylinders of appropriate dimension and thermal properties. These represent an inner core and an outer shell (Figure 15.10).

The controlling system is represented by a similar system to that used by Stolwijk and Hardy (1977) but with only two nodes to control (Figure 15.11).

An example of the model's predictions is provided in Figure 15.12. This model provides an integral part of the calculations to determine the new effective temperature (ET^*) and standard effective temperature (SET) thermal indices.

Table 15.4.i Values for surface area volumes and heat transfer coefficients

Segment %	Surface area		Volume		Heat transfer coefficients		
	m² S(I)	% of total	m³ ×10⁻³	% of total	Radiant heat transfer coeff. HR (I) W m⁻²°C⁻¹	Convective heat transfer coeff. HC (I) W m⁻²°C⁻¹	Combined coefficient H (I) W m⁻²°C⁻¹
Head sphere	0.1326	7.00	4.02	5.4	4.8	3.0	7.8
Trunk, cylinder	0.6804	36.02	41.00	55.1	4.8	2.1	6.9
Arms, cylinder	0.2536	13.41	7.06	9.5	4.2	2.1	6.3
Hands, cylinder	0.0946	5.00	0.67	0.9	3.6	4.0	7.6
Legs, cylinder	0.5966	31.74	20.68	27.8	4.2	2.1	6.3
Feet, cylinder	0.1299	6.86	0.97	1.3	4.0	4.0	8.0
Total	1.8877	100.00	74.4	100.00			

Table 15.4.ii Weight and heat capacity of the four layers in each segment

Segment	Wt kg	% of total	Core						Muscle		Fat		Skin	
			Skeleton		Viscera		Total							
			Wt kg	C(N) W h °C⁻¹	Wt kg	C(N) W h °C⁻¹	Wt kg	C(N) W h °C⁻¹	Wt kg	C(N) W h °C⁻¹	Wt kg	C(N) W h °C⁻¹	Wt kg	C(N) W h °C⁻¹
Head	4.02	5.4	1.22	0.71	1.79	1.86	3.01	2.57	0.37	0.39	0.37	0.26	0.27	0.28
Trunk	38.50	51.7	2.83	1.64	9.35	9.85	12.18	11.44	17.90	18.80	7.07	4.94	1.35	1.41
Arms	7.06	9.5	1.51	0.86	0.74	0.77	2.25	1.63	3.37	3.54	0.97	0.67	0.48	0.50
Hands	0.67	0.9	0.23	0.13	0.03	0.03	0.26	0.16	0.07	0.07	0.15	0.10	0.19	0.20
Legs	20.68	27.8	5.02	2.92	1.92	2.02	6.94	4.94	10.19	10.67	2.38	1.66	1.20	1.25
Feet	0.97	1.3	0.37	0.21	0.06	0.06	0.43	0.27	0.07	0.07	0.22	0.15	0.24	0.26
Central blood	2.50	3.4			2.50	2.60	2.50	2.60						
Total	74.40	100.0	11.18	6.47	16.39	17.19	27.57	23.66	31.97	33.60	11.16	7.79	3.73	3.90

Source: Stolwijk and Hardy (1977).

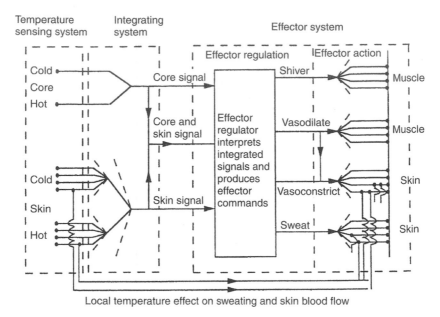

Local temperature effect on sweating and skin blood flow

Figure 15.8 Representation of the active (controlling) system of the 25-node model of Stolwijk and Hardy (1977).

Table 15.5.i Estimates for distribution of sensory input and effector output over the various skin areas

Segment	Surface area		SKINR(I)	SKINS(I)	SKINV(I)	SKINC(I)
	m²	%				
Head	0.1326	7.00	0.21	0.081	0.132	0.05
Trunk	0.6804	36.02	0.42	0.481	0.322	0.15
Arms	0.2536	13.41	0.10	0.154	0.095	0.05
Hands	0.0946	5.00	0.04	0.031	0.121	0.35
Legs	0.5966	31.72	0.20	0.218	0.230	0.05
Feet	0.1299	6.85	0.03	0.035	0.10	0.35
Total	1.8877	100.00	1.00	1.000	1.000	1.00

Table 15.5.ii Estimates for distribution of heat production in muscle compartments

Segment	Total Muscle Mass (%)	Work M (1)	Chil M (1)
Head	2.323		0.02
Trunk	54.790	0.30	0.85
Arms	10.525	0.08	0.05
Hands	0.233	0.01	0.00
Legs	31.897	0.60	0.07
Feet	0.233	0.01	0.00

Source: Stolwijk and Hardy (1977).

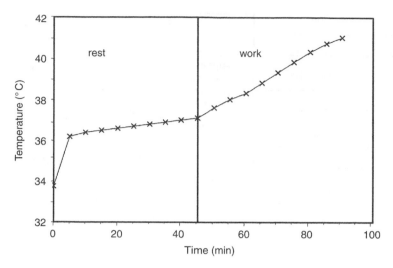

Figure 15.9 Example of the predicted mean skin temperature response of the model of Stolwijk and Hardy (1977) as modified by Haslam and Parsons (1988). Thermal conditions were for $t_a = t_r = 49.5\,°C$, $v = 0.1\,\mathrm{m\,s^{-1}}$, $rh = 32$ per cent, $I_{cl} = 0.1\,\mathrm{Clo}$, $f_{cl} = 1$, $i_m = 0.5$, $M = 306\,\mathrm{W\,m^{-2}}$, $W = 49\,\mathrm{W\,m^{-2}}$ for work period and $M = 58\,\mathrm{W\,m^{-2}}$, $W = 0\,\mathrm{W\,m^{-2}}$ for rest period.

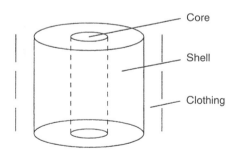

Figure 15.10 Representation of the passive (controlled) system of the two-node model of Nishi and Gagge (1977).

Wissler model

Wissler (1988) describes a model that computes 225 temperatures in 15 elements plus O_2 and CO_2 and lactate concentrations. The model is an order of magnitude larger than the Stolwijk and Hardy model and it is well validated for hot and cold environments and both one atmosphere and

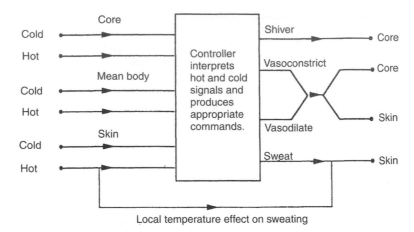

Figure 15.11 Representation of the active (controlling) system of the two-node model of Nishi and Gagge (1977).

hyperbaric environments. The detail provided allows its use in specialist areas such as for cold water immersion and diving, where it has been used in application. The passive system of the model is shown in Figure 15.13 and an example of its use in Figure 15.14.

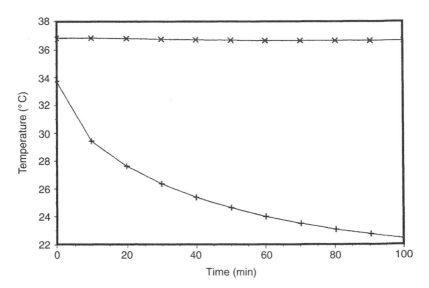

Figure 15.12 Example of the predicted mean skin temperature (+) and core temperature (×) response of the two-node model of Nishi and Gagge (1977). Thermal conditions were for $t_a = t_r = 5\,^\circ\mathrm{C}$, $v = 0.1\,\mathrm{m\,s^{-1}}$, $rh = 30$ per cent, $I_{cl} = 0.1\,\mathrm{Clo}$, $f_{cl} = 1$, $i_m = 0.5$, $M = 60\,\mathrm{W\,m^{-2}}$, $W = 0\,\mathrm{W\,m^{-2}}$.

Table 15.6 Set point values and initial condition temperatures (Stolwijk and Hardy, 1977)

Segment	Compartment	N	Temperature °C
Head	Core	1	36.96
	Muscle	2	35.07
	Fat	3	34.81
	Skin	4	34.58
Trunk	Core	5	36.89
	Muscle	6	36.28
	Fat	7	34.53
	Skin	8	33.62
Arms	Core	9	35.53
	Muscle	10	34.12
	Fat	11	33.59
	Skin	12	33.25
Hands	Core	13	35.41
	Muscle	14	35.38
	Fat	15	35.30
	Skin	16	35.22
Legs	Core	17	35.81
	Muscle	18	35.30
	Fat	19	35.31
	Skin	20	34.10
Feet	Core	21	35.14
	Muscle	22	35.03
	Fat	23	35.11
	Skin	24	35.04
Central blood		25	36.71

Werner model

Werner (1990) provides a three-dimensional model involving 63 types of tissue with the temperature grid in the body (1 cm for the trunk, 0.5 cm for other parts) represented by 400 000 points. 'An unresolved problem is the control strategy of the system, that is the question whether the inhomogeneous pattern of effector distribution is maintained in the cold and warmth or whether active modification and control is distributed' (Werner and Buse, 1988). A simple lumped parameter approach was therefore used.

$$Y = 0.8T_{core} + 0.1T_{skin} + 0.1T_{muscle} - 37,$$

where

Y = effector signal

T_{core} = mean model core temperature

T_{skin} = mean model skin temperature

T_{muscle} = mean model muscle temperature.

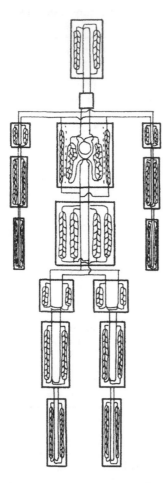

Figure 15.13 Representation of the passive (controlled) system of the Wissler model.

Source: Wissler (1985).

The model provides detailed results and has been supported by a programme of research which develops knowledge about both passive and controlling properties. A representation of the passive system of the model is shown in Figure 15.15 and an example of some results is shown in Figure 15.16.

Other models

A number of models, other than those described above, have been produced. Many of these have been based on the model of Stolwijk and Hardy

Experimental values of $(h_c/k_{m,w})$

System	Pressure Ata	$(h_c/k_{m,w})$ J/(mole °C)
Air-water	0.14–5.2	29.90 ± 1.5
Helium-Water	1.00–30.0	35.46 ± 1.3

Source: Zimmerman and Ramsey.

Values of the effective heat transfer coefficients for air and heliox mixtures at various pressures

Gas	Windspeed m/s	Pressure Ata	h_{eff} w/(m²°C)
Air	1.00	1	15.3
Air	0.75	2	17.7
Air	0.40	4	18.0
Heliox	0.40	4	21.7
Heliox	0.30	8	29.5
Heliox	0.25	16	38.0
Heliox	0.25	32	54.2

Summary of computed results for hyperbaric welding

Case	P Ata	T_{eff} °C	T_{insp} °C	Met Watts	$T_{a, 4hrs}$ °C	Sweat $_{4hrs}$ grams	$Q_{sh, 4hrs}$ Watts
1	21	32	32	145	37.06	132	0
2	31	32	32	145	37.05	90	0
3	31	30	30	145	36.51	0	6
4	31	34	34	145	37.65	694	0
5	31	34	20	145	37.14	154	0
6	31	34	15	145	37.10	59	0
7	31	32	32	233	37.67	794	0
8	31	32	32	133/233	37.50	290	0

Notes

1 $T_{a,4h}$ = arterial temperature at the end of 4h.
2 Sweat $_{4h}$ = amount of sweat produced during 4h.
3 $Q_{sh, 4h}$ = shivering rate at the end of 4h.
4 Case 8 involved 30min of rest and 60min of work at the metabolic rates shown.

Figure 15.14 Example of the output of the Wissler model.
Source: Wissler (1985).

(1977). The developments in the use of computers will increase the interest in this area and it is reasonable to expect that models will be used in many applications. Models include those of Montgomery (1974), Richardson (1985) and Tikuisis (1989, 1992, 1998) who consider cold and immersion in water, O'Neill *et al.* (1985) for heat stress in tractor cabs, Gordon (1974), Ringuest (1981) who modified the controlling system in the Stolwijk and Hardy model and there are others.

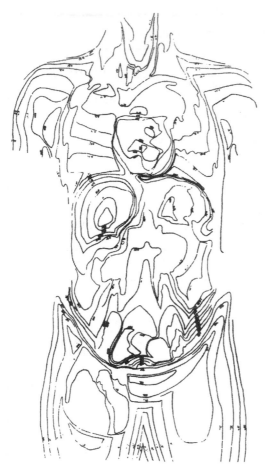

Figure 15.15 Photogrammetric analysis: contour lines of the trunk after the elimination of all organs.

Sources: Werner and Buse, 1988 and Mekjavic *et al.* (1988).

A series of models was produced by Haslam and Parsons (1989) from Loughborough University of Technology (LUT) in the United Kingdom. The models were essentially practical updates of existing models and were hence termed the LUTre (modified Givoni and Goldman), LUT 2-node, LUT 25-node and LUTISO (modified ISO DIS 7933) models. Lotens *et al.* (1989) describes a model of the human foot, and Oakley (1990) a model of the finger for investigation of the effects of cold exposure. Jones and Ogawa (1992) and Smith (1991) describe a model for use in transient conditions and Lotens (1989, 1990, 1993) presents a dynamic model of clothing systems as does Jones *et al.* (1994).

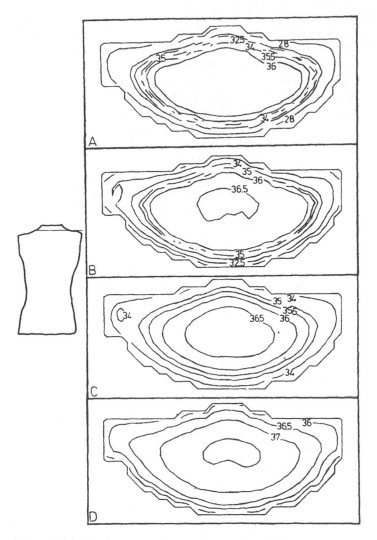

Figure 15.16 Isotherms on the trunk at the level of the collarbone (see left) for
$t_a = 10\,°C$ (A), $20\,°C$ (B), $30\,°C$ (C) and $40\,°C$ (D).

Sources: Werner and Buse, 1988 and Mekjavic *et al.* (1988).

One problem which has emerged with the use of models is that it becomes
a non-trivial problem to identify errors. In addition models often change as
they are updated. It is important to keep up-to-date with developments in
what rapidly becomes a specialist area. For a further discussion of models
and their practical use the reader is referred to Wissler (1988), Haslam and
Parsons (1989a), Werner (1996) and Tikuisis (1998).

Significant developments in models of human thermoregulation and heat exchange have been the use of sophisticated software traditionally used in engineering design and the increase in non-military applications of models, including the area of thermal comfort. Engineering design approaches have allowed the use of Computational Fluid Dynamics (CFD) including finite element and finite difference methods where equations of heat transfer have been solved by computer software written by the modeller or, more prevalently, where standard software packages for constructing thermal models have been used. Neale (1998) continued the work of Haslam and Parsons (1988) and Haslam (1989), and developed the LUT 25-node model to include anthropometric inputs (Neale *et al.*, 1995), acclimation and the distribution of clothing across the body (Neale *et al.*, 1996). Similar developments have also been described by Kraning (1995) and Candas *et al.* (1998). Neale (1998) also used a commercially available finite volume software package to simulate the LUT 25-node model in cold environments. Interestingly, although the number of layers of the model are greatly increased, it did not improve accuracy of prediction of human response.

The software is generally in three parts: a pre-processor which defines the physical and dynamic properties of the model; a processor which provides a mathematical solution to the problem; and a post-processor which presents the results in useable form. The use of such packages are becoming routine in environmental design and the work of Neale (1998) demonstrated their application to human thermal modelling.

Fu (1995) and Fu and Jones (1996) report a transient three-dimensional finite element thermal model for the human body (based on Smith, 1991) combined with a two-dimensional model of clothing (based on Jones and Ogawa, 1992). Tikuisis (1994) and Tikuisis and Young (1996) have developed a model for the prediction of responses to cold outdoor environments and Candas *et al.* (1998) describe a model for assessing the risk of discomfort due to transient environments.

Fiala (1998) and Fiala and Lomas (2001) describe a dynamic human thermal model for the prediction of thermal comfort responses (Dynamic Thermal Sensation – DTS). The passive system of the model has body elements for the head, face, neck, shoulders, arms, hands, thorax, abdomen, legs and feet. Each element consists of tissue layers with appropriate thermal properties (brain, lung, bone, muscle, viscera, fat and two layers of skin). The active system simulates the responses of the human thermoregulatory system ('cutaneous vasomotion', sweating, shivering). Regression analysis was conducted between predicted physiological responses and thermal comfort measurements from a database of 'comfort experiments' (including those of Rohles and Nevins, 1971 – see Chapter 8). It was concluded that skin temperature, core temperature and rate of change of skin temperature provide a valid comfort model. Although the study is extensive, it is debatable whether the regression model furthers understanding of the underlying mechanisms of thermal comfort response and the model should be reconciled with the effects of skin wettedness as described by Gagge *et al.* (1967).

Zhu (2001) describes a 37-node human body thermoregulatory model (HBTM) for the prediction of moderate transient building environments. The model is an extension of the Stolwijk and Hardy (1977) 25-node model to include two layers of clothing (six segments × six layers plus central blood = 37 nodes). The controlling system includes the effects of rate of change of temperature. Comparison of the models' predictions with measured head core and mean skin temperatures of subjects exposed to transient conditions shows good agreement although in changes from neutral to cool or cold environments, predicted body temperatures drop faster than those measured.

Tanabe *et al.* (2002) describe a 65-node model of human thermoregulation (based on Stolwijk and Hardy, 1977) combined with a thermal radiation model and Computational Fluid Dynamics (CFD). The original model has 16 body segments (head, chest, back, pelvis, left and right shoulder, left and right arm, left and right hand, left and right thigh, left and right leg, and left and right foot), four layers (core, muscle, fat, skin) and central blood (16 × 4 + 1 = 65 nodes). This was modified to 1542 elements (1542 × 4 (layers) + 1 body compartment). The integrated model is used to predict the physiological and physical state of the human body standing in a room exposed to direct solar radiation from a window and a cooling panel in the ceiling.

Other models based on the Stolwijk and Hardy (1977) model include Thellier *et al.* (1994a,b) who modelled physiological response during extravehicular activity in astronauts, thermal comfort responses in buildings and local thermal sensations of a car driver in winter conditions. They integrate the (clothed) model with other simulation models (e.g. buildings, vehicles) to provide an overall design tool. Wang (1990) used a simplified model to investigate thermal transients and developed a Dynamic Thermal Sensation index (DTS) based on the model predictions.

Other models include those of Höppe (2001) who considered indoor and outdoor comfort using the Instationary Munich Energy Balance Model (IMEM) which may contribute to the development of a Universal Thermal Climate Index (UTCI). A simulation model (INKA/TILL – finite difference solver) developed by vehicle manufacturers (BMW, Opel) is used commercially to simulate both vehicle environments and thermal comfort response. This was used in the Autoglaze European programme of research (see Chapter 9). Injuries CAused by Radiation Upon the Skin (ICARUS) is a three-layer skin model with clothing on an incorporated model of human thermoregulation (Green and Prescott, 1999). It has been developed and used to predict burn injuries to personnel subjected to intense thermal radiation.

Candas *et al.* (1998) use a computer model which predicts thermal dissatisfaction from skin temperature changes (including head, torso, back, arms, hands, legs and feet), the thermal differences between the extremities as well as the thermal state of the extremities (head, hands and feet) and the differences between front and back, left and right, and upper and lower parts of the body. Based upon a programme of physiological and thermal comfort experiments conducted over thirty years, a model is presented for interpretation of predicted physiological state to produce thermal comfort responses.

DO MODELS WORK?

The discussion at the beginning of this chapter suggested that the question 'do the models work?' is one of whether the imperfections of the model are significant in terms of the application to which they are put. Often this involves whether the predictions are of sufficient accuracy for use in making practical decisions. A validation or evaluation of a model can involve comparing predictions of human response from the model (e.g. in terms of body temperature, etc.) with those measured on humans who are exposed to those conditions. Even for rational thermal models, response data from human subjects are often used in the development of the model. Using these same data however, is not a true validation or evaluation of the model.

One of the first independent evaluations of models was conducted by Wissler (1982) in a symposium for the US Air Force. International experts were invited to present human response data and these were compared with the predictions of the Wissler, Stolwijk (25-node), Gordon and Goldman's models for both heat and cold (including the Givoni and Goldman model above).

It was found that for comparison with human response data within the range for which it was intended to be applicable, the Goldman heat model was 'quite good, although the model had a tendency to overestimate the increase in rectal temperature during heavy exercise in a hot environment'. The Goldman cold model provided 'very good results in some cases and totally erroneous results in others'. Gordon's model was found to require a large effort to prepare data for a particular run. Where it was possible to run the model results appeared to be similar to those for other models (see Gordon, 1974). Stolwijk's model appeared to work well for predictions in hot and moderate environments but not so well in the cold where central body temperatures appear to fall too quickly. The Wissler model was extensively tested and was found to work well over a wide range of conditions. When compared with individual subjects' responses however, there were significant differences between responses and predictions, especially in cold environments.

One of the most elaborate evaluations of models was conducted by Haslam (1989) and Haslam and Parsons (1989a). The study began in 1985 when they were interested in the possible use of the models in practical application. The models of ISO DIS 7933 (SW_{req} – heat balance equation), Givoni and

Goldman (1972, 1973), Stolwijk and Hardy (1977) and Nishi and Gagge (1977) were investigated. Each of the models was modified so that it could be used for practical application to provide predictions for work/rest cycles and persons wearing clothing (involving both dry and vapour heat transfer). The predictions of the physiological responses based on the four modified models (LUTre, LUTISO, LUT 25-node, LUT 2-node) were compared with an extensive database of experimental responses. The results are presented in Tables 15.7 and 15.8.

It was found that predictions were reasonable for at least one of the models tested for cool, neutral, warm and hot environments. Predictions of responses to cold environments were poor. The LUTre model was useful for predicting responses to very hot, heavy exercise conditions.

An interesting evaluation of the Stolwijk and Hardy (1977) model was conducted by Cooper *et al.* (1987). The body dimensions and composition, and responses to heat and cold, were experimentally determined for each of six male subjects. Each individual subject's dimensions and composition was substituted into the model replacing those of the standard man. Thus an individual's passive system was produced for the model. Comparison of the predicted responses using the individual's model with that of individual's measured responses did not however improve the accuracy of prediction.

Wissler (1988) assesses current modelling capabilities. He provides the following scale and gives ratings for his estimate of how useful present models would be with respect to different applications (Table 15.9).

The ratings take account of the extent to which models have been validated, importance of factors not properly accounted for and difficulty in simulation. Gonzalez *et al.* (1977) compared the LUT 25-node model with the USARIUM model (based upon Govoni and Goldman, 1972).

Many of the models presented above have now been independently validated although the criteria for validation (does it work?) have varied, are not always clear, and validation, albeit independent, has often been conducted by the authors of the model – often not physiologists. Malchaire *et al.* (2000) developed an extensive database of human responses to hot environments with 'strict' quality control of data. They split the data into two matched halves and used one half to aid development of the model and the other half for validation. The availability of global communication has made the development and availability of large databases possible and should enhance validation opportunities in the future.

Hybrid models

The data presented in Tables 15.7 and 15.8 provide an indication of which of the models tested provides 'most accurate' predictions for given conditions. For example, if a prediction were required for conditions involving US soldiers, during exercise in hot conditions then the LUTre model would

Table 15.7 Accuracy of thermal models over thermal conditions*

	$t_a \leq 5$		$5 < t_a \leq 15$		$15 < t_a \leq 25$		$25 < t_a \leq 35$		$35 < t_a$	
	t_{cr}	t_{sk}	t_{cr}	t_{sk}	t_{cr}	t_{sk}	t_{cr}	t_{sk}	t_{cr}	t_{sk}
nude/rest/ no-wind	NONE	lut2	lut2	lut25 lut2	luttre lut2 lut25	lut2 lut25	lut25 lut2 luttre	lut2† lut25†	lut25 lut2† luttre†	lut25 lut2
nude/rest/wind	–	–	lut2	lut25	luttre lut2	lut25	–	–	–	–
nude/work/ no-wind	–	–	–	–	lut25 luttre	lut2 lut25	lut25 luttre lut2	lut25† lut2†	lut25 lut2	lut2 lut25
nude/work/wind	–	–	–	–	–	–	lut2 lut25 luttre	–	lut25	lut2† lut25†
clothed/rest/ no-wind	NONE	lut2† lut25†	lut2	lut2 lut25	–	–	–	–	–	–
clothed/rest/wind	NONE	NONE	NONE	lut2	–	–	–	–	–	–
clothed/work/ no-wind	lut25	lut25	lut2 lut25	lut25	–	–	lut2‡	lut2 lut25	NONE‡	lut2 lut25
clothed/work/ wind	NONE	NONE	–	–	–	–	lut25	–	–	–

Source: Haslam and Parsons (1989a).

Notes
* The most accurate models for each environment category in descending order, where the mean rmsds for the models' predictions were within the maximum average standard deviations found for the observed data (i.e. 0.5 °C for t_{cr} and 1.6 °C for t_{sk}). Dashes indicate that no experimental data were available for that environmental category, 'NONE' indicates that none of the models' predictions were within the maximum average standard deviations observed.
† Mean rmsds were equal in this category.
‡ Data were available for Tac only.

Table 15.8 Accuracy of thermal models over thermal conditions*

	$t_a \leq 5$		$5 < t_a \leq 15$		$15 < t_a \leq 25$		$25 < t_a \leq 35$		$35 < t_a$	
	t_{cr}	t_{sk}	t_{cr}	t_{sk}	t_{cr}	t_{sk}	t_{cr}	t_{sk}	t_{cr}	t_{sk}
nude/rest/ no-wind	NONE	NONE	NONE	NONE	NONE	lut2	lut25 lut2	lut2† lut25†	lut25	lut25 lut2
nude/rest/wind	–	–	NONE	NONE	luttre	lut25	–	–	–	–
nude/work/ no-wind	–	–	–	–	lut25	NONE	lut25	lut25† lut2†	lut25	lut2
nude/work/wind	–	–	–	–	–	–	lut2 lut25	–	NONE	lut2† lut25†
clothed/rest/ no-wind	NONE	NONE	NONE	lut2 lut25	–	–	–	–	–	–
clothed/rest/wind	NONE	NONE	NONE	NONE	–	–	–	–	–	–
clothed/work/ no-wind	NONE	lut25	lut2	NONE	–	–	lut2†	lut2	NONE‡	lut2
clothed/work/ wind	NONE	NONE	–	–	–	–	NONE	–	–	–

Source: Haslam and Parsons (1989a).

Notes
* The most accurate models for each environment category in descending order, where the mean rmsds for the models' predictions were within 0.3 °C for t_{cr} and 0.8 °C for t_{sk}. Dashes indicate that no experimental data were available for that environmental category, 'NONE' indicates that none of the models' predictions were within the maximum average standard deviations observed.
† Mean rmsds were equal in this category.
‡ Data were available for Tac only.

Table 15.9 Current modelling capabilities

Difficulty	Probability of obtaining a useful result	
1	Excellent	(90%)
2	Good	(70%)
3	Fair	(50%)
4	Poor	(30%)
5	Unlikely	(10%)

Source: Adapted from Wissler (1988).

provide the best simulation. Haslam and Parsons (1989b) however, found that no one-model tested was best for all conditions. A model 'system' was then developed whereby the most appropriate model could be selected according to the human exposure conditions.

A further development would be to use a thermal model in combination with a database of human responses. A 'hybrid' model could be produced which would provide an indication of how appropriate a rational thermal model would be and of how good a 'match' had been found between conditions of interest and those in the database of human responses. Such a system is feasible, could be kept up-to-date, and may provide the best of all models. The concept of a simple model could also be challenged and may be the way forward is to provide tools and a software environment that allows the construction of a model or system to optimize predictions for a particular application.

Computer aided design (CAD)

Computer aided design (CAD) has been a specialist topic for experts using sophisticated computer equipment for particular applications usually involving relatively large expense. The widespread availability and use of microcomputers and the development of software for practical use has meant that CAD has become an integral part of the design and evaluation process. It is likely therefore that where there is a need to involve knowledge of the human sciences in design and assessment, so there will be a need for its involvement in CAD software. Parsons (1989) considers CAD of the thermal environment and demonstrates how a CAD system could be produced for human thermal environments.

The production of any product naturally requires a specification in terms of what it is intended to do, how well it should do it, who should be able to use it, etc. (Eason, 1989). The development of a CAD system for thermal environments requires a method of dividing the knowledge domain (i.e. all knowledge to be included on human response to the thermal environment) and a dynamic method of presenting and assessing the knowledge (i.e. the interface design). A number of studies have attempted to produce such systems. For example, Smith and Parsons (1987) divided the knowledge domain into knowledge for the area, simulation (e.g. the use of models) and standards and limits. Virk (1986) investigated so-called knowledge elicitation techniques in this area. He observed experts in the area of human response to the thermal environment to identify how they considered and investigated practical problems to provide expert advice. Keyson and Parsons (1990) considered the design of a 'user interface' for an environmental ergonomics knowledge-based system, using evaluations and rapid prototyping techniques. Chui and Parsons (1990) investigated the use of video in the system (KEES – Parsons, 1987). Knowledge of human thermal environments is divided into hot, moderate and cold environments and effects are considered in terms of health, comfort and performance. For each combination of these areas, knowledge (a summary of the area), literature (which an expert would recommend), analysis and simulation (allows calculation of appropriate thermal models and indices), standards and limits (ISO etc.) and video (a short video presentation of an expert providing advice on the topic) are provided. The interface for the whole environmental ergonomics system is shown in Figure 15.17.

Symington and Warren (1989) present a computer system which allows calculation of metabolic cost and heat stress for use in US military applications. This system integrates models for predicting metabolic rate for various activities (e.g. walking rapidly over rough terrain in military clothing, etc.) with the heat stress model of Givoni and Goldman (1972) to provide

1 Thermal	2 Noise	3 Vibration	4 Light	5 Air quality	6 Psychol	7 Spatial
1. Hot..........Office........... Hand-arm.... Office..........Particles...... Group..........Anthro.......
2. Moderate Traffic........... Aircraft.........Industry........Gases..........Personal......................
3. Cold.........Aircraft..........Ship...
4.................Neighbour.....Building...
5.................Domestic...... Roadvehicles...
6.................................... Off roadvehicles...

```
                    1.....Health............
                    2.....Comfort..........
                    3.....Performance...

                1.....Knoweledge......................
                2.....Literature.........................
                3.....Analysis and simulation.....
                4.....Limits and standards.........
                5.....video...............................
                                                        'X' to exit
        Enter selection.....                            'Q' to quit
```

Figure 15.17 Computer software interface for an environmental ergonomics system.

a practical system for use by military personnel. Similar larger scale computer systems have been developed for use by experts in applied physiology (e.g. MAPS – Modular Applied Physiology System (Wadsworth and Parsons, 1989, 1991)).

Galer and Taylor (1989) describe a system using hypertext (sophisticated computer software system for searching key and related words and topics rather like a dynamic book) to provide an electronic book or manual. This is not primarily concerned with thermal environments but is to support the implementation of human factors knowledge into design. The use of solely text-based systems (i.e. no facility for calculation) in the design and assessment of thermal environments is clearly limited; however there are possibilities.

The work of Symington and Warren (1989) was related to the MANPRINT programme developed in the USA. This programme has also been adopted in the UK and is concerned with ensuring that human factors issues and knowledge are considered from the design stage and throughout a programme for developing a product. Computer software tools, such as hypertext forms of military standards are under development. This activity is clearly not restricted to military applications. It is both feasible and likely that there will be increasing demand for CAD tools for designing and evaluating human thermal environments.

Example of the use of a simple CAD system

A simple computer support or CAD system is described by Parsons (1989). The interface for the system is shown in Figure 15.17 which also demonstrates the facilities and selections available to the user. For the thermal environment it can be seen that the user selects hot, moderate or cold environments and whether he is interested in health, comfort or performance. For each of the combinations of selections, utilities offered are knowledge, literature, simulation and analysis, standards and limits and video. With experience the user will use the system in a dynamic way to help provide solutions and advice to hypothetical and practical problems. It is impossible to present the dynamic features of such a system in textual form. However, a practical example is provided below with a description of how the system may be used.

Problem definition

Consider receiving the following request for advice:

> We are designing the environment of a clean room which requires a high ventilation rate to maintain fumes at an acceptable level. Workers are seated performing a task involving manual dexterity. We anticipate that the air moving across the worker will be about $1.6\,\mathrm{ft\,s^{-1}}$ and at

a temperature of around 71 °F. The workers wear light protective clothing and will carry out eight-hour shifts with fifteen-minute breaks every two hours. Before we engage in the expense of building the rooms and purchasing air conditioning and ventilation equipment we would like to know if the thermal environment would be acceptable. That is, in terms of thermal comfort, do they meet industrial limits and will the conditions maintain productivity and health?

Solution

The reader will see that there is no unique solution to the above problem and that the CAD system could be used in many ways. A possible use of the system is described below.

The system structure will help to analyze the problem in terms of the questions asked and the way in which solutions can be structured (see Figure 15.7).

Convert units

The conversion facility will convert $1.6\,\mathrm{ft\,s^{-1}}$ to $0.5\,\mathrm{m\,s^{-1}}$ and 71 °F to 21.7 °C. The units are now in a form appropriate for inputs to models and for comparison with limit values.

Select 'knowledge' utility

Knowledge for moderate environments in terms of health, comfort and performance are consulted. General principles and a single summary sheet are obtained. These are presented below.

Moderate environments are usually associated with thermal comfort. Health effects are usually related to individual components of the thermal environment. The response of persons will depend upon the air temperature, radiant temperature, air velocity, humidity and the type of work and clothing worn. Generally for office workers for air temperatures of around 20 °C, maintain r_h around 40–60 per cent, air velocity $<0.15\,\mathrm{m\,s^{-1}}$ and no major directional radiation components.

The mechanisms for health effects in moderate environments are not fully understood. Symptoms are both behavioural and physiological. Workers complain of drowsiness, general lack of well-being, and headaches. Skin rashes and musculoskeletal complaints can occur. This is often labelled sick building syndrome of which the thermal environment may be only one contributing factor.

Low humidity can cause problems, dryness, electrostatic, skin rashes, etc. However it is not low humidity alone but as an interaction with other components. For example, raising humidity levels will reduce deposition of dust or particles in the air onto the skin. Air movement over exposed skin (neck, ankles, etc.) may lead to discomfort and health effects.

Thermal comfort is defined as that condition of mind which expresses satisfaction with the thermal environment. It is a subjective reaction usually associated with moderate environments.

The comfort of a person will depend upon the air temperature, radiant temperature, air velocity and humidity of the environment and individual factors such as the type of work and clothing worn.

The work of Fanger (1970) is important. He used data from USA and Denmark to develop the *PMV*, predicted mean vote, and the *PPD*, predicted percentage dissatisfied, in an environment. The work of Gagge *et al.* (1971) and of ASHRAE (1993) have provided the main principles. The four conditions for thermal comfort are heat balance, mean skin temperature within comfort limits, sweat rate (or skin wettedness) within comfort limits and that there is no local thermal discomfort.

The *PMV* method is outlined in Fanger's book *Thermal Comfort* and in ISO 7730. The six basic parameters (above) are used in an equation to provide the *PMV*, a value between +3 (hot) and −3 (cold), with 0 (neutral) providing optimum comfort. The *PPD* is calculated from the *PPD* (see 'analysis and simulation' in this CAD system).

Local discomfort can be caused by air movement (best $<0.15\,\mathrm{m\,s^{-1}}$), asymmetric radiation (keep below 5–10 °C), etc. See B and K booklet on Local Thermal Discomfort. Draughts, etc. are a major source of dissatisfaction in moderate environments. Surveys have shown that thermal conditions are a major factor in overall worker's dissatisfaction.

Moderate environments are defined by air temperature, radiant temperature, humidity, and air velocity and effects depend upon work activity, clothing worn, and other individual and contextual factors.

Although a number of studies have investigated the effects of thermal environments on performance, no accurate prediction can be made. What we mean by performance is an important issue. A convenient division is into manual and cognitive (mental) performance. Laboratory studies have been inconclusive, some have shown improved performance in mild heat stress.

Motivation, level of skill and type of task are important. In extreme conditions where subjects collapse or behavioural changes occur (also dehydration, heat stroke, etc.) then performance will be affected. However, these are unlikely to occur in moderate conditions. Performance may be affected by the distracting effect of cold discomfort and cold hands may reduce ability to perform fine manual tasks.

The concept of arousal is often used to explain effects. Minimum arousal will occur in slightly warm conditions. Vigilance tasks will be most affected by low arousal. But how does one measure or predict arousal? It is often suggested that slightly cool environments will increase arousal and be optimal for performance at some tasks.

The work of Pepler (1964), Meese, Wyon, Kok *et al.* in the late 1970s in South Africa and the summaries in the books by McIntyre and by Chrenko are of interest. These include the work of Vernon, Warner *et al.* in the

British coal mines and the work of Wing (1965). Ramsey in Texas has considered limits for hot environments in terms of the *WBGT* index. Fox (1967) reviews the effects of cold and concludes that hand skin temperature is important.

The above provide a brief summary of the area and can act to provide basic expert guidance towards the solution of a problem.

Standards and limits

Selection of the 'Standards and Limits' utility in the areas of health, comfort and performance in moderate environments will provide references to relevant current standards and list guidelines and limits where available.

Analysis and simulation

The analysis and simulation utility provides a facility for analyzing the data and calculating thermal index values relevant to the problem (for comparison with limits for example). Interactive thermal modelling will also be useful. The knowledge utility advised calculation of the *PMV* and *PPD*. This is provided below. It will be seen that original information is insufficient for all inputs to the *PMV* model. Reasonable assumptions or further consultation with the client must therefore be made.

Input data

Metabolic rate	$W\,m^{-2}$	$= 70$
Clothing insulation	$m^2\,°C\,W^{-1}$	$= 0.155$
Air temperature	$°C$	$= 21.5$
Mean radiant temperature	$°C$	$= 21.5$
Relative air velocity	$m\,s^{-1}$	$= 0.5$
Partial vapour pressure	Pa	$= 1014$

Output

	PMV
Hot	3
Warm	2
Slightly warm	1
Neutral	0
Slightly cool	-1
Cool	-2
Cold	-3

Predicted mean vote	$PMV = -0.5$
Predicted percentage of dissatisfied	$PPD = 10.3$

In this case therefore, in terms of whole-body thermal comfort, workers are predicted to feel on average between 'neutral' and 'slightly cool'. This may be acceptable; however, it may also be useful to calculate how much one should raise air temperature (or change other combinations of parameters) to provide optimum conditions. The CAD model is then used again to find a simulation involving an air temperature of 23.5 °C that will provide comfort (*PMV* = 0).

In terms of predicted physiological response of workers (which can be related to discomfort) the two-node model of human thermoregulation will provide a prediction.

Parsons KCP2N – J. B. Pierce (two-node model of thermoregulation)

Thermal conditions

Air temperature (°C)	21.5
Mean radiant temperature (°C)	21.5
Air velocity (m s^{-1})	0.5
Relative humidity (fraction)	0.5
Clothing insulation (Clo)	1.0
Metabolic rate (W m^{-2})	70.0
External work (W m^{-2})	0.0

Predicted final conditions

Final skin wettedness (ND)	0.1
E_{max} (W m^{-2})	129.9
Final mean skin temperature (°C)	33.2
Final body core temperature (°C)	36.8
Skin heat loss (W m^{-2})	63.2
Dry heat loss (W m^{-2})	48.8
Total evaporative heat loss (W m^{-2})	20.0
Final body heat store (W m^{-2})	0.0

Simulation – 4-h exposure

Time (min)	Skin temp (°C)	Core temp (°C)	Wettedness (fraction)
0	33.70	36.80	0.06
30	33.21	36.83	0.11
60	33.20	36.83	0.11
90	33.20	36.83	0.11
120	33.20	36.83	0.11
150	33.20	36.83	0.11
180	33.20	36.83	0.11
210	33.20	36.83	0.11
240	33.20	36.83	0.11

It is difficult to present the dynamic characteristics of the CAD system, however it should be clear from the above that the simulations can be used to design and evaluate environments.

Literature

Selected literature for effects of moderate environments on health, comfort and performance will provide sources for further information in this area.

Reply to the problem

A comprehensive reply can be provided to the client based on the above. From the knowledge facility it was clear that local thermal discomfort will occur due to unacceptable levels of air movement across workers causing draught and possible health problems. Humidity should also be considered. From the analysis and simulation, whole-body discomfort was of little importance but air temperatures of 23.5 °C would improve comfort. The design of clothing will be important. Limits and guidelines can be provided as well as further relevant reading.

There are many issues, about the use of such CAD and advice-giving systems, which have yet to be resolved. The quality and integrity of information as well as methods for providing warnings for misuse of data are important. Such systems however, have clear practical value; issues will be resolved and they will play a major role for practitioners in the future.

Appendix   Fundamentals of heat transfer

Heat is a form of energy and is associated with the thermodynamic property of temperature. Heat flows from higher to lower temperatures and this is called heat transfer. The rate of heat flow depends upon a number of system properties and parameters and obeys the laws of thermodynamics.

The *first law of thermodynamics* states that the change in internal energy of a system is equal to the heat added to the system minus the work done by the system, that is, there is *conservation of energy*. The amount of heat (energy) given up by one system which is interacting with another system is equal to the amount taken up by the second system.

The *second law of thermodynamics* states that heat flows spontaneously from a hot body to a cold one but not the reverse.

Heat transfer is commonly considered as three types; conduction, radiation and convection and is related to evaporation.

CONDUCTION

Conduction of heat in stationary substances occurs due to internal temperature gradients causing vibrational motions of free electrons (solids) and molecules (liquids and gases) causing transfer of heat from higher to lower temperatures. All effects are at the microscopic level and there is no appreciable motion of the substance.

Fourier's law of conduction is given by:

$$Q = -kA \frac{T_2 - T_1}{d},$$

where

Q = rate of heat transfer by conduction $(\mathrm{W\,m^{-2}})$

k = thermal conductivity of medium (Table A.1) $(\mathrm{J\,s^{-1}\,m^{-1}K^{-1}})$

A = cross-sectional area normal to conductive direction $(\mathrm{m^2})$

$T_2 - T_1$ = temperature difference across the medium (K)

d = distance between points at temperatures T_2 and T_1 (m).

Conduction of heat is the only mechanism possible in opaque and stationary media.

Table A.1 Typical values of the thermal conductivity for solids, liquids and gases

Substance	Temperature (°C)	Thermal conductivity ($\mathrm{W\,m^{-1}\,^\circ C^{-1}}$)
Solids		
Silver (pure)	0	417
Copper (pure)	0	356
Aluminium (pure)	0	228
Iron (pure)	0	73.7
Stainless steel (18% Ni, 8% Cr)	0	16.2
Ice	0	2.21
Glass (soda-lime)	20	0.76
Bone (*in vivo*)	37	0.4–2.15
Liquids		
Water	0	0.57
Blood	37	0.51–0.53
Glycerine (pure)	0	0.28
Alcohol ethyl	20	0.18
Kerosene	20	0.15
Gases		
Hydrogen	0	0.168
Helium	0	0.142
Water vapour	0	0.025
Air	0	0.024
Nitrogen	0	0.024
Oxygen	0	0.015
Carbon dioxide	0	0.015

Source: Shitzer and Eberhart (1985a,b).

RADIATION

All bodies above a temperature of absolute zero emit thermal radiation, and heat transfer occurs in the form of electro-magnetic waves between two opaque solids at different temperatures. This is similar to light (0.39–0.78 µm) but heat rays are over the waveband 0.1–10 µm. The energy emitted (q) is proportional to the fourth power of the absolute temperature (T).

$$q = \varepsilon\sigma T^4,$$

where $\sigma = 5.67 \times 10^{-8}\,\mathrm{W\,m^{-2}\,K^{-4}}$, is the Stefan–Boltzmann constant. ε is the emissivity of the body (see Table A.2).

In reality, the emitted energy varies with the wavelength of the radiation such that there is an emission spectrum. The thermal emissivity is 1.0 for a black body but is typically less than 1.0 and depends upon the properties of the body surface.

For elementary calculations it is convenient to assume, as in Table A.2, that emissivity does not vary with wavelength, although this is not strictly true.

Table A.2 Typical values of normal total radiation emissivity for various surfaces

Substance	Temperature (K)	Emissivity
Aluminium, highly polished, 98.3% pure	500–850	0.039–0.057
Copper, polished	390	0.023
Steel, polished	373	0.066
Asbestos, board	297	0.96
Glass, smooth	296	0.94
Water	273–373	0.95–0.963
Skin	305–307	0.97–0.99

Source: Shitzer and Eberhart (1985a,b).

When two black bodies exchange heat by radiation:

$$Q_{1-2} = A_1 F_{12} \sigma T_1^4 - A_2 F_{21} \sigma T_2^4,$$

where

A_i = surface area involved in the exchange of energy (m^2)

F_{ij} = shape factor describing the proportion of the ith area viewed by the other (j) area.

The value of F_{ij} for a totally closed system is unity. For surfaces not black the expressions $A_i F_{ij} \sigma T_i^4$ are multiplied by grey body emissivity (ε).

When radiation energy strikes a medium, there is reflectivity, absorbtivity and transmissivity. The first law of thermodynamics requires that these add to unity.

Radiation occurs between non-contacting surfaces and requires no medium. It is the only energy transfer possible in a vacuum.

CONVECTION

Heat transfer by convection involves the bulk movement of fluid in four simultaneously occurring and interrelated processes:

1 conduction of heat from the solid surface to immediately adjacent fluid particles;
2 absorption and storage of the conducted energy by the particles with the elevation of their internal energy;
3 movement of these higher energy particles to regions of lower temperature, with mixing and transfer of part of the stored energy; and
4 transport of the energy by the bulk movement of the fluid.

Associated with convection (q) is a boundary layer over the surface of the solid. Heat transfer by convection is given by Newton's law of cooling:

$$q = h_c(T_1 - T_2).$$

Table A.3 Typical values of the convection heat transfer coefficient h_c

Mode	h_c $Wm^{-2}K^{-1}$
Free convection	5–25
Forced convection	
Gases	25–250
Liquids	50–20 000
Convection with phase change (boiling or condensation)	2 500–100 000

Source: Shitzer and Eberhart (1985a,b).

where

h_c = convective heat transfer coefficient (see Table A.3)

T_1 = surface temperature of the solid

T_2 = bulk temperature of fluid outside boundary layer.

Experimental data relating to heat transfer by convection are often reported in terms of dimensionless parameters. Some of these are presented with interpretation in Table A.4.

Natural convection occurs due to buoyancy forces resulting from temperature differences, e.g. warm air rises. Forced convection is a result of a positive displacement of fluid produced by mechanical means.

EVAPORATION

Evaporation is similar to heat transfer by convection but also requires an initial change of state from liquid to vapour at the skin surface and subsequent diffusion of vapour across the boundary layer into the ambient air. The driving force for diffusion is the concentration gradient and the quantity transferred is mass. The rate of mass transfer (m) is

$$m = h_d(C_s - C_a).$$

where C_s and C_a are water vapour concentrations (mass per unit volume) at the skin and ambient air respectively and h_d is the mass transfer coefficient, analogous to h_c and related by

$$h_d = \frac{h_c}{\rho c}.$$

This is the Lewis relation (Lewis, 1922) and is the mechanism that allows calculation of heat transfer from a sweating body.

Table A.4 Dimensionless parameters associated with the transport of energy in a biological tissue

Parameter	Name	Physical interpretation
$\dfrac{hd}{k}$	Biot (Bi)	Energy transferred by convection at the surface of a tissue divided by energy conducted through the tissue
$\dfrac{kt}{pcd^2}$	Fourier (Fo)	Energy conducted through tissue divided by heat capacity of tissue (thermal inertia)
$\dfrac{pcwd^2}{k}$		Energy transported by bloodstream (capillary perfusion) divided by energy conducted through tissue
$\dfrac{qd^2}{k\Delta T}$		Metabolic heat generation divided by energy conducted through the tissue
$\dfrac{pud}{\mu}$	Reynolds (Re)	Inertial force maintaining blood flow in a channel divided by viscous force restraining the flow
$\dfrac{\mu c}{k} = \dfrac{\nu}{\alpha}$	Prandtl (Pr)	Kinematic viscosity (ρ/ρ) divided by thermal diffusity ($k/\rho c$)
$\dfrac{pcud}{k}$	Peclet (Pe)	Thermal convection divided by thermal diffusivity
$\dfrac{pcud^2}{kL}$	Graetz (Gz)	Channel aspect ratio multiplied by convected energy in channel divided by conducted energy in blood channel
$\dfrac{hd}{k}$	Nusselt (Nu)	Energy transferred by convection at the surface of a tissue divided by energy conducted through the fluid
$\dfrac{\mu}{\rho D}$	Schmidt (Sc)	Kinematic viscosity divided by mass diffusity
$\dfrac{h_D d}{D}$	Sherwood (Sh)	Equivalent to Nu for mass transfer
$\dfrac{p^2 q\beta L^3 \Delta T}{\mu^2}$	Grashof (Gr)	A measure of thermally induced flow (natural convection)

Source: Shitzer and Eberhart (1985a,b).

References

ACGIH, 1992, American Conference of Governmental Industrial Hygienists, Threshold Limit Values and Biological Exposure Indices, Cincinnati, OH, USA.

ACGIH, 1996, TLVs and BEIs. Threshold Limit Values for chemical substances and physical agents. Biological Exposure Indices. American Conference of Governmental Industrial Hygienists, Cincinnati, OH, USA, ISBN 1–882417–13–5.

ACGIH, 1998, TLVs and BEIs. Threshold Limit Values for chemical substances and physical agents. Biological Exposure Indices. American Conference of Governmental Industrial Hygienists, Cincinnati, OH, USA, ISBN 1–88–2417–23–2.

Aitken, J., 1887, Addition to thermometer screens, *Proceedings of the Royal Society, Edinburgh*, **14**, 428.

Akimoto, T., Nobe, T., Tanabe, S. and Kimnra, K., 1999, Floor-supply displacement air conditioning: laboratory experiments, *ASHRAE Transactions*, **105** (Pt2), 739–749.

Allen, J. R., 1988, A technical basis for the development of thermal performance standards for immersion protection, in I. B. Mekjavik, E. W. Banister and J. B. Morrison (Eds), *Environmental Ergonomics*, London: Taylor & Francis, 205–220, ISBN 0–85066–400–4.

Ambler, H. R., 1955, Notes on the climate of Nigeria with reference to personnel, *Journal of Tropical Medicine and Hygiene*, **58**, 99–112.

Angel, H. A., 1995, Laboratory studies of clothing ventilation using a direct tracer gas technique and an indirect user performance test. MSc thesis, Loughborough University, UK.

Aptel, M., 1988, Comparison between required clothing insulation and that actually worn by workers exposed to artificial cold, *Applied Ergonomics*, **19**(4), 301–305.

Arnaud, M. J. (Ed.), 1998, *Hydration throughout life*, International Conference, Vittel (France), June 9–12, Montrouge: John Libbey Eurotext.

ASHVE, 1924, Thermal comfort conditions, ASHVE standard.

ASHRAE, 1966, Thermal comfort conditions, ASHRAE standard 55.66, New York.

ASHRAE, 1989a, Physiological principles, comfort and health, in *Fundamentals Handbook*, Atlanta.

ASHRAE, 1989b, Thermal environmental conditions for human occupancy, ANSI/ASHRAE standard, Atlanta.

ASHRAE, 1992, Thermal environmental conditions for human occupancy, ANSI/ASHRAE standard, Atlanta.

ASHRAE, 1993, Physiological principles and thermal comfort, ASHRAE Handbook of Fundamentals, Chapter 8, Atlanta, USA.

ASHRAE, 1997, Thermal comfort, ASHRAE Handbook of Fundamentals, Chapter 8, Atlanta, USA.

Astrand, P. and Rodahl, K., 1986, *Textbook of Work Physiology–Physiological Bases of Exercise*, 3rd edn, New York: McGraw-Hill.

Atha, J., 1974, Physical fitness measurements, in L. A. Larson (Ed.), *Fitness, Health and Work Capacity*, New York: Macmillan.

Auliciems, A., 1981, Towards a psychophysiological model of thermal perception, *International Journal of Biometerology*, 25, 109–122.

Baker, N. and Standeven, M., 1996, Thermal comfort for free-running buildings, *Energy and Buildings*, 23.

Ballantyne, E. R., Barned, J. R. and Spencer, J. W., 1967, Environment assessment of acclimatized caucasian subjects at Port Moresby, Papua, *Proceedings of the Third Australian Building Research Congress*.

Baron, R., 1972, Aggression as a function of ambient temperature and prior anger arousal, *Journal of Personality and Social Psychology*, 21, 183–189.

Baron, R. and Lawton, 1972, Environmental influences on aggression. The facilitation of modelling effects by high ambient temperatures, *Psychonomic Science*, 26, 80–86.

Bedford, T., 1936, *The warmth factor in comfort at work: A physiological study of heating and ventilation*, Industrial Health Research Board Report No. 76, London: HMSO.

Bedford, T., 1946, *Environmental warmth and its measurement*, Medical Research Memorandum, No. 17, London: HMSO.

Bedford, T. and Warner, C. G., 1934, The globe thermometer in studies of heating and ventilation, *Journal of Hygiene*, 34, 458–473.

Behmann, F. W., 1988, Field evaluation methods, in *Handbook on Clothing*, Research Group 7, Biomedical Research Aspects of Military Protective Clothing, NATO, Brussels.

Belding, H. S., 1970, The search for a universal heat stress index, in J. D. Hardy and C. C. Thomas (Eds), *Physiological and Behavioural Temperature Regulation*, Springfield IL.

Belding, H. S. and Hatch, T. F., 1955, Index for evaluating heat stress in terms of resulting physiological strain, *Heating Piping and Air Conditioning*, 27, 129–136.

Berglund, L. G., 1988, Thermal comfort: A review of some recent research, in I. B. Mekjavik, E. W. Banister and J. B. Morrison (Eds), *Environmental Ergonomics*, London: Taylor & Francis, pp. 70–86.

Berglund, L. G. and Cunningham, D. J., 1986, Parameters of human discomfort in warm environments, *ASHRAE Transactions*, 92(2), 732–746.

Berglund, L. G. and Fobelets, A. P., 1987, Subjective human response to low-level air currents and asymmetric radiation, *ASHRAE Transactions*, 93.

Berlin, H. M., Stroschein, L. and Goldman, R. F., 1975, *A computer program to predict energy cost, rectal temperature, and heart rate response to work, clothing and environment*, Department of the Army report No ED-SP-75011, Edgewood Arsenal, Maryland, USA.

Bethea, D. and Parsons, K. C., 1998a, The user-oriented design, development and evaluation of the clothing envelope of thermal performance, in M. A. Hanson (Ed.), *Contemporary Ergonomics*, London: Taylor & Francis, pp. 520–524, ISBN 0–7484–0811–8.

Bethea, D. K. and Parsons, K. C., 1998b, The validity of ISO 7933:1989 duration limit exposure (DLE) and predicted sweat rate (SWP) for clothed subjects, in J. A. Hodgdon, J. H. Heaney and M. J. Buono (Eds), *Environmental Ergonomics VIII*, pp. 301–305, ISBN 0–9666953–1–3.

Bethea, D. and Parsons, K. C., 2000, The development of a practical heat stress assessment methodology for use in UK industry, Final report in HSE contract 3589/R53.162, Loughborough University, UK.

Birnbaum, R. R. and Crockford, G. W., 1978, Measurement of the clothing ventilation index, *Applied Ergonomics*, 9(4), 194–200.

Black, A. E., Coward, W. A., Cole, T. J. and Prentice, A. M., 1996, Human energy expenditure in affluent societies: an analysis of 574 doubly labelled water measurements, *European Journal of Clinical Nutrition*, 50, 72–92.

Blazejczyjk, K., 2000, Absorption of solar radiation by man, in J. Werner and M. Hexamer (Eds), *Environmental Ergonomics IX*, pp. 43–46, Aachen, Germany: Shaker Verlag, ISBN 3–8265–7648–9.

Bligh, J. and Moore, R. (Eds) 1972, *Essays on temperature regulation*, North-Holland, Amsterdam, ISBN 0–7204–4103–X.

Bligh, J., 1978, Thermoregulation: What is regulated and how? in Y. Houdas and J. D. Guieu (Eds), *New Trends in Thermal Physiology*, Paris: Masson, pp. 1–10.

Bligh, J., 1979, The central neurology of mammalian thermoregulation, *Neuroscience*, 4, 1213–1236.

Bligh, J., 1985, Regulation of body temperature in man and other mammals, in A. Shitzer and R. C. Eberhart (Eds), *Heat Transfer in Medicine and Biology – Analysis and Applications*, Vol. 1, New York: Plenum Press, pp. 15–52.

BOHS, 1990, *The Thermal Environment*, British Occupational Hygiene Society Guide No. 8, by Youle, A., Collins, K. J., Crockford, G. W., Fishman, D. S., Parsons, K. C. and Sykes, J., Science Reviews Lid with H and H Scientific Consultants Ltd, Leeds.

BOHS, 1996, *The Thermal Environment*, 2nd edn, British Occupational Hygiene Society Technical Guide No. 12, by Youle, A., Collins, K. J., Crockford, G. W., Fishman, D. S., Mulhall, A. and Parsons, K. C. H. and H Scientific Consultants Ltd, Leeds, ISBN 0–948237–29–5.

Bonjer, F. H., Davies, C. J. M., Lange Andersen, K., Sargeant, A. J. and Shepherd, R. J., 1981, Measurement of maximum aerobic power, in J. S. Weiner and J. A. Lourie (Eds), *Practical Human Biology*, London: Academic Press.

Bonnes, M. and Secchiatoli, G., 1995, *Environmental psychology: A psycho-social introduction*. London: Sage, ISBN 0–8039–7905–1.

Borg, G. A. V., 1982, Psychophysical bases of perceived exertion, *Medical Science Sports Exc.*, 14(5), 377.

Borg, G., 1998, Borg's perceived exertion and pain scales, Human Kinetics, IL, USA, ISBN 0–88011–623–4.

Botsford, J. H., 1971, A wet globe thermometer for environmental heat measurement, *American Industrial Hygiene Journal*, 32, 1–10.

Boyce, P. R., 1981, *Human Factors in Lighting*, London: Applied Science.

Bouskill, L. and Parsons, K. C., 1996, Effectiveness of a neck cooling personal conditioning unit at reducing thermal strain during heat stress, in S. A. Robertson (Ed.), *Contemporary Ergonomics*, London: Taylor & Francis, pp. 196–201, ISBN 0–7484–0549–6.

Bouskill, L. M., Kuklane, K., Holmer, I., Parsons, K. C. and Withey, W. R., 1998a, The relationship between ventilation index and thermal insulation, in J. A. Hodgdon, J. H. Heaney and M. J. Buono (Eds), *Environmental Ergonomics VIII*, pp. 195–202, ISBN 0–9666953–1–3.

Bouskill, L., Livingston, R., Parsons, K. C. and Withey, W. R., 1998b, The effect of external air speed on the clothing ventilation index, in M. A. Hanson (Ed.), *Contemporary Ergonomics*, London: Taylor & Francis, pp. 540–544, ISBN 0–7484–0811–8.

Bouskill, L., Sheldon, N., Parsons, K. C. and Withey, W. R., 1998c, The effect of clothing fit on the clothing ventilation index, in M. A. Hanson (Ed.), *Contemporary Ergonomics*, London: Taylor & Francis, pp. 510–514, ISBN 0–7484–0811–8.

Bouskill, L. M., 1999, Clothing ventilation and human thermal response. PhD thesis, Department of Human Sciences, Loughborough University, UK.

Bouskill, L. M., Havenith, G., Kuklane, K., Parsons, K. C. and Withey, W. R., 2002, Relationship between clothing ventilation and thermal insulation, *American Industrial Hygiene Association Journal*, In Press.

Boutelier, C., 1979, Survival and protection of aircrew in the event of accidental immersion in cold water, AGARD Report No. AG211, Neuilly sur Seine, France, NATO.

Braun, T. L. and Parsons, K. C., 1991, Human thermal responses in crowds, in E. J. Lovesey (Ed.), *Contemporary Ergonomics*, London: Taylor & Francis, pp. 190–195.

Breslin, R., 1995, Gender differences and thermal requirements, Final year undergraduate report, Department of Human Sciences, Loughborough University, UK.

Brierley, C., 1996, Acclimation: familiarisation to hot, humid environments, and its effects on thermal comfort requirements. MSc thesis, Department of Human Sciences, Loughborough University, UK.

Brooke, S. and Ellis, H., 1992, Cold, in A. P. Smith and D. M. Jones (Eds), *Handbook of Human Performance*, 1, The Physical Environment, London: Academic Press, ISBN 0–12–650351–6.

Brooks, J. and Parsons, K. C., 1999, An ergonomics investigation into human thermal comfort using an automobile seat heated with encapsulated carbonised Fabric (ECF), *Ergonomics*, 42(5), 661–673.

Brouha, L., 1960, *Physiology in industry*, New York: Pergamon Press.

Brunstrom, J. M. and MacRae, A. W., 1997a, Effects of temperature and volume on measures of mouth dryness, thirst and stomach fullness in males and females, *Appetite*, 29, 31–42.

Brunstrom, J. M. and MacRae, A. W., 1997b, Mouth state: a nuisance variable, in preference tests, *Food Quality Preferences*, 8, 349–352.

Brunstrom, J. M., Tribbeck, P. M. and MacRae, A. W., 2000, The role of mouth state in the termination of drinking behaviour in humans, *Physiology and Behaviour*, 68, 579–583.

BS 4086, 1966, *Recommendations for maximum surface temperatures of heated domestic equipment*, London: BSI.

BS 4086, 1983, *Recommendations for maximum surface temperatures of heated domestic equipment*, revised, London: BSI.

BS 7915, 1998, Ergonomics of the thermal environment – Guide to design and evaluation of working practices for cold indoor environments, London: BSI.

BS 7963, 2000, Ergonomics of the thermal environment – Guide to the assessment of heat strain in workers wearing personal protective equipment, London: BSI.

BS EN ISO 7730, 1995, Moderate thermal environments – determination of the PMV and PPD indices and specification of the conditions for thermal comfort, London: BSI.

BS PD 6504, 1983, *Medical information on human skin contact with hot surfaces*, London: BSI.

Bull, J. P., 1963, Burns, *Postgraduate Medicine Journal*, 39, 717–723.

Burton, A. C. and Bazett, H. C., 1936, A study of the average temperature of the tissues of the exchanges of heat and vasomotor responses in man by means of a bath calorimeter, *American Journal of Physiology*, 117, 36.

Burton, A. C. and Edholm, O. G., 1955, *Man in Cold Environment*, London: Edward Arnold.

Cabanac, M., 1981, Physiological signals for thermal comfort, in K. Cena and J. A. Clark (Eds), *Bioengineering, thermal physiology and comfort*, Amsterdam: Elsevier, ISBN 0–444–99761–X.

Cabanac, M., 1992, Selective brain cooling in humans: Fancy or Fact, in Lotens and Havenith (Eds), *Proceedings of the Fifth International Conference on Environmental Ergonomics*, Maastricht, November, pp. 214–215.

Cabanac, M., 1995a, *Neuroscience intelligence unit: human selective brain cooling*, Heidelberg, Germany: Springer-Verlag, ISBN 3–540–59083–8.

Cabanac, M., 1995b, *Human selective brain cooling*, Germany: Springer-Verlag, ISBN 1–57059–223–3.

Cabanac, M., 2000, Human selective brain cooling during hyperthermia: does it remain a doubt? in J. Werner and M. Hexamer (Eds), *Environmental Ergonomics IX*, Aachen, Germany: Shaker Verlag, pp. 3–8, ISBN 3–8265–7648–9.

Candas, V. and Hoeft, A., 1988, Clothing assessment and effects on thermophysiological responses of man working in humid heat, *Proceedings of CEC Seminar on Heat Stress Indices*, Commission of the European Communities, Luxembourg.

Candas, V., Libert, J. P. and Brandenberger, G., 1985, Hydration during exercise – effects on thermal and cardiovascular adjustments, *European Journal of Applied Physiology*, 55(2), 113–122.

Candas, V., Sari, H. and Herrmann, C., 1998, Assessment of risk of discomfort due to thermal transients based on a computer model of thermoregulation, *Environmental Ergonomics VIII*, J. A. Hodgdon, J. H. Heaney and M. J. Buono (Eds), ISBN 0–9666953–1–3.

Canter, D., 1975, *Environmental Interaction*, London: Surrey University Press.

Canter, D., 1977, *The Psychology of Place*, Architectural Press.

Canter, D., 1985, *Applying Psychology*, Inaugural Lecture by Professor D. Canter, University of Surrey, May.

Carpenter, G. A. and Moulsley, L. H., 1972, A visualization technique for studying air movement in large enclosures over a wide range of ventilation rates, *JIHVE*, 39, 279–287.

CEC, 1988, Heat Stress Indices. Proceedings of a seminar held by the Commission of the European Communities, Luxembourg.

CEN, 1989, Safety of machinery; basic concepts; general principles for design, PrEN, Brussels, May.

CEN, 1990, Temperature of touchable surfaces. Ergonomics data to establish temperature limit values for hot surfaces, PrEN 563, Brussels.

Cena, K. and Clark, J. A., 1981, *Bioengineering, thermal physiology and comfort*, Vol. 10, Amsterdam: Elsevier, ISBN 0–444–99761–X.

Cena, K. and Spotila, J. R., 1984, *Thermal Comfort of the Elderly*, ASHRAE Research Project 421–RP.

Cena, K. and Spotila, J. R., 1986, *Thermal Comfort for the Elderly: Behavioural Strategies and Effort of Activities*, Final Report, ASHRAE, RP–460.

Cetas, T. C., 1985, Temperature measurement, in A. Shitzer and R. C. Eberhart (Eds), *Heat Transfer in Medicine and Biology*, New York: Plenum Press, pp. 373–391.

Chato, J. C., 1985, Measurement of thermal properties of biological materials, in A. Shitzer and R. C. Eberhart (Eds), *Heat Transfer in Medicine and*

Biology – Analysis and Applications, Vol. 1, New York: Plenum Press, pp. 167–192.

Chatonnet, J. and Cabanac, M., 1965, The perception of thermal comfort, *International Journal of Biometeorology*, 9, 183–193.

Chen, F., 1997, Thermal responses of the hand to convective and contact cold – with and without gloves. PhD thesis, Department of Industrial Ergonomics, Linkoping University, Sweden.

Chen, F., Lin, Z. and Holmér, I., 1996, Hand and finger temperature in convective and contact cold, *European Journal of Applied Physiology and Environmental Physiology*, 72, 372–279.

Chen, F., Nilsson, H. and Holmér, I., 1994, Cooling responses of finger in contact with an aluminium surface, *American Industrial Hygiene Association Journal*, 55, 218–222.

Chrenko, F. A. (Ed.), 1974, *Bedford's Basic Principles of Ventilation and Heating*, 3rd edn, London: H. K. Lewis.

Christensen, N. K., Albrechtsen, O., Fanger, P. O. and Trzeciakiewicz, A., 1984, Air movement and draught, *Proceedings of INDOOR AIR '84*, August 20–24, Stockholm, Sweden.

Chui, Y. P. and Parsons, K. C., 1990, The use of video in an environmental ergonomics knowledge based system, in E. J. Lovesey (Ed.), *Contemporary Ergonomics*, London: Taylor & Francis, pp. 321–326.

CIBSE, 1986, Environmental criteria for design, Section A1, *CIBSE Guide*, Vol A, London: The Chartered Institute of Building Services Engineers.

Clark, R. P. and Edholm, O. G., 1985, *Man and His Thermal Environment*, London: Edward Arnold.

Cohnheim, J., 1873, *Neue Untersuchungen uber die Entzundung*, Berlin.

Collins, K. J., 1983, *Hypothermia – The Facts,* Oxford: Oxford University Press.

Collins, K. J. and Hoinville, E., 1980, Temperature requirements in old age, *Building Services Engineering Research and Technology*, 1(4), 165–172.

Cooke, H. M., Wyndham, C. H., Strydom, N. B., Maritz, J. S., Bredell, G. A. G., Vileyn, V. W., Morrison, J. F., Peters, J. and Williams, C. G., 1961, The effects of heat on the performance of work underground, *Journal of the Mine Ventilation Society of South Africa*, October.

Cooper, S., Haslam, R. A. and Parsons, K. C., 1987, A model of human thermoregulation: the effects of body size on the accuracy of prediction, in E. D. Megaw (Ed.), *Contemporary Ergonomics*, London: Taylor & Francis, pp. 275–279.

Crawshaw, L. I., Nadel, E. R., Stolwijk, J. A. J. and Stamford, B. A., 1975, Effect of local cooling on sweating rate and cold sensation, *Pflügers Archive*, 354, 19–27.

Crockford, G., 1991, Communication to the Clothing Science Special Interest Group of the BOHS, and Ergonomics Society, UK.

Crockford, G. W. and Rosenblum, H. A., 1974, The measurement of clothing microclimate volumes, *Clothing Research Journal*, 2(3), 109–114.

Crockford, G. W., Crowder, M. and Prestidge, S.P., 1972, A trace gas technique for measuring clothing microclimate air exchange rates, *British Journal of Industrial Medicine*, 29, 378–386.

Danielsson, U., 1998, Risk of frostbite, in I. Holmér and K. Kuklane (Eds), *Problems with Cold Work*, pp. 133–135, NIWL Report Nr. 1998:18, Salna, Sweden.

Dasler, A. R., 1974, Ventilation and thermal stress, ashore and afloat, in Chapter 3, *Manual of Naval Preventive Medicine*, Washington: Navy Department, Bureau of Medicine and Surgery.

Dasler, A. R., 1977, Heat stress, work functions and physiological heat exposure limits in man, in *Thermal Analysis – Human Comfort – Indoor Environments*, NBS Special Publication 491, Washington: US Department of Commerce.

de Dear, R. J., 1998, A global database of thermal comfort field experiments, *ASHRAE Transactions*, **104**, part 1.

de Dear, R. J. and Brager, G. S., 1998a, Developing an adaptive model of thermal comfort and preference, *ASHRAE Transactions*, Atlanta, USA.

de Dear, R. J. and Brager, G. S., 1998b, A global database of thermal comfort field experiments, *ASHRAE Transactions*, Atlanta, USA.

de Dear, R. J. and Brager, G. S., 2002, Thermal comfort in naturally ventilated buildings: revisions to ASHRAE Standard 55, *Energy and Buildings*, **34**(6), pp. 549–572.

de Dear, R. J. and Fountain, M. E., 1994, Field experiments on occupant comfort and office thermal environments in a hot–humid climate, *ASHRAE Transactions*, **100**(2), 457–475.

de Dear, R. J., Brager, G. S. and Cooper, D. J., 1997, Developing an adaptive model of thermal comfort and preference – Final Report on ASHRAE RP–884, Sydney: MRL.

de Dear, R. J., Leow, K. G. and Ameen, A., 1991, Thermal comfort in the humid tropics – Part I. Climatic chamber experiments on temperature preferences in Singapore. *ASHRAE Transactions*, **97**(1), 874–879.

Dexter, E., 1904, *Weather Influences*, New York: Macmillan.

Dill, D. B. and Costill, D. L., 1974, Calculation of percentage changes in volumes of blood plasma and red cells in dehydration, *Journal of Applied Physiology*, **37**(2), 247–248.

Diller, K. R., 1985, Analysis of skin burns, in A. Shitzer and R. C. Eberhart (Eds), *Heat Transfer in Medicine and Biology*, Vol. 2, New York: Plenum Press.

DIN 33 403, 1994, Climate at workplaces and their environments. Part 5. Ergonomic design of cold workplaces, German Standard DIN Berlin.

Dodt, E. and Zotterman, Y., 1952, Mode of action of warm receptors, *Acta Physiology Scandinavia*, **26**, 345–357.

Donnini, G., Molina, J., Martello, C., Lai, D.H.C., Kit, L.H., Chang, C.Y., Laflamme, M., Nguyen, V.H. and Haghighat, F., 1996, Field study of occupant comfort and office thermal environments in a cold climate, Final report on ASHRAE RP – 821 Montreal, Quebec, Canada.

Drury, C. G., 1990, Methods for direct observation of performance, in J. R. Wilson and E. N. Corlett (Eds), *Evaluation of Human Work – A Practical Ergonomics Methodology*, London: Taylor & Francis, pp. 35–57.

DuBois, E. F., 1937, *Lane Medical Lectures. The Mechanism of Heat Loss and Temperature Regulation*, Stanford: Stanford University Press.

DuBois, D. and DuBois, E. F., 1916, A formula to estimate surface area if height and weight are known, *Archives of Internal Medicine*, **17**, 863.

Dufton, A. F., 1929, The eupatheostat, *Journal of Scientific Instruments*, **6**, 249–251.

Dufton, A. F., 1936, The equivalent temperature of a warmed room, *JIHVE*, **4**, 227–229.

Durnin, J. V. G. A. and Passmore, R., 1967, *Energy, Work and Leisure*, London: Heinemann Educational.

Eason, K. D., 1989, *Information Technology and Organizational Change*, London: Taylor & Francis.

Edholm, O. G., 1978, *Man – hot and cold*, London: Edward Arnold.

Edholm, O. G., 1981, Habitual activity and daily energy expenditure, in J. S. Weiner and J. A. Lourie (Eds), *Practical Human Biology*, London: Academic Press, pp. 227–239.

Edholm, O. G. and Weiner, J. S., 1981, *The Principles and Practice of Human Physiology*, London: Academic Press.

EEC, 1984, *The Treaty of Rome*, Brussels: The European Community.

Eley, C., Goldsmith, R., Layman, D., Tan, G. L. E., Walker, E. and Wright, B. M., 1978, A respirometer for use in the field for the measurement of oxygen consumption. The 'MISER', a miniature, indicating and sampling electronic respirometer, *Ergonomics*, 21, 253–264.

Ellis, F. P., 1953, Thermal comfort in warm and humid atmospheres. Observations on groups and individuals in Singapore, *Journal of Hygiene*, 51, 386.

Ellis, F. P., Smith, F. E. and Walters, J. D., 1972, Measurement of environmental warmth in SI units, *British Journal of Industrial Medicine*, 29, 361–377.

EN 563, 1994, Temperatures of touchable surfaces, Ergonomics data to establish temperature limit values for hot surfaces, CEN, Brussels.

Enander, A., 1987, Effects of moderate cold on performance of psychomotor and cognitive tasks, *Ergonomics*, 30, 1431–1445.

Environment Canada, 2001, Wind Chill science and equations, World Wide Web report by Environment Canada and DCIEM, Toronto, Canada.

Factories Act, 1961, London: HMSO.

Fanger, P. O., 1967, Calculation of thermal comfort: introduction of a basic comfort equation, *ASHRAE Transactions*, 73, Part 2.

Fanger, P. O., 1970, *Thermal Comfort*, Copenhagen: Danish Technical Press.

Fanger, P. O., 1972, Near future prospects of the meteorological environment in Developing Countries in deserts and tropical areas. Improvement of human comfort and resulting efforts on working capacity, *Biometeorology*, 5, 11.

Fanger, P. O., Hojbjerre, J. and Thomsen, J. O. B., 1977, Can winter swimming cause people to prefer lower room temperatures? *Int. J. Biometero*, 21(1), 44–50.

Fanger, P. O. and Pedersen, C. L. K., 1977, Discomfort due to air velocities in spaces, *Proceedings of Conference: Institut International du Froid*, Commissions B1, B2 and El, Belgrade, 289–296.

Fanger, P. O. and Christensen, N. K., 1986, Perception of draught in ventilated spaces, *Ergonomics*, 29, 215–235.

Fanger, P. O. and Toftum, J., 2002, Extension of the PMV model to non air-conditioned buildings in warm climates, *Energy in Buildings*, 34(6), 533–536.

Fanger, P. O., Melikov, A. K., Hanzawa, H. and Ring, J., 1989, Turbulence and draught, *ASHRAE Journal*, April.

Farmer, E., Brooke, R. S. and Chambers, E., 1923, A comparison of different shift systems in the glass trade, Report to the Industrial Fatigue Board No. 24, London: HMSO.

Fiala, D., 1998, Dynamic simulation of human heat transfer and thermal comfort. PhD thesis, de Montfort University, UK.

Fiala, S., 1998, Dynamic simulation of human heat transfer and thermal comfort PhD thesis, de Montfort University, UK.

Fiala, D. and Lomas, K. J., 2001, The dynamic effect of adaptive human responses in the sensation of thermal comfort, in *Conference Proceedings Moving Thermal Comfort Standards into the 21st Century*, Windsor, UK, ISBN 1–873640–33–1.

Fishman, D. and Jenne, B., 1981, Some like it hotter! An examination of preferred showering temperatures, *Watson House Bulletin*, 46(2), British Gas.

Fishman, D. S. and Pimbert, S. L., 1978, Survey of subjective responses to the thermal environment in offices, in *Indoor Climate: Effects on Human Comfort, Performance and Health in Residential, Commercial and Light Industry Buildings*, WHO conference, Copenhagen.

Fitts, P. M. and Posner, M. I., 1967, *Human Performance*, Belmont, CA: Brooks Cole.

Fleishman, E. A. and Ellison, Z., 1962, A factor analysis of fine manipulative tests, *Journal of Applied Psychology*, 46, 96–105.

Flemming, G. and Graveling, R. A., 1997, A series of working practices for cold work in the UK, Report to the BSI by Institute of Occupational Medicine, IOM, Edinburgh, UK.

Fobelets, A. P. R. and Gagge, A. P., 1988, Rationalization of the effective temperature, ET* as a measure of the enthalpy of the human indoor environment, *ASHRAE Transactions*, 94, Part 1.

Fœrevik, H., 2000, Protective clothing and survival at sea, in K. Kuklone and I. Holmér (Eds), Ergonomics of Protective Clothing, Proceedings of NOROBETEF 6 and 1st European Conference on Protective Clothing, Stockholm, Sweden, May, NIWL Nr 2000: 8.

Fountain, M. E. and Huizenga, C., 1996, A windows 3.1 thermal sensation model – User's manual, Berkeley, Calif: Environmental Analytics.

Fox, W. F., 1967, Human performance in the cold, *Human Factors*, 9, 203–220.

Fox, R. H., 1974, Temperature regulation with special reference to man, in R. J. Linden (Ed.), *Recent Advances in Physiology*, Vol. 9, London: Churchill Livingstone.

Fox, R. H., Goldsmith, R., Hampton, I. F. G. and Lewis, H. E., 1964, The nature of the increasing sweating capacity produced by heat acclimatization, *Journal of Physiology*, 171, 368–376.

Fox, R. H., Solman, A. J., Isaacs, R., Fry, A. J. and MacDonald, J. C., 1973, A new method for monitoring deep body temperature from the skin surface, *Clinical Science*, 44, 81–86.

Fu, G., 1995, A transient 3D mathematical thermal model for the clothed human. PhD thesis, Kansas State University, Manhattan, Kansas, USA.

Fu, G. and Jones, B., 1996, Combined finite element human thermal model and finite difference clothing model, In Y. Shapiro, D. S. Moran and Y. Epstein (Eds), *Environmental Ergonomics*, pp. 166–169, Freund, Tel Aviv. ISBN 965-294-123-9.

Fuller, F. H. and Brouha, L., 1966, New engineering methods for evaluating the job environment, *ASHRAE Journal*, 8(1), 39–52.

Fuller, F. H. and Smith, P. E., 1980, The effectiveness of preventive work produces in a hot workshop, in F. N. Dukes–Dobos and A. Henschel (Eds), *Procedures of a NIOSH Workshop on Recommended Heat Stress Standards*, DHSS (NIOSH) Publication No. 81–108: 32–45, Cincinnati: US Department of Health and Human Services, Public Health Service, Centers for Disease Control.

Fuller, F. H. and Smith, P. E., 1981, Evaluation of heat stress in a hot workshop by physiological measurements, *American Industrial Hygiene Association Journal*, 42, 32–37.

Fung, W. and Parsons, K. C., 1996, Some investigations into the relationship between car seat cover materials and thermal comfort using human subjects, *Journal of Coated Fabrics*, Vol. 26, October.

Gagge, A. P., 1937, A new physiological variable associated with sensible and insensible perspiration, *American Journal of Physiology*, 120, 277–287.

Gagge, A. P., Burton, A. C. and Bazett, H. C., 1941, A practical system of units for the description of the heat exchange of man with his thermal environment, *Science NY*, 94, 428–430.

Gagge, A. P., Stolwijk, J. A. J. and Hardy, J. D., 1967, Comfort and thermal sensations and associated physiological responses at various ambient temperatures, *Environmental Research*, 1, 1–20.

Gagge, A. P., Stolwijk, J. A. J. and Nishi, Y., 1971, An effective temperature scale based on a single model of human physiological temperature response, *ASHRAE Transactions*, 77, 247–262.

Gagge, A. P., Nishi, Y. and Gonzalez, R. R., 1972, Standard effective temperature index of temperature sensation and thermal discomfort, *Proceedings of the CIB Commission W45 (Human requirements) Symposium*, Building Research Station, UK.

Gagge, A. P., Nishi, Y. and Gonzalez, R. R., 1973, CIB Commission W45 Symposium Thermal comfort and moderate heat stress, Watford, UK, London: HMSO.

Gagge, A. P., Fobelets, A.P. and Bergland, L.G., 1986, A standard predictive index of human response to the thermal environment, *ASHRAE Transactions*, 92(1).

Galer, M. and Taylor, B., 1989, Human factors in information technology: Esprit project 385, in E. D. Megaw (Ed.), *Contemporary Ergonomics*, London: Taylor & Francis, pp. 82–86.

Garg, A., Chaffin, D. B. and Herrin, G. D., 1978, Prediction of metabolic rates for manual materials handling jobs, *American Industrial Hygiene Association Journal*, 39(8), 661–674.

General Board of Health, 1857, Report by the Commissioners, appointed to inquire into the warming and ventilation of dwellings.

Geng, Q., 2001, Hand cooling, protection and performance in cold environment. PhD thesis, Department of Human Work Sciences, Lulea University of Technology, Sweden.

Giancoli, D. C., 1980, *Physics: Principles with Applications*, London: Prentice-Hall.

Giorgi, G., Megri, A.C., Donnini, G. and Haghighat, F., 1996, Responses of disabled persons to thermal environments, ASHRAE Research Project 885-RP, AND Inc, Montreal, Canada.

Givoni, B., 1963, A new method for evaluating industrial heat exposure and maximum permissible work load, Paper submitted to the International Biometeorological Congress in Pau, France, September.

Givoni, B., 1976, *Man, Climate and Architecture*, 2nd edn, London: Applied Science.

Givoni, B. and Goldman, R. F., 1971, Predicting metabolic energy cost, *Journal of Applied Physiology*, 30(3), 429–433.

Givoni, B. and Goldman, R. F., 1972, Predicting rectal temperature response to work, environment and clothing, *Journal of Applied Physiology*, 2(6), 812–822.

Givoni, B. and Goldman, R. F., 1973, Predicting heart rate response to work, environment and clothing, *Journal of Applied Physiology*, 34(2), 201–204.

Glickman, N., Inouye, T., Keeton, R. W. and Fahnestock, M. K., 1950, Physiological examination of the effective temperature index, *ASHVE Transactions*, 56, 51.

Goldman, R. F., 1988, Standards for human exposure to heat, in I. B. Mekjavic, E. W. Banister and J. B. Morrison (Eds), *Environmental Ergonomics*, London: Taylor & Francis, pp. 99–136.

Gonzalez, R. R., 1979, Role of natural acclimatization (cold and heat) and temperature: Effect on health and acceptability in the built environment, Indoor Climate, P.O. Fanger.

Gonzalez, R. R., 1988, Biophysics of heat transfer and clothing considerations, in K. B. Pandolf, M. N. Sawka and R. R. Gonzalez (Eds), *Human Performance Physiology and Environmental Medicine at Terrestrial Extremes*, Brown and Benchmark, USA, pp. 45–96.

Gonzalez, R. R. and Gagge, A. P., 1973, Magnitude estimates of thermal discomfort during transients of humidity and operative temperature (ET*), *ASHRAE Transactions*, 79(1), 89–96.

Gonzalez, R. R., Mc Lellan, T. M., Cheung, S. K., Withey, W. R. and Pandolf, K. B., 1997. Heat strain models applicable for protective clothing systems: Comparison of core temperature response, Journal of Applied Physiology, 83: 1017–1032, 1997.

Gordon, R. G., 1974, The response of a human temperature regulatory system model in the cold. PhD thesis, University of California, Santa Barbara, CA, USA.

Grandjean, E., 1988, *Fitting the Task to the Man: A Textbook of Occupational Ergonomics*, 4th edn, London: Taylor & Francis.

Gravelling, R. and Fleming, G., 1996, Report to client, the development of a Draft British Standard on the thermal stress of working practices in cold indoor environments, Institute of Occupational Medicine Report, Edinburgh, p. 95.

Greger, R. and Windhorst, U., 1996, *Comprehensive human physiology*, Heidelberg, Germany: Springer-Verlag, ISBN 3–540–58109–X.

Griefahn, B., 1998, Cold – its interaction with other physical stressors, in I. Holmér and K. Kuklane (Eds), *Problems with Cold Work*, NIWL, Salna, Sweden, Nr. 1998:18, ISBN 91–7045–483–3.

Griefahn, B., Künemund, C. and Gehring, U., 2001, Annoyance caused by draught: The extension of the draught rating model (ISO 7730), in *Moving Thermal Comfort Standards into the 21st Century proceedings*, Windsor, UK, ISBN 1–873640–33–1.

Griefahn, B., Mehnert, P., Bröde, P. and Forsthoff, A., 1997, Working in moderate cold – a possible risk to health, *Journal of Occupational Health*, 39, 36 41.

Griffin, M. J., 1990, *Handbook of Human Vibration*, London: Academic Press.

Griffiths, I., 1975, The thermal environment, in D. Canter and P. Stringer (Eds), *Environmental Interaction*, Guildford: Surrey University Press.

Grivel, F. and Fraise, J. P., 1978, Field study of behavioural adjustments to the thermal environment in six underground railway stations, in *Indoor climate: effects on human comfort, performance and health in residential, commercial and light industry buildings*, WHO Conference, Copenhagen.

Guildford, J. P., 1954, *Psychometric Methods*, 2nd edn, New York: McGraw-Hill.

Guyton, A. C., 1969, *Function of the human body*, 3rd edn, Philadelphia: W. B. Saunders.

Halabi, L. and Parsons, K. C., 1995, Surface temperatures and the thermal sensation and discomfort of handrails, in S. A. Robertson (Ed.), *Contemporary Ergonomics*, London: Taylor & Francis, ISBN 07484–0328–0.

Hales, J. R. S. and Richards, D. A. B., 1987, *Heat Stress – physical exertion and environment*, Amsterdam: Excerpta Medica.

Hamlet, M. P., 1998, Human cold injuries, in K. B. Pandolf, M. N. Sawka and R. R. Gonzalez, *Human Performance Physiology and Environmental Medicine at Terrestrial Extremes*, USA: Brown and Benchmark, pp. 435–466.

Hamlet, M. P., 1998, Peripheral cold injury, in I. Holmer and K. Kuklane (Eds), *Problems with Cold Work*, pp. 127–131. National Institute for Working Life, Publication NR 1998:18, Solna, Sweden, ISBN 91–7045–483–3.

Hanson, M. and Graveling, R. A., 1999, Development of a draft British Standard, The assessment of heat strain for workers wearing personal protective equipment, IOM Research Report TM/99/03, IOM, Edinburgh, UK.

Hardy, J. D. and DuBois, E. F., 1938, The technique of measuring radiation and convection, *Journal of Nutrition*, 15, 461–475.

Hardy, J. D., Wolff, H. G. and Goodell, H., 1952, *Pain Sensations and Reactions*, Baltimore: Williams and Wilkins.

Haslam, R. A., 1989, An evaluation of models of human response to hot and cold environments, PhD thesis, Loughborough University.

Haslam, R. A. and Parsons, K. C., 1988, Quantifying the effects of clothing for models of human response to the thermal environment, *Ergonomics*, 31(12), 1787–1806.

Haslam, R. A. and Parsons, K. C., 1989a, Models of human response to hot and cold environments, *Human Modelling Group Final Report*, Vols 1 & 2, APRE, Farnborough.

Haslam, R. A. and Parsons, K. C., 1989b, Computer-based models of human responses to the thermal environments – are their predictions accurate enough for practical use? in J. B. Mercer (Ed.), *Thermal Physiology*, Amsterdam: Elsevier.

Havenith, G., 1997, Individual heat stress response. PhD thesis, Katholieke Universiteit Nijmegen, The Netherlands, ISBN 90–9010979–X.

Havenith, G., Heus, R. and Lotens, W. A., 1990, Resultant clothing insulation: a function of body movement, posture, wind, clothing fit and ensemble thickness, *Ergonomics*, 33, 67–84.

Havenith, G., Holmér, I., den Hartog, E. A. and Parsons, K. C., 1998, Representation of the effects of movement and wind on clothing vapor resistance in ISO standards, in J. A. Hodgdon, J. H. Heaney and M. J. Buono (Eds), *Environmental Ergonomics VIII*, pp. 297–300, ISBN 0–9666953–1–3.

Havenith, G., Holmér, I., de Hartog, E. A. and Parsons, K. C., 1999, Clothing evaporative heat resistance. Proposal for improved representation in standards and models, *Annals of Occupational Hygiene*, 43(5), 339–346.

Havenith, G., Holmér, I., Parsons, K. C., den Hartog, E. A. and Malchaire, J., 2000, Calculation of dynamic heat and vapour resistance, in J. Werner and M. Hexamer (Eds), *Environmental Ergonomics IX*, Aachen, Germany: Shaker Verlag, pp. 125–128, ISBN 3–8265–7648–9.

Havenith, G., Holmér, I. and Parsons, K. C., 2002, Personal factors in thermal comfort assessment: Clothing properties and metabolic heat production, *Energy and Buildings*, 34(6), 581–592, July.

Havenith, G. and Zhang, P., 2002, Comparison of different tracer gas dilution methods for the determination of clothing ventilation, in Y. Tochihara (Ed.), *Environmental Ergonomics X, Proceedings of the tenth conference on Environmental Ergonomics*, Japan: Fukuoka, ISBN 4–9901358–0–6.

Hawking, S., 1988, *A Brief History of Time*, London: Bantum Press.

Hayward, G. S., Eckerson, J. D. and Collis, M. L., 1975, Thermal balance and survival time prediction of man in cold water, *Can. J. Physiol. Pharmacol.*, 53, 21–32.

Heberden, W., 1826, An account of the heat of July, 1825; together with some remarks upon sensible cold, *Philosophical Transactions of the Royal Society, London*, II, 69.

Hellon, R. F. and Crockford, G. W., 1959, Improvements to the globe thermometer, *Journal of Applied Physiology*, 14, 649–650.

Henchel, A., 1980, Comparison of heat stress action levels, in F. Dukes-Dubos and A. Henshel (Eds), *Workshop on Recommended Heat Stress Standards*, 21–31, NIOSH publication No. 81–108, US Department of Health, Education and Welfare Centers for disease control.

Henriques, F. C. and Moritz, A. R., 1947, Studies of thermal injury, I. The conduction of heat to and through skin and the temperatures attained therein. A theoretical and experimental investigation, *American Journal of Pathology*, **23**, 531–549.

Henschel, A., 1980, Heat Stress Indices, In NIOSH (1986).

Hensel, H., 1981, *Thermoreception and Temperature Regulation*, London: Academic Press.

Hey, E. N., 1968, Small globe thermometers, *Journal of Physics E*, **1**, 954–957.

Hill, L., 1919, The science of ventilation and open air treatment. Part I. Report for the Medical Research Council, London, No. 32.

Hill, L., Griffith, O. W. and Flack, M., 1916, The measurement of the rate of heat loss at body temperature by convection, radiation and evaporation, *Philosophical Transactions of the Royal Society (B)*, **207**, 183–220.

Hill, I. D., Webb, L. H. and Parsons, K. C., 2000, Carers' view of the thermal comfort requirements of people with physical disabilities. Proceedings of the IEA 2000/HFES 2000 Congress, San Diego, USA, pp. 716–719.

HMSO, 1938, Report of the Building Research Board for 1937, in Chrenko, 1974, *Bedford's Basic Principles of Ventilation and Heating*, H. K. Lewis (Ed.), London.

HMSO, 1963, Offices, Shops and Railway Premises Act 1963, London: HMSO.

Hodder, S. G. and Parsons, K. C., 2001a, In Automotive Glazing, Task 2.4 – Thermal Comfort, Final Technical Report No. BE97-3020 T2.4 to Brite/Euram Programme, Contract No. BRPR-C797-0450, European Commission, Brussels.

Hodder, S. G. and Parsons, K. C., 2001b, Field trials in Seville, In Automotive Glazing: Task 2.4 – Thermal Comfort, Final Technical Report NO. BE96-3020, Brite/Euram, European Commission, Brussels.

Hollies, N. R. S., 1971, The comfort characteristics of next-to-skin garments, including shirts, Shirley International Seminar on Textiles for Comfort, 1–12.

Hollies, N. R. S., Custer, A. G., Morin, C. J. and Howard, M. E., 1979, A human perception analysis approach to clothing comfort, *Textile Research Journal*, **49**, 557–564.

Holmér, I., 1984, Required clothing insulation (IREQ) as an analytical index of cold stress, *ASHRAE Transactions*, **90**(1), 116–128.

Holmér, I., 1988, Personal Communication.

Holmér, I., 2000, Thermal manikins in research and standards, in H. Nilsson and I. Holmer (Eds), *Thermal Manikin Testing*, National Institute for Working Life, Publication NR 2000:4, Solna, Sweden, pp. 1–7, ISBN 91-7045-554-6.

Holmér, I. and Geng, O., 2000, Temperature limit values for cold touchable surfaces – final report by EU project SMT4-CT97-2149, European Commission, DGXII SMT.

Holmér, I., Nilsson, H., Havenith, G. and Parsons, K. C., 1998, Convective heat loss through clothing, in J. A. Hodgdon, J. H. Heaney and M. J. Buono (Eds), *Environmental Ergonomics VIII*, pp. 293–296, ISBN 0-9666953-1-3.

Holmér, I. Nilsson, H., Havenith, G. and Parsons, K.C., 1999, Clothing convective heat exchange. Proposal for improved representation in standards and models, *Annals of Occupational Hygiene*, **43**(5) 329–337.

Höppe, P., 2001, Different aspects of assessing indoor and outdoor thermal comfort, In Conference Proceedings, Moving Thermal Comfort Standards into the 21st Century, Windsor, UK, pp. 368–375, ISBN 1–873640–33–1.

Horvath, S. M. and Freedman, A., 1947, The influence of cold upon the efficiency of man, *Aviation Medicine*, **18**, 158–164.

Houdas, Y., Colin, J., Timbal, J., Boutelier, C. and Guien, J., 1972, Skin temperatures in warm environments and the control of sweat evaporation, *Journal of Applied Physiology*, **33**(1), 99–104.

Houdas, Y. and Guicu, J., 1975, Physical models of human thermoregulation, *Proceedings of the temperature regulation and drug addiction symposium*, Paris: Karger, pp. 11–21.

Houghton, F. C. and Yagloglou, C. P., 1923, Determining equal comfort lines, *Journal of ASHVE*, **29**, 165–176.

Houghton, F. C. and Yagloglou, C. P., 1924, Cooling effect on human beings produced by various air velocities, *Journal of ASHVE*, **30**, 193.

Houghton, F. C., Gutberlet, C. and Witkowski, E., 1938, Draft temperature and velocities in relation to skin temperature and feeling of warmth, *ASHRAE Transactions*, **44**, 289–308.

House, J. R., 1994, Hand immersion as a method of reducing heat strain during rest periods, in J. Frim, M. B. Ducharme and P. Tikuisis (Eds), *Proceedings of the sixth International Conference on Environmental Ergonomics*, Motebello, Canada, September 25–30, pp. 10–11, ISBN 0–662–21650–4.

House, J. R., 1996, Reducing heat strain with ice vests or hand immersion, in Y. Shapiro, D. S. Moran, Y. Epstein (Eds), *Environmental Ergonomics Recent Progress and New Frontiers*, Freund, Tel Aviv, pp. 347–350, ISBN 965–294–123–9.

Hudack, S. and McMaster, P. D., 1932, The gradient of permeability of the skin vessels as influenced by heat, cold and light, *Journal of Experimental Medicine*, **55**, 431–439.

Humphreys, M. A., 1972, *Clothing and thermal comfort of secondary school children in summertime*, CIB Commission W45 symposium thermal comfort and moderate heat stress, Watford, London: HMSO.

Humphreys, M. A., 1974, Environmental temperature and thermal comfort, *Building Services Engineer (RHVE)*, **42**, 77–81.

Humphreys, M. A., 1976, Field studies of thermal comfort compared and applied, *Journal of the Institute of Heating and Ventilating Engineers*, **44**, 5–27.

Humphreys, M. A., 1977, The optimum diameter for a globe thermometer for use indoors, *Annals of Occupational Hygiene*, **20**(2), 135–140.

Humphreys, M. A., 1978, Outdoor temperatures and comfort indoors, Building Research and Practice, **6**(2), 92–105.

Humphreys, M. A. and Nichol, J. F., 1970, An investigation into the thermal comfort of office workers, *Journal of the Institution of Heating and Ventilation Engineers*, **30**, 181–189.

Humphreys, M. A. and Nicol, J. F., 1995, An adaptive guideline for UK office temperatures, in J. F. Nicol, M. A. Humphreys, O. Sykes and S. Roaf (Eds), Standards for thermal comfort: Indoor air temperature standards for the 21st century. London: E&FN Spon.

Humphreys, M. A. and Nicol, J. F., 2002, The validity of ISO – PMV for predicting comfort rates in everyday thermal environments, *Energy and Buildings*, **34**(6), 667–684.

Humphreys, N., Webb, L. H. and Parsons, K. C., 1998, A comparison of the thermal comfort of different wheelchair seating materials and an office chair, in M. A. Hanson (Ed.), *Contemporary Ergonomics*, London: Taylor & Francis, pp. 525–529, ISBN 0–7484–0811–8.

Humphrey, S. J. and Wolff, H. S., 1977, *The Oxylog Journal of Physiology*, 267, 12.

Huntington, E., 1915, *Civilization and Climate*, New Haven, CT: Yale University Press.

IDECG, 1990, The doubly labelled water method for measuring energy expenditure. Technical recommendations for use in humans. International Dietary Energy Consultancy Group, in A.M. Prentice (Ed.), Vienna: IDECG/IAEA, NAHRES-4.

Ilmarinen, R., Tammela, E. and Korhonen, E., 1990, Design of functional work clothing for meat cutters, *Applied Ergonomics*, 21(1), 2–6.

ISO 7243, 1982, Hot environments – estimation of the heat stress on working man, based on the WBGT-index (wet bulb globe temperature), Geneva: International Standards Organization.

ISO 7726, 1998, Ergonomics of the thermal environment. Instruments for measuring physical quantities. Geneva: International Standards Organization.

ISO 7730, 1984, Moderate Thermal Environments – determination of the PMV and PPD indices and specification of the conditions for thermal comfort, Geneva: International Standards Organization.

ISO 7730, 1994, Moderate Thermal Environments – Determination of the PMV and PPD indices and specification of the conditions for thermal comfort, Geneva: International Standards Organization.

ISO 7243, 1989, Hot Environments – Estimation of the heat stress on working man, based on the WBGT-index (wet bulb globe temperature), Geneva: International Standards Organization.

ISO 7726, 1985, Thermal Environments – instruments and methods for measuring physical quantities, Geneva: International Standards Organization.

ISO 7933, 1989, Hot Environments – Analytical determination and interpretation of thermal stress using calculation of required sweat rate, Geneva: International Standards Organization.

ISO 8996, 1990, Ergonomics of the Thermal Environment: Estimation of metabolic heat production, Geneva: International Standards Organization.

ISO 9886, 1992, Evaluation of thermal strain by physiological measurements, Geneva: International Standards Organization.

ISO 9920, 1995, Ergonomics of the thermal environment – estimation of the thermal insulation and evaporative resistance of a clothing ensemble, Geneva: International Standards Organization.

ISO 10551, 1995, Ergonomics of the thermal environment – Assessment of the influence of the thermal environment using subjective judgement scales, Geneva: International Standards Organization.

ISO 12894, 2001, Ergonomics of the thermal environment – Medical supervision of individuals exposed to extreme hot or cold environments, Geneva: International Standards Organization.

ISO 11092, 1993, Textiles – Physiological effects – Measurement of thermal and water vapour resistance under steady-state conditions (sweating guarded–hotplate test), Geneva: International Standards Organization.

ISO TS 13732-2, 2001, Method for the assessment of human responses to contact with surfaces (ISO DTR 13732) Part 2 : Human contact with surfaces at moderate temperature, London: BSI.

ISO CD 13732-3, 2001, Ergonomics of the thermal environment – Touching of cold surfaces, Berlin: DIN.

ISO TS 14415, 1999, Ergonomics of the thermal environment : The Application of international standards for people with special requirements, Geneva: International Standards Organization.

ISO CD 15265, 2002, Strategy for risk assessment and management and working practice in cold environments, London: BSI.

ISO CD 7933, 2001, Ergonomics of the thermal environment – Analytical determination and interpretation of heat stress using calculation of the Predicted Heat Strain, London: BSI.

ISO TR 11079, 1993, Evaluation of cold environments: Determination of required clothing insulation, IREQ, Geneva: International Standards Organization.

Ittleson, W. H., Proshansky, H. M., Rivlin, L. G. and Winkel, G. H., 1974, *An Introduction to Environmental Psychology*, New York: Holt, Rinehart and Winston.

Jay, O., 2002, Short-term fingertip contact with cold materials. PhD thesis, Department of Human Sciences, Loughborough University, UK.

Jay, O. and Havenith, G., 2000, Skin contact with cold materials: A comparison between male and female responses to short term exposures, in J. Werner and M. Hexamer (Eds), *Environmental Ergonomics IX*, Aachen, Germany: Shaker Verlag, pp. 185–188, ISBN 3–8265–7648–9.

Jessen, C. and Kuhnen, G., 2000, Selective brain cooling: A current appraisal, in J. Werner and M. Hexamer (Eds), *Environmental Ergonomics IX*, Aachen, Germany: Shaker Verlag, pp. 9–14, ISBN 3–8265–7648–9.

Johanessen, K., 1985, A new thermal anemometer probe for indoor air velocity measurements, in Bruel and Kjaer, *Technical Review*, Vol. 2, ISBN 007–2621, Copenhagen.

Jones, B. W. and Ogawa, Y., 1992, Transient interaction between the human and the thermal environment, *ASHRAE Transactions*, Part 1.

Jones, B. W., Ito, M. and McCullough, E. A., 1990, Transient thermal response of clothing systems, *Proceedings of the International Conference on Environmental Ergonomics IV*, Austin, Texas.

Jones, P. R. M., West, G. M., Harris, D. H. and Read, J. D., 1989, The Loughborough Anthropometric shadow scanner (LASS), *Endeavour, New Series*, 13(4), 162–168.

Jones, B., Sipes, J., He, Q., and McCullough, E., 1994, The transient nature of thermal loads generated by people, *ASHRAE Transactions*, 100(2).

Kampmann, B., Malchaine, J. and Piette, A., 1999, Comparison between the PHS model and the ISO 7933 Standard, in J. Malchaine (Ed.), *Evaluation and control of warm working conditions*, Proceedings of BIOMED 'Heat Stress' Research Project Conference, Barcelona, June.

Kampmann, B. and Piekarski, 2000, The evaluation of workplaces subjected to heat loss: can ISO 7933 (1989) adequately describe heat strain in industrial workplaces? *Applied Ergonomics*, 31(1), 59–72.

Keatinge, W.R. and Donaldson, G.C., 1998, Differences in cold exposures associated with excess winter mortality, in I. Holmér and K. Kuklane (Eds), *Problems with Cold Work*, NIWL, Salna, Sweden, Nr 1998:18, ISBN 91–7045–483–3.

Keatinge, W. R., 1969, *Survival in cold water*, Oxford: Blackwell Scientific.

Keele, C. A. and Neil, E., 1971, *Pain. Applied Physiology* (by S. Wright, revised by Keele and Neil), Oxford: Oxford University Press.

Kelly, G. A., 1955, *The Psychology of Personal Constructs*, New York: Norton.

Kenney, W. L., Lewis, D. A., Hyde, D. E. *et al.*, 1988, Physiologically derived critical evaporative coefficients for protective clothing ensembles. *Journal of Applied Physiology*, **63**, 1095–1099.

Kenney, W. L., Mikita, D. J., Havenith, G., Puhal, S. M. and Crosby, P., 1993, Simultaneous derivation of clothing-specific heat exchange coefficients, Medicine and Science in Sports and Exercise, Special Communications, pp. 283–289.

Kenshalo, D. R., 1968, The skin senses, *Proceedings of the 1st International Symposium on the Skin Senses*, Tallahassee, Florida: Florida State University.

Kenshalo, D. R., 1970, Psychophysical studies of temperature sensitivity, in W. D. Neff (Ed.), *Contributions to Sensory Physiology*, New York: Academic Press.

Kenshalo, D. R., 1979, *Sensory Functions of the Skin of Humans*, New York: Plenum Press.

Kerslake, D. M., 1972, *The Stress of Hot Environment*, Cambridge: Cambridge University Press.

Keyson, D. K. and Parsons, K. C., 1990, Designing the user interface using rapid prototyping, *Applied Ergonomics*, **21**, 3.

Koch, W., Jennings, B. H. and Humphreys, C. M., 1960, Environmental study II – sensation responses to temperature and humidity under still air conditions in the comfort range, *ASHRAE Transactions*, **66**, 264.

Kraning K. K., 1995, Validation of mathematical models for predicting physiological events during work and heat stress, Technical Report from US Army Research Institute of Environmental Medicine, Natick, Massachusetts, Report Number T95-18.

Krasner, L., 1980, *Environmental Design and Human Behaviour – A Psychology of the Individual in Society*, Oxford: Pergamon.

Lang, F., 1996, The body compartments of water and electrolytes, in R. Greger and U. Windhorst (Eds), *Comprehensive Human Physiology*, Heidelberg, Germany: Springer-Verlag, ISBN 3–540–58109–X.

Langkilde, G., 1977, Thermal comfort for people of high age, *INSERIM*, **77**, 187–194.

Larson, L. A., 1974, Fitness, Health, and Work Capacity: International Standards for assessment, International Committee for the Standardization of Physical Fitness Tests, London: Macmillan.

Lawrence, J. C. and Bull, J. P., 1976, Thermal conditions which cause skin burns, *IMechE*, **5**(3), 61–63.

Leach, E. H., Peters, R. A. and Rossiter, R. J., 1943, Experimental thermal burns, especially the moderate temperature burn, *Quarterly Journal of Experimental Physiology*, **32**, 67.

Lee, T., 1976, *Psychology and the Environment*, London: Methuen.

Legg, S. J. and Pateman, C. M., 1984, A physiological study of the repetitive lifting capabilities of healthy young males, *Ergonomics*, **27**(3), 259–272.

Leithead, C. S. and Lind, A. R., 1964, *Heat Stress and Heat Disorders*, London: Cassell.

Leslie, J., 1804, *An Experimental Inquiry into the Nature and Propagation of Heat*, London: J. Mawman.

Lewin, K., 1936, *Principles of Topological Psychology*, New York: McGraw-Hill.

Lewis, W. K., 1922, The evaporation of a liquid into gas, *ASME Transactions*, **44**, 325–335.

Lewis, T., 1930, Observations upon the reactions of the vessels of the human skin to cold, *Heart*, **15**, 177–208.

Liddell, D. K., 1963, Estimation of energy expenditure from expired air, *Journal of Applied Physiology*, **18**, 25–29.

Lind, A. R., 1963, A physiological criterion for setting thermal environmental limits for everybody's work, *Journal of Applied Physiology*, **18**, 51–56.

Lind, A. R. and Bass, D. E., 1963, The optimal exposure time for the development of acclimatisation to heat, *Federal Proceedings*, **22**, 704–708.

Lind, A. R., Hellon, R. F., Jones, R. M., Weiner J. S. and Fraser, D. C., 1957, Reactions of Mines-rescue personnel to work in hot environments, *Medical Research Bulletin*, **1**.

Lotens, W. A., 1988, Comparison of the thermal predictive models for clothed humans, *ASHRAE Transactions*, **94**, Part 1.

Lotens, W. A., 1989, A clothing model, *Proceedings of the International Conference on Environmental Ergonomics – IV*, Austin, Texas.

Lotens, W. A., 1990, Clothing thermal evaluation using heat balance techniques, *Proceedings of the International Conference on Environmental Ergonomics – IV*, Austin, Texas.

Lotens, W. A., 1993, Heat transfer from humans wearing clothing. PhD thesis, University of Delft, The Netherlands.

Lotens, W. A. and Havenith, G., 1988, Ventilation of rainwear determined by a trace gas method, in I. B. Mekjavic, E. W. Bannister and J. B. Morrison (Eds), *Environmental Ergonomics*, London: Taylor & Francis, pp. 162–176.

Lotens, W. A. and Van De Linde, E. J. G., 1983, Insufficiency of current clothing description, Conference of Medical Biophysics, Aspects of Protective Clothing.

Lotens, W. A., Heus, R. and Van de Linde, E. J. G., 1989, A 2-node thermoregulatory model for the foot, in J. B. Mercer (Ed.), *Thermal Physiology*, Amsterdam: Excerpta Medica, pp. 769–775.

Loveday, D. L., Parsons, K. C., Taki, A. H., Hodder, S. G. and Jeal, L. D., 1998, Designing for thermal comfort in combined chilled ceiling and displacement ventilation environments. *ASHRAE Transactions*, **104** (Part 1B), 901–911.

Loveday, D. L., Parsons, K. C., Taki, A. H., Hodder, S. G. and Jeal, L., 2000, Designing for thermal comfort in combined chilled ceiling/displacement ventilation environments, *ASHRAE Transactions*.

Loveday, D. L., Parsons, K. C., Taki, A. H. and Hodder, S. G., 2001, Thermal comfort design of chilled ceiling/displacement ventilation environments based on ISO 7730: v Validation, adaptive context and influence on system configurations, in K. McCartney (Ed.), *Moving thermal comfort standards into the 21st century*, *Proceedings of a conference held on 5th–8th April*, Cumberland Lodge, Windsor, UK, pp. 19–30, ISBN 1-873640.

Loveday, D. L., Parsons, K. C., Taki, A. H. and Hodder, S. G., 2002, Displacement ventilation environments and chilled ceilings: thermal comfort design within the context of the BS EN ISO 7730 versus adaptive debate, *Energy and Buildings*, **34**(6) July, 573–580.

Lukaski, H. C., 1987, Methods for the assessment of body composition: traditional and new, *American Journal of Clinical Nutrition*, 46, 437–456.

Lumley, S. H., Story, D. L. and Thomas N. T., 1991, Clothing ventilation – update and applications, *Applied Ergonomics – technical note*, 22(6), 390–394.

Lynn, R., 1991, *Personality and National Character*, Oxford: Pergamon Press.

Mackworth, N. H., 1950, Researches on the measurement of human performance, Medical Research Council Special Report, No. 268, London: HMSO.

Mackworth, N. H., 1952, Some recent studies of human stress from a marine and naval viewpoint, *Journal of Institute of Marine Engineers*, 64, 123–138.

Mackworth, N. H., 1953, Finger numbness in very cold winds, *Journal of Applied Physiology*, 5, 533.

Macpherson, R. K., 1960, Physiological responses to hot environments, Medical Research Council Special Report Series No. 298, London: HMSO.

Madsen, T. L. and Olesen, B. W., 1986. A new method for evaluation of the thermal environment in automotive vehicles. *ASHRAE Transactions*, Vol. 92, Part 1B, pp. 38–54.

Madsen, T. L., 1999, Development of a breathing thermal manikin, in Nilsson and Holmer (Eds), *Proceedings of the third international meeting on Thermal Manikin Testing 3IMM*, NIWL, Salna, Sweden, pp. 74–78, NR 2000: 4.

Maggio, R. C., 1956, A moulded skin simulant material with thermal and optical constants approximating those of human skin, Naval Material Laboratory, New York Naval Shipyard, Laboratory Project 5046–3, Part 105.

Mairiaux, P. and Malchaire, J., 1988, Comparison and validation of heat stress indices in experimental studies, in *Heat Stress Indices Seminar Proceedings*, Commission of the European Communities, Luxembourg, pp. 81–110.

Mairiaux, P. and Malchaire, J., 1990, Work in hot environments (Le travail en ambiance chaude) *Collection de Monographies de Medicine du Travail*, 7, Paris: Masson.

Mairiaux, P., Davis, P. R. and Stubbs, D. A., 1983, Intra-abdominal pressure relationship to forces exerted in an erect posture, in K. Coombes (Ed.), *Proceedings of the Ergonomics Society's Conference 1983*, London: Taylor & Francis, p. 200.

Malchaire, J., Gebhardt, H. J. and Piette, A., 1999, Strategy for evaluation and prevention of risk due to work in thermal environments. *The Annals of Occupational Hygiene*, 43(5), 367–376.

Malchaine, J., Kampmann, B., Gelshardt, H. J., Mehnert, P. and Alfano, G., 1999, The Predicted Heat Strain Index: Modifications brought to the required sweat rate index, In Evaluation and control in Warm Working Conditions.

Malchaire, J., Piette, A., Kampmann, B., Havenith, G., Mehnert, P., Holmer, I., Gebhardt, H., Griefahn, B., Alfano, G. and Parsons, K. C., 2000, Development and validation of the Predictive Heat Strain (PHS) model, in J. Werner and M. Hexamer (Eds), *Environmental Ergonomics IX*, Aachen, Germany: Shaker Verlag, pp. 133–136, ISBN 3–8265–7648–9.

Malchaine, J., Piette, A., Kampmann, B., Mehnert, P., Gebhardt, H., Havenith, G., de Hartog, E., Holmér, I., Parsons, K., Alfano, G. and Griefahn, B., 2001, Development and validation of the predicted heat strain model, *Annals of Occupational Hygiene*, 45(2), 123–135.

Markham, S., 1947, *Climate and the Energy of Nations*, New York: Oxford University Press.

Markus, T. A. and Morris, E. N., 1980, *Buildings, Climate and Energy*, London: Pitman.

502 References

Marzetta, L. A., 1974, *Engineering and construction manual for an instrument to make burn hazard measurements in consumer products*, US Department of Commerce, National Bureau of Standards Technical Note 816, Washington.

Mawby, F. D. and Street, P. J., 1985, Development of fire fighting apparel, in D. J. Oborne (Ed.), *Contemporary Ergonomics*, London: Taylor & Francis, pp. 181–186.

McArdle, B., Dunham, W., Holling, H. E., Ladell, W. S. S., Scott, J. W., Thomson, M. L. and Weiner, J. S., 1947, The prediction of the physiological effects of warm and hot environments, Medical Research Council, London, *RNP Rep. 47/391*.

McArdle, W. D., Katch, F. I., Katch, V. L., 1996, Exercise Physiology: Energy, Nutrition and Human Performance, Baltimore: Williams and Wilkins, ISBN 0–683–05731–6.

McCaig, R. H., 1992a, Medical supervision of individuals exposed to hot or cold environments, Working document of ISO/TC 159/SC5 WG1.

McCaig, R. H., 1992b, Ergonomics of the thermal environments: medical supervision of individuals exposed to hot or cold environments, ISO/TC 159/SC5 N231, Geneva.

McCartney, K. J. and Nicol, J. F., 2002, Developing an adaptive control algorithm for Europe, *Energy in Buildings*, **34**(6), 623–635.

McCullough, E. A., 1990, Physical techniques for determining the resistance to heat transfer provided by clothing. In *Proceedings of the International Conference on Environmental Ergonomics, IV*, Austin, USA.

McCullough, E. A. and Hong, S., 1992, A database for determining the effect of walking on clothing insulation, in Lotens and Havenith (Eds), *Proceedings of the Fifth International Conference on Environmental Ergonomics*, Maastricht, The Netherlands.

McCullough, E. A. and Jones, B. W., 1984, A comprehensive database for estimating clothing insulation, Institute for Environmental Research, Kansas State University, *IER Technical Report 84–01*, December.

McCullough, E. A., Jones, B. W. and Huck, J., 1985, A comprehensive database for estimating clothing insulation, *ASHRAE Transactions*, **91**(2A), 29–47.

McCullough, E. A., Jones, B. W. and Tamura, T., 1989, A database for determining the evaporative resistance of clothing, *ASHRAE Transactions*, **95**.

McIntyre, D. A., 1980, *Indoor Climate*, London: Applied Science.

McNall, P. E., Ryan, P. W., Rohles, F. W., Nevins, R. G. and Springer, W. E., 1968, Metabolic rates at four activity levels and their relationship to thermal comfort, *ASHRAE Transactions*, **74**, Part I. IV. 3. 1.

McNeill, M. and Parsons, K. C., 1996, Heat stress in night-clubs, in S. A. Robertson (Ed.), *Contemporary Ergonomics*, Taylor & Francis, pp. 208–213, ISBN 0–7484–0549–6.

McNeill, M. B. and Parsons, K. C., 1999, Appropriateness of international heat stress standards for use in topical agricultural environment, *Ergonomics*, **42**(6), 779–797.

Mecheels, J. and Umbach, K. H., 1977, Themophysiologische Eigenschaften von Kleidunfssystemen (Thermophysiological properties of clothing systems), *Melliand Textilberichte*, **58**, 73–81.

Meese, G. B. and Schiefer, R. E., 1984, Thermal effects on the performance of schoolchildren. Outline of a proposed research project, *South African Journal of Science 1983*, **79**, 172.

Meese, G. B, Kok, R., Lewis, M. I. and Wyon, D. P., 1982, Effects of moderate cold and heat stress on factory workers in Southern Africa. 2, Skill and performance in the cold, *South Africa Journal of Science*, **78**, 189–197.

Meese, G. B., Kok, R., Lewis, M. I. and Wyan, D. P., 1984, A laboratory study of the effects of moderate thermal stress on the performance of factory workers, *Ergonomics*, **27**, 1, 19–43.

Meinander, H., 1992, Coppelius – A sweating thermal manikin for the assessment of functional clothing, Nokobetef IV: Quality and usage of protective clothing, Kittila, Finland, pp. 157–161.

Meinander, H., 1999, Extraction of data from sweating manikin tests, in O. Nilsson and I Holmér (Eds), *Proceedings of the Third International Meeting on Thermal Manikin Testing*, 3IMM NIWK, Salna, Sweden, NR 2000 4.

Mekjavic, I. B., Bannister, E. W. and Morrison, J. B., 1988, *Environmental Ergonomics: Sustaining Human Performance in Harsh Environments*, London: Taylor & Francis.

Mekjavic, I. B., Sun, J., Lun, V. and Giesbrecht, G., 1992, Evaluation of an infra red tympanic thermometer during cold water immersion rewarming, in Lotens and Havenith (Eds), *Proceedings of the Fifth International Conference on Environmental Ergonomics*, Maastricht, The Netherlands, November, pp. 42–43.

Mehnert, P., Malchaire, J., Kampmann, B., Griefahn, B. and Piette, A., 2000, Prediction models for the mean skin temperature, in J. Werner and M. Hexamer (Eds), *Environmental Ergonomics IX*, Aachen, Germany: Shaker Verlag, pp. 121–124, ISBN 3–8265–7648–9.

Mercer, J. B., 1989, Thermal Physiology 1989, *Proceedings of the International Symposium on Thermal Physiology*, Amsterdam: Excerpta Medica.

Meredith, A. E., 1978, Manual dexterity tests and real military tasks, *APRE Report No. 20/77*, Farnborough, UK.

Metz, B., 1988, *Proceedings of CEC Seminar on Heat Stress Indices*, Commission of the European Communities, Luxembourg.

Michel, J. M. and Vogt, J. J., 1972, Measurement of skin temperature. Part I. Contact thermometers and fluxmeters, *Thermique et Aeraulique* (07), 691–714.

Mills, C., 1939, *Medical Climatology: Climatic and Weather Influences in Health and Disease*, Baltimore: Charles C. Thomas.

Missenard, A., 1935, Théorie simpliféed Thermomètre Résultant, *Chauffage Ventilation*, 347–352.

Missenard, A., 1948, A thermique des ambiences: équivalences de passage, équivalences de séjours, *Chaleur et Industrie*, **276**, 159–172 and **277**, 189–198.

Missenard, A., 1959, On thermally equivalent environments, *JIHVE*, **27**, 231–237.

Mitchell, D., 1974, Convective heat loss from man and other animals, in J. L. Monteith and L. E. Mount (Eds), *Heat Loss from Animals and Man*, London: Butterworths.

Molnar, R. W., 1960, An evaluation of wind-chill, in S. M. Harvath (Ed.), *Cold Injury*, New York, J. Macy Jr. Foundation, pp. 175–222.

Montagna, W. and Parakkal, P. F., 1974, *The Structure and Function of Human Skin*, New York: Academic Press.

Monteith, J. L. and Unsworth, M. H., 1990, *Principles of Environmental Physics*, 2nd edn, London: Edward Arnold.

Montgomery, L. D., 1974, A model of heat transfer in immersed man, *Annals of Biomedical Engineering*, **2**, 19–46.

Moos, R. H., 1976, *The Human Context – environmental determinants of behaviour*, Chichester: John Wiley and Sons.

Moritz, A. R. and Henriques, F. C., 1947, Studies in thermal injury II. The relative importance of time and air surface temperatures in the causation of cutaneous bums, *American Journal of Pathology*, **23**, 695–720.

Morris, L. A. and Graveling, R. A., 1986, *Responses to intermittent work in hot environments*, Institute of Occupational Medicine, TM/88/13.

Morrissey, S. J. and Liou, Y. H., 1984, Metabolic cost of load carriage with different container sizes, *Ergonomics*, 27(8), 847–853.

Moser, C. and Kalton, G., 1971, *Survey Methods in Social Investigation*, 2nd edn, London: Heinemann.

Murgatroyd, P. R., Shetty, P. S. and Prentice, A. M., 1993, Techniques for the measurement of human energy expenditure: a practical guide, *International Journal of Obesity*, 17, 549–568.

Nadel, E. R., Bergh, U and Saltin, B., 1972, Body temperatures during negative work exercise, *Journal of Applied Physiology*, 33, 553–558.

Nadel, E. R., Mitchell, J. W. and Stolwijk, J. A. J., 1973, Differential thermal sensitivity in the human skin, *Pflügers Archives*, 340, 71–76.

Neal, M. S., 1998, Development and application of a clothed thermoregulatory model. PhD thesis, Department of Human Sciences, Loughborough University, UK.

Neal, M. S., Withey, W. R., Misham, C. and Parsons, K. C., 1996, Theoretical maxima for clothing insulation values, in Y. Shapiro, D. S. Moran, Y. Epstein (Eds), *Environmental Ergonomics, Recent Progress and New Frontiers*, Freund Publishing House, pp. 389–392, ISBN 965–294–123–9.

Neiburger, M., Edinger, J. G. and Bonner, W. D., 1982, *Understanding Our Atmospheric Environment*, San Francisco: W. H. Freeman.

Neilsen, M., 1938, Die Regulation der Kopertemperatur bei Muskelerbeit, *Skandinavian Archives of Physiology*, 79, 193–230.

Nevins, R. G., 1971, Thermal comfort and draughts, *Journal de Physiologie*, 13, 356–358.

Nevins, R. G. and Feyerherm, A. M., 1967, Effects of flow surface temperature on comfort. Part IV, Cold Floors, *ASHRAE Transactions*, 73(2), III 2.1.

Nevins, R. G., Michaels, K. B. and Feyerherm, A. M., 1964a, The effect of floor surface temperatures on comfort, Part I: College-age males, *ASHRAE Transactions*, 70, 29.

Nevins, R. G., Michaels, K. B. and Feyerherm, A. M., 1964b, The effect of floor surface temperatures on comfort, Part II: College-age females, *ASHRAE Transactions*, 70, 37.

Nevins, R. G., Rohles, F. H., Springer, W. and Feyerherm, A. M., 1966, A temperature humidity chart for thermal comfort of seated persons, *ASHRAE Transactions*, 72(I), 283–291.

Nevola, V. R., 1998, Commanders' Guide: Drinking for optimal performance during military operations in the Heat. Defence Evaluation and Research Agency Centre for Human Sciences, Farnborough, DERA/CHS/PP5/CR980062/1.0.

Nicol, F. and McCartney, K., 1997, Modelling temperature and human behaviour in buildings. Thermal comfort field studies: 1996–1997. Sustainable Building. *Proceedings of a BEPAC and EPSRC Conference*, Oxford, UK.

Nicol, J. F. and Humphreys, M. A., 1972, Thermal comfort as part of a self-regulating system. In Thermal comfort and moderate heat stress, *Proceedings of the CIB commission*, BRE, Watford, UK, pp. 263–274, ISBN 0–11–670520–5.

Nicol, J. F. and Humphreys, M. A., 2002, Adaptive thermal comfort and sustainable thermal standards for buildings, *Energy and Buildings*, 34(60), 563–572.

Nicol, F. and Parsons, K. C., 2002, Editorial. Special issue on thermal comfort standards, *Energy and Buildings*, 34(6), July, 529–532.

Nicol, J. F. and Raja, I. A., 1997, Thermal comfort, time and posture. Explanatory studies in the nature of adaptive thermal comfort, Oxford Brookes University, UK.

Nilsson, H., Holmér, I., Bohm, M. and Noren, O., 1999, Definition and theoretical background of the equivalent temperature. Paper 99A4082, *Proceedings of the ATA Conference*, Florence, Italy.

Nilsson, H. O. and Holmér, I., 2000, *Proceedings of the 3rd International Meeting on Thermal Manikin Testing 3IMM*, NIWL report Nr 2000:4, Salna, Sweden, ISBN 91-7045-554-6.

NIOSH, 1972, Occupational exposure to hot environment. National Institute for Occupational Safety and Health, HSM 72-10269, Department of Health, Education and Welfare, USA.

NIOSH, 1986, *Occupational exposure to hot environments*, National Institute for Occupational Safety and Health, DHHS (NIOSH) Publication No. 86-113, Washington DC, USA.

Nishi, Y. and Gagge, A. P., 1970, Moisture permeation of clothing – A factor governing thermal equilibrium and comfort, *ASHRAE Transactions*, 76, Part 1, 137–145.

Nishi, Y. and Gagge, A. P., 1977, Effective temperature scale useful for hypo- and hyperbaric environments, *Aviation Space and Environmental Medicine*, 48, 97–107.

Oakley, E. H. N., 1990, A new mathematical model of finger cooling used to predict the effects of windchill and subsequent liability to freezing cold injury, *Proceedings of International Conference in Environmental Ergonomics – IV*, Austin, Texas, USA.

Oakley, E. H. N. and Lloyd, C. J., 1990, Investigations into the pathophysiology of mild cold injury in human subjects, *Proceedings of International Conference in Environmental Ergonomics – IV*, Austin, Texas, USA.

O'Brian, N. V., 1996, An assessment of heat stress on workers in tunnels using compressed air compared to those in free air conditions. MSc thesis, Department of Human Sciences, Loughborough University, UK.

O'Brian, N. V., Parsons, K. C. and Lamont, D. R., 1997, Assessment of heat strain on workers in tunnels using compressed air compared to those in free air conditions. In *Tunnelling 97*. The Institution of Mining and Metallurgy, London, pp. 341–352. ISBN 1870706-34X.

O'Leary, C. M., 1994, An investigation into the role of the IREQ index in the Design of working practices for cold environments. MSc thesis, Department of Human Sciences, Loughborough University, UK.

O'Leary, C. O. and Parsons, K. C., 1994, The role of the IREQ index in the design of working practices for cold environments, *Annals of Occupational Hygiene*, 38(5), 705–719.

Ohnaka, T., Hodder, S. G. and Parsons, K. C., 2002, The effects of simulated solar radiation to the head and trunk on the thermal comfort of seated subjects, in *Environmental Ergonomics, Proceedings of the 10th meeting*, Fukuoka, Japan.

Olesen, B. W., 1977, Thermal comfort requirements for floors occupied by people with bare feet, *ASHRAE Transactions*, 83, Part 2.

Olsen B. W., 1982, Thermal comfort, in Brüel and Kjaer, *Technical Review No. 2*, Copenhagen, Denmark.

Olesen, B. W., 1985a, Local Thermal Discomfort, in Bruel and Kjaer, *Technical Review No. 1*, Copenhagen.

Olesen, B. W., 1985b, Heat Stress, in Bruel and Kjaer, *Technical Review No. 2*, Copenhagen.

Olesen, B. W., 2001, Introduction to the new revised draft of EN ISO 7730, in *Conference Proceedings Moving Thermal Comfort Standards into the 21st Century*, Windsor, UK, pp. 31–44, ISBN 1–873640–33–1.

Olesen, B. W. and Dukes–Dubos, F. N., 1988, International standards for assessing the effect of clothing on heat tolerance and comfort, in S. Z. Mansdorf, R. Sager and A. P. Nielson (Eds), *Performance of Protective Clothing*, Philadelphia: ASTM, pp. 17–30.

Olesen, B. W. and Parsons, K. C., 2002, Introduction to thermal comfort standards and to the proposed new version of EN ISO 7730, *Energy and Buildings*, 34(6), July, 537–548.

Olesen, B. W., Scholer, M. and Fanger, P. O., 1979, Vertical air temperature differences and comfort, in P. O. Fanger and O. Valbjorn (Eds), *Indoor Climate*, Copenhagen: Danish Building Research Institute, pp. 561–579.

Olesen, B. W., Sliwinska, E., Madsen, T. L. and Fanger, P. O., 1982, Effects of body posture and activity on the thermal insulation of clothing: measurements by a movable thermal manikin, *ASHRAE Transactions*, 88(2), 791–805.

O'Neill, D. H. and Whyte, R. T., 1985, Predicting heat stress in complex thermal environments, *Report of National Institute of Agricultural Engineering*, Silsoe, UK.

O'Neill, D. H., Whyte, R. T. and Stayner, R. M., 1985, Predicting heat stress in complex thermal environments, in D. J. Oborne (Ed.), *Contemporary Ergonomics*, London: Taylor & Francis.

Oohori, T., Berglund, L. G. and Gagge, A. P., 1988, Simple relationships among current vapour permeability indices of clothing with a trapped-air layer, in I. B. Mekjavic, E. W. Banister and J. B. Morrison (Eds), *Environmental Ergonomics*, ISBN 0–85066–400–4.

Oohori, T., Berglund, L. G. and Gagge, A. P., 1984, Comparison of current two parameter indices of vapour permeation of clothing – as factors governing thermal equilibrium and human comfort, *ASHRAE Transactions*, 90(2A), 85–101.

Osczevski, R. and Bluestein, M., 2001, A new wind chill index. In wind chill science and equations, World Wide Web report by Environment Canada and DCIEM, Toronto, Canada.

Oseland, N. A., 1997, Thermal Comfort: A comparison of observed occupant requirements with those predicted and specified in standards. PhD thesis, Cranfield University, UK.

Oseland, N. A. and Humphreys, M. A., 1994, Thermal comfort: past, present and future, Building Research Establishment Report, BR 263, ISBN 0–85125–633–3, Watford, pp. 184–197.

Oseland, N. A., Humphreys, M. A., Nicol, J. F., Baker, N. V. and Parsons, K. C., 1998, Building design and management for thermal comfort, BRE report CR 203/98, Garston, UK.

Osgood, C. E., Suci, G. J. and Tannenbaum, P. H., 1957, *The Measurement of Meaning*, University of Chicago: Chicago Press.

Pacink, M., 1990, The role of personal control of the environment in thermal comfort and satisfaction at the workplace. Coming of Age, EDRA 21/1990, in R.I. Selby, K. H. Anthony, J., Choi and B. Orland (Eds), Environmental Design Research Association, Okla, IS, pp. 303–312.

Pandolf, K. B., Givoni, B. and Goldman, R. F., 1977, Predicting energy expenditure with loads while standing or walking very slowly, *Journal of Applied Physiology*, 43(4), 577–581.

Pandolf, K. B., Sawka, M. N. and Gonzalez, R. R., 1988, *Human Performance Physiology and Environmental Medicine at Terrestrial Extremes*, USA: Brown and Benchmark.

Parker, R. D. and Parsons, K. C., 1990, Computer based system for the estimation of clothing insulation and metabolic heat production, in E. J. Lovesey (Ed.), *Contemporary Ergonomics*, London: Taylor & Francis, pp. 473–478.

Parsons, K. C., 1987, Human response to hot environments: a comparison of ISO and ASHRAE methods of assessment, *ASHRAE Transactions*, 93(1), 1027–1038.

Parsons, K. C., 1988, Protective Clothing: Heat exchange and physiological objectives, *Ergonomics*, 31(7), 991–1007.

Parsons, K. C., 1989, Computer aided design of the thermal environment, *Proceedings of Human Factors Society 33rd Annual Meeting*, 1, 512–516.

Parsons, K. C., 1990, Human response to thermal environments: principles and methods, in J. R. Wilson and E. N. Corlett (Eds), *Evaluation of Human Work*, London: Taylor & Francis, pp. 387–405.

Parsons, K. C., 1991, User performance tests for determining the thermal properties of clothing, in Y Quéinnec and F. Daniellou (Eds), Designing for everyone, *Proceedings of the 11th Congress of the International Ergonomics Association*, Paris, London: Taylor & Francis.

Parsons, K. C., 1992a, The Thermal Audit, in E. J. Lovesey (Ed.), *Contemporary Ergonomics*, London: Taylor & Francis, pp. 85–90.

Parsons, K. C., 1992b, Contact between human skin and hot surfaces: Equivalent Contact Temperature (Tceq), in W. A. Lotens and G. Havenith (Eds), *Proceedings of the Fifth International Conference on Environmental Ergonomics*, Maastricht, The Netherlands, November 2–6, pp. 144–146, ISBN 90–6743–227–X.

Parsons, K. C., 1992c, Thermal stress in compressed air work, in F. M. Jardine and R. I. McCallum (Eds), *Engineering and Health in Compressed Air Work*, London: E&FN Spon, ISBN 0–419–18460–0.

Parsons, K. C., 1993, Safe Surface Temperatures, in E. J. Lovesey (Ed.), *Contemporary Ergonomics*, Taylor & Francis, ISBN 0–7484–0072–2.

Parsons K. C., 1995, Computer models as tools for evaluating clothing risks and controls, *Annals of Occupational Hygiene*, 39(6), 827–840.

Parsons, K. C., 1998a, Case study of cold work in a hospital 'plating area', in I. Holmer and K. Kuklane (Eds), *Problems with cold work*, National Institute for Working Life, Publication NR 1998:18, Solna, Sweden, pp. 66–68, ISBN 91–7045–483–3.

Parsons, K. C., 1998b, Working practices in the cold : measures for the alleviation of cold stress, in I. Holmér and K. Kuklane (Eds), Problems with Cold Work, pp. 48–57, NIWL Report Nr. 1998:18, Salna, Sweden.

Parsons, K. C., 2000, An adaptive approach to the assessment of risk for workers wearing protective clothing in hot environments, in K. Kuklane and I. Holmer (Eds), *Ergonomics of Protective Clothing*, National Institute for Working Life, Publication NR 2000:8, Solna, Sweden, pp. 34–37, ISBN 91–7045–559–7.

Parsons, K. C., 2001a, Introduction to thermal comfort standards, in K. McCartney (Ed.), *Moving thermal comfort standards into the 21st century*, Proceedings of a conference held on 5th–8th April, Cumberland Lodge, Windsor, UK, pp. 19–30, ISBN 1–873640.

Parsons, K. C., 2001b, The estimation of metabolic heat for use in the assessment of thermal comfort, in K. McCartney (Ed), *Moving thermal comfort standards into the 21st century*. Proceedings of a conference held on 5th–8th April, Cumberland Lodge, Windsor, UK, pp. 301–308, ISBN 1–873640.

Parsons, K. C., 2002, The effects of gender, acclimation state, the opportunity to adjust clothing and physical disability on requirements for thermal comfort, *Energy and Buildings*, 34(6) July, 593–600.

Parsons, K. C. and Egerton, D. W., 1985, The effect of glove design on manual dexterity in neutral and cold conditions, in D. J. Oborne (Ed.), *Contemporary Ergonomics*, London: Taylor & Francis, pp. 203–209.

Parsons, K. C. and Hamley, E. L., 1989, Practical methods for the estimation of human metabolic heat production, in J. B. Mercer (Ed.), *Thermal Physiology*, pp. 777–781.

Parsons, K. C. and Bishop, D., 1991, A data base model of human responses to thermal environments, in E. J. Lovesey (Ed.), *Contemporary Ergonomics*, London: Taylor & Francis, pp. 444–449.

Parsons, K. C. and Webb, L. H., 1999, Thermal comfort design conditions for indoor environments occupied by people with physical disabilities, Final report to EPSRC research grant GR/K71295, Loughborough University, UK.

Parsons, K. C., Havenith, G., Holmer, I., Nilsson, H. and Malchaire, J., 1998, A proposed method for quantifying the effects of wind and human movement on the thermal and vapor transfer properties of clothing, in J. A. Hodgdon, J. H. Heaney and M. J. Buono (Eds), *Environmental Ergonomics VIII*, pp. 287–291, ISBN 0–9666953–1–3.

Parsons, K. C., Havenith, G., Holmér, I. Nilsson, H. and Malchaine, J., 1999, The effects of wind and human movement on the heat and vapour transfer properties of clothing, *Annals of Occupational Hygiene*, 43(5), 367–376.

Påsche, A., 2000, Current and future standards of survival suits and diving suits, in K. Kuklone and I. Holmér (Eds), *Ergonomics of Protective Clothing, Proceedings of NOKOBETEF 6 and 1st European Conference on Protective Clothing*, Stockholm, Sweden, May, NIWL Nr 2000: 8.

Payne, R. B., 1959, Tracking proficiency as a function of thermal balance, *Journal of Applied Physiology*, 14, 387–389.

Pepler, R. D., 1964, Psychological effects of heat, Chapter 12, in C. S. Leithead and A. R. Lind (Eds), *Heat Stress and Heat Disorders*, London: Cassell.

Pfeiffer, J., 1969, *The Cell*, New York: Time Life Books.

Pierce, F. T. and Rees, W. H., 1946, The transmission of heat through textile fabrics, Part II, *Journal of Textile Institute*, 37, 181–204.

Piette, A. and Malchaire, J., 1999, Validation of the PHS model, in J. Malchaine (Ed.), *Evaluation and control of warm working conditions, Proceedings of BIOMED 'Heat Stress' Research Project Conference*, Barcelona, June.

Piette, A. and Malchaire, J., 2000, Exposure duration limits for cold grip on different materials, in J. Werner and M. Hexamer (Eds), *Environmental Ergonomics IX*, Aachen, Germany: Shaker Verglag, pp. 193–196, ISBN 3–8265–76489.

Pitts, D. R. and Sissom, E. L., 1977, *Heat Transfer*, Schaum's Outline Series, London: McGraw-Hill.

Poulton, E. C., 1976, Arousing environmental stresses can improve performance, whatever people say, *Aviation Space and Environmental Medicine*, 47, 1193–1204.

Powell, S. and Havenith, G., 2000, The effects of contact cooling on manual dexterity and cooling of the hand, in J. Werner and M. Hexamer (Eds), *Environmental Ergonomics IX*, Aachen, Germany: Shaker Verglag, pp. 205–208, ISBN 3–8265–76489.

Preston-Thomas, H., 1968–76, International practical temperature scale of 1968, *Metrologia*, **5**, 35; 1969, amended Edn of 1975, *Metrologia*, **12**, 7; 1976.

Provins, K. A., 1958, Environmental conditions and driving efficiency. A review, *Ergonomics*, **2**, 97–107.

Provins, K. A., 1966, Environmental heat, body temperature and behaviour: An hypothesis, *Australian Journal of Psychology*, **18**, 118–129.

Raja, I. A., Nicol, J. F., 1997, A technique for postural recording and analysis for thermal comfort research, *Applied Ergonomics*, **27**(3), 221–225.

Ramsey, J. D. and Kiron, Y. C., 1988, Simplified decision rules for predicting performance loss in the heat, in *Heat Stress Indices, Proceedings of a seminar held by the Commission of the European Communities*, Luxembourg, pp. 337–372.

Ramsey, J. D., Burford, C. L., Beshir, M. Y. and Jensen, R. C., 1983, Effects of workplace thermal conditions of safe work behaviour, *Journal of Safety Research*, **14**, 105–114.

Randle, I. P. M., 1987, Predicting the metabolic cost of intermittent load carriage in the arms, in E. D. Megaw (Ed.), *Contemporary Ergonomics*, London: Taylor & Francis, pp. 286–291, ISBN 0–85066–386–5.

Randle, I. P. M., Legge, S. J. and Stubbs, D. A., 1989, Task-based prediction models for intermittent load carriage, in E. D. Megaw (Ed.), *Contemporary Ergonomics*, London: Taylor & Francis, pp. 380–385.

Rao, M. N., 1952, Comfort range in tropical Calcutta. A preliminary experiment, *International Journal of Medical Research*, **40**, 45.

Ray, R. D., 1984, The theory and practice of safe handling temperatures, *Applied Ergonomics*, **15**(1), 55–59.

Reid, D. B., 1844, *Illustrations of the Theory and Practice of Ventilation*, London: Longman, Brown, Green and Longmans.

Reinertsen, R., 2000, The effort of the distribution of insulation in immersion suits on thermal responses, in K. Kuklone and I. Holmér (Eds), *Ergonomics of Protective Clothing, Proceedings of NOKOBETEF 6 and 1st European Conference on Protective Clothing*, Stockholm, Sweden, May, NIWL Nr 2000: 8.

Reischl, U., Spand, W. A. and Dakes-Dobos, 1987, Ventilation analysis of industrial protective clothing in AsFour S. S. (Ed.), *Trends in Ergonomics/Human Factors IV*, Amsterdam: Elsevier, pp. 421–427.

Renbourn, E. T. (Ed.), 1972, *Materials and Clothing in Health and Disease*, H. K. Lewis & Co. Ltd.

Richardson, G., 1985, A review of human thermoregulation and its simulation, RAF Institute of Aviation Medicine, Farnborough, Hampshire: *IAM Report 643*.

Ringuest, J. L., 1981, A statistical model of the controller function of the human temperature regulating system. PhD thesis, Clemson University, USA.

Rintamaki, H. and Parsons, K. C., 1998, Limits for cold work, in I. Holmer and K. Kuklane (Eds), *Problems with cold work*, National Institute for Working Life, Publication NR 1998:18, Solna, Sweden, pp. 72–74, ISBN 91–7045–483–3.

Rissanen, S. and Rintamäki, H., 2000, Individual variation during slow and rapid contact cooling, in J. Werner and M. Hexamer (Eds), *Environmental Ergonomics IX*. Aachen, Germany: Shaker Verlag, pp. 189–191, ISBN 3–8265–76489.

Ritz, P., 1998, Methods of assessing body water and body composition, in M. J. Arnaud (Ed), Hydration throughout life, John Libbey, Eurotxt ISBN 2 7420 0226 X

Rivolier, J., Goldsmith, R., Logg, D. J. and Taylor, A. J. W., 1988, *Man in the Antarctic*, London: Taylor & Francis.

Robertson, A. S., Burge, P. S., Hedge, A., Sims, J., Gill, F. S., Finnegan, M., Pickering, C. A. C. and Dalton, G., 1985, Comparison of health problems related to work and environmental measurements in two office buildings with different ventilation systems, *British Medical Journal*, **291**, 373–376.

Rohles, F. H., 1969, Preference for the thermal environment by the elderly, *Human Factors*, **11**(1), 37–41.

Rohles, F. H., 1970, Thermal sensations of sedentary man in moderate temperature, *Institute for Environmental Research: Special Report*, Kansas State University, USA.

Rohles, F. H. and Nevins, R. G., 1971, The nature of thermal comfort for sedentary man, *ASHRAE Transactions*, **77**(1), 239–246.

Rohles, Jr. F. H. and Milliken, G. A., 1981, A scaling procedure for environmental research, *Proceedings of Human Factors Society – 25th Annual Meeting*.

Rohles, F.H. and Wallis, P., 1979, Comfort criteria for air conditioned automotive vehicles, Society of Automotive Engineers, SAE INC, 790122.

Rolls, B. J., Wood, R. J., Rolls, E. T., Lind, H., Lind, R. W. and Ledingham, J. G. G., 1980, Thirst following water deprivation in humans, *American Journal of Physiology*, **239**, R476–R482.

Rosenblad-Wallin, E. and Karholm, M., 1987, Environmental mappings as a basis for the formulation of clothing demands, *Applied Ergonomics*, **18**(2), 103–110.

SAE, 1993, Equivalent temperature, SAE Information Report, J2234 1993, Society of Automotive Engineers, USA.

Santee, W. R. and Gonzalez, R. R., 1988, Characteristics of the thermal environment, in K. B. Pandolf, M. N. Sawka and R. R. Gonzalez (Eds), *Human Performance Physiology and Environmental Medicine at Terrestrial Extremes*, USA: Brown and Benchmark, pp. 1–44.

Sawka, M. N., 1988, Body fluid responses and hypohydration during exercise-heat stress, in K. B. Pandolf, M. N. Sawka and R. R. Gonzalez (Eds), *Human Performance Physiology and Environmental Medicine at Terrestrial Extremes*, USA: Brown and Benchmark, pp. 227–266.

Schiller, G., Arens, E., Bauman, F., Benton, C., Fountain, M. and Doherty, T., 1988, A field study of thermal environments and comfort in office buildings, *ASHRAE Transactions*, **94**(2), 280–308.

Sevitt, S., 1949, Local blood-flow changes in experimental burns, *Journal of Pathological Bacteriology*, **61**, 427–442.

Shiraki, K., Sagawa, S., Tajima, F., Yokota, A., Hashimoto, M. and Brengelmann, G. L., 1988, Independence of brain and tympanic temperatures in an unanesthetized human, *Journal of Applied Physiology*, **65**, 482–486.

Shitzer, A. and Eberhart, R. C., 1985a, *Heat Transfer in Medicine and Biology – Analysis and Applications*, Vol. 1, New York: Plenum Press.

Shitzer, A. and Eberhart, R. C., 1985b, *Heat Transfer in Medicine and Biology – Analysis and Applications*, Vol. 2, New York: Plenum Press.

Siekmann, H., 1989, Determination of maximum temperatures that can be tolerated on contact with hot surfaces, *Applied Ergonomics*, **20**(4), 313–317.

Siekmann, H., 1990, Recommended maximum temperatures of touchable surfaces, *Applied Ergonomics*, **20**(1), 69–73.

Sinclair, M. A., 1990, Subjective assessment, in J. R. Wilson and E. N. Corlett (Eds), *Evaluation of Human Work – A Practical Ergonomics Methodology*, London: Taylor & Francis, pp. 58–88.

Singleton, W. T., 1974, *Man-Machine Systems*, London: Penguin.

Siple, P. A., 1939, Adaptation of the explorer to the climate of Antarctica, Thesis of Clark University, USA.

Siple, P. A. and Passel, C. F., 1945, Measurements of dry atmosphere cooling in sub-freezing temperatures, *Proceedings of American Philosophical Society*, 89, 177–199.

Slonim, A. D., 1952, *Fundamentals of the General Ecological Physiology of Mammals*, Moscow and Leningrad: Academic Press of USSR.

Smallhorn, E. A., 1988, A new immersible thermal manikin, in I. B. Mekjavic, E. W. Banister and J. B. Morrison (Eds), *Environmental Ergonomics*, London: Taylor & Francis, pp. 195–202.

Smith, C. E., 1991, A transient three-dimensional model of the human thermal system. PhD thesis, Kansas State University, USA.

Smith, T. A. and Parsons, K. C., 1987, The design, development and evaluation of a climatic ergonomics knowledge based system, in E. D. Megaw (Ed.), *Contemporary Ergonomics*, London: Taylor & Francis, pp. 257–262.

Spitzer, H. and Hettinger, Th., 1976, Caloricentafels. Tabellen voorhet omzetten van fysissche activiteiten in Calorisch Waarden, Acco, Leuven.

Sprague, C. H. and Munson, D. M., 1974, A composite ensemble method for estimating thermal insulation values of clothing, *ASHRAE Transactions*, 80(2), 120–129.

Stammers, R. B., Carey, M. S. and Astley, J. A., 1990, Task analysis, in J. R. Wilson and E. N. Corlett, *Evaluation of Human Work*, London: Taylor & Francis.

Stevens, S. S., 1960, The psychophysics of sensory function, *American Science*, 48, 226.

Stevens, J. C., Marks, L. E. and Gagge, A. P., 1969, The quantitative assessment of thermal discomfort, *Environmental Research*, 2, 149–165.

Stirling,. M., 2000, Assessment, prevention and rehydration procedures. PhD thesis, Loughborough University, UK.

Stirling, M. and Parsons, K. C., 1998a, Physiological and biochemical changes in heat acclimation for subjects with a controlled diet, in J. A. Hodgdon, J. H. Heaney and M. J. Buono (Eds), *Environmental Ergonomics VIII*, pp. 85–88, ISBN 0–9666953–1–3.

Stirling, M. and Parsons, K. C., 1998b, Measures of human hydration state, in J. A. Hodgdon, J. H. Heaney and M. J. Buono (Eds), *Environmental Ergonomics VIII*, pp. 125–128, ISBN 0–9666953–1–3.

Stoll, A. M., Chianta, M. A. and Piergallini, J. R., 1979, Thermal conduction effects in human skin, *Aviation Space and Environmental Medicine*, August.

Stolwijk, J. A. J. and Hardy, J. D., 1965, Skin and subcutaneous temperature changes during exposure to intense thermal radiation, *Journal of Applied Physiology*, 20, 1006–1013.

Stolwijk, J. A. J. and Hardy, J. D., 1966, Temperature regulation in man – a theoretical study, *Pflüger Archives*, 291, 129–162.

Stolwijk, J. A. J. and Hardy, J. D., 1977, Control of body temperature, in *Handbook of Physiology, section 9: reaction to environmental agents*, Bethesda, Maryland: American Physiological Society, pp. 45–68.

Stubbs, D. A., Buckle, P. W., Hudson, M. P., Burton, P. E. and Rivers, P. M., 1983, Back pain in the nursing profession III. Uniform evaluation, Poster abstract to K. Coombes (Ed.), *Proceedings of the Ergonomics Society's Conference*, London: Taylor & Francis, p. 193.

Sullivan, P. J., Mekjavic, I. B., Kakitsuba, N., 1987, Determination of clothing micro-environment volume, *Ergonomics*, 30(7), 1043–1052.

Symington, L. E. and Warren, P. H., 1989, Exploratory MANPRINT research: a Load carrying expert system, in E. D. Megaw (Ed.), *Contemporary Ergonomics*, London: Taylor & Francis, pp. 146–151.

Tanabe, S., Kimura, K. and Hara, T., 1987, Thermal comfort requirements during the summer season in Japan, *ASHRAE Transactions*, 93(1), 564–577.

Tanabe, S. I., Kobayashi, K., Nakano, J., Ozeki, Y. and Konishi, M., 2002, Evaluation of thermal comfort using combined multi-node thermoregulation (65MN) and radiation models and computational fluid dynamics (CFD), *Energy and Buildings*, 34(6), 637–646.

Teichner, W. H., 1954, Recent studies of simple reaction time, *Psychological Bulletin*, 51, 128–150.

Teichner, W. H., 1967, The subjective response to the thermal environment, *Human Factors*, 5, 497–510.

Thellier, F., Althabégoity, F. and Cordier, A., 1994a, Modelling of local thermal sensations of a car driver in winter conditions, in J. Frim, M. B. Ducharme and P. Tikuisis (Eds), *Environmental Ergonomics*, Montebello, Canada, pp. 190–191, ISBN 0–662–21650–4.

Thellier, F., Monchoux, F., Bonin, J.-L. and Clément, G., 1994b, Model of physiological responses during spatial extra-vehicular activity, in J. Frim, M. B. Ducharme, and P. Tikuisis (eds), *Environmental Ergonomics*, Montebello, Canada, pp. 174–175, ISBN 0–662–21650–4.

Thellier, F., Serin, G., Monchoux, F. and Cordier, A., 1992, Clothing effort on thermal sensations. Evaluation by transient modelling, in W. A. Lotens and G. Havenith (Eds), *Proceedings of the Fifth International Conference on Environmental Ergonomics*, Maastricht, The Netherlands.

Tikuisis, P., 1989, Prediction of the thermoregulatory response for clothed immersion in cold water, *European Journal of Applied Physiology*, 59, 334–341.

Tikuisis, P., 1992, Modelling of heat transfer, in W. A. Lotens and G. Havenith (Eds), *Environmental Ergonomics*, TNO, The Netherlands.

Tikuisis, P., 1994, Prediction of survival time for cold exposure, in J. Frim, M. B. Ducharme and P. Tikuisis (Eds), Proceedings of the sixth International Conference on *Environmental Ergonomics*, Motebello, Canada, September 25–30, pp. 160–161, ISBN 0–662–21650–4.

Tikuisis, P., 1998, Prediction of cold responses, in I. Holmér and K. Kuklane (Eds), *Problems with Cold Work*, NIWL Report Nr 1998: 18, Salna, Sweden, pp. 101–107.

Tikuisis, P. and Young, A. J., 1996, Prediction model development of tolerance/survival times for semi-immersed cold/wet work, in Y. Shapiro, D.S. Moran, Y. Epstein (Eds), *Environmental Ergonomics*, Freund, Tel Aviv, pp. 186–189, ISBN 965–294–123–9.

Tochihara, Y., Kimura, Y., Yadoguchi, I. U. and Nomura, M., 1998, Thermal responses to air temperature before, during and after bathing, in. J. A. Hodgdon, J. H. Heaney, M. J. Buono (Eds), *Environmental Ergonomics VIII*, pp. 309–313, ISBN 0–9666953–1–3.

Todd, S., 1988, Can performance of a manual task be predicted from hand skin temperature in cold conditions? Final Year Undergraduate Ergonomics Project, Loughborough University, UK.

Toftum, J., 2002, Human response to combined indoor environment exposures, *Energy and Buildings*, **34**(6), 601–606.

Toftum J. and Nielsen, R., 1996a, Draught sensitivity is influenced by general thermal sensation, *International Journal of Industrial Ergonomics*, **18**(4), 295–305.

Toftum, J. and Nielsen, R., 1996b, Impact of metabolic rate on human response to air movements during work in coal environments, *International Journal of Industrial Ergonomics*, **18**(4), 307–316.

Toner, M. M. and McArdle, W. D., 1988, Physiological adjustments of man to the cold, in K. B. Pandolf, M. N. Sawka and R. R. Gonzalez (Eds), *Human Performance Physiology and Environmental Medicine at Terrestrial Extremes*, USA: Brown and Benchmark, pp. 97–152, ISBN 0-697-14823-8.

Towle, J. A., Parsons, K. C. and Haisman, M. F., 1989, Design, Development and Implementation of a biomechanics knowledge base, in E. D. Megaw (Ed.), *Contemporary Ergonomics*, London: Taylor & Francis, pp. 152–157.

Turk, J., 1974, Development of a practical method of heat acclimatization for the army, in J. L. Monteith and L. E. Mount (Eds), *Heat Loss From Animal and Man*, London: Butterworth.

Umbach, K. H., 1984, Testing of textiles – Determination of physiological properties – Measurement of stationary thermal and water vapour resistance by means of a thermoregulatory model of human skin, ISO/TC 38/SC 8.

Umbach, K. H., 1988, Physiological tests and evaluation models for the optimization of the performance of protective clothing, in I. B. Mekjavic, E. W. Banister and J. B. Morrison (Eds), *Environmental Ergonomics*, London: Taylor & Francis, pp. 139–161.

Underwood, C. R. and Ward, E. J., 1966, The solar radiation area of man, *Ergonomics*, **10**, 399–410.

Van der Held, E. F. M., 1939, *Temperaturer hohung von Handgriffen Elektrowarme*, **9**(2), 31–34.

Vernon, H. M., 1919a, The influence of hours of work and of ventilation on output in tinplate manufacture, Report to Industrial Fatigue Research Board, No. 1, London: HMSO.

Vernon, H. M., 1919b, An investigation of the factors concerned in the causation of industrial accidents, Health of Munitions Workers Committee Memo No. 21, CD 9046.

Vernon, H. M., 1920, Fatigue and efficiency in the iron and steel industry, Report to Industrial Fatigue Research Board, No. 5, London: HMSO.

Vernon, H. M., 1930, The measurement of radiant heat in relation to human comfort, *Journal of Physiology*, **70**, 15.

Vernon, H. M. and Warner, C. G., 1932, The influence of the humidity of the air on capacity for work at high temperatures, *Journal Hygiene Cambridge*, **32**, 431–462.

Vernon, H. M., Bedford, T. and Warner, C. G., 1927, The relations of atmospheric conditions to the working capacity and the accident rate of miners, Report to Industrial Fatigue Board, No. 39, London: HMSO.

Virk, G., 1986, A thermal comfort expert system in an expert system shell. MSc thesis, Loughborough University.

Vogt, J. J., Candas, V., Libert, J. P. and Daull, F., 1981, Required Sweat Rate as an index of thermal strain in industry, in K. Cena and J. A. Clark (Eds), *Bioengineering, Thermal Physiology and Comfort*, Amsterdam: Elsevier, pp. 99–110.

Vogt, J. J., Meyer, J. P., Candas, V., Libert, J. P. and Sagat, J. C., 1983, Pumping effects on thermal insulation of clothing worn by human subjects, *Ergonomics*, **26**(10), 963–974.

Wadsworth, P. M. and Parsons, K. C., 1986, Laboratory evaluation of ISO/DIS 7933 (1983), Analytical determination of heat stress, in D. J. Oborne (Ed.), *Contemporary Ergonomics*, London: Taylor & Francis, pp. 193–197.

Wadsworth, P. M. and Parsons, K. C., 1989, The design, development, evaluation and implementation of an expert system into an organisation, *Proceedings of the 3rd International Conference on Human–Computer Interaction*, Amsterdam: Elsevier.

Wadsworth, P. M. and Parsons, K. C., 1991, The MAPS system, *Applied Physiology APRE Report*, Farnborough, UK.

Wagstaff, M. A., 1983, An evaluation of gloves and mitts with respect to thermal protection and their effects on manual dexterity and strength in cold environments. MSc thesis in Ergonomics, Loughborough University, UK.

Wang, X., 1990, A dynamic model for estimating thermal comfort, Climate and Building, **2**, Stockholm.

Webb, C. G., 1959, An analysis of some observations of thermal comfort in an equatorial climate, *British Journal of Industrial Medicine*, **16**, 297.

Webb, L. H. and Parsons, K. C., 1997, Thermal comfort requirements for people with physical disabilities, in *sustainable buildings, Proceedings of the BEPAC and EPSRC mini conference*, Oxford, UK.

Webb, L. H., Bailey, A. D. and Parsons, K. C., 2000, A software tool that provides guidance on thermal comfort conditions for people with physical disabilities, *Proceedings of the IEA 2000/HFES 2000 Congress*, San Diego USA, pp. 708–711.

Webb, P., 1964, Bioastronautics data book, NASA SP-3006.

Webb, P., Annis, J. F. and Troutman, S. J., 1978, Heat flow regulation, in Y. Houdis and J. D. Guieu (Eds), *New Trends in Thermal Physiology*, Paris: Masson, pp. 29–32.

Weddel, G. and Miller, S., 1962, Cutaneous sensibility, *Annual Review of Physiology*, **24**, 199–222.

Weiner, J. S., 1982, The measurement of human workload, *Ergonomics*, **25**(11), 953–965.

Weiner, J. S. and Hutchinson, J. C. D., 1945, Hot human environments: its effect on the performance of a motor coordination test, *British Journal of Industrial Medicine*, **2**, 154–157.

Weiner, J. S. and Lourie, J. A., 1981, *Practical Human Biology*, London: Academic Press.

Weir, J. B. de V., 1949, New methods for calculating metabolic rate with special reference to protein metabolism, *Journal of Physiology*, **109**, 1–9.

Werner, L., 1990, Properties of the human thermostat: results from mathematical and experimental analyses, *Proceedings of the International Conference on Environmental Ergonomics – IV*, Austin, Texas.

Werner, 1996, Computer modeling of physiological responses: Trends for the future, in Y. Shapiro, D. S. Moran and Y. Epstein (Eds), *Environmental Ergonomics*, Freund, London, Tel Aviv, ISBN 965-294-123-9.

Werner, J. and Buse, M., 1988, Three dimensional simulation of cold and warm defence in man, in I. B. Mekjavic, E. W. Banister and J. B. Morrison (Eds), *Environmental Ergonomics*, London: Taylor & Francis, pp. 286–296.

Weston, H. C., 1922, A study of efficiency in fine linen weaving, Report to Industrial Fatigue Research Board, No. 20, London: HMSO.

Weston, J. C., 1951, Heating research in occupied houses, *Journal of the Institute of Heating and Ventilating Engineers*, **19**, 47–108.

WHO, 1969, Health factors involved in working under conditions of heat stress, *Technical Report 412*, Geneva.

White, M. D., Johnston, C. E., Wu, M-P., Bristow, G. K. and Giesbrecht, G. G., 1994, Orthostatic intolerance during 63° head-up tilt following hot bath immersion is longer after ethanol ingestion, in J. Frim, M. B. Ducharme and P. Tiknisis (Eds), *Proceedings of the 6th International Conference on Environmental Ergonomics*, ISBN 0–662–21650–4.

Wilkinson, R. T., 1974, Individual differences in response to the environment, *Ergonomics*, **17**(6), 745–756.

Williams, S., 1996, Behavioural responses to maintain thermal comfort in office environments. MSc thesis, Department of Human Sciences, Loughborough University, UK.

Wilson, J. R. and Corlett, E. N., 1995, *Evaluation of Human Work*, 2nd Edn, London: Taylor & Francis, ISBN 0–7484–0084–2.

Wilson, O. and Goldman, R. F., 1970, Role of air temperature and wind in the time necessary for a finger to freeze, *Journal of Applied Physiology*, **29**(5), 658–664.

Wing, J. F., 1965, A review of the effects of high ambient temperature on mental performance, *Aerospace Medical Research Laboratories*, AMRL–TR–65 102, (NTIS AD 624144).

Winslow, C. E. A., Herrington, L. P. and Gagge, A. P., 1936, A new method of partitional calorimetry, Ibid., 116, 669.

Wissler, E. H., 1982, An evaluation of human thermal models, a report based on a workshop held at the University of Austin, Texas, USA, for the USAF.

Wissler, E. H., 1985, Mathematical simulation of human thermal behavior using whole body models, in A. Shitzer and R. C. Eberhart (Eds), *Heat Transfer in Medicine and Biology*, Volume 1, New York: Plenum, pp. 325–373.

Wissler, E. H., 1988, A review of human thermal models, in I. B. Mekjavic, E. W. Banister and J. B. Morrison (Eds), *Environmental Ergonomics*, London: Taylor & Francis, pp. 267–285.

Withey, W. R. and Parsons, K. C., 1994, A systematic basis for thermal risk assessment, in J. Frim, M. B. Ducharme and P. Tikuisis (Eds), *Proceedings of the sixth International Conference on Environmental Ergonomics*, Motebello, Canada, September 25–30, pp. 244–245, ISBN 0–662–21650–4.

WMA, 1985, The declaration of Helsinki and its subsequent revisions. The World Medical Association Handbook of Declaration.

Wood, E. J. and Bladon, P. T., 1985, The human skin, *Studies in Biology*, **164**, London: Edward Arnold.

Woodcock, A. H., 1962, Moisture transfer in textile systems, *Textile Research Journal*, 8, 628–633.

Wu, Y. C., 1972, Material properties. Criteria for thermal safety, *Journal of Materials*, JMSLA, December, **17**(4), 573–579.

Wu, Y. C., 1977, Control of thermal impact for thermal safety, American Institute of Aeronautics and Astronautics, *AIAA Journal*, **15**(5), 674.

Wyatt, S., Fraser, J. A. and Stock, F. G. C., 1926, Fan ventilation in a human weaving shed, Report to Industrial Fatigue Research Board, No. 37, London: HMSO.

Wyndham, C. H., 1963, Thermal comfort in the humid tropics of Australia, *Brit. J. Industr. Med.*, **20**, 110.

Wyndham, C. H., 1974, The physiological and psychological effects of heat, *Proceedings of the Mine Ventilation Society of South Africa*.

Wyndham, C. H. and Strydon, N. B., 1969, Acclimatising man to heat in climatic rooms or mines, *Journal of South African Institute of Mining and Metallurgy*, 60–64.

Wyon, D. P., 1969, The effects of moderate heat stress on the mental performance of children, National Swedish Building Research Document, No. 8.

Wyon, D. P., 1970, Studies of children under imposed noise and heat stress, *Ergonomics*, **13**(5), 598–612.

Wyon, D. P., 1986, The effects of indoor climate on productivity and performance: a review (in Swedish), *WS and Energi*, **3**, 59–65.

Wyon, D. P., 1989, The use of thermal manikins in environmental ergonomics, *Scandinavian Journal of Work, Environment and Health*, **15** (supplement), 84–94.

Wyon, D. P., 1996, Individual microclimate control: required range, probably benefits and current feasibility, *Proceedings of Indoor Air '96*, **1**, 1067–1072.

Wyon, 1999, Dr Wyon puts micro climates under the microscope, *M&E Design, The journal for mechanical and electrical consulting engineers and designers*, February.

Wyon, D. P. and Halmberg, I., 1972, Systematic observation of classroom behaviour during moderate heat stress, in *Thermal Comfort and Moderate Heat Stress*, Proceedings of CIB, W45 Symposium, Watford, London: HMSO.

Wyon, D. P. and Sandberg, M., 1996, Discomfort due to vertical thermal gradients, *Indoor Air*, **6**, 48–54.

Wyon, D. P., Andersen, I. and Lundqvist, G. R., 1979, The effects of moderate heat stress on mental performance, *Scandinavian Journal of Work, Environment and Health*, **5**, 352–361.

Wyon, D. P., Tennstedt, J. C., Lundgren, I. and Larsson, S., 1985, A new method for the detailed assessment of human heat balance in vehicles, Volvo's thermal manikin VOLTMAN, SAE paper No. 850042.

Yaglou, C. P. and Minard, D., 1957, Control of heat casualties at military training centers, *American Medical Association Archives of Industrial Health*, **16**, 302–316 and 405 (corrections).

Yagloglou, C. P. and Miller, W. E., 1925, Effective temperature with clothing, *ASHVE Transactions*, **31**, 89.

Yagloglou, C. P. and Drinker, P., 1929, The summer comfort zone: climate and clothing, *ASHVE Transactions*, **35**, 269.

Yoshida, A., Matsui, I. and Taya, H., 1989, *Technical report references on contact with hot and cold surfaces, tactile warmth and warmth of floors*, ISO/TC 159/SC5/WG1 N164, Geneva.

Yoshida, J. A., Banhidi, L. Polinezky, T., Kintses, G., Hachisu, H., Imai, H., Sato, K. and Nonaka, M., 1993, A study on thermal environment for physically handicapped persons. Results from Japanese – Hungarian joint experiment in 1990, *Journal of Thermal Biology*, **18**, 363–375.

Yoshida, J.A. Nomura, M., Mikami, K. and Hachisu, H., 2000, Thermal comfort of severely handicapped children in nursery schools in Japan, *Proceedings of the IEA 2000/HFES 2000 Congress*, San Diego, USA, pp. 712–715.

Youle, A., Collins, K. L, Crockford, G., Fishman, D., Parsons, K. C. and Sykes, J., 1990, *The Thermal Environment*, BOHS Technical Guide No. 8, Scientific Reviews Ltd.

Zhou, G., 1999, Human perception of air movement, Impact of frequency and airflow direction on sensation of draught. PhD thesis, Technical University of Denmark.

Zhou, G. H., Loveday, D. L., Taki, A. H. and Parsons, K. C., 2002a, Design and testing of a garment for measuring sensible heat transfer between the human body and its environment, *ASHRAE Transactions*, June, Atlanta, USA.

Zhou, G. H., Loveday, D. L., Taki, S. A. H. and Parsons, K. C., 2002b, Measurement of the air flow and temperature fields around live subjects and the evaluation of human heat loss. In Proceedings of Indoor Air Conference, July, Monterey, USA.

Zhu, F., 2001, A human thermoregulatory model for predicting thermal sensation in transient building environments. MSc thesis, Department of Architecture, University of Cambridge, UK.

Index

Printed and bound by CPI Group (UK) Ltd, Croydon, CR0 4YY

01/11/2024

01782635-0016